Cambridge Studies in Biological Anthropology 5

Somatotyping – development and applications

SOMATOTYPING – DEVELOPMENT AND APPLICATIONS

Cambridge Studies in Biological Anthropology

Series Editors

G. W. Lasker
Department of Anatomy, Wayne State University, Detroit, Michigan, USA

C. G. N. Mascie-Taylor
Department of Biological Anthropology,
University of Cambridge

D. F. Roberts
Department of Human Genetics, University of Newcastle upon Tyne

Somatotyping – development and applications

J. E. LINDSAY CARTER
Department of Physical Education
San Diego State University, San Diego, California, USA

and

BARBARA HONEYMAN HEATH
Department of Anthropology
University of Pennsylvania, Philadelphia, Pennsylvania, USA

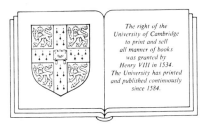

The right of the
University of Cambridge
to print and sell
all manner of books
was granted by
Henry VIII in 1534.
The University has printed
and published continuously
since 1584.

CAMBRIDGE UNIVERSITY PRESS

Cambridge
New York   Port Chester
Melbourne   Sydney

Published by the Press Syndicate of the University of Cambridge
The Pitt Building, Trumpington Street, Cambridge CB2 1RP
40 West 20th Street, New York, NY 10011, USA
10 Stamford Road, Oakleigh, Melbourne 3166, Australia

First published 1990

Printed in Great Britain by Redwood Press Limited, Melksham, Wiltshire

British Library cataloguing in publication data

Carter, J. E. L. (J. E. Lindsay)
Somatotyping – development and applications
1. Man. Body types
I. Title.   II. Heath, Barbara Honeyman
612

Library of Congress cataloguing in publication data

Carter, J. E. L. (J. E. Lindsay)
Somatotyping – development and applications / J. E. Lindsay Carter
and Barabara Honeyman Heath
   p.   c. – (Cambridge studies in biological anthropology)
Includes index.
ISBN 0-521-35117-0.
1. Somatotypes.   2. Physical anthropology.   3. Man–Constitution.
I. Heath, Barbara Honeyman.   II.Title.   III. Series.
GN66.5.C37   1990
573′.6 – dc20   89-35775   CIP

ISBN 0 521 35117 0

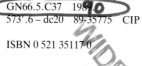

Contents

Foreword

How man varies in physique has been an important topic in the courses on human population biology that I have given for the last 30 or more years. Each year I have wished for an authoritative work on somatotyping to which to refer the students. Similarly when young researchers have come to discuss with me investigations in which they require to incorporate physique as a variable, the same lack has been painfully present. It has been necessary to refer them to the early books by Sheldon (with all their faults) and to the classic research papers on the subject by, e.g., Tanner, Dupertuis, Parnell, and of course the authors of the present volume, Barbara Heath Roll and Lindsay Carter. Some years ago, discussing this difficulty with Barbara Heath, at her lovely home in California, the seed was sown and the pages that follow are the harvest.

The somatotype story has been lengthy, sometimes stormy, always interesting. Barbara Heath has been actively involved throughout its history, right from the earliest days of her work with Sheldon. She alone can, and does, explain at first hand why it was that Sheldon's innovative concept provoked so much antagonism. Yet it was a concept that revolutionized the classification and description of physique, and as such its technique deserved to be objectively appraised and tested. Thanks to the work of those who did this, and found it acceptable for particular problems while appreciating its limitations, it has now, in its modified form, been applied world wide. The survey of these applications and the array of data that this book includes is not a mere literature review, for Barbara Heath has herself been associated as a consultant or rater for nearly every major sample of descriptive somatotype photographs in existence. Lindsay Carter, attracted to the field independently, has been largely responsible for disseminating the sports applications of the method, quite apart from his own notable contributions to refining it with his co-author. The combination of names, Heath–Carter, is justifiably as well known to those working on physique as is that of Hardy–Weinberg to those in populations genetics.

I for one enjoyed reading this book. It makes a contribution to the subject that is long overdue. No book has been written by two authors more distinguished in their own field. Besides illustrating the concept and technique, and so helping those wishing to apply them in their own studies, the book will be a useful source of reference for those wishing to ask questions of existing somatotype data.

D. F. Roberts
Department of Human Genetics
University of Newcastle upon Tyne

Preface

In the dialogue of 'Phaedrus' Plato observed that 'we are bound to our bodies like an oyster to its shell', and his contemporary, Hippocrates, proposed an early system of classification of our inescapable bodies. None of the many systems proposed to classify human physique has provoked so many controversies as the '*somatotype*' procedures that Sheldon described in his 1940 book, *The Varieties of Human Physique*. Although the early controversies over the Sheldon somatotype system are now a matter of history, the nature of the dissension and the roles of the figures chiefly involved are relevant to present attitudes to somatotyping and its applications.

This book was written to meet the need for a comprehensive text that brings up to date details of somatotype method and a summary of its applications. Since 1940 there have been many articles and several books relating to limited aspects of somatotyping, but there has been no comprehensive text since Parnell's *Behaviour and Physique* (1958). With the benefit of hindsight we have supported the idea that the phenotypic method of somatotyping is the most viable approach. The accumulated evidence supports this view and extensive application of the phenotypic method has significantly broadened our appreciation of human variation.

This book examines somatotyping, the method and its history, and describes a modified somatotype method constructed to correct the problems inherent in the Sheldon system. It reviews and re-evaluates the somatotype concept and argues its utility as a research tool for describing and understanding variations in human physique. It is intended to serve as a reference volume on somatotyping for those who wish to work in this research field. It contains the details of the method and a summary of its applications. It summarizes present knowledge, including much previously unpublished data available to the authors. It separates evidence from opinion regarding controversial methodology. It presents a phenotypic approach to somatotyping that is in keeping with current thinking in human biology. And it provides a framework for future investigations.

The book is arranged as follows:

Chapter One gives a brief history of systems of morphological taxonomy, of Sheldon and his development of the somatotype concept, and of the evolution of the Heath–Carter somatotype method from Sheldon's system. Inasmuch as the history of human concerns derives from individual histories, this chapter relates some of the history of the individuals centrally involved. However, this requires fairly detailed reference to the procedures themselves, so the reader unfamiliar with somatotyping may want to skim this chapter and return to it after reading the body of the book.

Chapter Two gives a description and comparison of the most important somatotype methods, showing the differences among them, with particular emphasis on the Sheldon and Heath–Carter methods. Chapter Three reviews the present knowledge of adult somatotype variation as reflected in studies throughout the world. Chapter Four reviews the development of somatotypes in childhood and changes in adult life, emphasizing that the phenotypic method is especially adapted to showing change in somatotypes. Chapter Five describes the limitations in knowledge of genetic factors in somatotypes. Chapters Six and Seven review the great number of studies of somatotypes in different kinds of athletes, in physical performance tests, in the fields of health, disease and behavioural variables, in occupational choice, and other applications of somatotyping. Chapter Eight recapitulates present knowledge and discusses promising areas for future research in somatotyping.

The Appendices and Glossary give technical details: Appendix I gives definitions and procedures in application of the Heath–Carter somatotype method; Appendix II presents details of the analyses of somatotype data; Appendix III lists data for rating by Heath of previously published somatotype photographs.

The Reference list contains about 80% of the complete listing of somatotype references, together with relevant non-somatotype references. A special effort has been made to cite previously little known studies from colleagues in Eastern Europe, India and Latin America.

The book has been arranged for the reader to use the chapters and appendices independently, reading only those of immediate interest. Naturally, some cross referencing will be necessary.

Acknowledgements

The friends and colleagues who have generously contributed time, materials, imaginative ideas and constructive criticism in the course of four decades deserve a chapter of their own. Some of the stories of historically significant contributions are included in the text. The tables and figures include acknowledgment of data and photographs. We are immeasurably grateful to these many people.

We owe special debts of gratitude to William Sheldon, who created the original concepts of somatotyping; to Albert Behnke, Marcel Hebbelinck, Carl Hopkins, William Laughlin, Margaret Mead, Richard Parnell, Derek Roberts, William Ross, Philip Smithells, Jiři Štěpnička, and James Tanner, for the ideas and dialogue that encouraged us and helped greatly in shaping and developing the concepts and conclusions presented in this book.

The many colleagues, research assistants and students who have participated in the collection and analyses of data will see that their labours, both voluntary and paid, played significant roles in the overall task of completing this book. Contributions of students and colleagues are noted in citations of published articles, theses and dissertations. We want to express special thanks to Stephen Aubry, William Duquet and Richard Shoup who began as enthusiastic students labouring 'in the trenches' and emerged as valued colleagues.

We are most grateful to colleagues who read one or more chapters and gave important suggestions for improvements: Claude Bouchard, Bruce Copley, Otto Eiben, Fred Hulse, Wilton Krogman, Gabriel Lasker, William Ross, Richard Shoup, L. S. Sidhu, Roger Simmons, and David Sleet. We hope they will recognize their contributions.

During the past forty years we have benefited from the support and interest of many friends and colleagues, including Nancy Bayley, Harrison Clarke, Luther Cressman, Albert Damon, Paul Fejos, Stanley Garn, Menard Gertler, William Howells, Edward Hunt, Frances Ilg, John King, Nathan Kline, Arnold Labby, Howard Lewis, Carey Miller, Lita Osmundsen, Carl Seltzer, Phillip Tobias, Charles Torrance, and Frederic Wulsin.

We deeply appreciate the technical assistance of Jerry Takigawa and his associates for art work and design of most of the final figures, and the help of Harry King in assembly of the reference list.

We owe immeasurable thanks to Derek Roberts for unflagging interest, encouragement, and editorial guidance.

We owe a special debt to the love, tolerance, and cooperation of our spouses, Lolita Diȳoso Carter and Frederick Roll, during the exceedingly long evolution of this book. We are also grateful to Fred Roll for photographic skill in improving some of the prints used.

J. E. Lindsay Carter
Barbara Honeyman [Heath] Roll

1 *History of somatotyping*

*Any scientist worthy of his salt labours to bring about the obso-
lescence of his own work.*

(*Theodosius Dobzhansky*)

Early physique classifications

Somatotyping is the most recent development in the twenty-five
century history of morphological taxonomy and constitutional investigation.
Tucker & Lessa (1940*a,b*) in their review of the history of human classification defined constitution as the sum total of the morphological, physiological and psychological characters of an individual, in large part determined by heredity but influenced in varying degrees by environmental
factors or, simply, the total biological make-up of an individual.

Physicians have been prominent in the history of constitutional investigation, particularly in studies of interrelations of morphology and susceptibility to disease. In the fifth century BC the Greek physician, Hippocrates,
described people with long thin bodies as of *habitus phthisicus*, and observed
that they were susceptible to tuberculosis. He called those with short thick
bodies *habitus apoplecticus*, and said they were susceptible to vascular
disease and apoplexy. In the first century AD Celsus, a Roman medical
encyclopaedist, wrote that above all things one should know the nature of
constitution, why some people are fat and some thin.

Other early physicians turned their attention to relationships between
temperament and susceptibility to disease. In the second century AD Galen,
a Greek physician, applied the idea of four bodily 'humours', which had
persisted for the seven centuries since Pythagoras and Empedocles. He
suggested that the physician needed to ascertain a patient's temperament in
terms of humours in order to diagnose and treat disease. In the fourth
century BC Aristotle observed that a specific body always involves a specific
character. Avicenna, early eleventh century Arab physician and philos-
opher, recommended the study of temperament as it related to character.

In the late eighteenth and early nineteenth century typologies following
the pattern of Hippocrates were popular in France. Halle in 1797, followed
by Rostan in 1828, described three types of physical constitution as *type
digestif*, *type musculaire* and *type cerebrale*. Here for the first time the
important distinction of predominant muscular development was recognized.

The development of anthropometry added new dimensions to the study of morphology. Anthropometry was first used in studies of morphology in the seventeenth century when Elsholtz, at the University of Padua, established a method for taking measurements on the body. Two hundred years later Quetelet, a Belgian mathematician and astronomer, was the first to study the measurements of man statistically. Unaware that Elsholtz had used the word anthropometria, he thought he had invented it.

Late in the nineteenth century di Giovanni carried out a long series of anthropometric studies in the school of clinical anthropology that he founded at the University of Padua. His pupil, Viola, influenced by Beneke at the University of Marburg, differentiated three morphological types. He called subjects with large, heavy bodies and relatively short limbs macrosplanchnic, those with small trunks and relatively long limbs microsplanchnic, and those with intermediate variation normosplanchnic. He himself observed that microsplanchnic was approximately the same as the old term *habitus phthisicus*, and that macrosplanchnic was approximately the same as *habitus apoplecticus*. In order to distinguish between the long-limbed, small-trunked microsplanchnics and the large-trunked, short-limbed macrosplanchnics, Viola derived a measure of trunk volume and morphological indices from manipulation of eight trunk measurements and the length of one arm and one leg.

An important trichotomous classification, formulated by Huter about 1880, divided people into three types: cerebral (with ectodermic structures predominating), muscular (with mesodermic structures predominating) and digestive (with endodermic structures predominating).

In *Körperbau und Charakter* (1921) Ernst Kretschmer (1888–1964) described four physical and psychic types, deriving from his acute clinical observations and a minimum of measurements. These he called the athletic, pyknic, asthenic and dysplastic physiques. Later he substituted the word leptosomic for asthenic and made a distinction between the linearity and the slenderness, fragility and gracility of the leptosomic type. Kretschmer's types, which resembled the 'groupings' of di Giovanni, recognized a gradual gradation from psychoses to 'normality'. He had a significant and far-reaching influence on constitutional investigation despite criticism for limited sampling, scanty measurements, lack of indices, subjective estimates, and failure to classify data according to age, sex and social status.

The twentieth century produced considerable interest in constitutional investigation in the United States as well as in Italy and Western Europe. There is an excellent detailed review of these early workers in Tucker & Lessa (1940*a,b*) and another by Tittel & Wutscherk (1972). Pearl, Ciocco and Draper were primarily interested in relationships between physical type, temperament and susceptibility to various diseases. Those best known for their classifications of constitutional types were Bean, Bryant, Goldthwaite, Mills, Davenport and Stockard. There is a similarity of overall picture of variations of physique in such classifications as hypersthenic, sthenic and asthenic; fleshy, medium and slender biotype; lateral, intermediate and linear; and so on.

Early in this century biologists and anthropologists readily accepted the existence of discrete types and tried to find them in what we recognize today as the complex continuum of human variation. Sheldon's somatotype concept of continuous variation was a striking advance over previous systems of classification. He recognized that every individual instead of being of a particular type, was a mixture of all the three basic components of physique, but that these were present in varying degrees in different individuals. These three components, according to Tucker & Lessa (1940*b*), he originally called *pyknosomic, somatosomic,* and *leptosomic,* but then adopted the names *endomorphy, mesomorphy,* and *ectomorphy,* which are strikingly reminiscent of Huter and Von Rohden. However, Sheldon makes no allusion to his inspiration for these words. Without comment he merely lists Van Rohden in the bibliography.

The contributions of W. H. Sheldon

W. H. Sheldon (in collaboration with S. S. Stevens and W. B. Tucker) in 1940 introduced his concept of 'somatotype' in *The Varieties of Human Physique.* The subtitle was *An Introduction to Constitutional Psychology.* Sheldon declared that his purpose was 'to provide a three-dimensional system for description of human physique', but also that morphological classification was merely a means to the end of creating 'an analogous schema for the description and classification of temperament' (pp. xi, xii). It was a promising and innovative departure from constitutional systems that classified the human species into between two and five 'types'. (The 'analogous schema' appeared in *The Varieties of Temperament,* 1942.)

Sheldon's books evoked markedly antithetical responses from the academic community and from the general public. Meredith (1940) was among the first of the scientists who responded to and bitingly criticized the methodological weaknesses. On the other hand Aldous Huxley popularized the words 'somatotype', 'endomomorphy', 'mesomorphy' and 'ectomorphy' and so publicized the idea of easily recognized relationships between physique and temperament. In *Ends and Means* and in articles in periodicals he adopted the new words and welcomed the prospect of using them to describe and match physiques and temperaments. The words have become part of our vocabulary.

The facts of Sheldon's early biography do not foretell that temperamental perversity, and an unfortunate predilection for deliberately antagonizing the Establishment, which would jeopardize success and brilliant achievement. William Herbert Sheldon was born on 19 November 1898 in Warwick, Rhode Island, the youngest of the three children of William Herbert Sheldon and Mary Abby Greene. He was proud of his parents' pre-Revolutionary New England ancestry. He liked to talk of his mother's descent from a Revolutionary General Greene and of his parents' friendship with such contemporary luminaries as William James, whom he claimed as his godfather. This may well be one of his several, probably apocryphal whimsies, like that of his avowed descent from Benjamin Franklin.

By his own account, William was his mother's favourite. A precocious, gifted, versatile son, a voracious reader, thirsty for knowledge, avidly curious, he liked to say that he was raised with a litter of Irish setter pups. He referred to his father as a locally renowned naturalist and hunter, who brought him up on Rhode Island's then unpolluted and sparsely populated shores and in the nearby New England forests. He had affectionate memories of his father, who taught him the migratory and nesting habits of the shore birds. He also showed him the trees and shrubs where the cecropia, the luna, the promethea, and the polyphemus silk moths laid their eggs, and encouraged him to watch for the hatching of the eggs, to see them grow into exotic caterpillars and to marvel at the hundreds of feet of fine silk thread they spun for snug cocoons to pupate in until the long winter gave way to spring. Sometimes he watched the evanescent moth push the silk threads apart and struggle on to a twig to spread its wings to dry on an early warm day of spring, then fly away to find a mate – and having mated, die. In the family library he learned natural history in William Hamilton Gibson's self-illustrated *Sharp Eyes* and *Eye Spy*, Gene Stratton Porter's *Moths of the Limberlost*, and Hornaday's *American Natural History*.

From early boyhood Sheldon collected and traded the large copper early American cents, the 'old pennies'. By the time he was at Brown University he was a formidable expert and trader, familiar with most of the collectors and collections in the Boston–Providence area. According to him, he paid for most of his education from his numismatic profits. He continued through life to collect and trade the early coppers. By 1950 he had amassed a collection second only to that of George Clapp, the philanthropist founder of ALCOA (Aluminum Company of America), who had bequeathed his own collection to the American Numismatic Society in New York.

Sheldon made a hobby of applying rating scales and the principles of taxonomy to the grading and classification of coins. He established scales for rating condition, value and rarity. In his book *Early American Cents* (Harper & Bros., 1949) – which he said he wrote because 'ever since childhood I have wanted to read it' – he brought order and system to the identification, grading and evaluation of every known variety of early cent.

He provided a catalogue, plates and tables, which enable amateurs to identify accurately any large cent. He published a sequel, *Penny Whimsey*, in 1958. These two scholarly, authoratitive and charming out-of-print collectors' items are his only non-controversial writings.

Sheldon attributed his combined interests in morphology, psychology, philosophy and the philosophical concerns of religion to his boyhood and youthful environment. He formalized his interests in his undergraduate programme at Brown University (where he graduated in 1918), at the University of Colorado (where he received a masters degree in psychology in 1923), and at the University of Chicago (where he completed his PhD in psychology in 1925). While he was a graduate student at the University of Chicago he met Sante Naccarati, a young Italian anthropologist, who was on a fellowship in the United States investigating possible relationships among morphology, temperament and intelligence. Naccarati introduced him to the teachings of his master, Viola, and of di Giovanni and Pende, of the Padua School of Clinical Anthropology. Sheldon and Naccarati found much in common and planned to join in collaborative research, which they hoped would lead to a technique for combining some of the findings and methods of Viola with those described in Kretschmer's study, *Physique and Character* (*Körperbau und Charakter*, 1921). For his PhD dissertation ('Morphologic Types and Mental Ability', University of Chicago, 1925), Sheldon repeated the study that Naccarati had completed. While he was beginning his teaching career at the University of Chicago, he and Naccarati continued to ponder possible ways of investigating morphology, intelligence and behaviour. Their collaboration ended prematurely when Naccarati was killed in an automobile accident during a summer holiday in Italy in 1929. Twenty years later Sheldon referred to the loss of Naccarati as a singular personal tragedy that delayed his progress toward new techniques for describing human morphology and analysing its relationships with behaviour and function.

Meanwhile Sheldon decided he needed a medical education in order to proceed effectively with his investigations of human morphology and behaviour. When he entered medical school at the University of Wisconsin in 1929 he also taught psychology part-time. He graduated from medical school in 1933 and completed his internship in 1934.

Following medical school Sheldon spent two years (1934–1936) in England and Europe, financed by a travelling fellowship awarded by the National Council on Religion in Higher Education, and a grant from Dorothy Whitney Elmhirst of Darlington, England. He was committed to write *Psychology and the Promethean Will*, a philosophical statement of his conception of the 'religious problem' (as opposed to 'theological problems') and the need for what he chose to call a 'biological humanics'. He said he visited Kretschmer, Freud, Adler and Jung, and discussed with them his ideas about what he would call 'somatotyping'. He was disappointed that Kretschmer was not ready to embrace a three-dimensional concept of human physique, despite its obvious close relationship to his own studies of physique and behaviour. Sheldon's talks with the four greats of European psychiatry were exciting and stimulating, but did not lead to the collaborative research that he had hoped for.

In England he became acquainted with prominent British intellectuals such as Aldous Huxley, Julian Huxley, Gerald Heard, Christopher Isherwood and Bertrand Russell. It was Sheldon's most exciting, provocative, stimulating, golden time. When he finished the promised book he returned to the United States, filled with plans for setting up a Constitution Project as the umbrella for research and writing that would encompass somatotype, psychology, psychiatry, philosophy and religion.

Sheldon's golden dreams came to an abrupt end in 1936, in an emotional crisis involving a girl he called 'Starlight'. He said he was 'engaged' to her. When she unexpectedly married another man, he wrote to the new husband an emotional, ill-advised and threatening letter. Its outraged recipient widely distributed copies of the offending document in high academic circles. The repercussions from this incident prejudiced Sheldon's opportunities for academic appointment for the rest of his life. Several loyal colleagues at various institutions arranged for him to share space for carrying out his research and for writing his books. But after 1936 he held no formal, salaried academic posts. Thereafter, he was dependent upon his own resources and privately obtained funds to pay research assistants and to meet other expenses.

In 1936 the Divinity School of the University of Chicago made informal arrangements for Sheldon to lecture and to pursue his interest in religion and psychology. During this period he persuaded the departments of physical education at the University of Chicago, Oberlin College, the University of Wisconsin and several other institutions to cooperate in collecting the first somatotype photographs, as extensions of their 'posture picture' programmes.

Between 1938 and 1940 Sheldon worked with Professor Smith S. Stevens on the conceptual rationale of the somatotype method. Stevens, trained as a physicist, was Director of the Psycho-Acoustic Laboratory in the Harvard Psychology Department. His interests centred on hearing as a physical, psychological and physiological phenomenon. He was devoted to good scientific method, and found the logic of Sheldon's schema attractive. Despite reservations about seemingly premature inferences he collaborated with Sheldon as co-author of *The Varieties of Human Physique* (1940) and *The Varieties of Temperament* (1942).

Earnest A. Hooton, widely known and popular Harvard professor of anthropology, was an early supporter of Sheldon's somatotype research. Sheldon's originality and flair appealed to Hooton, who was openly enthusiastic about the prospects of somatotyping. Sheldon dedicated *The Varieties of Human Physique* (1940) to 'Earnest A. Hooton whose studies in physical anthropology have vitalized constitutional research'. In this book he defined somatotyping as the quantification of three components, which he called *endomorphy*, *mesomorphy* and *ectomorphy*. He rated each component on a 7-point scale. The ratings were made from standardized photographs and were expressed as a series of three numerals, to sum to no less than 9 and no more than 12. He endeavoured to relate his ratings to a series of measurements and indices, and gave guidance on descriptive features to assist rating. He accumulated his primary data from studies of 4000 college men. But he made a number of assertions for which he provided *no* evidence. Ensuing controversies over various aspects of Sheldon's methodology cooled Hooton's enthusiasm. Although he applauded Sheldon's creativity and originality, he deplored his intransigence in the turbulent academic debates over the permanence of the somatotype, the validity and validation of the method, the claims for embryological origins of the three somatotype components, and the appropriateness of using the same rating scale for males and females.

In 1938 Sheldon met Dr Emil Hartl, an ordained minister with a degree in psychology, the Director of the Hayden Goodwill Inn. Hartl was so impressed by the idea of a close relationship between physique and temperament that he invited Sheldon to set up a project to study the 200 delinquent boys in residence at the Goodwill Inn. During much of the study Sheldon lived at the Inn. World War II interrupted this study in 1942, but Sheldon completed it when he returned to Boston after his military service.

In 1942, soon after he had completed *The Varieties of Temperament*, Sheldon joined the United States Air Force as a Major in the Medical Corps. Thus he temporarily solved his continuing employment problem. He was stationed at Kelley Field, where he successfully set up a somatotype research project. He took somatotype photographs of several thousand enlisted men and other Air Force personnel. Some of the findings from the study were published as reports of the School of Aviation Medicine (Sheldon, 1943*a,b*, 1944*a*).

During his tour at Kelley Field, Sheldon fell ill with what was at first believed to be brucellosis, traced to a dairy supplying the military. When his illness persisted he was hospitalized several times for a series of tests, and at length the diagnosis of Hodgkin's disease was made. Following a course of deep radiation therapy, Sheldon was retired from the Air Force with the permanent rank of Major on full disability pension. The prognosis was poor, and he suffered from the unpleasant side effects and malaise associated with radiation therapy. He returned to Chicago, where he slowly convalesced, with no recurrence of Hodgkin's disease symptoms. During this period he married Milancie Hill, a former student, who was now a graduate student at the University of Chicago. (His first marriage, to Louise Steger in Chicago in 1925, had ended in divorce in 1928.) He and Milancie went to Boston, where Sheldon resumed and completed the study of delinquent boys at the Hayden Goodwill Inn. (Milancie Hill's marriage also ended in divorce about 1948.)

At Harvard before the War, Hooton's graduate student, C. Wesley Dupertuis, became Sheldon's most devoted student and champion of somatotype method. When Dupertuis became the resident anthropologist in Dr George Draper's Constitution Laboratory in the Columbia Presbyterian Medical Center in New York, he and Dr J. R. Caughey (Draper's associate) set up a somatotype study in the outpatient clinics. Draper retired in 1945, Caughey left to become Dean of Case Western Reserve Medical School, and in 1946 Dupertuis persuaded Sheldon to join him in the Constitution Laboratory, with the nominal title of Director, without salary or formal academic appointment. After he moved to Columbia Medical Center, Sheldon completed his book *Varieties of Delinquent Youth* (Sheldon, 1949), the report on the Hayden Goodwill Inn study.

During Sheldon's post-War residence in Boston he became acquainted with Eugene McDermott, a geophysicist, who was president and founder of Geophysical Services, Inc., which later became Texas Instruments. Fascinated by Sheldon's somatotype concepts, particularly by his hypothesized relationships between physique and temperament, McDermott made several modest contributions to help with the progress of the delinquency study at the Hayden Goodwill Inn. He continued his support of Sheldon's ongoing research at Columbia Medical Center. Sheldon included McDermott as co-author of *Varieties of Delinquent Youth*, although his collaboration consisted solely of financial and moral support.

Sheldon and Heath

Barbara Heath (then Barbara Honeyman Hirsch) in Washington, DC, in 1944 met Frederick Wulsin, a professor of anthropology at Tufts University, and Hooton's former student, long-time associate and friend. Wulsin introduced her to Sheldon's *The Varieties of Human Physique* and *The Varieties of Temperament* and kindled her interest in somatotype research. He told her he had been familiar with Sheldon's pre-publication research, and found his ideas exciting and persuasive. He said Sheldon had written two exceptionally great books which 'made good anthropological sense' and added that 'his analysis of temperament sheds more light on human conduct than any other point of view I know'.[1] Inadvertently Wulsin pointed Heath down the road that was to lead to her five-year association with Sheldon.

Heath promptly bought the Sheldon books, read them, re-read and studied them, and gave her serious attention to the possibility that she herself might take part in somatotype research, and continue her formal education beyond her Smith College bachelor's degree in history. Wulsin encouraged her and arranged meetings with his academic colleagues who were interested in somatotype research. Following correspondence, Hooton asked his graduate student, William Laughlin, on holiday in Oregon, to get in touch with Heath in Portland. Laughlin, fresh from a Columbia summer seminar on somatotype with Sheldon, was enthusiastic about the inherent potential in somatotype research.

Later that year Heath visited Boston and talked at some length with Hooton. He expressed intense interest in somatotype research and confirmed his sense of its potential value. At the same time he warned that anyone who worked with Sheldon should be aware of his temperamental difficulties, which had consistently interfered with harmonious relations with his peers, and of his chequered career in academia. He then arranged for Heath to spend a day with Stanley Garn, another graduate student, who was working on two somatotype projects – one a cardiac study at Massachusetts General Hospital, the other a growth study at the Forsyth Dental Clinic. He also arranged a meeting with Dr James Andrews, Assistant Curator of Somatology at the Peabody Museum, Harvard University, who was supervisor and statistical analyst of the somatotype study of Army personnel which Hooton had initiated at demobilization centres at the end of World War II (Hooton, 1959). Andrews told Heath that he himself preferred to call the three components *fat*, *muscle* and *length*, and rejected outright Sheldon's requirement that the sum of the three components in a rating should be no less than 9 and no greater than 12, but said nothing about the 7-point scale and other Sheldon strictures.

Heath also met Smith Stevens at the Harvard Psycho-Acoustic Laboratory. He referred to the controversies Sheldon had generated and expressed serious reservations about the course of somatotype research. He suggested that 'the best possible idea would be to save random somatotype photographs for twenty years, then dig them out and start drawing conclusions'.[2] Stevens seemed to have lost his original interest in somatotype, and had little contact with Sheldon after the publication of the first two books. In Boston, Heath visited the Hayden Goodwill Inn, where she talked at some length with the two Sheldon associates who remained loyal friends and collaborators to the end of Sheldon's life. These were Emil Hartl, the Director of the Inn, and Roland Elderkin, a social worker, who carried out much of the case work for Sheldon's study.

A little later Heath visited Sheldon at the Constitution Laboratory in Columbia Presbyterian Hospital in New York. There she met Wesley Dupertuis, who showed her the procedures of somatotype photography and data collection in the Presbyterian Hospital clinics. At the time Sheldon was completing the final draft of *Varieties of Delinquent Youth*, and gave her a copy of the draft chapter 'The Psychiatric Variables', which impressed her greatly. Heath returned to Portland intensely excited about somatotype research. Despite their various reservations, the people she had talked with encouraged her to pursue her interests. She had the impression that Dupertuis' imminent departure, to take up his post as clinical anthropologist at Case Western Reserve Medical School, would leave something of a vacuum in the Constitution Laboratory and would substantially slow down Sheldon's data gathering. She accepted Sheldon's apparent view of himself as a 'misunderstood genius' who had the bad luck not to attract competent and loyal associates.

In the year that followed Heath stirred up a good deal of interest in the potential of somatotype research among the teaching staff at the University of Oregon Medical School, where she worked as research secretary in the Psychiatric Clinic. In May 1948 Howard Lewis, the professor of medicine, invited Sheldon to visit the medical school and to give a series of lectures there. Tentative plans were made for seeking funds to support a somatotype research project in the medical school clinics.

At the same time, in a lively correspondence, Heath and her friend Frederick Wulsin carried on long discussions about somatotyping and William Sheldon. Wulsin cautioned Heath that she might find herself 'riding herd more or less indefinitely on a prima donna'. When she sent him a copy of the chapter from the *Varieties of Delinquent Youth* manuscript he wrote: 'Sheldon takes a perverse delight in saying things that will infuriate some of his readers. Why do it?' What is the use? It is not a skilful method of persuasion. So likewise he waxes lyrical about humor. It is clear that he is painting his own philosophy and point of view. It is all right for him to feel that way, but this is not the place to tell the world about it. Rather this volume should give the *impression* of complete objectivity, even though that be an impossible ideal.'[3]

For about a year Heath wrestled with the conflict between lifelong habits of conventional conformity and the desire to take part in somatotype research. She weighed the risks and faced up to the inevitable disapproval of much of the community. Then in the autumn of 1948 she moved to New York. She became acquainted with the routines and activities of the

Constitution Laboratory – two rooms in the Presbyterian Hospital Department of Medicine. Soon she was filling a number of roles as an associate of Sheldon. She rapidly learned the techniques of making somatotype ratings and soon made independent ratings of hundreds of somatotype photographs, which correlated almost perfectly with Sheldon's ratings. She took charge of the measuring and photographing of subjects in the clinics, acted as clinic secretary, often typing Sheldon's letters as he dictated them, and in the evenings took courses in biology at Columbia University. Dr Paul Fejos, Director of the Viking Fund (now the Wenner-Gren Foundation), encouraged and assisted her in improving her photographic skills.

Pleased with Heath's contributions to the work of the Constitution Laboratory and apparent prospects for continued progress, Eugene McDermott set up a fund of $100 000 to support the research of the laboratory. He and Sheldon called the fund the Biological Humanics Foundation. Sheldon was its nominal president. Heath was secretary and treasurer, and was now referred to as the Executive Secretary of the Constitution Laboratory. At the same time Sheldon named McDermott co-author, with Hartl, of *Varieties of Delinquent Youth*, which was then in press, and proposed that McDermott and Heath be co-authors of the forthcoming *Atlas of Men*.

In 1949 Sheldon persuaded the departments of physical education in half a dozen midwestern colleges and universities to cooperate in a project for taking somatotype photographs of the incoming autumn classes of young women. A few months later he made similar arrangements with east coast Ivy League women's colleges. With a team of three young women, Heath obtained somatotype photographs of almost 4000 college women – the first (and only) large collection of female somatotype photographs. Sheldon announced that an *Atlas of Women* based on these data was forthcoming, a companion volume to his *Atlas of Men*. During this period Heath also helped to photograph and measure about 1800 women in mental institutions in New York and Massachusetts. She made somatotype ratings of all these photographs. She helped to set up a longitudinal study of children at the Gesell Institute in New Haven, Connecticut. Within two years she had studied and rated at least 12 000 somatotype photographs. She had somatotyped more subjects by Sheldon's method than anyone except Sheldon himself.

When *Varieties of Delinquent Youth* was completed Sheldon began to write the text and to prepare the somatotype photographs for the proposed *Atlas of Men*. Despite the fact that Sheldon's data were wholly cross-sectional, with one photograph of each subject, he proposed to show each of the known somatotypes at ages 18 to 65 and over at five-year intervals. He prepared tables of somatotypes distributed according to the criterion of height divided by the cube root of weight, which showed that the height–weight ratios of a given somatotype varied with age (usually due to increases in weight). As indicated in Chapter 3, in the absence of longitudinal data the tables were based upon extrapolations from the arbitrary base of age 18 years. Usually the somatotypes of subjects were based on height–weight ratios calculated from the heights and weights at age 18, which they reported for themselves from memory.

The Viking Fund gave a $13 000 subsidy toward the production of the expensive volume, which Harper & Brothers had agreed to publish. Heath helped with the selection and page layout of the 1161 photographs to be published in the *Atlas*. The agreed format of six photographs on each 9 × 12-inch page called for some reduction of the original 5 × 7 prints and closer positioning of the three views of each subject. Sheldon suggested that the demands of the format be met by remounting the photographs on 4 × 5 cards, which could be reduced to the 3.5 × 4.5 dimensions required, and Heath found that with skilful use of scissors she could do this without mutilating or altering the size of the original images. She reassured the Harper editor and the Viking Fund on this point.

When Sheldon chose examples of the various somatotypes at successive ages he found that there were discrepancies between the height–weight ratio indicated for the 4-4-4 somatotype at age 40, for example, and the photograph he felt was the best example; or, he found that the subject who met the height–weight criterion was 50 years old instead of 40. Heath was astonished to see that Sheldon simply altered the age or height–weight ratio to meet the criteria of his extrapolated tables! Sheldon's insistence upon the reality and actual existence of the polar extremes of somatotype (e.g. 1-1-7, 7-1-1) led to really serious difficulties. He could have presented existing examples close to the extremes, and described the polar extreme as conceptual, but as yet not encountered. Instead, when he could find no subjects who met the exact criteria for 1-1-7, he asked Heath to trim a little from each view of a somatotype rated as $1\frac{1}{2}$-$2\frac{1}{2}$-7 (see photograph no. 1 in *Atlas of Men*, as an example). Shocked at this suggestion, Heath felt honour-bound to inform Dr Fejos of this lapse from agreed procedures, despite Sheldon's announced intention to make her a co-author of *Atlas of Men*. Fejos had already expressed misgivings about the 'cutting and pasting' technique, and agreed with Heath's decision to forego the co-authorship. Soon afterwards Sheldon found a woman medical student who was skilful with scissors and followed his instructions without apparent reservation.

Sheldon's handling of photographs and data for *Atlas of Men* thus presented Heath with a serious ethical problem. In addition her concern about Sheldonian somatotype methodology was already growing. She had established that she could match his somatotype ratings almost perfectly when she applied his criteria and ignored her contradictory anthroposcopic impressions. She was confident that the underlying schema for somatotyping was sound and that somatotyping was potentially a valuable research tool. But she also knew that the potential could not be realized without methodological modifications. Because she could see no immediate prospect of continuing somatotype research independent of the Constitution Laboratory, she postponed a crucial confrontation with Sheldon.

As Heath had hoped, in 1951 the Rockefeller Foundation made a five-year grant to the University of Oregon Medical School for a study of possible relationships between somatotype and disease in the population of the outpatient clinics. This gave her the wished-for opportunity to return to Oregon to continue full-time somatotype research and work with her colleagues of many years. In Oregon she was virtually independent of Sheldon and had time to think through the methodological problems of somatotyping. She was the Executive Secretary of the project and was paid a salary from the grant. She was in charge of recruitment, interviewing and somatotyping of subjects. All photographs and data sheets became the property of the medical school. Sheldon was given desk space and the nominal title of Clinical Professor, without salary and without financial benefit under the grant, spending two or three months at a time in Portland and the remainder of the time in New York.

Dr Carl E. Hopkins, Associate Professor of Public Health and Preventive Medicine at the medical school, who taught biostatistics and epidemiology, was intrigued by the idea of somatotype. Heath explored with him the flaws in Sheldon's method and discussed possible modifications. They agreed that the chief problems were: (1) the closed system and 7-point scale; (2) the insistence that the somatotype itself did not change despite radical shifts in weight; and (3) the non-linear relationship of somatotype and height–weight ratio.

On a visit to the Institute of Child Welfare in Berkeley in 1951, Sheldon and Heath reviewed together the serial somatotype photographs in the Berkeley Growth Study, directed by Dr Harold Jones. During the review Heath, for the first time, semi-publicly disagreed with Sheldon. She maintained that in some cases it was impossible to predict a subject's somatotype at age 18 (or other ages) from earlier somatotype photographs and data. Sheldon insisted, without evidence, that the young adult somatotype could be predicted from any age, and declined to consider the need to re-think the issue of permanence of somatotype indicated by the Berkeley longitudinal data.

By early 1953 it was clear to Heath that any further work with Sheldon was impossible. He had failed to take advantage of a singularly good research setting at the University of Oregon Medical School. He seemed to avoid working in harmony with the medical school faculty and personnel, who had been Heath's friends and colleagues for almost 15 years. He was intransigent about the slightest modification of his method. He resisted use of currently accepted statistical analyses of his data. Plainly the man who mutilated and manipulated the materials for his forthcoming *Atlas of Men* was not a 'misunderstood genius'.

In June 1953 Heath left the University of Oregon Medical School–Rockefeller project and, without regret, permanently severed her connection with Sheldon and projects connected with him. She enrolled in a graduate programme at New York University, where under the sponsorship and guidance of Professors Leonard Larsen and Roscoe Brown, she worked out the basic modifications of the Sheldon somatotype method. In late 1954 she and her ophthalmologist husband, Scott Heath (deceased 1974), moved to Carmel, California, where she has lived since then, continuing her work with somatotype research and methodology as a freelance consultant to projects in the United States and many other countries.

Sheldon after 1953

In 1953 Columbia Medical Center notified Sheldon that the space he used was no longer available. He moved his somatotype materials to the apartment of his friend and associate, Dorothy Iselin Paschal, on Riverside Drive in New York City. He continued his informal association with Dr Nolan Lewis, Dr Nathan Kline and his assistant, Dr Ashton Tenney, at Rockland State Hospital in New York. Sheldon also returned to Portland, Oregon, for some time each year, ostensibly analysing the data from the Rockefeller project, but he published no reports of the findings. The Rockefeller grant expired in 1956.

There were occasional reports that Sheldon still talked about publishing an *Atlas of Women*, using the somatotype photographs of college women collected by Heath and her assistants. By the mid 1950s there were data on nearly 4000 college women and on a similar number of women from outpatient and other hospital populations. These were a more than adequate sample to show patterns of somatotype distribution for adult women in the United States. However, it was apparent that Harper & Brothers and other publishers had no interest in undertaking the proposed *Atlas of Women*.

In the late 1950s Sheldon developed a new somatotype method, which he called the Trunk Index method. A brief first description of the system appeared in a paper Sheldon (1961) presented at the Children's Hospital in Boston. He apparently gave occasional lectures about the new direction he had taken. One of these was the Maudsley Bequest Lecture (Sheldon,

1965), presumably written by Sheldon, but read by his friend, Emil Hartl. Sheldon published only one major paper on somatotyping (Sheldon *et al.*, 1969) after *Atlas of Men* (1954).

In 1966 Dorothy Iselin Paschal bought a house in Cambridge, Massachusetts, where she installed Sheldon's somatotype materials and made an informal headquarters for him to live and work in. Sheldon and those who had remained close to him (Emil Hartl, Dorothy Paschal, Roland Elderkin, Edward Monnelly) referred to this headquarters as the Biological Humanics Foundation. But Sheldon the individualist so isolated himself from all but this handful of dedicated friends that there was no one who would provide legitimate validation or practical application of his ideas. Apparently he wholly abandoned his original somatotype method in favour of his new classification, the Trunk Index, though little has been published about it.

Sheldon was a gifted man. His conceptual insights were stunning. His personal insights were often dulled by his incapacity for redeeming empathy. His self-insights suffered from his lack of humility. He had the gift of charm without the warmth and generosity for friendship. He knew the answers without completing the research, and was unwilling to ask the appropriate questions. His writing shows his flair for natural history, for philosophy, and for scientific enquiry. He could envisage a synthesis of constitutional psychology, psychiatry and medicine and the possibility of emerging with new philosophical and religious concepts. He seemed to have a romantic image of himself as a tragic Arthurian knight destined to be victimized by those less cultivated and less sensitive, by prosaic intellects who referee scientific journals and deny space to 'original thought'. He was a 70-inch (178-cm) 3-3½-5 who saw himself as a 72-inch (183-cm) 2−4−5. In conversation he showed open contempt for all the human species except those of certified Anglo-Saxon lineage. His racism and male chauvinism are shockingly evident in his writing. William Herbert Sheldon died in Cambridge, Massachusetts in 1977 at age 78. His faithful friend, Dorothy Iselin Paschal, died in Cambridge in 1981 at age 76.

Further developments by Heath and Carter

Meanwhile there were further developments in somatotyping, which led to the evolution of the Heath–Carter method. The Heath–Carter method defines somatotype as a quantitative description of the present shape and composition of the human body. It is expressed in a three-number rating, representing three components of physique: (1) endomorphy refers to relative fatness; (2) mesomorphy refers to musculoskeletal robustness relative to height; and (3) ectomorphy refers to relative linearity. The somatotype can be used to record changes in physique and to estimate gross biological differences and similarities among human beings. It is an anthropological identification tag. This method of somatotyping is sensitive to changes in physiques over time and is used for rating both sexes at all ages.

The method is a modification of the original one in Sheldon (1940). It is a dynamic and more useful physique classification. It seems to have evolved from a series of singularly lucky incidents. Divers investigators, sceptical of Sheldon's method but interested in somatotyping, enlisted Heath to rate series of somatotype photographs, many of which were exactly suited to her proposed modifications. Others elsewhere pursued independent investigations which later came together. Finally, Lindsay Carter's interest led to the collaborative project that produced the Heath–Carter somatotype method. (Heath (1963) had proposed the following modifications: (1) Redistribution of the somatotype ratings so that there would be a linear relationship between somatotype and the height–weight ratio. For this she used as a reference framework Sheldon's table of somatotype distribution according to the criterion of height divided by the cube root of weight. (2) Elimination of the distribution tables that extrapolated height–weight ratios according to age. (3) Adoption of the modified table for both sexes at all ages. (4) Adoption of an open-ended rating scale.)

When Harold Jones and Harold Stoltz initiated the Berkeley Growth Study in 1928, they adopted the idea of photographs together with anthropometry from the posture pictures which had been used in undergraduate physical education since the late nineteenth century. (A few years later Sheldon would borrow the same idea from the physical educators and would call the photographs 'somatotype' photographs.) It was in rating these photographs that Heath expressed disagreement with Sheldon in 1951.

Harold Jones, Director of the study, procured a grant under which Heath could re-rate the photographs. The series consisted of approximately 300 boys and girls photographed and measured at 6-month intervals from ages 11 to 18 years. In addition there were photographs and measurements of about 65% of the original subjects at age 33, fifteen years after the conclusion of the study. Heath re-rated the series without reference to previous and subsequent data and photographs of a given subject. Jones supplied her with several hundred photographs at a time, picked at random as from a well-shuffled deck of cards.

Subsequent comparison of the results confirmed Heath's hypothesis that in many cases subjects could change so dramatically between ages 11 and 18 years that accurate prediction of somatotype at age 18 was out of the question. In many cases subjects changed as unpredictably between ages 18 and 33 as between 11 and 18.

When she rated Dr Carey Miller's material on 200 Hawaii-born Japanese male and female students at the University of Hawaii, Heath found that there are ethnic differences among somatotype distributions. The report of this analysis (Heath *et al.*, 1961) contained the first published reference to Heath's modified table for somatotype distribution and height–weight ratios.

For six or eight years after Dr James Tanner's 1952 visit to the Rockefeller project at the University of Oregon Medical School he and Heath corresponded regularly concerning somatotype series which Tanner sent to her for rating. They also carried on a simultaneous written discussion of the pros and cons of modifying Sheldon's somatotype system. When Heath and Tanner applied Sheldon's criteria their ratings corresponded closely, as Tanner (1954a) showed in a report on the reliability of anthroposcopic somatotyping, based on comparisons of Tanner's, Heath's and Dupertuis' ratings.

Tanner[4] resisted departure from Sheldon's criteria and discouraged Heath's suggested modifications, while at the same time he talked of modifications of his own.[5] Several times he mentioned regression equations for somatotyping young men, and a paper on 'anthropometric somatotyping'.[6]

In 1956 Tanner sent the first of several samples of women subjects to Heath. He admitted that he did not 'feel tremendously confident at somatotyping women'.[7] Tanner expressed surprise that Heath found so many 5s and $5\frac{1}{2}$s in mesomorphy in the first sample, and noted that his ratings were more than 1 unit lower on the average. He asked Heath if she had changed her standards and was giving mesomorphy a higher rating. She assured him she had not done so. He then sent a sample of 150 women in drama and music schools in London, who turned out to be lower in mesomorphy and closely resembled the distribution for American college women. In the correspondence about women's somatotypes Heath mentioned that she had talked with Albert Behnke about the need to redefine somatotype components, so as to take account of Behnke's measurements of total body fat and lean body mass. She suggested that each individual has a 'constellation of possible somatotypes'[8], a concept which would eliminate the fiction of somatotype 'permanence'.

In 1957 Tanner sent to Heath a somatotype sample of English Channel swimmers, which included subjects up to age 37. Heath pointed out that Tanner's previous samples had fallen in the 18–23-year age group, which did not 'introduce any radical rating problems'; and that 'you are for the first time seeing my ratings based on judging the somatotype as it is, and not trying to reconcile the present with a hypothetical reconstruction of the past'.[9] Tanner replied: 'I am not sure I agree with you about letting endomorphy rise with fat content.'[10]

In December 1957 Tanner wrote to Heath saying that he had asked Margaret Mead to send to her the 'complete file of your pictures of the New Guinea people (with relevant heights and weights). How Barbara and I will agree on the ratings I don't know.'[11] Tanner asked Heath to 'rate the children in terms of what somatotype you *guess* [authors' italics] they are going to be at eighteen, as well as putting on a figure for their immediate phenotype'.[12]

In 1957 Harrison Clarke, Research Professor in the School of Health and Physical Education at the University of Oregon, asked Heath to be the somatotype consultant to a growth study he had set up in the Medford (Oregon) schools. After a year's experience with rating the somatotype photographs of boys from age 7 to 18 years, Heath wrote to Tanner about the problem of rating children as they are, versus predictions of somatotypes at age 18. She pointed out that it is particularly difficult to predict somatotypes of pre-adolescent boys.[13] She also raised questions about Sheldon's distribution of somatotypes on the criterion of height divided by the cube root of weight. Tanner continued to raise objections to Heath's suggested modifications. It is apparent that he resisted, at least partly, because he was confident that he himself would develop modifications of a different kind. At the same time he perceived the logic of Heath's proposals.[14]

Tanner's suggestion that Margaret Mead send the Manus somatotype photographs to Heath led to her long association with Mead. It also led to longitudinal study of the somatotypes and growth patterns of a unique ethnic population, and provided Heath with solid evidence that mesomorphy could not be described adequately on a closed 7-point scale.

Heath's association with Margaret Mead began with Mead's letter saying, 'We are waiting most anxiously to see what you will make of these' [the somatotype photographs of the Manus of the Territory of Papua New Guinea].[15] Heath replied: 'The physiques are so different from any I have seen that I am going through them several times before attempting to somatotype them.'[16] Heath wrote to Tanner that the consistently low male endomorphy and high mesomorphy in the Manus series reinforced her conviction that the rating scales should be opened; that the distribution of somatotypes and height–weight ratios should fall at equal intervals; and that re-scaling for age ought to be eliminated. She commented: 'I think you should know that I opened up the scale to allow for ratings not given by WS [i.e. Sheldon] to my knowledge, to allow for subjects who are $8\frac{1}{2}$ in mesomorphy and $\frac{1}{2}$ in endomorphy.'[17]

In late 1958 Tanner sent to Heath somatotype photographs of students at Loughborough College, England, and of athletes in the British Empire Games in Cardiff, Wales (Tanner, 1964). Heath compared the Manus males with the most mesomorphic subjects in these series, and compared all of them with the 7s in mesomorphy in *Atlas of Men*. Her conclusion that several in each series were in fact more mesomorphic than any of the 7s in the *Atlas* strengthened the case for extending the rating scale.

However, the decisive evidence that the rating scales should be extended came slowly. Dr Carl C. Seltzer, Research Fellow in anthropology at the Peabody Museum, Harvard University, asked Heath to do somatotype ratings on a series of photographs from a study of obesity. These data on obese females demonstrated that the 7-point scale was inadequate for rating endomorphy. In her report to Seltzer[18] she pointed out the important differences between her criteria and Sheldon's, and sent him a copy of the Sheldon table of somatotype distribution according to the criterion of height–weight ratios together with a copy of her modified table. She called to his attention her open-ended rating scale and elimination of corrections for age, emphasizing that in such a selected sample the distribution is necessarily constricted in comparison with a more random sample. Rating by Sheldon's 7-point scale yielded ratings from 5 to 7 in endomorphy, while the open scale yielded a range of 5 to 14 in endomorphy. She commented: 'I believe the Sheldon ratings in the second component would be almost the same as mine, ranging from $2\frac{1}{2}$ to $5\frac{1}{2}$. In the third component there is little variation in samples of obese subjects, but I do make the distinction between the rating of 1 and $\frac{1}{2}$ in ectomorphy.'[19]

Each of the somatotype series Heath worked with confirmed the need for modifications of the somatotype method for describing the human species of all races at all ages and of both sexes. The Berkeley and Medford series showed the difficulties of using somatotypes to shed light on variations in children's patterns of growth. The Manus series and the Tanner series of athletes strengthened the argument for open rating scales. The Manus series and the University of Hawaii Japanese series called attention to ethnic variations in somatotype. All the series called attention to the narrow range of Sheldon's solely cross-sectional data.

In a roundabout way Dr Lindsay Carter's letter of June 1962 from the University of Otago in Dunedin, New Zealand, led to the Heath–Carter collaboration.[20] Carter had first learned about somatotyping when he was an undergraduate student of Philip Smithells, Director of the School of Physical Education at the University of Otago. Smithells had met Sheldon and Heath when he visited the Constitution Laboratory in 1949. When Carter was a research assistant and lecturer at the University of Otago in 1954–1955 Smithells encouraged him to initiate some somatotype research. From 1956 to 1959 Carter was a graduate student at the University of Iowa, Iowa City, where he completed his MA and PhD degrees. At Iowa he was greatly influenced by C. Harold McCloy and Frank Sills in physical education and by Howard Meredith in child development. Some years previously McCloy and Sills had established their photoscopic reliability with Sheldon on a test series of somatotype photographs. They regarded the somatotype concept as useful in their field of research. Meredith, on the other hand, was sceptical of somatotyping and had been a severe, outspoken critic (Meredith, 1940). When Carter arrived at Iowa he regarded himself as a devout disciple of the Sheldon method. He had studied all four of the Sheldon books. Based on blind ratings of the photographs in *Varieties of Delinquent Youth*, he regarded himself as a good rater.

For his MA thesis Carter studied the somatotypes of 123 boys aged 12 to 17 years, with Sills as his adviser. Carter tried to apply Sheldon's concept that somatotypes of subjects under age 18 should be estimates of a somatotype appropriate for age 18. Sills insisted that this was impossible, that the present somatotype was of greater importance. Sills' thinking prevailed – and sowed some doubts in Carter's mind. When he returned to the University of Otago in 1960 he had dropped somatotyping, after his confidence in the Sheldon system had been shaken and he had found no acceptable alternative. In Otago he came across Parnell's book, *Behaviour and Physique* (Parnell, 1958). He re-read Parnell's paper 'Somatotyping by physical anthropometry' (Parnell, 1954). Parnell's anthropometric approach was designed to apply to both males and females, and to a certain extent to children. This method was not precisely what Carter was looking for, but it provided him with a framework for further studies and for possible modifications. He was disappointed that Parnell had retained the age-adjusted scales and the concept that a somatotype could not change. He saw no solution to the problem of applying Sheldon criteria to photographs of samples of different ages and both sexes. However, he carried out a somatotype study of students and a large sample of New Zealand Air Force personnel, using photography and anthropometry, and applying Parnell's method (Carter & Rendle, 1965).

Communications between Carter and Heath got off to a slow start. Heath mislaid Carter's 1962 letter, and answered it on 11 April 1963. Her reply reached Carter six months later when it was forwarded to him by surface mail from New Zealand to San Diego, California, where he had accepted a post at San Diego State College (now San Diego State University). Carter's second letter[21] was forwarded to Heath in Moscow, where she was visiting the Institute of Anthropology of Moscow State University. He listed the somatotype samples he had studied, emphasized his great interest in somatotype research, and discussed Parnell's M.4 deviation chart somatotype method, which was to make an important contribution to modifications later incorporated in the Heath–Carter method.

Heath and Carter finally met in March 1964. After several hours of discussion of their interests in somatotyping, their reservations about

Sheldon's methodology, and Heath's suggested modifications, they agreed to explore the possibilities of collaborating on future research. They outlined a project for continuing the modifications of somatotype method that Heath (1963) had suggested. Heath and Carter were greatly pleased and excited to discover they shared a common interest in a research area given to generating more negative controversy than positive findings. Both believed appropriate modifications could produce a useful research tool. It was apparent that they had similar objectives and would each contribute different ideas. Heath had come from 'within the system'. Her proposed modifications seemed to provide a useful first step towards solution of the somatotype dilemma. Carter had come from outside, and had already found the importance of anthropometry in somatotype method.

The contribution of anthropometric measurements to the rating of the three somatotype components was obvious to Heath. In their first joint paper, 'A comparison of somatotype methods' (1966), Heath and Carter showed how Heath's modifications and an adaptation of Parnell's M.4 method could improve Sheldon's method.

Dr Albert Behnke's studies of body composition influenced both Heath and Carter. Heath had become familiar with his monograph 'The relation of lean body weight to metabolism and some consequent systematizations' (Behnke, 1953) when she was writing the prospectus for her PhD dissertation at New York University. Later she had collaborated with Behnke in a somatotype project when he was director of the US Naval Radiation Laboratory at Hunters Point in San Francisco. Carter was greatly impressed with Behnke's 'assessments of lean body mass',[22] which he heard about in Behnke's paper on body composition at the April 1964 meeting of the American College of Sports Medicine in Los Angeles. Heath had found Behnke's concept of lean body mass invaluable when she was thinking through possible redefinitions of the somatotype components. The lean body mass idea led her to try to visualize each subject in a somatotype photograph as he or she would appear with the least imaginable subcutaneous fat. The minimal measurable skinfolds of the Manus males demonstrated the appearance of a body with this 'least imaginable subcutaneous fat'.

The Roberts & Bainbridge (1963) paper 'Nilotic physique' provided the evidence that a 7-point scale could not adequately describe ectomorphy in the human species as a whole.[23] Heath wrote to Carter that this paper dramatically supported the open-ended scale for ectomorphy,[24] in the same way that the Manus had for mesomorphy and the obese women had for endomorphy. The most mesomorphic subjects in Tanner's British Empire Games series also supported the need for an open rating scale.

Heath obtained valuable advice and reassuring encouragement from her friends, Dr Carl E. Hopkins (now a professor in the School of Public Health, University of California, Los Angeles) and the late Dr Charles Torrance. Torrance examined Parnell's standard scales and corrections for age and skewness and concluded 'there was no real purpose to be achieved by these manipulations'.[25] Heath was increasingly interested in Tanner's, as well as Parnell's, anthropometric data, which illuminated the relationships of anthropometry to somatotype ratings.[26]

Three papers – 'Need for modification of somatotype methodology' (Heath, 1963), 'A comparison of somatotype methods' (Heath & Carter, 1966) and 'A modified somatotype method' (Heath & Carter, 1967) – encompassed the proposed modifications, the validation of the modifications, and a presentation of the Heath–Carter modified somatotype method. The third paper included the adoption of an anthropometric somatotype procedure (adapted from Parnell's M.4 chart) to conform with the modifications already adopted.

The validation process revealed to Heath and Carter that the relationship between ratings for endomorphy and skinfold values were highly related. They also found that 'relatively small increments of fat (skinfolds) are easily recognized visually in somatotype photographs of subjects at the low end of the endomorphic scale, and that the same absolute increment is not discernible in subjects high in the first component. The first component increases out of proportion to the other two components, so that proportionately larger decreases in height–weight ratio are necessary to be compatible with exceedingly high skinfolds.'[27] In the course of writing the third paper presenting the Heath–Carter method, Heath re-rated all the somatotype photographs in her own and in Carter's files, applying the modified criteria.[28]

Preliminary findings from Heath's first field trip to Manus in 1966 confirmed those of the first examination of Manus somatotype data in 1958. Subsequent follow-ups amplified the adult data and provided modified longitudinal growth data, reported in 'Growth and somatotype patterns of Manus children' (Heath & Carter, 1971).

Heath's and Carter's joint and individual published papers have included applications of their somatotype method plus several innovative adjuncts to it (Carter & Phillips, 1969; Carter, 1970, 1971, 1974; Carter et al., 1971; Carter et al., 1973; De Garay et al., 1974). Their modifications and applications of somatotype method have attracted attention in a number of countries and have been translated into several other languages. To meet the growing demands a teaching handbook, *The Heath–Carter Somatotype Method* was published by Carter in 1972, with new editions in 1975 and 1980 (Carter, 1980*a*).

Other contributions to somatotyping

From the beginning of somatotype research, investigators in anthropology, biology, medicine, physical education and sports sciences have shown the greatest interest. Primarily their research has concerned methodology, population studies, growth studies, and the effects of exercise and sports on somatotype. In physical anthropology in the United States, studies were made by Hooton, Howells, Lasker, Hunt, Garn, Newman, Bullen and Kraus. Damon, Gertler, Seltzer and Spain applied somatotyping in medicine and constitutional anthropology.

The physical educators were the most active and cooperative of Sheldon's early supporters at the time that he was collecting somatotype data on college men and women. Several prominent physical educators included somatotyping in their research and encouraged their students to do likewise. Among these were Peter Karpovich and Harrison Clarke (later Research Professor at the University of Oregon) at Springfield College, Carl Wilgoose at Boston University, Thomas Cureton at Springfield and the University of Illinois, and Sills and McCloy at the University of Iowa. Frederick Rand Rogers was an early visitor to the Constitution Laboratory and continued his support of somatotyping for many years although he made no published contributions.

Sheldon often made special reference to the debt he owed to the physical educators. In the preface to *Atlas of Men* (Sheldon, 1954) he wrote: 'grateful acknowledgment is due . . . to the faculties of the departments of physical education in 31 colleges and universities where students were photographed'. As mentioned, somatotype photographs were based upon the tradition of posture pictures common in college physical education programmes. Ben W. Miller, in a discussion of Sheldon's paper delivered before the American Academy of Physical Education (Sheldon, 1952, p. 79), said: 'Our profession has long been occupied with attempts to classify body build. Posture pictures and their analysis were really forerunners of somatotype photographs. So also were efforts to use anthropometry to define constitutional differences.'

In England, James Tanner used somatotyping in much of his research. Tanner first became interested in somatotype studies in the mid 1940s in his postgraduate medical studies at the University of Pennsylvania and later at Johns Hopkins University. He began collecting measurements and somatotype photographs of students and athletes while working at the Department of Human Anatomy at Oxford University. At the Institute of Child Health of the University of London he initiated the Harpenden Growth Study, a childhood to maturity project in which he included somatotype analysis. As mentioned earlier, in connection with his association with Heath, he explored somatotype methods and reliability of somatotype rating. Many of his numerous publications from 1947 onward include somatotype data. His most notable contributions to somatotyping are *The Physique of the Olympic Athlete* (1964) and *Atlas of Children's Growth* (Tanner & Whitehouse, 1982).

Another Englishman, Richard Parnell, who was trained in general medicine and worked in tropical medicine and emergency medicine during World War II, carried out substantial somatotype research. From 1947 to 1951 he was at the Student Health Centre at Oxford University, supported by a Nuffield grant. While there he included somatotypes in evaluations of student health and performance. He realized that combining anthropometry with photography would overcome some of the disadvantages in Sheldon's technique, and would provide more comprehensive and objective data, and that he could derive a preliminary 'phenotype or somatotype estimate' from selected measurements. He devised what he called 'the M.4 method', first published in 1954 and later elaborated in his excellent work *Behaviour and Physique* (1958). From 1952 for eleven years medical research grants supported his work at the Warneford Hospital, Oxford, where he was primarily concerned with constitutional psychiatry. He also was associated with Tanner in a project photographing and measuring students at Oxford. Parnell checked his photoscopic somatotype ratings with Tanner. Although Tanner and Parnell measured some of the subjects at Oxford and at Loughborough (physical education students), they worked independently. After 1963, when research money was no longer available, Parnell returned to practise in the North Birmingham area as a consultant physician specializing in geriatric and psychiatric problems, until his retirement in 1976. Parnell's major contribution was the adaptation of anthropometric measurements to the assessment of somatotype. This adaptation made a basic contribution to Heath's and Carter's 1967 modifications.

Parnell made corrections for age in the scales of total skinfold measurements and height–weight ratio, although he had reservations about the practicability of Sheldon's definition of the somatotype as 'the best available evidence of what the *morphogenotype* is believed to be'. This was an unfortunate concession to the unvalidated Sheldon concept of somatotype permanence. While Parnell recognized some of the valid criticisms of the theory and technique of Sheldon's method, he was unwilling to modify the closed 7-point rating scale, to eliminate the corrections for age, and to restrict somatotype ratings to phenotypic estimates based upon the combined evidence of anthropometry and photoscopy.

The scope of Parnell's work is best reflected in his book on behaviour and physique, in which he investigated a wide range of topics and made a clear statement of hypotheses and research methodology, accompanied by appropriate statistical analyses. Parnell's work included somatotype data for children, family relationships, disease entities, and psychiatric/mental disease problems. He continued this line of enquiry in *Family Physique and Fortune* (Parnell, 1984).

Also in England Roberts & Bainbridge (1963) and Bainbridge & Roberts (1966) made a unique contribution in their study of the Nilotes. Both were initially critical of the somatotype method, as stressed in the (unpublished) paper by Roberts & Weiner to the 1950 Anatomical Congress. But they perceived its potential for population comparisons and were convinced by the concordance they achieved when analysing independently the same subjects after rationalizing the scoring criteria. As mentioned earlier, their work provided new insights into variation in somatotypes and helped to justify important modifications of somatotype method. Surprisingly, their method of assessing dysplasia in physique (1966) has not been applied as widely as it deserves. Peter Bale, Peter Jones, Thomas Reilly and Henry Robson investigated somatotypes in relation to physical performance and John King conducted a large study of somatotypes in industry.

In continental Europe in the 1940s and 1950s a few investigators used Sheldon's system. There were F. Alexander and Pierre Swalus in Belgium, and Peter Verdonck and G. Petersen in The Netherlands. In general Sheldon's somatotype method was not widely accepted in Europe. In part this probably was due to strong traditions of anthropometry in physical anthropology and to wide acceptance of the classical biotypologies of the European schools of constitutional anthropology. Also, many used the approaches of Conrad (1941, 1963) and Lindegard (1953).

In Czechoslovakia somatotyping has been used in studies of children's growth and in relating growth to physical performance. In particular, Jiri Štěpnička used somatotyping in Prague from the late 1960s. By 1970 he had converted to using the Heath–Carter method. His comprehensive work stimulated other users in Czechoslovakia and elsewhere in Eastern Europe. His lead was followed by Jana Pařízková and Vladimir Bok in Prague, and by Anton Zrubák and Eva Chovanová in Bratislava. Otto Eiben, Ivan Szmodis, Jano Mészáros and Istvan Farmosi carried out extensive work in Hungary. The Heath–Carter technique has been used in Poland and the USSR, and to a limited extent in West Germany, Austria and The Netherlands.

Somatotyping has been used fairly extensively in Belgium. During the 1970s Marcel Hebbelinck together with his colleagues Jan Borms, William Duquet and Jan Clarys conducted large studies of children's growth at the Free Univeristy of Brussels. Also, there were extensive studies at the Catholic University of Leuven, directed by Pierre Swalus, and recent studies by Gaston Beunen and his colleagues, Jan Simons and Albrecht Claessens.

Somatotyping has been used to a limited extent in other parts of the world. In South Africa, Phillip Tobias, Marcus Fredman and, recently, Paul Smit and Bruce Copley included somatotypes in their investigations. In Japan, S. Yonemura and K. I. Hirata; in New Zealand, Lindsay Carter, James Hay, Samuel Lewis, Robert Leek and Leslie Williams; in India, J. N. Berry, D. K. Kansal, S. C. D. Rangan, L. S. Sidhu, S. P. Singh and H. S. Sodhi; and in Turkey, Pervin Olgun and C. Gurses, conducted somatotype studies. In Canada, William Ross, Claude Bouchard, Donald Bailey and Robert Mirwald applied somatotyping in large studies. In a number of Latin American countries there have been notable somatotype studies: in Mexico, by Alfonso De Garay and colleagues, and by Luis Vargas and Maria Villanueva; in Venezuela, by Betty Pérez and Fritzi Brief; in Chile, by Bernardo Chernilo; in Brazil, by Claudio Araújo, Paolo Gomes, Eduardo de Rose, Antonio Guimarães and Victor Matsudo; and in Cuba by Carlos Rodríguez.

Somatotyping has played an important role as a method of evaluation in growth studies. It was used in studies in Berkeley, California; in Medford, Oregon; in Toledo, Ohio; in Saskatoon, Saskatchewan; in Trois Rivières, Quebec; in Brussels and in Leuven, Belgium; in Nijmegen, The Netherlands; in Prague, Czechoslovakia; in the Harpenden Growth study in England; and the Kormend Growth Study in Hungary.

Summary

As Hunt (1981) noted, physical anthropology, having been frozen in a typological paradigm for the first half of the twentieth century, shifted to the dynamic, phenotypical viewpoint. In the 1940s Sheldon moved away from strict typology by introducing the concept of three continuous variables to describe human physical variation. However, his promising somatotype method came to be virtually abandoned because of its rigidity of technique and insistence upon an immutable somatotype. But important and essentially sound concepts underlay Sheldon's method, and Heath–Carter modifications of it have produced a useful, widely used research tool for the study of human somatic variation.

The history of somatotyping reflects the importance of listening to the dissenting and doubting voices, who called attention to the importance of recognizing the obvious changes in somatotypes in the course of growth and aging, the irrefutable somatotype changes wrought by dietary deprivation (as in the Minnesota starvation study), the often unpredictable somatotypic changes during growth, and the important insights gained from research in other related areas.

[1] Frederick Wulsin to Barbara Heath, 6 May 1947.

[2] Personal records, Barbara Heath.

[3] Frederick Wulsin to Barbara Heath, 1 May 1948.

[4] Letter from James Tanner to Barbara Heath, 15 April 1953.

[5] 'I should be strongly against endeavouring to rescale somatotype ratings. I don't think there is any point in doing so, except that of tidying up slightly.' Letter from James Tanner to Barbara Heath, 21 October 1953.

[6] Letter from James Tanner to Barbara Heath, 11 November 1953.

[7] 'I have had, as you know, very little experience in somatotyping women and quite obviously set my sights in mesomorphy wrong in this group.' James Tanner to Barbara Heath, 25 June 1956.

[8] Barbara Heath to James Tanner, 17 July 1956.

[9] '. . . it is probably more useful to equate total body fat and endomorphy, and so to rate endomorphy as it is even if we thereby create ratings not previously used'. Barbara Heath to James Tanner, 17 January 1957.

[10] James Tanner to Barbara Heath, 22 January 1957.

[11] James Tanner to Margaret Mead, 23 December 1957.

[12] James Tanner to Barbara Heath, 23 December 1957.

[13] 'If the intervals are equal, why the present arrangement? If we assume 13.20 is approximately the correct index for 4-4-4, 3-4-3 and 2-5-3 (and from experience it seems to fit), then why isn't 4-4-3 at 13.00 with 3-5-3 and 2-5-2? If you apply this rationale to the entire table considerable rearrangement of the table follows. 7-4-1 is at 12.00, not at 11.80, and so on. I tried this out on the Medford series and found they slide into slots far better.' Barbara Heath to James Tanner, 7 February 1958.

[14] '. . . as to your new system of somatotyping I am not so sure I agree with it, solely on the basis that somatotyping is an empirical approach and as yet has little in the way of theoretical foundation, consequently I feel it should be kept an absolutely constant technique in the interests of uniformity of procedures between investigators. When we do have a good theoretical reason (and I don't really feel the notion that height/weight intervals should be equal is a good one), then will be the time to re-do the somatotyping concept. Michael Healey's and my studies here on the amount of bone, muscle and fat in the body related to somatotype are gradually pushed forward . . . I really think that in a year's time we shall have a genuine basis for re-thinking the somatotype in terms partly of shape, but chiefly on tissue component . . . I have stuck fairly close to the Sheldon tables for height/weight and have not been led to assign any somatotypes to Manus that don't exist in the European group. I must confess that I am tempted to do so . . . in the direction of giving a 7.5 or 8.0 in mesomorphy.' James Tanner to Barbara Heath, 2 September 1958.

[15] Margaret Mead to Barbara Heath, 13 February 1958.

[16] Barbara Heath to Margaret Mead, 3 March 1958.

[17] Barbara Heath to James Tanner, 4 April 1958.

[18] Barbara Heath to Carl Seltzer, 15 November 1962.

[19] '. . . what we might call "anthroposcopic acuity" is a most important gift in somatotyping . . . the human eye is able to measure with great accuracy. In medicine doctors "measure" in very much the same sense that one does in anthroposcopy – and record their measurements in scaled ratings. The ophthalmologist "measures" millimeters of retinal arteries by eye, not by tape.' Barbara Heath to Carl Seltzer, 17 January 1963.

[20] 'I read your paper published in the *American Journal of Physical Anthropology* June 1961, with great interest. I have been interested in somatotyping for many years and have somatotyped a number of small samples both in the United States and in New Zealand.' Lindsay Carter to Barbara Heath, 15 June 1962.

[21] 'I would be very interested to hear your views of Parnell's method . . . I plan to keep on collecting data here but I am a little unsure as to whether or not I should attempt to stay with Sheldonian anthroposcopy of the somatotype rating or combine the best of Parnell's method and Sheldon. Perhaps it would be possible to get together sometime.' Lindsay Carter to Barbara Heath, 11 November 1963.

[22] '. . . in fact it seems to me that there is a considerable link between some of his work and the concept of somatotype, or at least the practical application of rating it. Parnell used some of the measurements but did not get as far as talking about lean body mass and percentage fat composition.' Lindsay Carter to Barbara Heath, 20 April 1964.

[23] Heath met both authors later. Roberts sent her part of his Nilote sample, which she used to validate the extension of the scale for ectomorphy. In 1975 Heath visited Roberts' Department of Human Genetics at the University of Newcastle upon Tyne, where she somatotyped the entire Nilote series.

[24] '. . . it almost says a good deal of what I said. Thank heaven my paper preceded it by one issue! I would love to somatotype that series – and hope to meet one or both authors some time.' Barbara Heath to Lindsay Carter, 27 September 1964.

[25] '. . . skewness is not a contraindication to linear relationships. He [Torrance] thinks Parnell boxed himself in, in exactly the same way Sheldon did, by confining the scale to a 1 to 7 range and making corrections for age.' Barbara Heath to Lindsay Carter, 24 May 1965.

[26] 'I am fascinated by the skinfold measurements. I am strongly inclined to think that neither Tanner nor Parnell realize how significant they are. Actually it is impossible to use these measures effectively within the restricted ground rules of the 7-point scale and corrections for age . . . Let's assume that Tanner and Parnell do not have samples of men and/or women representative of a population as a whole. If true, any distribution in their samples will be skewed. It is certain that the Olympic athletes are highly selected. Their somatotype distribution is certainly skewed; and so is their distribution for total fat. Enclosed is a tally for total fat and for endomorphy. Do you see what I see? I didn't see the pattern until I divided the range of total fat measurements into approximately the same number of intervals as the range of ratings in endomorphy . . . It looks as though the skewness almost disappears when men and women are considered together. Is this true? . . . I think if we had a sufficiently large series of skinfold measurements we would find there is a linear relationship between ratings in endomorphy and total fat as reflected in the three skinfold measurements.' Barbara Heath to Lindsay Carter, 24 June 1965.

[27] Lindsay Carter to Barbara Heath, 24 August 1966.

[28] 'I have tabulated my 1958 ratings of 98 of the British Empire Games athletes for comparison with my 1966 corrections of these same ratings. (You will recall that I rated all the photographs in Tanner's book, which added 70 subjects not included in the BEG series.) In 1958 I had expanded the scale, thrown out the age corrections, and established the linear relationship between HWRs and somatotype rating points – but I had not studied skinfold measurements . . . You will note that in 1958 I have ratings of $7\frac{1}{2}$, 8 and $8\frac{1}{2}$ for nine subjects,

but in 1966 I gave 9 in one case, and gave 18 ratings of 7, 8 and 8½. In 1958 I rated 14 as sevens, and in 1966 I rated 23 as sevens. Tanner rated 11 subjects as seven – all the rest were 6½ or less. In 1958 I rated 23 subjects as one in ectomorphy; in 1966 I gave ratings of one-half to 15 subjects, and one to 27 subjects. So you can see a marked influence upon my ratings after I looked again, taking into consideration the total skinfolds.' And: 'It seems to me it means more to say that Olympic athletes and Manus men show more mesomorphy than has been reported in Sheldon's Atlas, and that the degree of increment is reflected in our ratings. The same is true of Seltzer's samples selected for obesity, and for the Nilote study.' Barbara Heath to Lindsay Carter, 13 August 1966.

2 Review of somatotype methods

Introduction

Although Sheldon pioneered the origins and development of somatotyping (he had worked out the basic ideas in the mid 1930s), he, his associates and others modified and changed somatotype method in a variety of ways. Some somatotype methods, although they evolved from Sheldon's method, were based on different premises and used quite different procedures. This chapter reviews methods used in various studies, and compares their criteria, reliability and limitations.

Methods of somatotyping

Sheldon's methods

Sheldon introduced the concept and the word 'somatotype' in *The Varieties of Human Physique* (1940). He recognized the need for classifying the variations of human physique on continuous scales expressed in simple numerical values. Earlier systems of classification characterized a total physique as belonging to a broad category or type based on some anthropometric measurements plus visual impressions. Sheldon defined somatotyping as 'a quantification of three primary components determining the morphological structure of an individual expressed as a series of three numerals, the first referring to endomorphy, the second to mesomorphy, and the third to ectomorphy'.

Endomorphy, according to him, means relative predominance of soft roundness throughout the various regions of the body. When endomorphy is dominant the digestive viscera are massive and tend relatively to dominate the bodily economy. The digestive viscera are derived principally from the endodermal embryonic layer.

Mesomorphy means the relative predominance of muscle, bone and connective tissue. The mesomorphic physique is normally heavy, hard, and rectangular in outline. Bone and muscle are prominent and the skin is made

30

thick by heavy underlying connective tissue. The entire bodily economy is relatively dominated by tissues derived from the mesodermal embryonic layer.

Ectomorphy means relative predominance of linearity and fragility. In proportion to mass, the ectomorph has the greatest surface area and hence relatively the greatest sensory exposure to the outside world and the largest brain and central nervous system. In a sense, therefore, his bodily economy is relatively dominated by tissues derived from the ectodermal embryonic layer.

Following his dubious assertion that the three components are derived solely from specific embryonic layers, Sheldon stated that the somatotype is a trajectory along which an individual under average nutritional conditions and absence of major illness is destined to travel. He used the word 'morphophenotype' to refer to the present physique, and 'morphogenotype' to refer to the genetically determined physique. He maintained that the somatotype does not change through life; that 'the deposit or removal of fat does not change the somatotype, because it does not change significantly any of the measurements, except where the fat is deposited'. He stated that 'a 4-4-4 is not changed by nutritional disturbances to a 4-4-3, or to anything else. He only becomes a fat or a lean 4-4-4.'

From the outset Sheldon was sensitive and defensive about the objectivity and validation of his method. In *The Varieties of Human Physique* (p. 113) he said: 'the problem of somatotyping is to discover "objective" anthropometric correlates for the "subjective" discriminable aspects'. He chose an 'arbitrary' scale that permitted no rating more than 7 or less than 1 in any one component. He stated that 'the variables are independent but their sum is limited by the boundaries given by the numbers 9 and 12'. There is an inconclusive discussion on the rationale of the restricted range of 9 to 12 for sums of the three components of somatotype ratings.

For purposes of validation Sheldon and his associates constructed a series of tables based on ratios of 17 transverse measurements (taken from the 5×7-inch photographic negatives) to height, and entered these in a machine for determining somatotypes from the indices. The procedure was laborious and obviously not feasible for general use. (Tanner (1964, p. 37) said, 'This system does not work, and has never, in fact, been used.') Sheldon himself used somatotype photographs, the record of present height and weight (plus weight history, when available), a table of somatotypes distributed on the criterion of height divided by the cube root of weight, and anthroposcopy. He was justifiably confident of the reliability of his inspectional judgment. He also knew that high levels of skill required long practice. He expected, and received, widespread criticism of the subjectivity of his method. Nonetheless, despite such acerbic and intelligent critiques as Howard

Meredith's (1940), Sheldon had laid the groundwork for future modifications in a useful research area.

Sheldon (1940) summarized his photoscopic (he called it anthroposcopic) somatotype method as follows:

1. Calculation of height/$\sqrt[3]{}$weight ratio (HWR).
2. Calculation of ratios of 17 transverse measurements (taken from photographic negatives) to stature. The transverse measurements, selected from factor analysis of 32 measurements, were: four on the head and neck, three on the thoracic trunk, three on the arms, three on the abdominal trunk, and four on the legs.
3. Inspection of the somatotype photograph, referring to a table of known somatotypes distributed against the criterion of HWR, comparing the photograph with a file of correctly somatotyped photographs, and recording the estimated somatotype.
4. Comparison of the 17 transverse measurement ratios with a range of scores for each ratio, to give a final score.

Sheldon reasoned that the somatotype, derived from scores for the 17 measurements, accurately reflected deviation from the inspectional rating of step 3. This method was described, with an accompanying photograph (Sheldon, 1940, opp. p. 104), but in practice was discarded in favour of reference to a checklist of inspectional criteria (Sheldon, 1940, pp. 35–45), a file of photographic examples, and reference to the distribution of known somatotypes according to HWR.

In *Atlas of Men*, Sheldon (1954) described in some detail a procedure for somatotype rating for men at ages from 18 into the sixties. He introduced age-adjusted weight scales, extrapolated so that a given somatotype at age 18 would be rated the same through life, irrespective of weight gains (or losses) and decreases in stature. To match the hypothetical curves he constructed somatotype–HWR tables, one for each five-year interval from 18 years. A total of 1175 somatotype photographs were arranged to show the 'same' somatotype at different ages for 88 somatotypes. The photographs clearly showed that he gave the same somatotype to physiques differing in appearance. He gave no clear reason for and cited no data to support his persistent idea that a somatotype is permanent. Sheldon *et al.* (1969) later noted that 'otherwise the somatotype would be a mere phenotypic concept subject to change with any change in nutritional status' (p. 840), and that 'It had become very clear that despite nutritional changes the fundamental constitutional pattern remains stubbornly constant through life' (1969, p. 846). Defining the somatotype as permanent and constitutionally based, Sheldon *et al.* produced a system that supported the definition, and effectively resisted reasonable modification. To others it seemed obvious that weight variations change apparent fatness and apparent muscle tissue, and that changes in either of these also change the ratings in ectomorphy, i.e. in relative linearity. In short, individuals may change so that in the course of a lifetime they have a cluster of somatotypes, that may or may not be predictable.

There were indeed four persistent criticisms: (1) the somatotype changes, (2) somatotyping is not objective, (3) there are two, not three, primary components, for endomorphy and ectomorphy are essentially the inverse of each other, and (4) somatotyping omits the factor of size. To meet these criticisms and the negative reactions consistently provided by his somatotype system as presented in *The Varieties of Human Physique* and in the *Atlas of Men*, Sheldon (1961, 1965; Sheldon *et al.*, 1969) described a 'new' Trunk Index method.

The Trunk Index
The Trunk Index somatotype is calculated from the following data:

1. The Trunk Index, obtained from planimetry of standard somatotype photographs.
2. Maximal and minimal weight and stature, reported by the subject.
3. Table of HWRs and trunk indices.
4. A table of somatotypes plotted against maximal stature.
5. 'Basic Tables for Somatotyping', which combine the Trunk Index, maximal stature, and the SPI (somatotyping ponderal index, i.e. the lowest ponderal index on record). The tables are corrected for age and are read differently for men and women. Nothing is said about using them for children, or for subjects shorter than 137 cm (males) or 127 cm (females).

The 7-point rating scale is retained. The sum of the three components may range from 7 to 15.

Instead of presenting new data to defend his concept of permanency of somatotype, he changed tactics, abandoning the by now well-known techniques of his previous method, and substituting one based on even more questionable criteria – planimetry of photographic areas. He wrote, 'We are searching for a set of parameters which both reflect the basic pattern accurately and remain constant.' Among the anthropometric measurements and ratios he considered, he reported that the ratio of the area of the thoracic trunk to the area of the abdominal trunk was relatively constant. He said that after a long search he found by chance one ratio among hundreds that showed a consistent pattern. He implied that he adopted this particular ratio merely because it fitted his preconceived requirements; and that he radically changed his somatotype system merely to make use of a newly discovered 'consistent' ratio.

In this new method, the appropriate ratio was calculated from planimetric measurements of a standard somatotype photograph:

> The trunk is divided at the nearest possible approximation to the plane of the anatomical waist – this is the plane midway between the lowermost plane of the ribs and the uppermost plane of the pelvis, with the individual standing fully erect. For the purpose of deriving and standardising this index – now called the Trunk Index – the TT (thoracic trunk) is the numerator, and the AT (abdominal trunk) is the denominator. The upper limits of TT are defined by a line connecting the points where the sternocleidomastoids cross the trapezii. The lower limit of AT is the photographic center of the gluteal fold.   (Sheldon, 1965, p.8)

Instructions for drawing the areas on the photographs are reasonably clear, but it is by no means certain that even a good anatomist could accurately identify on nude photographs of all kinds of physiques such landmarks as lumbar angle, pubic tubercle, anterior superior iliac spine, and bottom of the ribs.

There is no indication that there are theoretical or statistical relationships between his two methods of calculating a somatotype. It appears that Sheldon merely stated that the Trunk Index differentiates quantitatively endomorphy and mesomorphy, that it is constant (at least from the third year of life to old age), that it is wholly independent of nutritional state, and that it allows for complete objectification of somatotyping. Walker (1974*a,b*) was not altogether successful in his attempt to use the Trunk Index method for somatotyping children and predicting adult height and SPI from children's growth data.

The 'Basic Tables for Somatotyping' show the somatotype rating and Trunk Index appropriate for given heights, weight and SPIs at ages from 20 to over 50 at ten-year intervals. Men and women with the same somatotypes have the same trunk indices, but heights and weights differ in that women are shorter and lighter at the same row in the table. The tables can be manipulated objectively, but the question of their meaning remains.

Thus it appears that Sheldon responded to continued criticism of his method with a quantum leap from an original, albeit moderately subjective system of somatotyping to a method that he said was objectively determined but has little or no apparent relationship to his previous method.

He reported (Sheldon *et al.*, 1969) that the Trunk Index method was used in a review of the longitudinal study of children at the Institute of Child Welfare, University of California, Berkeley; the starvation experiment at the University of Minnesota; a study of United States Military Academy (West Point) students; a study of patients in the outpatient clinic of

Columbia Presbyterian Hospital, New York City; a study of 46 pairs of identical twins; and a study of men and women attending a nutrition clinic. He reported that in all these studies the Trunk Index remained constant. However, except for one correlation coefficient ($r = 0.98$, the correlation between trunk indices before and after starvation) he provided no data or analyses to support these findings.

It is posssible that in the succession of life phases the area ratio of thoracic trunk to abdominal trunk remains constant, that the two areas increase and decrease in size in relation to one another. However this is not evidence that the somatotype remains constant. Moreover, the age and sex limited tables appear to fit a preconceived idea, as did the extrapolated tables of the age-corrected tables of somatotype distribution. Sheldon's new method had bases, premises and substance which differed from his previous systems. He asserted that the new system provided a measure of massiveness (HWR), a separator for the kinds of mass into endomorphy and mesomorphy (the Trunk Index) and, finally, a measure of the degree of stretched-outness into space (height). 'When the other two parameters are known, this is precisely what ectomorphy is' (Sheldon *et al.*, 1969, p. 848).

This statement on ectomorphy is a major departure from the original concept that somatotyping is a measure of shape, not of size. In his new method Sheldon used height as a measure of size, and substituted it for ectomorphy. He stated that, as a result of these changes, the sum of the somatotype components no longer need be limited to sums of 9 to 12, but in fact now extend to sums of 7 to 15. The original (1940) matrix of 76 somatotypes expanded to 88 (*Atlas of Men*, 1954); the Trunk Index matrix gives 267.

In his new components Sheldon was unclear on a fundamental point. He said: 'The somatotype is a simple point within cubical or three-dimensional space where three distributions cross. The three are: trunk index, height, and the somatotyping ponderal index' (Sheldon *et al.*, 1969, p. 870). He does not explain what became of the components or say whether this is a definition or a re-definition. Endomorphy is defined as the predominance of the abdominal trunk area over the thoracic trunk area as determined from the Trunk Index; mesomorphy is the dominance of the thoracic trunk area over the abdominal trunk area as determined from the Trunk Index; and ectomorphy is equal to height.

In a series of statements, for which he presented no scientific evidence, Sheldon tried to dispose of the four critical objections to his original somatotype method. He asserted that the somatotype cannot change, since maximal stature and maximal massiveness are simply items of historical fact, and Trunk Index is constant throughout life. It appears he reached this conclusion by manipulating the material simply to make a point.

He announced that the somatotype is completely objective and can be derived on a computer. But this only says that the Trunk Index is objective in the sense that little judgment and skill are required to delineate the areas on the trunk for planimetric measurement. It is so difficult to locate the required anatomical sites on photographs that there are serious variations among observers. Obviously computer derivation of the somatotype from planimetric data does not make it objective. It simply yields consistent calculations.

He said that the high negative correlations among the components had been corrected, thus ending the controversy over two versus three primary components. Actually this may not have been a valid objection. The point came up in factor analysis studies, some of which were designed with inappropriate measurements for distinguishing between endomorphy and mesomorphy, while in others sampling was so limited that the ranges of somatotype component ratings were incomplete and biased the statistical results.

In answer to the objection that somatotype omits the factor of size, he said that size had been restored by using stature as a determining parameter. That objection is not valid, because originally he defined the somatotype as a measure of shape. There are many established measures for body size, which need not be incorporated in this system. By introducing size he abandoned a useful concept in response to an invalid criticism.

Although the Trunk Index bore little resemblance to his previous method, it clearly did not meet all the objections to the original Sheldon method. It, like its predecessor, lacked scientific clarity, and was not based on tested hypotheses or normal statistical procedures. Finally, the requirement of maximal stature and minimal HWR ruled out its usefulness in growth studies.

Anthrotyping

There is an undated manuscript in the Tozzer Library, Harvard University, written by Sheldon and Tucker at the University of Chicago in about 1937 or 1938, entitled 'The anthrotyping technique – a method for studying human variation'. The manuscript describes a method that uses tables and ratios, defines primary and secondary components, and presents some material on the relationships between disease, psychological 'levels' and somatotype. The three elemental structural components are defined as exhibiting softness, ruggedness and linearity/fragility. The authors labelled these components 'endosomic', 'somatosomic' and 'leptosomic' and called the technique 'anthrotyping'. The secondary components are labelled '*g*' (for gynandromorphy) and '*d*' (for dysplasia, which for the first time is defined as 'basic structural disharmony'). The paper also reports that one of the writers had developed a technique for measuring, at the psychological level of personality, what seemed to be reflections of the same three elemental components, called viscerotonia, somatotonia and neurotonia (to be changed later to cerebrotonia). The forerunner of the somatochart first appears in this paper, with 24 tables showing the distributions of anthrotypes according to various anthropometric criteria, and the correlation matrices between the components for males and females. This early somatochart is referred to as 'a two-dimensional approximately proportional representation of the anthrotype, based on 4000 men and 2500 women'.

It is obvious that the material in this manuscript was expanded and modified into *The Varieties of Human Physique* (1940). In the latter, the terminology was changed to endomorphy, mesomorphy and ectomorphy. The somatochart 'triangle' is inverted, so that the 1-7-1 is at the top. The words endomorphy, mesomorphy and ectomorphy appear along each axis instead of endosomia, somatosomia and leptosomia.

Except for an abstract of a paper by Sheldon & Tucker (1938) at a meeting of the American Association of Physical Anthropologists, there had been no publications on somatotyping. *The Varieties of Human Physique* was essentially new to the scientific community when it was published in 1940.

Hooton's method

During the 1940s Hooton (1951) modified Sheldon's somatotype method for rating a large series of United States Army men. He used the term 'body build', not somatotype, and made ratings based on the height–weight ratio and inspection of standard somatotype photographs. He defined the first component as a concept dealing with fat development, the second component as a concept of bone size and muscle relief, and the third component as a concept of relative elongation or attenuation of the body. He rated the first and second components by appraising the areas of the thorax, abdomen, and upper and lower extremities, and averaged separately the four regional ratings to give a total body rating for fat and bone–muscle development. He made third component ratings from 1 to 7 from a table of distribution of HWRs. The HWRs were calculated from both metric units and Imperial units. Hooton used metric ratios, and Sheldon used Imperial ratios. Hooton used a whole-unit 7-point scale, and Sheldon used a 13-point scale (7 whole units and their half-unit intervals).

Dupertuis & Emanuel (1956) elaborated on the criteria of the Sheldon method, defining endomorphy as soft roundness, mesomorphy as muscularity and solidity, and ectomorphy as linearity and delicacy. They used extensive key files of somatotype photographs from Sheldon's laboratory, and his HWR tables, to aid visual inspection. They commented, 'According to Sheldon, the procedure in somatotyping is always to somatotype the whole body and then to look at different regions for dysplasias and other secondary components' (p. 3).

According to Dupertuis and Emanuel (1956), a report from Hooton says F. L. Stagg was 'the chief somatotyper' for his laboratory, and the final ratings were made in collaboration with himself. Hooton adds that 'the ratings of the first component on a seven point scale are, at best, approximations and the body types determined by rating the three components are in no sense biological eternal verities'. He also stated (p. 24) that he did not believe it possible to determine with accuracy whether a fat man should be rated 7-1 or 7-2 (endomorphy and mesomorphy ratings) by inspection of a photograph. He said that first component ratings were based on strict averaging of independent assessments of the four different regions. 'We do not believe in glancing at the whole anatomy and giving it an overall rating on the basis of general impressions (usually too much influenced by the muscular development in a few places, e.g. the trapezii, the pectorals, the muscles of the upper arms).' Others might argue that the body is a single construct better judged by overall scanning, plus the anthropometry now required, than by commanding the eye to break the body up into discrete segments.

Bullen and Hardy's method

When they made a study of physiques of 175 college women, Bullen & Hardy (1946) found no published data on female somatotypes, and only passing speculative references in four or five paragraphs in *The Varieties of Human Physique* (1940). Therefore they derived from Sheldon's criteria a checklist of 105 specific points based on observable criteria for predominance of a component. They believed that their checklist reduced to a minimum the problems of a rating continuum, could be used for comparisons between all age and sex groups, and could serve as a universal scale to be applied to all groups. This method was applied in the studies of Bullen (1952), Danby (1953), Kraus (1951) and Roberts & Bainbridge (1963).

Bullen and Hardy calculated the final somatotype from five regional ratings. Their endomorphy ratings ranged from 1 to 6, mesomorphy from 2 to 7, and ectomorphy from 1 to 7. They did not report whether they gave the same absolute rating values for women as for men. However, Bullen (1952) raised the question of relative versus absolute rating scales and pointed out that in *The Varieties of Human Physique* (1940) Sheldon used relative scales for women, with 7 as the maximum for female mesomorphy. Thereafter he applied absolute scales. In general the literature reflects use of absolute scales.

When Roberts and Bainbridge (1963) found in their study of Nilotic physique that Bullen's and Hardy's criteria, modified by Danby (1953), were inadequate for discrimination among higher levels of ectomorphy in their sample, they adopted additional criteria of length and breadth for each of the five regions (two in each region), and extended the HWR table to accommodate physiques that they rated as 1-2-8, 1-2-9, 1-1-9, and so on.

Cureton's method

Cureton (1947) devised a somatotype method that combined inspectional ratings of the photographs, palpation of the musculature, skinfold measurements, HWRs, and assessments of strength and vital capacity. He also developed an abbreviated checklist for somatotype rating. Cureton's (1947) simplified physique rating method rates '(1) External Fat, (2) Muscular Development and Condition, (3) Skeletal Development' on a scale of 1 to 7. The rating scales appear to be based upon body composition criteria, and the third component seems to have been redefined to combine characteristics previously associated with mesomorphy as being part of ectomorphy. It appears that sometimes Cureton used Sheldon's original criteria plus photoscopic ratings, and sometimes his own simplified method. He believed the rating system consisted primarily of assessments of bone, muscle and fat. He did not use age-adjusted scales because he worked with college men and young athletes. He seemed to treat all ratings as phenotypic.

In 1951 he added an objective scale, which was normally distributed, as the horizontal base of the somatotype distribution triangle. He also used various physical performance tests to explain the vertical distribution in terms of strength. He justified this by factor analysis, which showed a four-item strength index almost on the vertical axis of the somatotype distribution triangle, and the horizontal base of the triangle corresponding to 'thinness and heaviness in fat', or to 'vital capacity residuals' increasing with ectomorphy and decreasing with endomorphy, or to large girth measures of endomorphy compared with small girth measures corresponding to ectomorphy. He arranged these objective distributions horizontally across the bottom of the triangle as standard scores, and argued that the Sheldon, Stevens and Tucker somatotype method could be reduced to a two-dimensional system that could be scored objectively (Cureton, 1951).

Parnell's M.4 deviation chart method

Parnell (1954, 1958) made the first substantial contribution to making the somatotype method more objective. He developed a scoring method for using anthropometric measurements, which he recorded on the 'M.4 Deviation Chart to use in conjunction with somatotype photographs'. The M.4 deviation chart (Fig. 2.1) includes tables for obtaining an anthropometric somatotype. Parnell substituted the terms *fat, muscularity* and

ADULT DEVIATION CHART OF PHYSIQUE (Male Standards)

NAME .... AGE .... DATE ....

OCCUPATION ....Married/Single Ch.: M .... F .... REF. No. ....

Fat: (Skinfold)	Age	Total 3 Skinfold Measurements												
(mms.) Over triceps	16 – 24	10	12	14	17	20	24	29	36	45	57	73	93	114
Subscapular	25 – 34	12	14	17	20	24	30	38	48	60	74	94	114	+
Suprailiac	35 – 44	13	16	19	22	27	35	44	55	68	87	109	+	+
Total fat	45 – 54	14	17	20	23	29	37	47	61	74	95	118	+	+
ENDOMORPHY Estimate		1	1½	2	2½	3	3½	4	4½	5	5½	6	6½	7

Height (ins.)	55.0	56.5	58.0	59.5	61.0	62.5	64.0	65.5	67.0	68.5	70.0	71.5	73.0	74.5	76.0	77.5	79.0	80.5
Bone: Humerus	5.34	5.49	5.64	5.78	5.93	6.07	6.22	6.37	6.51	6.65	6.80	6.95	7.09	7.24	7.38	7.53	7.67	7.82
(cms.) Femur	7.62	7.83	8.04	8.24	8.45	8.66	8.87	9.08	9.28	9.49	9.70	9.91	10.12	10.33	10.53	10.74	10.95	11.16
Muscle: Biceps	24.4	25.0	25.7	26.3	27.0	27.7	28.3	29.0	29.7	30.3	31.0	31.6	32.2	33.0	33.6	34.3	35.0	35.6
(cms.) Calf	28.5	29.3	30.1	30.8	31.6	32.4	33.2	33.9	34.7	35.5	36.3	37.1	37.8	38.6	39.4	40.2	41.0	41.8

First estimate of mesormorphy	1	1½	2	2½	3	3½	4	4½	5	5½	6	6½	7	
Correction for fat (T.F. mms)	12	15	18	22	27	33	40	48	57	68	83	100	120	140
Age: 16 – 24	+½	+½	+¼	+¼	0	−¼	−½	−1	−1½	−2	−2½	−3	−4	
25 – 34	(+½)	+½	+¼	+¼	0	−¼	−½	−¾	−1¼	−1¾	−2¼	−2¾	−3½	−4
35 +	(+½)	(+½)	+¼	+¼	0	−¼	−¼	−½	−1	−1½	−2	−2½	−3	−3½
MESOMORPHY (corrected estimate)	1	1½	2	2½	3	3½	4	4½	5	5½	6	6½	7	

Weight Wt. lb. H.W.R.	Age.													
Present	18	12.1	12.3	12.5	12.7	12.9	13.1	13.3	13.5	13.7	13.8	14.0	14.2	14.4
H.K.W.	23	11.7	12.0	12.2	12.5	12.8	13.0	13.2	13.4	13.6	13.8	14.0	14.2	14.4
Usual	28	11.5	11.8	12.1	12.4	12.6	12.8	13.0	13.3	13.5	13.7	13.9	14.2	14.4
At 18 years	33	11.3	11.7	12.0	12.3	12.5	12.7	12.9	13.2	13.4	13.6	13.9	14.1	14.4
At 23 years	38	11.2	11.5	11.8	12.1	12.4	12.6	12.8	13.1	13.3	13.6	13.9	14.1	14.4
Recent change	43+	11.1	11.4	11.7	12.0	12.3	12.6	12.8	13.1	13.3	13.6	13.9	14.1	14.4
ECTOMORPHY		1	1½	2	2½	3	3½	4	4½	5	5½	6	6½	7

Fig. 2.1. The Parnell M.4 adult Deviation Chart for determining the somatotype. This chart is based on Parnell's (1958, p. 21) chart with the values for the Harpenden skinfold caliper inserted for the determination of the endomorphy estimate.

linearity (abbreviated to F, M and L) for Sheldon's component names. The estimate for F, or endomorphy, is derived from the skinfold measurements. The estimate for M, or mesomorphy, is derived from height, bone diameters and limb girths. The estimate for L, or ectomorphy, is derived directly from the HWR. Each of the three component scales is corrected for the several age groups shown on the M.4 adult deviation chart. For rating muscularity (Sheldon's mesomorphy) the M.4 chart is based upon the assumption that a rating of 4 in muscularity bears a constant proportional relationship to stature.

Parnell intended that his M.4 ratings should correspond as closely as possible with Sheldon's somatotype ratings. However, his change of the component names to F, M and L was to emphasize that the ratings are derived from physical measurements and therefore do not describe exactly the same entities that Sheldon rated photoscopically. That is, he emphasized that his ratings are phenotypic and describe a physique at a given time. But his provisions for age correction give the impression that in fact he tried to preserve the constancy of the somatotype over time. He provided a separate M.4-style deviation chart for 11-year-olds (Parnell, 1958).

Damon's anthropometric method

Damon *et al.* (1962) used a multiple regression technique for predicting somatotypes from anthropometric measurements of Black and White soldiers. Their 49 anthropometric measurements included weight, lengths, breadths, depths, circumferences, skinfolds, grip strengths, and pulmonary function. On a 7-point scale 80% of the predictions came within one-half rating unit of the photoscopic ratings made by Damon (an experienced Sheldonian rater). Multiple correlation coefficients (R) for endomorphy, mesomorphy and ectomorphy were: 0.78, 0.66 and 0.90 for Whites; and 0.83, 0.84 and 0.88 for Blacks. As many as ten different measurements were used in some of the equations for predicting a given component. Grip strength scores and pulmonary function were not used in any of the equations.

Medford equations

Clarke (1971) used equations for predicting somatotype components for boys aged 9 to 17 years in the Medford Growth Study, conducted by the University of Oregon and directed by him. In the same study Sinclair (1966, 1969) and Munroe *et al.* (1969) derived regression equations from anthropometric and performance measures for predicting somatotype components rated by Heath. A variety of equations with multiple regression correlations were quite high for the first and third components, but lower for the second component. The predictions for the second component improved when values for the first and third components were used in a regression equation. The regression equations were specific to the age at which the measurements were taken.

Petersen's method for children

Petersen's (1967) atlas contains 560 somatotype photographs of children aged 6 to 15 years, selected from a study of 12 000 school children in Holland at the time of routine school medical examinations. The author gave the children's ages and somatotype ratings, but included no measurements or other data; for somatotype ratings he relied upon what he variously called 'somatoscopy', 'photoscopy' and 'scopy'. He used no objective criteria. The photographs are presented in order of ascending endomorphy, beginning with the 1s and progressing to 7s. Although well produced, the book is not really useful to those interested in growth and other studies of children. It has no objective criteria for rating children's somatotypes, and ignores the fact that Sheldon himself published no criteria for children's somatotypes.

The Leuven method

The Leuven Growth study (Ostyn *et al.*, 1980) is a mixed longitudinal study of growth and motor performance in 12–17-year-old Belgian boys, started in 1967. The investigators developed a modified method of somatotyping, based on comparisons of the methods of Sheldon, Parnell, and Heath and Carter (Swalus and Van der Maren, 1968–9; Swalus, 1969; Swalus *et al.*, 1970), and called it the Leuven method. They derived a phenotypical somatotype from: (1) a first estimate of endomorphy, based on the sum of three skinfolds, according to Parnell (1957); (2) a first estimate of ectomorphy, based on the HWR, according to Heath & Carter (1967); and (3) an estimate of mesomorphy (and the final rating) based on comparisons with the photographs of 16–24-year-olds in Sheldon (1954).

The Heath–Carter somatotype method

Heath (1963) described certain limitations in Sheldon's method and suggested the following modifications to overcome them:

1. Replace the arbitrary 7-point scale with a rating scale of equal appearing intervals, beginning theoretically with zero (in practice beginning with one-half) and having no arbitrary end point.
2. Eliminate the unjustified restrictions of sums of components to between 9 and 12.
3. Construct a table that preserves a logical linear relationship between somatotype ratings and HWRs.

4. Adopt a single table of HWRs and somatotypes suitable for both sexes at all ages.

Heath's (1963) table (Table 2.1) was based on Sheldon's (1954) table for males at age 18 years, which showed the distribution of somatotypes according to HWR. She found that 15 somatotypes (marked with asterisks in Table 2.1) were placed so that there was a difference of 0.20 in HWR with an increase or decrease of one unit in one component rating. For example, the HWRs for the somatotypes 3-6-1 and 2-6-1 are 12.40 and 12.60, and the HWRs for the somatotypes 3-5-2 and 2-5-2 are 12.80 and 13.00. That is, on Sheldon's table there is a logical linear relationship between HWR and ratings for 15 somatotypes, which is compatible with Heath's photoscopic ratings. She therefore redistributed the remaining somatotypes so that for each increase or decrease of one rating unit in one component there is a corresponding increase or decrease of 0.20 in HWR.

She gave ratings to present somatotypes, or morphophenotypes, without estimates of the somatotype at age 18 and without predictions of future somatotypes. The concept of a series of somatotypes for each individual replaced that of one somatotype for a lifetime. It became possible to study the evolution of adult somatotypes from a succession of pre-adult somatotypes. And it became possible to compare directly subjects of different ages in a given reference population, because they were measured on the same scale. In the course of the decade following 1963 she applied the foregoing criteria to rating approximately 15 000 somatotype photographs, which were part of the data in more than 30 published studies.

Heath & Carter (1967) further objectified Heath's system by incorporating a modified form of Parnell's M.4 technique. They defined somatotype as 'present morphological conformation', expressed in a three-numeral rating of primary components of physique that identify individual features of morphology and body composition. The first component, or endomorphy, refers to relative fatness and leanness. The second component, or mesomorphy, refers to relative musculoskeletal development for the individual's height. It may be thought of as relative lean body mass. The third component, or ectomorphy, represents relative linearity of individual physiques. Its ratings, derived largely but not entirely from HWRs, evaluate body form and longitudinal distribution, or 'stretched-outness' of the first and second components. Extremes in each component are found at both ends of the scales. That is, low first component ratings signify physiques with little non-essential fat, while high ratings signify high degrees of non-essential fat. Low second component ratings signify light skeletal frames and little muscle relief, while high ratings signify marked musculoskeletal development. Low third component ratings signify great mass for a given

Table 2.1. *Distribution of somatotypes according to HWR [height/cube root of weight] for all ages*

HWR (Imperial units)	Somatotypes					
11.40	9-5-1					
11.60	9-4-1	8-5-1				
11.80	8-4-1	7-5-1				
	4-8-1					
12.00	7-4-1	4-7-1				
	6-5-1	5-6-1				
12.20	7-3-1	3-7-1	5-5-1			
	6-4-1	4-6-1				
12.40	7-2-1	2-7-1	7-3-2	3-7-2		
	*3-6-1	6-3-1	6-4-2	4-6-2		
	5-4-1	4-5-1	5-5-2			
12.60	7-1-1	1-7-1	*3-6-2	5-4-2		
	*2-6-1	6-2-1	4-5-2	7-2-2		
			2-7-2	6-3-2		
12.80			7-1-2	1-7-2	6-3-3	3-6-3
			6-2-2	5-3-2	5-4-3	4-5-3
			*3-5-2	4-4-2		
			*2-6-2			
13.00			6-1-2	1-6-2	*2-6-3	6-2-3
			*2-5-2	5-2-2	*3-5-3	5-3-3
13.20			*2-5-3	5-2-3	*4-4-4	3-5-4
			*4-3-3	6-1-3	5-3-4	
			*3-4-3	1-6-3		

13.40			*2-5-4	5-2-4					
			6-1-4	4-3-4					
			*3-4-4						
13.60			4-2-4	3-3-4	*3-4-5	4-3-5			
			5-1-4	1-5-4	5-2-5				
			2-4-4						
13.80					3-3-5	2-4-5			
					4-2-5	5-1-5			
					1-4-5	4-1-5			
14.00					2-2-5		3-3-6	4-2-6	
							2-3-6	3-2-6	
							4-1-6		
14.20							3-1-6		
							2-2-6		
14.40							1-2-6	1-3-6	
14.60									3-2-7
14.80								2-1-6	2-2-7	3-1-7
									1-2-7	2-1-7
15.00									1-1-7

Adapted from Heath (1963).

*These somatotypes are in the same locations as in Sheldon's (1940) table.

height, while high ratings signify linearity of body segments and of the body as a whole with little mass for a given height, together with high HWRs. (See Appendix I and the Glossary.)

The following data are needed for Heath–Carter ratings: height, weight, four skinfolds (triceps, subscapular, suprailiac, calf), two bone diameters (humerus, femur), two muscle girths (calf, flexed arm), age and a revised HWR table. The Heath–Carter anthropometric somatotype, which correlated highly with criterion ratings by Heath, can be calculated directly from these data recorded on the rating form. The final somatotype rating is based on the anthropometric somatotype, the somatotype photograph, and reference to the distribution table for somatotypes and HWRs. Skilled somatotypers are able to make ratings from the photograph and the HWR table alone (see Appendix I).

Comment on somatotype methods and modifications

Hunt (1949) called attention to the pitfalls in idiosyncratic somatotype rating methods that arise when different raters (simply because they are different people) differently interpret the component criteria for subjective ratings of photographs. Following the publication of Sheldon's first somatotype books some investigators modified in various ways what they understood to be his method. Some made minor modifications and some produced separate methods. The modifications of Dupertuis & Emanuel (1956), Bullen & Hardy (1946), Damon *et al.* (1962), Petersen (1967), Preston & Singh (1972), Clarke (Medford equations, 1971) and Roberts & Bainbridge (1963) were minor.

Some modifications were designed to interpret data which did not appear amenable to Sheldon's method and criteria. Modifications such as Preston's & Singh's (1972) 'redintegrated somatotyping' were merely aberrant and created confusion. They integrated photographic diameters of limbs and torso with height, using photoelectronic computation and analysis, to produce graphs. They produced exact (if meaningless) quantifications, more appropriate to problems of body surface measurement than to somatotyping.

Somatotype is a generic term that embraces a number of methods derived from the original concepts of Sheldon, who coined the word. However it is defined, a body build method is not a somatotype method unless it includes a three-component rating of physique based upon the components endomorphy, mesomorphy and ectomorphy.

Comparisons of somatotype methods

In the following review 'Sheldon's method' refers to his 1940 or 1954 method and 'Trunk Index' refers to his 1969 method. The 'Heath method' refers to ratings based on her 1963 modifications, and the 'Heath–Carter method' refers to their 1967 modifications, with which anthropometric, photoscopic, or anthropometric plus photoscopic somatotypes may be rated.

Hooton versus Sheldon

Dupertuis & Emanuel (1956) compared the somatotypes of 500 Air Force flying personnel rated according to Sheldon's method and Hooton's method. Hooton and Stagg made the Hooton photoscopic ratings, and Dupertuis made the Sheldon photoscopic ratings. Comparison of the systems (Table 2.2) shows moderately good correlations between the two sets of ratings for all three components, but the differences between the component means are significant. For the first component the Hooton ratings are 0.5 units greater than the Sheldon ratings, for the second component 1.0 units less, and for the third component 0.7 units greater. Dupertuis and Emanuel point out that they chose a sample of extremes of physique, which may have yielded higher correlations than could be expected from a sample selected randomly from a total sample of Air Force personnel.

Parnell versus Sheldon

To determine the validity of his deviation chart, Parnell (1954) compared his somatotype ratings obtained by anthropometry with ratings by Sheldon's photoscopic method. He used males at ages 16 to 20 (N = 154). He found agreement within one-half unit in 90% of the ratings. He also found that 87.3% of the ratings agreed within one-half unit when he compared 282 Oxford males rated by the deviation chart method with photoscopic ratings by Tanner (1952). Tanner's mesomorphy ratings were 0.22 units lower than the deviation chart ratings. Parnell (1958) reported further comparisons of ratings of 2063 healthy young men by the M.4 method and by Sheldon's methods. He found discrepancies exceeding one-half unit in 3.9% of the ratings of endomorphy, in 10.6% of the ratings of

Table 2.2. *A comparison of the mean values of Hooton and Dupertuis ratings for the primary body build components (N = 500)*

	Hooton		Dupertuis			
Component	Mean	SD	Mean	SD	Critical ratio	*r*
First	3.86	1.17	3.34	1.31	15.38	0.82
Second	3.43	1.12	4.44	1.17	33.74	0.83
Third	3.69	1.33	3.02	1.46	20.01	0.88

From Dupertuis & Emanuel (1956).

mesomorphy, and in 9.8% of the ratings of ectomorphy. He commented that his (1954) M.4 method and photoscopic ratings differed little for the first and third components, but for the second component about one-third of the discrepancies exceeded one-half unit.

Cureton versus Sheldon

In a comparison of ratings of 50 males that Sheldon had somato-typed, Cureton (1947) had small average errors for endomorphy (+0.13) and mesomorphy (−0.06), and a large error for ectomorphy (+1.51). He suggested that the large error could be due to rating tall, large-boned men too high in ectomorphy. However, in a sample of Danish athletes, with a mean HWR of 12.43 (or its metric equivalent 41.09), his mean somatotype was $2\frac{1}{2}$-$5\frac{1}{2}$-4 (Cureton, 1951, p. 26). According to Sheldon's distribution table a somatotype of 3-6-1 is most probable for that HWR, and a $2\frac{1}{2}$-$5\frac{1}{2}$-4 would have a HWR of 13.10 (43.31). Such disparities may be due in part to using different rating criteria and in part to including performance scores as criteria for somatotype rating. Moreover, when Cureton constructed a somatotype triangle with ectomorphy on the left side and endomorphy on the right side, he failed to meet his own criterion that somatotypes should be placed at the location indicated on the Sheldon 'triangle'.

Trunk Index versus Parnell M.4

In a study of 102 normal male university students (out of an initial 126 volunteers) Haronian & Sugarman (1965) described and compared the Sheldon Trunk Index and the Parnell M.4 methods, but did not report who made the measurements and the somatotype ratings, or the reliability of the techniques used. They did report that they had removed from the study 20 subjects (which still leaves 4 subjects unaccounted for) whose somatotypes were near the means for the group, because they thought that 'to use these subjects for the purpose of correlating their performances on the psychologi-cal tests with their somatotype ratings would have been wasteful effort'. Of course actually they altered the true nature of the sample and significantly altered the somatotype distribution.

Later Haronian & Sugarman asked Heath to rate the same series (oddly, the number now was 101, not 102). She applied the revised Heath–Carter (1967) criteria, referred to the skinfold data, the HWR data and the photographs. Her mean somatotype was 3.4-4.6-2.9, the Trunk Index mean was 3.8-4.7-3.0, and the Parnell M.4 mean was 3.5-3.7-3.3. Correlations for the components could not be calculated without individual Trunk Index and Parnell ratings, but it is clear that the Heath and Trunk Index means are quite comparable and the Heath and Parnell means for mesomorphy markedly different. Haronian & Sugarman reported that the Trunk Index and Parnell methods produced significantly different means: higher for endomorphy (+02.7) and mesomorphy (+0.98), and lower for ectomorphy (−0.23).

Heath versus Parnell M.4

A comparison of Parnell's M.4 and Heath's photoscopic somatotype ratings of 59 adult males and 61 adult females (Heath & Carter, 1966) showed that Heath's first component means were significantly lower than the M.4 means for males, higher for the second component for females, and lower for the third component for both sexes.

Product–moment correlations between the two methods were low for the first component ($r = 0.74$), but relatively high for the second ($r = 0.89$) and the third ($r = 0.92$). Some of the differences are doubtless due to the Parnell age-scaled corrections versus the Heath–Carter universal reference scale for both sexes at all ages.

Heath versus Trunk Index in boys

Morton (1967) compared the Heath photoscopic and the Trunk Index methods in a study of 106 boys, aged 9 to 16, in the Medford Growth Study at the University of Oregon. The Trunk Index mean for endomorphy is approximately 0.8 units higher than the Heath mean (Table 2.3) and the majority of the Trunk Index means for mesomorphy and ectomorphy are 0.25–0.35 units higher than the Heath means (Clarke, 1971). For endomorphy, correlation between the two methods is 0.48 at age 9 and between 0.69 and 0.76 at other ages; correlations are 0.39–0.58 for mesomorphy and 0.55–0.70 for ectomorphy at all ages. This study showed that ratings by these methods differed in all of the components, that somatotype distributions were different at all ages, and associations between the methods no more than moderate.

In a study of 68 boys, aged 7.1 to 12.6, in a Sports Fitness programme at the Champaign–Urbana University of Illinois, Slaughter & Lohman (1977) obtained means of 2.6-4.1-3.3 for Heath–Carter ratings and 3.8-4.5-3.5 for Trunk Index ratings. The correlations between the two methods, $r = 0.82$ for endomorphy, 0.36 for mesomorphy and 0.80 for ectomorphy, are similar to Morton's (1967) values.

Heath versus Sheldon

Zuk (1958) compared Heath's and Sheldon's ratings of the subjects in the Berkeley Growth Study at ages 17 and 33. For males the correlations between raters ranged from 0.61 for endomorphy at age 17 to 0.89 for ectomorphy at age 33; for females the correlations were 0.24 for endomorphy and 0.83 for ectomorphy.

Table 2.3. *Differences between means of somatotype components when assessed by Trunk Index and anthroposcopic techniques*

Age	Means		Mean difference
	Trunk Index	Anthroposcopy	
Endomorphy			
9	4.27	3.38	0.89 ⎤
10	4.36	3.55	0.81 ⎥
11	4.23	3.41	0.82 ⎥
12	4.29	3.49	0.80 ⎥
13	4.31	3.51	0.80 ⎬ $p > 0.01$
14	4.30	3.29	1.01 ⎥
15	4.28	3.29	0.99 ⎥
16	4.31	3.42	0.99 ⎥
MP[a]	4.29	3.95	0.34 ⎦
Mesomorphy			
9	4.43	4.22	0.21 ⎤
10	4.36	4.15	0.21 ⎥
11	4.41	4.17	0.24 ⎥
12	4.39	4.05	0.34 ⎬ $p > 0.01$
13	4.37	4.03	0.34 ⎥
14	4.40	4.08	0.32 ⎥
15	4.40	4.12	0.28 ⎦
16	4.36	4.20	0.16
MP[a]	4.34	4.37	−0.03
Ectomorphy			
9	3.43	2.90	0.53 $p > 0.01$
10	3.47	2.85	0.62 $p > 0.01$
11	3.41	3.18	0.23 $p > 0.05$
12	3.45	3.20	0.25 $p > 0.01$
13	3.40	3.17	0.23 $p > 0.05$
14	3.40	3.35	0.05
15	3.39	3.47	−0.08
16	3.51	3.43	0.07
MP[a]	3.48	2.51	0.97 $p > 0.01$

From Clarke (1971).
[a] MP, age of minimum inverse ponderal index.

Zuk said that 'Dr Sheldon did his set of age 33 ratings by inspection of the photographs and supervised the age 17 ratings, which were turned out "mechanically" by the Harvard Somatotyping Machine.' It seems improbable that the age 17 ratings were done by machine, because Sheldon had abandoned the machine at Harvard when he rated these photographs in Heath's presence about 1950. Unfortunately, the correlations between Sheldon's and Heath's ratings do not show that Sheldon insisted that there had been no changes in somatotypes from age 12 to 17. When Heath re-rated the series in the mid-1950s, applying her modifications (see Heath, 1963), she found significant and often startling somatotypic changes during growth.

Sheldon versus Heath–Carter

Štěpnička (1970, 1974a) compared his Sheldon photoscopic ratings with his Heath–Carter photoscopic plus anthropometric ratings of more than 600 Czechoslovakian male athletes and approximately 300 male university students. For reference he used the 16–24-year-old subjects in Sheldon's (1954) *Atlas of Men*. Means for his two sets of ratings are shown in Table 2.4.

Štěpnička found the Heath–Carter scale improved the differentiation of degrees of mesomorphy, particularly as shown in the gymnasts and body builders.

In their *Atlas of Children's Growth* Tanner & Whitehouse (1982) gave Sheldon (1940, 1954) somatotype ratings to 'the normal subjects when they reached young adulthood'. Heath gave photoscopic ratings to the subjects at all ages, which ranged from 14 to 22 years. Tanner's mean somatotype for the 23 males (average age 19.1) was 2.9-4.2-3.6 and Heath's was 2.9-4.0-3.4; intercorrelations by component were 0.91 for endomorphy, 0.78 for mesomorphy and 0.86 for ectomorphy. There were no differences by somatotype mean, components or category (Carter & Heath, 1986).

For the 31 girls (average age 18.2) Tanner's mean somatotype was 4.7-2.8-3.1 and Heath's was 4.6-3.6-2.7, with intercorrelations by component of 0.80 for endomorphy, 0.46 for mesomorphy and 0.84 for ectomorphy. There were differences by somatotype means, category, mesomorphy and ectomorphy. The somatotype attitudinal distance (SAD) between the mean ratings of the girls was three times greater (SAD = 0.90) than for the boys (SAD = 0.28). (See Appendix III-1 for Heath ratings.)

For a comparison of Sheldon and Heath–Carter photoscopic ratings (Carter, 1985a), Heath rated a 20% age-stratified sample ($N = 232$) of the photographs in *Atlas of Men* (Sheldon, 1954). Figures 2.2 and 2.3 show that the Heath–Carter somatotypes have moved to the west and northwest sectors of the chart, emphasizing that Sheldon's 7-point scale confined the somatotypes within the borders.

Table 2.4. *Means for Sheldon and Heath–Carter ratings*

	Sheldon ratings	Heath–Carter ratings
University students	3.4-4.0-3.3	3.4-4.3-2.9
Gymnasts	1.4-6.4-2.0	1.4-6.9-2.1
100–200 metre sprinters	2.4-4.6-3.6	1.8-5.2-3.0
Body builders	1.8-6.8-1.5	1.8-7.9-1.4
Volleyball players	2.8-4.8-3.0	2.5-5.4-2.6

Fig. 2.2. Somatotypes of adult males, 18–59 years old, in *Atlas of Men*. In this and subsequent somatocharts numbers in the circles are the number of subjects with the same somatotype. ▲ = mean somatotype. Ratings by Sheldon, 1954 method. (Redrawn from Carter, 1985*a*.)

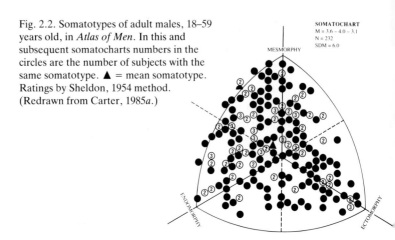

Fig. 2.3. Somatotypes of adult males, 18–59 years old, in *Atlas of Men*. ▲ = mean somatotype. Ratings by Heath, Heath–Carter photoscopic method. (Redrawn from Carter, 1985*a*.)

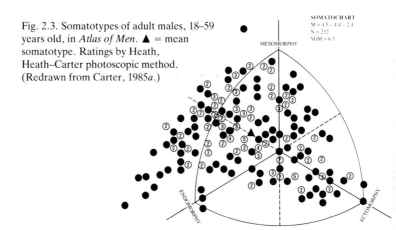

Fig. 2.4. Somatotypes of adult males in *Atlas of Men* whose HWR was less than 11.00 (36.37 metric). ▲ = mean somatotype. Ratings by Sheldon, 1954 method. (Redrawn from Carter, 1985*a*.)

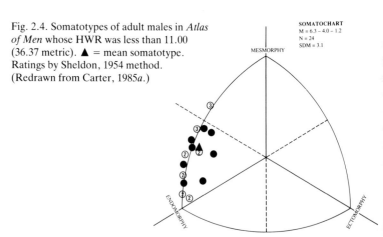

Fig. 2.5. Somatotypes of adult males in *Atlas of Men* whose HWR was less than 11.00 (36.37 metric). ▲ = mean somatotype. Ratings by Heath, Heath–Carter photoscopic method. (Redrawn from Carter, 1985*a*.)

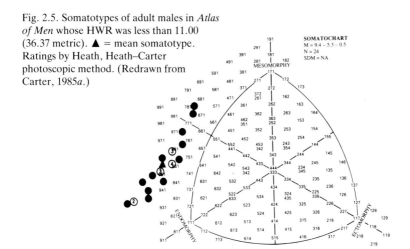

The Heath–Carter mean for ratings of a subsample of 24 *Atlas* subjects with HWRs lower than 11.00 (36.37) was $9\frac{1}{2}$-$5\frac{1}{2}$-$\frac{1}{2}$, and for the Sheldon ratings the mean was $6\frac{1}{2}$-4-1. The dramatic differences between individuals and means rated by the two methods are illustrated in Figs. 2.4 and 2.5, which show that no two subjects have the same rating and that the two distributions have no overlap. (See Appendix III-3 for Heath ratings.)

For comparison with her 1950 ratings by the Sheldon system, Heath re-rated by the Heath–Carter photoscopic method a sample of 48 females (Mount Holyoke College). The differences between the 1977 means (4.7-3.5-2.8) and the 1950 means (4.4-3.7-3.2) were +0.4 for endomorphy, −0.2 for mesomorphy, and −0.4 for ectomorphy. None is statistically significant, but all are in the same direction as in other re-rated samples.

Sheldon versus Parnell M.4 and Heath–Carter

In a somatotype study of 132 males, aged 17 to 19, Swalus *et al.* (1970) compared ratings by Parnell's M.4, Heath–Carter's anthropometric and Sheldon's photoscopic methods. The means for the Parnell method were 3.4-3.4-3.8, for Heath–Carter 2.5-3.6-3.4, and for Sheldon photoscopic 3.1-4.1-3.4. For the three methods, differences between the means were significant, except for the Sheldon and Heath–Carter methods. However, the correlation coefficients between the components were 0.92 and above for the Parnell and Heath–Carter methods.

Trunk Index versus Heath, Parnell M.4 and Heath–Carter

In a somatotype study of 300 urban Mexican males, aged 15 to 65 years, Villanueva (1976) compared Parnell (1958), 'Carter method' 'Heath

Table 2.5. *Comparison of Sheldon's Trunk Index, Parnell's M.4, Heath's photoscopic and Carter's anthroposcopic methods (N = 300)*

	$\overline{X}$	SD	SE
Endomorphy			
Parnell	3.82	0.93	0.05
Heath	3.33	1.29	0.07
Carter	3.31	1.40	0.08
Sheldon	3.65	0.80	0.05
Mesomorphy			
Parnell	4.33	1.00	0.06
Heath	4.46	0.95	0.06
Carter	4.77	1.06	0.06
Sheldon	4.10	0.90	0.05
Ectomorphy			
Parnell	3.48	1.20	0.07
Heath	2.57	1.24	0.07
Carter	2.77	1.28	0.07
Sheldon	2.79	1.21	0.07

From Villanueva (1976).

method' and Trunk Index ratings; Villanueva apparently intended 'Carter method' to refer to anthropometric ratings, and 'Heath method' to refer to Heath's photoscopic plus skinfold ratings. In general, the Trunk Index ratings do not correlate highly with the Carter, Heath and Parnell ratings in endomorphy and mesomorphy ($r = 0.50$–0.60), but are higher for ectomorphy ($r = 0.70$–0.80). The Carter, Heath and Parnell ratings correlate from 0.80 to 0.90 in all three components, but the means differ significantly. (See Table 2.5 for means, standard deviations and standard errors of means by component, and Table 2.6 for analysis of variance.)

Table 2.6. *Analysis of variance for each component*

					F-ratio
Endomorphy:	Carter	Heath	Sheldon	Parnell	
	3.31	3.33	3.65	3.82	14.5**
Mesomorphy:	Sheldon	Parnell	Heath	Carter	
	4.10	4.33	4.46	4.77	24.2**
Ectomorphy:	Heath	Carter	Sheldon	Parnell	
	2.57	2.77	2.79	3.48	31.3**

Calculated from data in Villanueva (1976).
**F-ratio significant beyond the 1% level of confidence.

Parnell versus Heath–Carter

In a somatotype study of 40 women, aged 19 to 69, in a physical fitness class, De Woskin (1967) compared his Parnell M.4 and his Heath–Carter anthropometric ratings. The Parnell means were 4.7-3.8-3.1, the Heath–Carter means were 5.9-4.8-1.7, and the correlations between components were 0.85 for endomorphy, 0.86 for mesomorphy, and 0.88 for ectomorphy. There were significant *t*-ratios for the paired differences between component means. The chi-square test and comparative ratios showed differences in somatotype distributions.

J. F. Climie (unpublished, San Diego State University) calculated Parnell M.4 somatotypes of 20 female physical education students and compared them with Heath–Carter anthropometric plus photoscopic ratings by Carter. The means for the Parnell ratings were 4.7-3.5-3.0 and for the Heath–Carter ratings 4.4-4.2-2.3; the differences between methods were significant for each component.

In Carter & Heath's (1971) somatotype study of 35 San Diego State College US football players, the mean for Parnell M.4 ratings was 4.7-5.5-2.1 and 4.2-6.3-1.4 for the Heath–Carter ratings. The Parnell restriction to maximum ratings of 7 in any component produced a conspicuous graphic difference in the somatoplots for a sample of football players with a good number of Heath–Carter ratings of 7, 7½ and 8 in mesomorphy. De Woskin (1967), with endomorphy ratings higher than 7, obtained a similar effect.

Sheldon photoscopic versus Sheldon Trunk Index

It is surprising that there is in the literature little evidence of the relationship between the Sheldon photoscopic and Trunk Index methods. Indirect comparisons by inference can be made. Parnell's (1954) M.4 ratings agreed fairly closely with Sheldon's photoscopic ratings. Theoretically, then, if Parnell's M.4 ratings and Sheldon's photoscopic ratings agree, Sheldon's Trunk Index and photoscopic ratings also should agree fairly well. However, Haronian & Sugarman's (1965) and Villanueva's (1976) studies indicate that there are considerable differences between M.4 and Trunk Index ratings.

In a direct comparison of the two Sheldon methods, Livson & McNeill (1962) reported that for the 'old and present' component values the correlations were 0.70 for endomorphy, 0.73 for mesomorphy and 0.79 for ectomorphy, and that for the Trunk Index the mean endomorphy was 1.0 unit higher, mesomorphy was the same, and the mean ectomorphy was 0.3 units higher.

In a somatotype study of 59 males in Northern India, Berry (1972) compared Dupertuis' Sheldon photoscopic ratings with his own Trunk Index ratings. Percentage agreements, plus or minus one-half unit, were 85%

(+0.19 units) for endomorphy, 83% (+0.09 units) for mesomorphy and 57% (−0.90 units) for ectomorphy.

Sheldon Trunk Index method versus Heath photoscopic method

In *Varieties of Delinquent Youth* Sheldon (1949) presented somatotype photographs, ages, heights (no weights!) and photoscopic somatotype ratings for 200 boys. In their 30-year follow-up, *Physique and Delinquent Behavior*, Hartl *et al.* (1982) re-published the original somatotype photographs with current ages, heights, weights and Trunk Index somatotypes. It can be inferred that the ratings were derived from an arcane application of the Trunk Index formula: planimetry on the 30-year-old photographs, maximum stature and the somatotype ponderal index. Heath photoscopically rated the photographs in *Varieties of Delinquent Youth* (larger than those in *Physique and Delinquent Behavior*) for comparison with the Sheldon photoscopic and Trunk Index ratings. The Heath–Carter means were 3.7-4.5-2.6, Sheldon means were 3.5-4.6-2.7 and Trunk Index means were 3.8-4.5-2.9. Component correlations between Sheldon photoscopic and Trunk Index methods were 0.70–0.87, and 0.89–0.96 between Sheldon photoscopic and Heath–Carter. As expected, there were high correlations for Sheldon and Heath–Carter photoscopic ratings in a sample of young adults with no extreme somatotypes. Absence of 30-year follow-up somatotype photographs is regrettable. A comparison of Trunk Index and Heath–Carter ratings on such a follow-up would be particularly interesting. (See Appendix III.2 for Heath ratings.)

Comparisons of somatotype ratings

Comparisons of whole somatotype ratings can be more clearly expressed graphically on somatocharts or as calculated distances than by component means, percentages and correlations. For example, if an empirical difference of one-half unit (for one component or for a combination of three components) is chosen as the criterion, the differences can be expressed as somatotype dispersion distance (SDD) = 1.0 and SAD = 0.5 (see Appendix II). SDD (somatotype dispersion distance) is the distance in two dimensions between any two somatoplots. The SDD is calculated in the *Y*-units of the two-dimensional coordinate system (Ross & Wilson, 1973). SAD (somatotype attitudinal distance), in three dimensions, is the distance between any two somatopoints (the plot of the somatotype in 3D, three dimensions), calculated in component units.

It is evident that the majority of SDDs and SADs for comparisons in this chapter (see Table 2.7 for sources of data, subject numbers, techniques and comparisons of methods) are greater than the criterion values of 1.0 and 0.5.

Table 2.7. *Comparison of somatotype methods by means of the Somatotype dispersion distance (SDD) and the Somatotype attitudinal distance (SAD)*

Method	Versus	SDD	SAD	Source; subjects; technique
Parnell's M.4 Method (1954, 1958)	Sheldon photo- scopic (1940, 1954)	—	0.19[a]	Parnell (1954); N = 154 Oxford Univ. males; Sheldon's photo- metric tables
		—	0.22[a]	Parnell (1954); N = 282 Oxford Univ. males; photoscopic rat- ings by Tanner
		2.01	0.82	Swalus *et al.* (1970); N = 132 males 17–19 yrs; photoscopic by Swalus
	Sheldon Trunk Index (1961, 1965; Sheldon *et al.*, 1969)	2.11	1.04	Haronian & Sugarman (1965); N = 102 Princeton Univ. males; (techniques by authors?)
		0.99	0.75	Villanueva (1976); N = 300 urban Mexican males, 15–65 yrs
	Heath photo- scopic (1963)	0.87	0.58	Heath & Carter (1966); N = 59 New Zealand PE males; photo- scopic by Heath
		1.31	0.55	Heath & Carter (1966); N = 61 New Zealand PE females; photoscopic by Heath, M.4 by Carter
	Heath & Carter (1967)	2.23	0.95	Heath (unpubl.); N = 101 Prince- ton Univ. males; photoscopic + skinfolds by Heath, M.4 by Haronian & Sugarman
	Heath & Carter photoscopic + anthropo- metric	2.36	0.97	J. F. Climie (unpublished); N = 20 San Diego State University PE females; anthropometric + photoscopic ratings by Carter, measurements by Climie
		1.81	1.04	Villanueva (1976); N = 300 urban Mexican males, 15–65 yrs; photoscopic + skinfolds by Heath
	Heath & Carter (1967) anthro- pometric	1.92	1.03	Swalus *et al.* (1970); N = 132 males 17–19 yrs; (measure- ments by authors?)
		3.12	0.98	Villanueva (1976); N = 300 urban Mexican males, 15–65 yrs; tech- niques by author
		4.85	2.03	De Woskin (1967); N = 40 fe- males, 19–69 yrs; techniques by author
		2.82	1.17	Carter & Heath (1971); N = 35 male US football players; tech- niques by Carter

(*continued*)

Table 2.7. (*Continued*)

Method	Versus	SDD	SAD	Source; subjects; technique
Sheldon Trunk Index (1961, 1965)	Heath & Carter (1967) anthropometric + photoscopic	1.27	0.53	Villanueva (1976); N = 300 urban Mexican males, 15–65 yrs; photoscopic + skinfolds by Heath
		0.38[a]	0.35[a]	Heath (unpubl.); N = 101 Princeton Univ. males; photoscopic + skinfolds by Heath
	Heath & Carter (1967) anthropometric	1.79	0.75	Villanueva (1976); N = 300 urban Mexican males, 15–65 yrs; techniques by author
Heath & Carter (1967) anthropometric	Sheldon (1940, 1954) photoscopic	1.23	0.81	Swalus *et al.* (1970); N = 132 males, 17–19 yrs; photoscopic by Swalus
	Heath photoscopic	0.58[a]	0.37[a]	Villanueva (1976); N = 300 urban Mexican males, 15–65 yrs; photoscopic + skinfolds by Heath
Sheldon (1940)	Hooton photoscopic	3.22	1.32	Dupertuis & Emanuel (1956); N = 500 US Air Force males, 20–34 yrs; photoscopic ratings by Dupertuis, Hooton and Stagg
Sheldon Trunk Index (1961, 1965)	Heath (1963) photoscopic	1.20 1.04 2.39	1.06 (9 yrs) 0.90 (12 yrs) 0.91 (16 yrs)	Morton (1967); N = 106 Medford Growth Study males, aged 9, 12 and 16 yrs, photoscopic by Heath, Trunk Index by Sheldon
	Heath & Carter anthropometric	1.83	1.28	Slaughter & Lohman (1977); N = 177 boys aged 7.1–12.5 yrs; measurements by author
	Sheldon (1940) photoscopic	—	1.05	Livson & McNeill (1962); N = 177 boys aged 11–16; Sheldon photoscopic by Dupertuis, Trunk Index by Sheldon's staff (Ann Turner)

[a] SDD < 1.0; SAD < 0.5.

reflecting the substantial differences in the whole somatotypes rated by different methods. However, Parnell's and Sheldon's photoscopic techniques are not significantly different in two out of three comparisons.

The summary (Table 2.7) of comparisons in this chapter shows that the SDDs and SADs are greater than the criterion values of 1.0 and 0.5, and the somatocharts show that complete somatotypes emphasize the differences among rating methods. However, in two of the three comparisons, the SADs between Parnell M.4 and the Sheldon photoscopic ratings are not significantly different. The SADs between Heath photoscopic and Parnell M.4 ratings are barely different for male or female physical education teachers, and the SDD for males is not different. As expected, there are no differences between Heath–Carter anthropometric and photoscopic ratings in Villanueva's (1976) study.

In general, rating differences are due to the use by Sheldon and Parnell of closed, age-adjusted scales versus the Heath–Carter use of open-ended, universal scales. This is especially evident in samples such as Carter & Heath's (1971) US football players, De Woskin's (1967) study of fitness programme women, Carter's (1985a) US men, and Štěpnička's (1974a) male athletes. However, findings are questionable when investigators have not established their rating reliability, and others, although they have used more 'objective' methods, have not established their reliability with anthropometry.

Heath & Carter (1966) proposed the following criteria for rating reliability: (1) a mean difference less than one-fifth of a rating unit in any

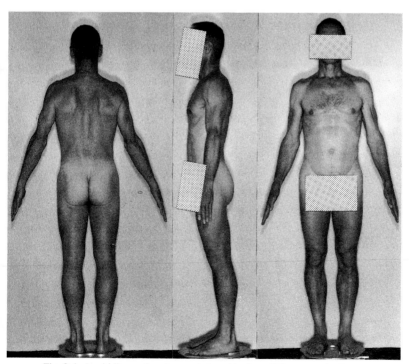

Fig. 2.6. A middle-aged male somatotyped by four different methods. Age = 54.3 yr; height = 171.7 cm; weight = 66.0 kg; HWR = 42.29. Heath–Carter anthropometric somatotype = 2½-6½-2½; Heath–Carter photoscopic plus anthropometric somatotype = 2-6-2½; Sheldon's Trunk Index somatotype = 2½-4-3½; and Parnell's M.4 Deviation Chart somatotype = 2½-6½-4. (From Carter & Heath, 1971.)

component; (2) statistically insignificant differences; (3) 90% or better agreement (plus or minus one-half unit); (4) correlation 0.90 or higher. Procedures for the SDD and SAD were added to these criteria later.

Although one-half unit for a component is the smallest possible photoscopic rating error, mean differences can be less and statistically meaningful and significant.

The somatotype photographs in Fig. 2.6 of a middle-aged male show the differences among Heath–Carter anthropometric, Heath–Carter photoscopic, Trunk Index and Parnell M.4 ratings. Only the two Heath–Carter ratings are similar for all three components. The other ratings are dissimilar

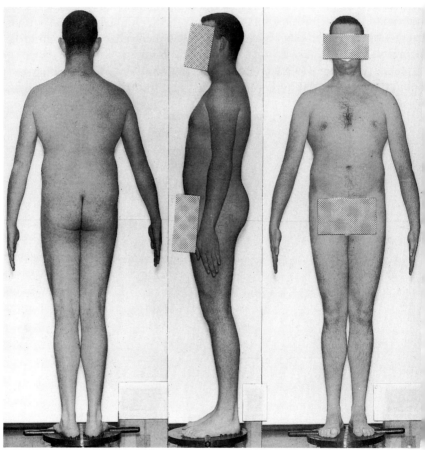

Fig. 2.7. A young male somatotyped by four different methods. Age = 22 yr; height = 190.8 cm; weight = 97.5 kg; HWR = 41.46. Heath–Carter anthropometric somatotype = 8-?-2; Heath–Carter anthropometric plus photoscopic somatotype = 6-4-2½; Sheldon's Trunk Index somatotype = 5-5-3; Parnell's M.4 Deviation Chart somatotype = 6½-1½-2½. (From Haronian & Sugarman, 1965.)

from one another as well as from the Heath–Carter ratings, especially in the second and third components. There are similar differences among methods in the ratings of a young male (Fig. 2.7).

Reliability of raters

The plotted data in *Varieties of Human Physique* (1940, p. 101) do not support Sheldon's reported correlations of 0.94–0.95 for himself and Tucker. He says that in 'the series of 600 specific estimates Tucker's somatotypes differed from Sheldon's by more than one degree [*sic*] in only one instance'. Presumably he means one rating unit. It appears that approximately 5% of the component ratings differed by $1\frac{1}{2}$ and 2 units. Also the plots indicate that the ratings were decimalized; presumably the raters averaged five regional ratings.

At the University of Illinois Sheldon showed a group of ten physical education instructors, who were familiar with somatotyping, how to use rating scales, and how to judge the three components, illustrated with examples of extreme dominance in each component. Cureton (1947) reported that, with limited training, the group made satisfactory judgments for rough somatotype classification of 50 photographs. Their mean error, irrespective of direction, was less than one-third of a rating unit for each component. Nine judges rated endomorphy and ectomorphy slightly high, and one rated ectomorphy $+1\frac{1}{2}$. Their high and low mesomorphy ratings were equally distributed.

In a study to establish reliability in Sheldon (1940) somatotype rating, Sills (1950) and two other members of the Physical Education faculty of the University of Iowa obtained a correlation of 0.88 or greater with the reference ratings for each of the three components. The reliability among the three faculty members for 158 subjects was 0.92 for endomorphy, 0.90 for mesomorphy and 0.95 for ectomorphy.

Tanner (1954*a*) found 90% (plus or minus one-half unit) overall agreement in the Sheldon photoscopic somatotype ratings of three experienced raters (Dupertuis, Heath and Tanner), and reliability coefficients of about 0.83 for the first and second components and 0.92 for the third, when the ratings cover the full range of the scale. Absolute mean differences among the raters ranged from 0.00 to 0.28 and reliability coefficients from 0.82 to 0.93, while percentage differences greater than one-half unit ranged from 5% to 17% for the first component, from 12% to 17% for the second, and from 3% to 13% for the third. Ectomorphy appears to be the easiest component to rate and mesomorphy the most difficult.

In a comparison of Hooton and Sheldon ratings of 28 somatotype photographs (Hunt & Barton, 1959) five raters obtained first component correlations from 0.66 to 0.89 and second component correlations from 0.66

to 0.82. (Hooton values for ectomorphy were calculated from a HWR table, without rating.)

In a comparison of Hooton and Sheldon methods, expert raters (Dupertuis & Emanuel, 1956) had correlations of 0.82 and 0.86. Sheldon's and Damon's (Damon *et al.*, 1955; Damon, 1960) ratings of two series of photographs of 'normal' White males (N = 124 and N = 146) were 'virtually identical', with component mean differences from 0.02 to 0.08. Roberts & Bainbridge (1963) had an average difference of 0.15 units per component. In a reliability check against Tanner's ratings on 28 subjects, Berry (Berry & Deshfukh, 1964) agreed to plus or minus one-half unit in 81% of the cases, by one unit in 15%, and to one and one-half units in 3.5%; absolute mean differences ranged from 0.03 to 0.19. Heath & Carter (1966), in ratings of 120 male and female subjects, showed differences of one-sixth to one-fifth of a rating unit on endomorphy and mesomorphy, 93% to 100% agreement (plus or minus one-half unit), and correlations of 0.88–0.92 for all three components.

Hunt and Barton (1959) reported a correlation of 0.96 when Hunt repeated the first component ratings of 30 photographs of army personnel after a 2-year interval. When Damon *et al.* (1962) re-rated the photographs of 199 White and 65 Black soldiers after an interval of 6 months, Damon's percentage agreement (plus or minus one-half unit) for the White soldiers was 92.4%, and his correlations for the components ranged from 0.83 (second component for Whites) to 0.93 (third component for both Whites and Blacks). For the Whites only, his mean differences were −0.01 for all three components. Damon (1960) also re-rated 174 females, aged 22 to 75, after a 6- to 8-week interval. His percentage agreement (plus or minus one-half unit) averaged 93.7%; his absolute mean differences for the components ranged from 0.05 to 0.14; and his correlations ranged from 0.79 for the second, to 0.90 for the first and third components.

It is apparent that when raters study somatotype criteria carefully, they generally obtain good inter-observer reliability, and that single observers can repeat their own assessments with a high degree of objectivity.

However, photoscopic rating *is* partially subjective. And, although Sheldon's Trunk Index method, Parnell's M.4 method and the Heath–Carter anthropometric somatotype are more objective, they depend primarily on accurate anthropometric measurements – or, in the case of Sheldon's Trunk Index method, on planimetric measurements on photographs. Obviously, investigators using these methods should always establish reliability. As Heath & Carter (1967) pointed out, anthropometric somatotypes are accurate in rating ranges from 1 to $7\frac{1}{2}$ in endomorphy, and from $\frac{1}{2}$ upward in ectomorphy. But photographs increase the validity of ratings in endomorphy higher than $7\frac{1}{2}$, ratings in ectomorphy of 1 and $\frac{1}{2}$, and all ratings in mesomorphy. Also although initial measurements may be accurate, calculation of the somatotype is often subject to error, and sometimes to misinterpretation. Investigators should always check their procedures and ratings with an expert in the application of the method. (There have been incorrect applications of Sheldon's *Atlas of Men* tables, and wrong uses of Parnell's M.4 rating form and Heath–Carter's anthropometric rating form.)

Rating scales

Sheldon's rating scales were constructed so that 1 represented the minimum and 7 represented the maximum value, with 4 apparently at the midpoint of the scale. Later he expanded them to 13-point scales by allowing for half-unit ratings. Others extended the scales beyond 7 (Roberts & Bainbridge, 1963; Heath, 1963; Seltzer & Mayer, 1964). Because of the relatively large intervals between points on a 7-point scale, some investigators have asked whether it should be regarded as an ordinal, interval or ratio-level measurement.

Experienced raters find that they can discern half-unit rating differences, but that differences less than that are unreliable. That is, a scale's accuracy has little to do with its continuity. Although the Heath–Carter anthropometric somatotype can be calculated to the nearest tenth of a rating unit, its accuracy may be spurious. The precision of the rating method, not the nature of the scale, produces the wide intervals. The scale applied in the Heath–Carter method can be regarded as a continuous ratio-level scale with its points marked at every half-unit.

In some samples the distribution of values of the components is skewed either positively or negatively. Positive skewness seems most likely to occur for endomorphy and ectomorphy. Sometimes a truncated distribution occurs at the low end of ectomorphy.

Heath and Carter (1967) resolved the problem of relative and absolute scales in favour of an absolute scale for both sexes at all ages. In this respect the scales should be regarded in the same way as height and weight scales.

Independence of components

The tables of somatotype distribution according to HWR, whether by Heath & Carter (1967), Carter (1980*a*), Duquet (1980) or Sheldon (1940, 1954), show that for a given HWR there are often several possible combinations of components and, when the values of two components are known, there is often more than one possible rating for the remaining component. However, there are biological constraints that limit the reasonable alternatives for ratings of a third component when the other two are known. Although it could be argued that adding a fourth or fifth factor would give the sum of possible variation, to date even elaborate studies with large numbers of subjects, many measurements and analyses have failed to yield factors that unequivocally identify all aspects of physique.

In short, two components without the third tell little about the whole somatotype and are of little value. Knowing the breadth and depth of a box without knowing its length reveals little about its appearance and potential use.

Component correlations

Zero-order intercorrelations among the three components vary according to age, sex, sample and somatotype method. In effect they are population specific.

A summary of intercorrelations for Heath–Carter ratings (Table 3.3) shows the following: (1) between endomorphy and mesomorphy correlations are generally low and positive, with occasional extremely low negative correlations; (2) between endomorphy and ectomorphy correlations are generally moderate and negative, usually higher for females; (3) between mesomorphy and ectomorphy correlations are generally moderate to moderately high and negative. The negative correlations between endomorphy or mesomorphy and ectomorphy do *not* signify a simple continuum from ponderosity to linearity. Ponderosity can be due to widely different endomorphy and mesomorphy rating combinations. For example, although the somatotypes 6-2-1, 2-6-1 and 4-4-1 have the same HWR (12.60, or 41.66), they have the same ponderosity.

Characteristically, according to most somatotyping methods ectomorphy ratings decrease as values for endomorphy and mesomorphy rise. Conversely, the first and second components decrease as ectomorphy increases. Furthermore, given ratings for two components, the rating of the remaining component can vary, although in some cases the third rating is relatively predictable. The ratings of ectomorphy can vary most when combined with ratings of 3 and 4 in endomorphy and mesomorphy. Thus, there is a degree of dependence among the values of the components.

In reality, the biological restrictions on the variations of human form and the restraints implied in the component definitions seem to limit the range of one component when we know the values of the other two components. Still, we should also note that there is moderate interdependence amongst most anthropometric variables. For example, although we use height and weight as independent variables, in some samples they correlate between 0.40 and 0.60. Despite possible weaknesses, the concept of somatotype scaling and ratings has proved feasible for describing human physique.

In the case of anthropometric ratings, the value of each component is based on separate sets of scaled measurements and calculated independently. In the case of photoscopic ratings, the somatotype as a whole is rated within the restrictions of the HWR table and all available anthropometry is taken into account.

It is impossible to determine 'true' correlations between somatotype components, because some methods use whole-unit ratings and some one-tenth units; some use a closed and some an open-ended rating scale; and characteristics vary from sample to sample.

There are many sample-specific characteristics of somatotype distributions, which differ from the theoretical model in which component ratings are on three orthogonal axes in space, with origins at 0-0-0. Thus differences between a sample and the model are a measure of the departure from the orthogonal.

Factor analysis

Hammond (1957) stressed the importance in body type research of factor analysis of metrical data and somatotype. He observed that the former, without preconceived ideas about what types might exist, allowed the data to demonstrate whatever clusters or groups there might be. But somatotypers accepted a three-component system empirically, taking the position that the three components account for most of the gross variation in shape and composition of human physique. Some factorists believed that body build can be derived from two, three or four factors; some identified muscle and bone factors separately; some identified general size as a factor. The Heath–Carter method holds that size is not relevant because somatotypes are size-dissociated.

In some factor analyses the anthropometric measurements were so limited that it was impossible to determine whether such factors as endomorphy, mesomorphy and ectomorphy were identifiable. Many investigators chose breadths and circumferences that were primarily skeletal dimensions, unsuitable for distinguishing between endomorphy and mesomorphy, and did not include size-dissociated measurements (Adcock, 1948; Howells, 1952; Burdick & Tess, 1983). But in general, when adequate measurements are available, the factors remaining after exclusion of general size contrast linearity with girth and mass (Burt, 1944; Roberts, 1977). But no factor analysis has yet identified a combination of measurements that equates with any component loading.

Since factor analysis rests on empirical science as well as on mathematics, factor elucidation may be theoretically satisfactory but not useful. Some investigators do not understand the basic concepts of somatotyping. Also, lack of clarity of Sheldon's (1940) definitions has led to some confusion. The three-numeral ratings, and the separateness of the three components and the unpredictability of the third component when two are known, also present special problems in analysis by traditional statistical techniques. Factor analysis has not proved successful as a reasonable alternative to

somatotype. Except for identifying groups of body measurements potentially important in describing human variation, it has led to no clearer
understanding of body type.

Checklists

Kraus (1951) applied Sheldon's list of criteria for somatotype
components of males to a study of Japanese males. In a study of American
college women, Bullen & Hardy (1946) modified Sheldon's list of inspectional criteria for males (Sheldon, 1940, pp. 37–46). In her study of East
Africans, Danby (1953) further adapted the Sheldon list to suit African
physique, and Roberts & Bainbridge (1963) also modified it. There were
seven criteria for each component in the five classified regions of the body in
Bullen & Hardy's (1946) and Danby's (1953) lists.

Apparently Bullen & Hardy (1946) designed their checklist to provide a
partial solution to criticisms of Sheldon's method; also they eliminated
secondary sexual characteristics (*g* component) and references to fineness of
hair and texture of skin (*t* component). Others (Hooton, 1946, 1951;
Cureton, 1947; Danby, 1953; Roberts & Bainbridge, 1963) designed checklists specifically to describe present or phenotypic somatotype, not to predict
a 'genotypical' somatotype. Danby's list appears to be more anatomically
based than others, but is subject to individual interpretation and is not
quantified. None of the lists softens Meredith's criticism of Sheldon's lack of
scientific rigor.

Sheldon's definitions emphasized the characteristics of component extremes. Dupertuis (1950), Humphreys (1957) and Hammond (1957), studied subjects with extreme somatotypes, and other authors have written little
about what to look for in moderate or low values of a component. However,
Cureton (1947) and Carter (1980*a*) describe characteristics over the range of
the rating scales, and in their definitions of components, Heath & Carter
(1967) describe what low as well as high levels of components look like. (See
Appendix I.)

Checklists that emphasize traits of a particular group, such as gymnasts,
may well distinguish between the extremes, but are less useful when applied
to subjects who are not extremes.

Second-order variables in somatotyping

Sheldon (1940) defined endomorphy, mesomorphy and ectomorphy as first-order variables of physique, and dysplasia, gynandromorphy,
texture and hirsutism as second-order variables as follows:

Dysplasia. The aspect of disharmony between different regions of
the body; that is, the body may be of one somatotype in the region of the
head and neck and another in the legs and trunk (Sheldon, 1940, p. 7).

Gynandromorphy (g). The bisexuality of a physique; the degree to which each sex exhibits some of the secondary sex characteristics (Sheldon, 1940, p. 7).

Texture (t). A description of physical harmony, symmetry and beauty; somatotype photographs can be rank ordered according to 'fineness' or 'coarseness' (Sheldon, 1940, p. 7).

Hirsutism (h). The hairiness of a physique (Sheldon, 1940, p. 7).

Except by Sheldon, little use has been made of second-order variables in somatotype studies. Zubin & Taback (1941) suggested that Sheldon's method for estimating dysplasia was unnecessarily complex, and provided a simpler equation. Damon (1942) replied that Sheldon's method is simple enough. Dupertuis & Emanuel (1956) reported no relationship between the dysplasia ratings of Dupertuis and Hooton ($r = 0.05$). Dupertuis' and Hooton's comparison of gynandromorphy ratings yielded a correlation of 0.60.

Dysplasia appears to be a useful aspect of somatotype, but Sheldon's concept of it is suitable only for somatotype ratings by body region. Bainbridge & Roberts (1966) defined dysplasia as 'variations among the body regions about the overall somatotype' (p. 251). They pointed out that Sheldon's method of determining dysplasia was unwieldy and that the method of recording the regions as somatotypes 'concealed more than it revealed'. Instead, for their Nilote sample they used regional deviations from the mean somatotype sample. They found that dysplasia differed according to the level of the components, and argued that dysplasia is of biological significance.

Carter (1980*a*) suggested rating the upper segments of the body (above the iliac crest) and the lower segments (below the iliac crest) separately. This simple schema is valuable in rating the overall somatotype when there appear to be discrepancies between the upper and lower parts of the body. Such discrepancies are particularly noticeable in some male gymnasts, whose upper bodies are in many cases considerably more mesomorphic than their lower bodies.

Judging physiques of body builders sometimes includes overt, sometimes intuitive, recognition of all four second-order variables. Muscular development of all regions of the body is favoured over uneven development. Skin texture, overall body harmony, symmetry and beauty are highly regarded. Absence of body hair is also a positive judging point.

Although gynandromorphy is usually obvious in a general way, in extreme cases it is more easily rated from the somatotype photograph and by anthropometric techniques such as those of Bayley & Bayer (1946) and of Tanner (1951*a*).

There appears to be no positive reason for trying to relate hirsutism to somatotype. Dupertuis and Sheldon showed a low ($r = 0.27$) correlation between ratings on hirsutism and somatotype components, with only a slight tendency for endomorphic mesomorphs to be more hirsute than more ectomorphic subjects (Sheldon, 1940, p. 79).

A simple, precise, quantifiable scale is needed for dysplasia, which seems to be the only potentially useful second-order variable. When dysplasia is not essential for the somatotype rating, the proportionality procedures of Behnke's 'somatogram' and Ross & Wilson's 'phantom' are more appropriate (Behnke & Wilmore, 1974; Ross & Wilson, 1974).

Photoscopic appraisal of physique

Brožek (1961), noting that the trained eye of the human morphologist is impressively accurate, emphasized the need for a visual guide analogous to a radiographic atlas of skeletal development. He suggested that such an atlas include rating scales, a system of quantitative appraisal of photographs, objective criteria of body composition, and equations for predicting body composition from photoscopic and photogrammetric data. He added that Parnell's modifications of Sheldon's system, which combined physical anthropology and photography, was a step in the right direction.

Hunt (1961) suggested that somatotypers should 'aim squarely at estimating fat content of the body from photographs', and could improve their skill in rating endomorphy if they concentrated on the most representative locations of adipose thickness.

Somatotype and body composition

The Heath–Carter somatotype is related to but not synonymous with body composition. Endomorphy is an estimate of relative fatness. Mesomorphy is an estimate of musculoskeletal robustness per unit of height. The purposes of the investigator determine the choice of body composition estimates or somatotype. The somatotype provides a general or *gestalt* summary of body shape, from which estimates of body composition can be inferred. Body composition models yield quantitative estimates of body composition, but do not describe body shape.

Brožek (1965) said that *in vivo* methods of estimating body composition are at best estimates of complete *in vitro* analysis, which are estimated to have individual prediction errors of plus or minus 2–10%. Bakker & Struikenkamp (1977), Lohman (1981) and Martin et al. (1985) confirmed these errors.

In the commonly used two-compartment model for body composition estimates, the body is partitioned into fat and fat-free mass (FFM) compartments. The compartments may be quantified as absolute or relative values.

In somatotype terms endomorphy corresponds to the fat compartment, mesomorphy to the fat free (FFM) compartment relative to height. Ecto-morphy, which is not included in the two-compartment model, adds a third dimension, i.e. the distribution of endomorphy and mesomorphy in space, or, tissue distribution relative to height. The somatotype rating expresses relative, not absolute, values.

Consider two subjects, A and B, with the same height and the same relative body composition (percentage fat and percentage FFM) but whose somatotypes are almost polar opposites, and assume that a rating of 1 in endomorphy is the equivalent of 5% body fat.

Subject A	*Subject B*
Height = 175.0 cm	Height = 175.0 cm
Fat = 5%	Fat = 5%
FFM = 95%	FFM = 95%
Weight = 71.0 kg	Weight = 48.0 kg
HWR = 43.00	HWR = 48.27
Somatotype = 1-6-2	Somatotype = 1-2-6

Subject A is highly mesomorphic and low in endomorphy and ectomorphy. Subject B is highly ectomorphic and low in endomorphy and mesomorphy. Although the percentage FFM is the same for both subjects, in absolute terms it is significantly different (67.5 vs. 45.6 kg). The somatotype rating shows how the difference 'looks'. The somatotype rating tells even the novice rater the difference between the two physiques, while relative body composition values do not distinguish between them. If both subjects were to have the same relative body composition and HWR, subject B would require a stature of 156.3 cm. Neither relative body composition values nor somatotype ratings give information about body size.

In general, studies of relationships between body composition and soma-totype show moderately high associations between percentage fat and endomorphy, and low to moderate associations between mesomorphy and lean body weight (absolute or relative). Usually there are negative associ-ations between percentage fat and lean body weight on the one hand and ectomorphy on the other (Dupertuis *et al.*, 1951*a*; Carter & Phillips, 1969; Slaughter & Lohman, 1976; Slaughter *et al.*, 1977*a,b*; Wilmore, 1970).

The statement by Slaughter & Lohman (1976, 1977), Slaughter *et al.* (1977*a,b*) and Wilmore (1970), who used anthropometric estimates of somatotype, that mesomorphy *is* lean body mass (LBM), is a misinterpret-ation of the Heath & Carter (1967) definition ('Mesomorphy can be thought of as lean body mass' (p. 70)). Heath and Carter implied that the rating of mesomorphy is related to LBM. Slaughter and Wilmore, using LBM to determine relationships with mesomorphy and equating LBM (kg) and mesomorphy, found low associations between LBM and mesomorphy. However, when Slaughter *et al.* 1977*a*,*b*) used independent anthropometry (height, elbow and knee widths, corrected biceps and calf girths) in multiple regression equations to predict LBM and mesomorphy, their multiple *r* values were 0.88 for 7–12-year-old boys, and 0.94 and 0.95 for college women. They did not explain their conflicting results.

Complete prediction of somatotype values from body composition variables is not to be expected; the somatotype components are not completely independent, and the relationships among the component values differ for different samples. The relationships which do exist are likely to be sample specific and can be expected to show the contribution of body composition to the somatotype ratings. The variance not accounted for may be due to the form and shape aspects of the somatotype (as seen in the photograph) and the errors associated with the body composition and somatotype procedures, as well as the contribution of the interrelationships among the components. Taking the somatotype components one at a time for analysis may destroy the somatotype concept and lead to misinterpretation of findings. For example, the somatotypes 4-5-1 and 1-5-4 have the same mesomorphy, but in both absolute and relative terms they differ greatly in fat and fat-free tissues. The first may have approximately 20% fat weight and 80% fat-free weight, while the second may have approximately 5% fat weight and 95% fat-free weight. And absolute weights can differ greatly, depending on the size of the subjects.

Bulbulian (1984) found that body density could be predicted better in young women of different somatotype groups by using somatotype-specific regression equations. These equations used component ratings as well as additional anthropometry.

Sterner & Burke (1986) compared skinfolds and visual ratings as estimates of percentage fat calculated from densitometry in men. Their finding that the percentage fat estimates of raters, experienced in body composition assessment, were as accurate as skinfold estimates, is no surprise to somatotypers, who have asserted this for several decades.

Summary

Methods
The various methods described in this chapter support one or the other of these basic premises: (1) a somatotype rating attempts to assess the constitutional, unchanging pattern of somatotype; (2) a somatotype rating estimates the phenotypic (i.e. present) somatotype. The methods considered are based on various combinations of photoscopic estimates, planimetry, anthropometry and functional performance. The Sheldon, Parnell

and Heath–Carter methods attempted to provide systems suitable for both sexes at all ages. Hooton combined Sheldon's genetic concept with a phenotypic concept (especially for the third component). He did not include corrections for age for his presumably young sample of Army personnel. Bullen & Hardy, Kraus, Danby, Roberts & Bainbridge, Cureton, Damon, Clarke, and Ostyn and colleagues all attempted to adhere as closely as possible to Sheldon's concept, except for phenotypic ratings. Their modest modifications were suited to special samples with characteristics not taken into account in Sheldon's criteria. The Heath–Carter method, derived from modifications and simplifications of Sheldon's system and designed to provide an objective phenotype method, has become the most used somatotype method.

Whatever method is used, its validity, and the reliability and objectivity of measurements and ratings derived from it, should be integral parts of the study.

Comparisons
The findings from comparisons of methods are as follows:
1. There are major differences among methods.
2. Not all of the differences can be accounted for by rater bias or unreliable measurement techniques, although these certainly can contribute.
3. Comparisons using the same methods may produce variable degrees of difference according to the sex and age of subjects in the study.
4. The differences between methods for samples of young adult males are the smallest. Differences are greater for children, older adults and females.
5. The two largest sources of differences between methods are (*a*) genotype versus phenotype ratings, and (*b*) open versus closed component scales.
6. Although the differences between some means may appear to be small, the overall shift in the somatotype distributions can cause significant differences in the numbers in the somatotype categories on the somatochart.
7. Because of the differences among the methods the word somatotyping must be regarded as a generic term embracing a number of different methods.

Conclusions
1. Appropriately trained photoscopic somatotypers reach high levels of reliability and objectivity.

2. Those who use anthropometry should demonstrate test/re-test reliability.

3. Somatotype component scales are ratio measurement scales that begin at zero, are open at the upper end, and can be recorded to one-half unit or calculated as decimal ratings.

4. Intercorrelations between components range from low to moderate. Their magnitudes depend on the characteristics of the sample.

5. Factor analysis has not resulted in a better method of overall physique description, and has contributed little to our conceptual understanding of the somatotype.

6. Checklists, which were designed to make somatotyping more objective, tended to describe extreme traits rather than to provide for evaluation on continuous scales.

7. With the exception of dysplasia, the second-order variables appear to be of no value in rating somatotype components, and gynandromorphy, hirsutism and texture do not add information relevant to the somatotype. They are best used as independent variables.

8. Phenotypic visual appraisal of physique based on body shape and apparent composition is desirable and with appropriate training is feasible and reproducible.

9. The somatotype and body composition are complementary and overlapping characteristics of physique. They are neither interchangeable nor synonymous.

3 *Human variation in adult somatotypes*

Introduction

This review of present knowledge of adult somatotype variation within and among different populations draws both on samples rated by the Heath–Carter method (Heath and Carter participated in many of these studies) and on samples rated by other methods.

In Part I the samples were cross-sectional, and in some cases grouped according to age. So far as was known the subjects were essentially healthy, and had no overt disabilities. If known, ethnicity, occupation, socio-economic status and other data are indicated. The original authors usually reported this information. Adults are defined as 18 years of age and older, except in a few samples that include some subjects aged 16 and 17.

Individual somatotypes, available for most samples, are plotted on somatocharts. Sample characteristics are presented in catalogue form. Somatocharts and descriptive statistics are found in tables and figures. Examples of somatotype photographs and ratings are shown in Appendix I.

In Part II, the studies collated mainly for the historical record begin with Sheldon's series of US men from *The Varieties of Human Physique* and *Atlas of Men*. Tables summarize the source, descriptive statistics and rating methods used.

PART I. HEATH–CARTER SOMATOTYPES

Variation in somatotypes of adult men

Thirty-nine samples, consisting of 21 814 subjects from different parts of the world, show the variations in male somatotypes. Table 3.1 catalogues the characteristics of each sample, the rating method and references, and Table 3.2 the statistics from each group. Figures 3.1 to 3.30 are somatocharts showing means, and wherever possible individual distributions.

Nationality samples

The nationality samples are the samples (Table 3.1) of Canadians (N = 8970), USA Federal Aviation Administration (FAA) control tower

Table 3.1. *Catalogue of characteristics of adult male samples*

Nationality samples

Canada	National sample
Sample	Participants in YMCA–LIFE programme, adult males. Collected under direction of Donald Bailey, 1976–78. Broad geographic and socio-economic spectrum. Probably relatively healthy sample. Height and weight similar to national Canadian and USA surveys
Age range	15–69 years, in six age groups, by decade
Ethnicity	Predominantly White
Number	8970
	Ages 15–19 = 161
	Ages 20–29 = 2259
	Ages 30–39 = 2985
	Ages 40–49 = 2031
	Ages 50–59 = 1159
	Ages 60+ = 375
Rating method	Heath–Carter anthropometric
References	Bailey *et al.* (1982)
USA	Norman, Oklahoma (University of Oklahoma)
Sample	Candidates at centre for training of Federal Aviation Administration control tower operators. Data from Clyde Snow, late 1960s
Age range	20–49 years
Ethnicity	Approx. 90% White, 10% Black
Number	345
	Ages 20–29 = 244
	Ages 30–39 = 59
	Ages 40–49 = 42
Rating method	Heath–Carter photoscopic plus skinfolds, by Heath in 1970
References	Unpublished data
USA	Primarily northeast and midwest
Sample	A 20% age-stratified sample taken from 1166 photographs of males in *Atlas of Men* (Sheldon, 1954). Photographs by Sheldon and colleagues in the late 1930s and the 1940s. Includes students, military personnel, and a variety of subjects from labourers to professionals
Age range	18–66 years
Ethnicity	Whites and a few Blacks
Number	232
	Ages 18–29 = 141
	Ages 30–39 = 32
	Ages 40–49 = 25
	Ages 50–66 = 34
Rating method	Heath–Carter photoscopic, by Heath in 1978
References	Sheldon (1954), Carter (1985*a*); see also Chapter 2
England	Dartford tunnel
Sample	Workers on second Dartford Tunnel under the River Thames. Men from a wide range of occupations, e.g. pony boys, miners, fitters, electricians, engineers, subcontractors and factory inspectors. Measured and photographed by John D. King, Medical Director, London Hyperbaric Medical Service. Collected 1974–76

(continued)

Table 3.1. (*Continued*)

Age range	18–64 years
Ethnicity	White, from Britain; a large number of Irish and Welsh
Number	775
	Ages 18–29 = 442
	Ages 30–39 = 226
	Ages 40–49 = 79
	Ages 50–64 = 28
Rating method	Heath–Carter anthropometric plus photoscopic, by Heath
References	J. D. King, personal communication
Mexico	Mexico, DF
Sample	Urban university students, technical and clerical workers. Measured and photographed by J. Faulhaber and co-workers, as a reference group for athletes measured at 1968 Mexico City Olympic Games
Age range	14–54 years
	(Age positively skewed by a few subjects over 35 years of age)
Ethnicity	102 Caucasian, 162 Mestizo
Number	264
Rating method	Heath–Carter anthropometric plus photoscopic, by Heath. (In a comparison of methods (see Chapter 3) Villanueva presented somatotype means for same group plus 35 additional subjects, extending the age range to 65 years. The mean anthropometric somatotype was 3.3-4.8-2.8. The mean anthroposcopic plus photoscopic by Heath was 3.3-4.5-2.6)
References	De Garay *et al.* (1974), Villanueva (1976)
India	Northwest
Sample	Various ethnic groups. Data collected 1974–80
Age range	18–45 years
Ethnicity	Gaddis, Rajput and Brahmins (N = 291), Punjabis (N = 59), Jat-Sikhs (N = 245), Banias (N = 221), Spitians (N = 52)
Number	868
Rating method	Heath–Carter anthropometric
References	Sidhu & Kansal (1974), Sidhu & Wadhan (1975), Sodhi (1976), Singh & Sidhu (1980), Singh (1981), Kansal (1981), Singh *et al.* (1985)
Ethnic samples	
USA	Wainwright, Alaska
Sample	Village with population of about 300 (approx. 40 households), about 145 km SW of Point Barrow. Photographs by C. Dotter of Univ. Oregon Medical School, 1958; anthropometry by Jamison & Zegura
Age range	16–75 years
Ethnicity	Eskimos and a few Eskimo hybrids
Number	81
Rating method	40 ratings: photoscopic, plus skinfolds for some subjects
	41 ratings: anthropometric somatotypes
	Ratings by Heath
Note	Somatotype distributions by both methods are essentially the same. Two sets of statistics are provided in Table 3.2, one for N = 81 and one for N = 30 (photographs and skinfolds were available for the latter)
References	Heath (1973, 1977), Heath *et al.* (1968)
Brazil	Caingang Indians

(*continued*)

Table 3.1. (*Continued*)

Sample	Caingang Indians, who live on Indian reservation at Rio das Cobras, a northern tributary of the Ignacu, in the western part of the state of Parana, about 450 km west of Curitiba. Altitude 610 m. Humid, subtropical forest
Age range	18–60 years
Ethnicity	Caingang Indians
Number	48
Rating method	Heath–Carter photoscopic, by Heath in April 1975
References	Roberts (personal communication)
Papua New Guinea	Pere Village, Manus Province
Sample	Adult males measured and photographed by Theodore Schwartz and Barbara Heath, 1966
Age range	22–72 years
Ethnicity	Melanesian
Number	100
Rating method	Heath–Carter photoscopic, by Heath
References	Heath *et al.* (1968), Heath (1973)
Sudan	
Sample	Northern Nilotes in southern Sudan in Nile Province. Shilluks, agricultural, with a few cattle; Dinka, pastoral, with cattle, and some agriculture. No obvious pathology. Data collected by Derek Roberts, 1953–4
Age range	18–45 years
Ethnicity	Nilotes of the Sudan
Number	Shilluk = 48
	Northern Dinka = 227 + 52 (279)
Rating method	Heath–Carter photoscopic, by Heath
References	Roberts & Bainbridge (1963), Heath (unpubl.)
Military samples	
USA	West Point, New York (United States Military Academy)
Sample	Officer cadets in their first year at West Point. Measured and photographed 1947; re-measured and re-photographed 1950
Age range	18–22 years
Ethnicity	(Predominantly White)
Number	424
Rating method	Rated by Sheldon method by Heath in 1948 and 1950. Re-rated by Heath–Carter criteria in 1978 by Heath (without photographs)
References	(Some of subjects included in *Atlas of Men*, but no reference to them)
USA	Hawaii
Sample	US Navy personnel stationed in Hawaii. Data and photographs provided by A. R. Behnke. Two samples: 19 submariners, 29 non-submariners
Age range	18–42 years
Ethnicity	25 White, 23 Black
Number	48
Rating method	Heath–Carter photoscopic, by Heath
References	Behnke *et al.* (1957)
England	Sandhurst (Royal Military Academy)

(*continued*)

Table 3.1. (*Continued*)

Sample	One complete entering class at Sandhurst (April 1952). Data and photographs collected by J. M. Tanner, 1952
Age range	18–20 years
Ethnicity	White
Number	286 (Tanner reported 287)
Rating method	Heath rated series in mid-1950s, photoscopic, based on age, height and weight. Heath re-rated series in 1978, using Heath–Carter revised HWR table
Reference	Tanner (1964)
England	Portsmouth
Sample	British naval personnel. Data collected by Derek Roberts, 1953
Age range	18–45
Ethnicity	White, British
Number	110
Rating method	Heath–Carter photoscopic, by Heath, 1975
Reference	Derek Roberts (personal communication)
Belgium	
Sample	Belgian military personnel. Data collected by Jan Clarys, 1975–6
Age range	17–30 years
Ethnicity	White
Number	283
Rating method	Heath–Carter anthropometric
References	None
Brazil	
Sample	Brazilian soldiers. Data and photographs by Derek Roberts, 1966
Age range	Approximately 17–35
Ethnicity	Brazilian White, Mulatto, Black
Number	45
Rating method	Heath–Carter photoscopic, by Heath, April 1975
References	Derek Roberts (personal communication)
Turkey	All regions
Sample	Random selection of recruits at entrance to compulsory military service. Included only those having same birthplace (by province) as their parents
Age range	18–20 years
Ethnicity	Not given
Number	6000
Rating method	Heath–Carter anthropometric method, by teams trained by authors
Reference	Olgun & Gürses (1986)

University student samples

USA	Princeton, New Jersey (Princeton University)
Sample	College, university and seminary students. Data collected 1962
Age range	17–28 years
Ethnicity	Predominantly White
Number	101
Rating method	Heath–Carter anthropometric plus photoscopic, by Heath
Reference	Haronian & Sugarman (1965)
USA	Monterey, California (Monterey Peninsula College)

(*continued*)

Table 3.1. (*Continued*)

Sample	Student volunteers at Monterey Peninsula College. Majority students in physical anthropology class. Data collected by Heath, 1972
Age range	16–52 years
Ethnicity	Majority White
Number	66
Rating method	Heath–Carter anthropometric plus photoscopic, by Heath
References	Unpublished data
USA	San Diego, California (San Diego State University)
Sample	Students, voluntarily enrolled in five physical conditioning classes at SDSU. Data collected 1973–5
Age range	17–32 years (one at 54 years)
Ethnicity	Over 90% White or Mexican-American. A few Black and Oriental
Number	328
Rating method	Heath–Carter anthropometric
Reference	Aubry (unpublished study, 1976)
USA	Honolulu, Hawaii (University of Hawaii)
Sample	Hawaii-born males, both parents of Japanese ancestry. 55% first generation, 20% second generation, remainder one parent born in Asia the other in Hawaii. Students at University of Hawaii, volunteers from physical education classes. Data collected by Carey D. Miller in 1955
Age range	18–22 years
Ethnicity	Japanese
Number	104
Rating method	Heath photoscopic, by Heath
Reference	Heath *et al.* (1961)
USA	Iowa City, Iowa (University of Iowa)
Sample	Undergraduate students in physical activity classes at University of Iowa, 1958–59
Age range	18–24 years
Ethnicity	Mostly White
Number	50
Rating method	Heath–Carter photoscopic, by Heath, 1982
References	Jennett (1959)
USA	Kirskville, Missouri (Northeast Missouri State University)
Sample	Students
Age range	18–22 years
Ethnicity	Not given
Number	161
Rating method	Heath–Carter anthropometric, by authors
References	Bale *et al.* (1984)
USA	Berkeley, California (University of California)
Sample	Volunteers attending University of California, Berkeley. Considered normal and healthy
Age	22.0 years
Ethnicity	Presumably the majority were White
Number	133
Rating method	Heath–Carter anthropometric, by Wilmore (with estimated medial calf skinfold)
Reference	Wilmore (1970)

(*continued*)

Table 3.1. (*Continued*)

Canada	British Columbia
Sample	Recruited from a general education exercise class, a campus residence and a physical education class for elementary school teachers at Simon Fraser University, University of British Columbia and University of Victoria. Measured by W. D. Ross and colleagues
Age range	17–33 years
Ethnicity	98% White
Number	153
Rating method	Heath–Carter anthropometric
References	Ross & Ward (1982), Carter *et al.* (1982)
England	London University, St Thomas' Medical School, St Thomas' Hospital
Sample	Medical students at St Thomas' Medical School. Data and photographs by J. M. Tanner, 1950–6
Age range	Not given
Ethnicity	White
Number	274
	(subsamples = 46, 46, 77, 105)
Rating method	Heath–Carter photoscopic, by Heath, 1978
References	Tanner (1951*b*, 1954*b*), Tanner *et al.* (1959)
Czechoslovakia	Prague
Sample	Students at Charles University, Prague. Data collected in late 1960s and early 1970s
Age range	17–30 years
Ethnicity	White, Central European
Number	302
	Partial sample = 267, plotted sample means = 3.4-4.3-3.0
	Total sample = 302, means = 3.4-4.3-2.9
Rating method	Early subjects: Heath–Carter photoscopic, by Štěpnička
	Later subjects (after 1970): Heath–Carter anthropometric, by Štěpnička
References	Štěpnička *et al.* (1979*a*), J. Štěpnička, personal communication
South Africa	Johannesburg (University of Witwatersrand)
Sample	Second-year medical students at University of Witwatersrand, Johannesburg, 1975–8
Age range	18–25 years
Ethnicity	White
Number	425
Rating method	Anthropometric and photoscopic by author. Photoscopic proficiency obtained by studying criteria and a criterion set of photographs rated by Heath
Reference	Gordon (1984)
Hungary	Budapest (Technical University)
Sample	Students at a technical university
Age	20 years
Ethnicity	Hungarian
Number	569
Rating method	Heath–Carter anthropometric, by author
Reference	Gyenis (1985)
South Korea	Pusan (Dong-A University)
Sample	Students in a variety of areas

(*continued*)

Table 3.1. (*Continued*)

Age	21.2 years
Ethnicity	Korean
Number	30
Rating method	Heath–Carter anthropometric, by author
Reference	Shin (1985)

Older men

USA	San Diego, California
Sample	Business and professional men enrolled in Adult Fitness Program at San Diego State University. Occupations included insurance, real estate, sales, dentistry. Voluntary participation, some referred by physicians. All had medical clearance. Some were 'overweight', some had mild hypertension
Age range	35–61 years
Ethnicity	White (one Oriental)
Number	56
Rating method	Heath–Carter anthropometric plus photoscopic, by Carter
References	Carter *et al.* (1965), Rahe & Carter (1976)
USA	Boston, Massachusetts (Veterans Administration Outpatient Clinic)
Sample	American veterans of Spanish-American War, who volunteered for military service in 1898. Agreed in 1959 to be photographed and measured at VA Outpatient Clinic. Ambulatory, not ill at time (about one-half previously known medically in clinic). Measured and photographed by Albert Damon
Age range	72–91 years
Ethnicity	96% White, 4% Black. Predominantly northwest European descent. Former occupations ranged from labourer to professional. More craftsmen and foremen and fewer labourers than expected
Number	130
Rating method	Heath–Carter photoscopic, by Heath, 1977
Reference	A. Damon (personal communication)

trainees (N = 345), a USA sample selected (SEL) to show wide variation (N = 232), British tunnel workers (N = 775) and urban Mexicans (N = 265). The somatoplot (Fig. 3.1) shows the means of the samples, the majority being above or to the left of the centre of the somatochart, that is, to the north and northwest. The Indians and Nilotes are to the near and far east of the centre. The somatoplots for the different age groups of the Canadians (Fig. 3.2), USA (FAA) (Fig. 3.3) and British (Fig. 3.4) show that the means move to the northwest. That is, they become more endo-mesomorphic with age. (The adjective endo-mesomorphic is an abbreviated form of the words endomorphic mesomorph which refers to a category of somatotypes with similar relative component rating values. There are 13 categories for grouping somatotype. The bases for the categories are described in Appendix II; they are defined in Table II.1, and their areas on the somatochart are

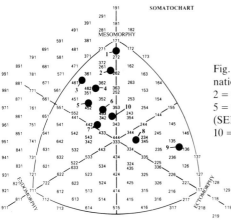

Fig. 3.1. Mean somatotypes of male nationality and ethnic samples. 1 = Manus; 2 = Caingang; 3 = Eskimo; 4 = British; 5 = Canadian; 6 = USA (FAA); 7 = USA (SEL); 8 = Indians; 9 = Nilotes; 10 = Mexican.

illustrated in Fig. II.7.) When compared by age group, the British tunnel workers are consistently the most mesomorphic. Endo-mesomorphy predominates in each of these samples.

The USA (SEL) sample is both non-random and different in composition from the other North American samples, and was specially selected from 1175 somatotype examples in *Atlas of Men* (Sheldon, 1954). This was the sample Sheldon selected to represent variation in American male somatotypes. The means in no way represent the US male population. The component standard deviations, the SDMs and SAMs (Table 3.2), are large compared with other samples. Endomorphy ratings range from 1 to $11\frac{1}{2}$, mesomorphy from 1 to 9, and ectomorphy from $\frac{1}{2}$ to 7 (Carter, 1985a). The somatoplots of the 18- to 29-year-old subsample (Fig. 3.5) shows the wide range of somatotypes among young US males. See Fig. 2.3 for the overall distribution of this sample (N = 232). High mesomorphy, balanced mesomorphy and endo-mesomorphy, found in other samples, are conspicuously absent from this sample.

Although there were many students in the urban Mexican sample, there were enough other subjects to qualify it as a nationality group (Fig. 3.6). The sample was plotted as a single distribution because there were no significant differences in means between the Mexicans of European descent and the Mestizos (De Garay *et al.*, 1974). There is no age breakdown in this group, because the majority were in their twenties plus a few subjects in their thirties and forties. The greatest number of balanced mesomorphs are among the Mexicans, the youngest group in the Canadian sample, and the USA (FAA) sample.

The Indian samples, aged 18–25 years, were drawn from seven ethnic subsamples from northwest India. The overall mean was approximately 2.6-3.3-4.0. The Indians, although they overlap with the meso-ectomorphs

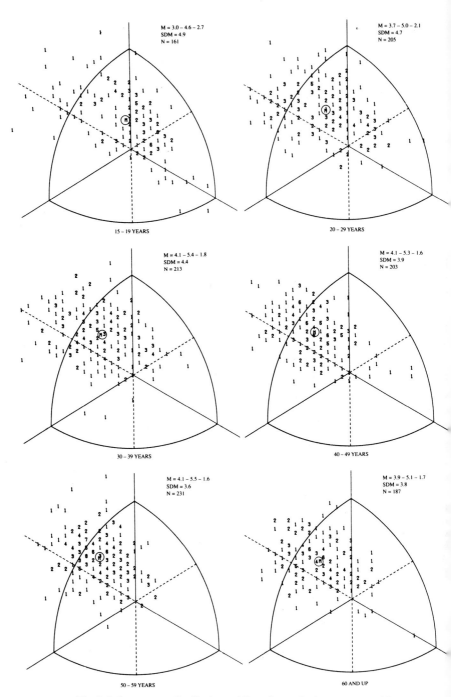

Fig. 3.2. Somatotype distributions of Canadian males by age groups. M = mean; 1, 2, 3 = number of subjects at a point. (From Bailey *et al.*, 1982.)

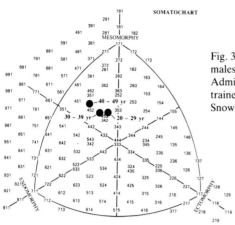

Fig. 3.3. Mean somatotypes of United States males by age. Federal Aviation Adminstration control tower operator trainees. (Photographs provided by C. P. Snow.)

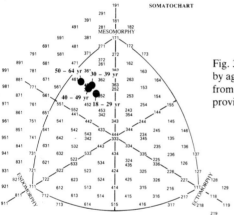

Fig. 3.4. Mean somatotypes of British males by age. Workers on Dartford tunnel project from various occupations. (Photographs provided by J. D. King.)

Fig. 3.5. Somatotypes of United States males aged 18–29 years, selected to show a variety of somatotypes. ▲ = mean somatotype. Ratings of photographs in *Atlas of Men* (Sheldon, 1954) by Heath using the Heath–Carter photoscopic method.

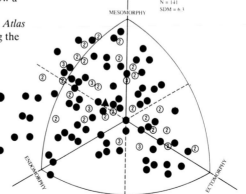

Table 3.2. Somatotypes of adult males

Sample/year	N	Statistic	Age (yr)	Height (cm)	Weight (kg)	Somatotype	HWR	SDM	SAM
Nationality samples									
Canada (YMCA–LIFE) (1976–78)	8970	X̄	38.3	177.1	79.2	3.9-5.2-1.9	41.25		
		SD	11.5	6.8	11.2	1.4 1.3 1.1			
15–19 yrs	161	X̄	18.2	175.5	70.9	3.9-4.7-2.8	42.40	4.9	2.1
		SD	1.4	9.5	12.7	1.4 1.5 1.4	—	2.9	1.3
20–29 yrs	2259	X̄	26.0	177.6	77.4	3.6-5.0-2.2	41.67	[a]4.7	[a]2.1
		SD	2.7	7.0	11.5	1.4 1.3 1.1	—	2.3	1.0
30–39 yrs	2985	X̄	34.5	177.5	80.0	4.0-5.2-1.8	41.19	[b]4.4	[b]2.0
		SD	2.8	6.8	11.1	1.4 1.2 1.0	—	2.3	1.0
40–49 yrs	2031	X̄	44.9	177.3	80.9	4.1-5.3-1.7	40.99	[c]3.9	[c]1.8
		SD	2.9	6.5	11.4	1.3 1.2 1.0	—	2.2	1.0
50–59 yrs	1159	X̄	54.3	176.0	79.8	4.1-5.4-1.6	40.88	[d]3.6	[d]1.6
		SD	2.7	6.7	10.1	1.2 1.2 0.9	—	1.9	0.9
60+ yrs	375	X̄	65.3	174.2	75.9	4.0-5.1-1.8	41.14	[e]3.8	[e]1.7
		SD	4.7	6.3	9.4	1.2 1.1 1.0	—	2.0	0.9
USA (FAA) (1968?)	345	X̄	28.5	177.0	73.5	3.4-4.8-2.4	42.10	—	—
		SD	6.2	6.1	10.2	1.3 0.9 1.2	1.79		
20–29 yrs	244	X̄	25.1	176.8	72.2	3.2-4.7-2.5	42.64	—	—
		SD	1.9	6.3	10.7	1.4 0.8 1.2	1.80		
30–39 yrs	59	X̄	33.6	177.9	75.4	3.5-4.8-2.4	42.26	—	—
		SD	3.5	5.7	10.2	1.3 0.8 1.1	1.65		
40–49 yrs	42	X̄	41.6	176.5	78.7	4.0-5.3-1.8	41.28	—	—
		SD	1.3	3.4	9.2	1.3 0.8 0.9	1.46		
USA (northeast and midwest)	232	X̄	18–66	—	—	4.5-4.4-2.4	—	6.3	2.7
		SD				2.0 1.5 1.7		3.1	1.3
18–29 yrs	141	X̄				4.1-4.4-2.7		6.3	2.7
		SD				1.9 1.5 1.7		3.1	1.3

Sample	n		Age	Height	Weight				
		SD				1.8 1.6 1.2		3.1	1.2
40–49 yrs	25	X̄				4.9-4.4-2.5		7.0	3.0
		SD				2.3 1.5 2.0		3.5	1.5
50–66 yrs	34	X̄				5.1-4.4-1.9		5.3	2.3
		SD				1.8 1.3 1.4		2.8	1.1
England (tunnel) (1974–76)	775	X̄	30.1	174.1	73.7	3.1-5.5-1.7	41.67	—	—
		SD	8.7	6.7	9.8	1.1 1.0 1.2	1.88		
18–29 yrs	442	X̄	24.1	174.7	72.6	3.0-5.4-1.9	42.01	—	
		SD	3.2	6.3	9.5	1.1 1.0 1.0	1.83		
30–39 yrs	226	X̄	33.9	173.6	75.0	3.3-5.6-1.6	41.29	—	
		SD	2.8	7.2	10.1	1.2 0.9 0.9	1.86		
40–49 yrs	79	X̄	43.8	173.6	75.0	3.4-5.5-1.5	40.38	—	
		SD	2.5	7.0	9.4	1.1 1.1 1.1	1.79		
50–64 yrs	28	X̄	54.7	171.7	77.8	3.9-5.9-1.3	40.38	—	
		SD	4.3	7.4	11.1	1.0 1.1 0.9	1.69		
Mexico (urban) (1968)	265	X̄	22.3	170.4	63.0	3.4-4.6-2.9	42.94	—	
		SD	4.9	7.0	8.6	1.4 0.9 1.3	1.80		
Caucasian	102	X̄	22.3	172.9	64.8	3.3-4.5-3.1	43.21	—	
		SD	5.6	7.2	10.1	1.5 1.0 1.3	1.75		
Mestizos	162	X̄	22.2	168.7	62.0	3.3-4.7-2.7	42.75	—	
		SD	4.5	6.4	7.7	1.3 0.9 1.2	1.75		
India (northwest) except Spitians (1974–80)	816	X̄	18–45	167.0	52.5	2.6-3.3-4.0	44.61	—	
		SD		5.3	5.6	1.0 0.9 1.2			
Ethnic samples									
USA (Eskimos) (1958) (villagers)	30	X̄	16–41	167.7	69.0	4.0-5.1-1.6	41.01		
		SD	17.0	6.1	9.0	1.0 4.0 0.6	1.38		
USA (Eskimos) (1958, 1968–69)	81	X̄	16–75	166.3	68.4	3.4-5.9-1.3	41.13		
Brazil (Caingang)	48	X̄	—	—	—	1.9-6.0-1.6	—	—	
		SD				0.5 0.9 0.6			

(continued)

Table 3.2. (*Continued*)

Sample/year	N	Statistic	Age (yr)	Height (cm)	Weight (kg)	Somatotype	HWR	SDM	SAM
Papua New Guinea (Manus) (1966)	100	$\bar{X}$	38.9	161.3	58.6	1.7-6.7-1.7	41.66	2.6	—
		SD	11.1	4.4	6.2	0.6 0.9 0.7	1.26	1.7	
Sudan (Nilotes) (1953–54)	52	$\bar{X}$	18–45	181.9	58.0	1.6-3.5-6.2	47.07	—	—
		SD		6.7	6.2	0.4 0.9 1.3	1.56		
India (Gaddis, Rajput and Brahmins) (1979)	200	$\bar{X}$	25–45	163.1	51.2	2.0-3.5-3.5	43.93	2.5	
		SD		5.0	3.3	0.6 1.2 0.9	—	1.5	
(Gaddis, Rajput and Brahmins, 1974–75)	91	$\bar{X}$	18–20	161.8	48.2	1.8-3.3-4.0	44.53	—	
		SD		4.6	3.9	0.5 0.4 0.9	1.19		
(Punjabi, 1976)	59	$\bar{X}$	24.0	170.0	58.0	3.5-3.3-3.6	44.03	1.8	
		SD	3.8	6.1	6.7	1.4 0.9 1.3	1.65	0.9	
(Jat-Sikh, 1975)	100	$\bar{X}$	21.7	170.4	54.5	3.5-3.5-4.0	44.95	—	
		SD	(18–25)	5.0	6.2	0.9 0.7 1.3	—		
(Jat-Sikh, 1976–77)	145	$\bar{X}$	18–21	170.7	54.0	2.4-3.0-4.5	45.17	—	
		SD		5.6	6.2	1.0 0.8 1.2	—		
(Bania, 1974)	71	$\bar{X}$	—	167.4	52.4	3.0-3.0-4.0	44.74	—	
		SD		5.5	7.5	—			
(Bania, 1976–77)	150	$\bar{X}$	18–21	168.3	52.4	3.0-3.1-4.4	44.98	—	
		SD		5.7	6.7	1.3 0.8 1.4	—		
(Spitians, 1980)	52	$\bar{X}$	33.9	158.9	52.5	3.0-4.8-2.5	42.46	—	
		SD	5.2	5.3	2.9	0.7 0.8 0.7			
Military samples									
USA (West Point) (1947)	424	$\bar{X}$	18–24	177.8	68.9	3.3-4.3-3.3	43.40	3.4	1.5
		SD		5.7	7.7	0.8 0.9 1.1	1.47	1.8	0.7
USA (Navy Hawaii) (1942)	48	$\bar{X}$	28.8	176.5	74.4	3.7-5.0-2.6	41.80	2.9	1.2
		SD	6.6	5.8	8.3	0.8 0.7 1.0	1.31	1.4	0.6

Sample	N		Age	Height	Weight				
England (Sandhurst) (1952)	286	X̄	18.5	NA	NA	3.4-4.4-3.2	43.43	3.0	1.3
		SD	0.6	NA	NA	0.6 0.8 1.0	1.24	1.6	0.6
England (Portsmouth)	110	X̄	—	NA	NA	2.9-4.8-2.3	—	—	—
		SD		NA	NA	0.8 0.8 1.0			
Belgium (soldiers)	283	X̄	20.1	174.5	68.3	2.4-4.3-2.8	42.81	3.9	1.7
		SD	2.1	6.7	9.0	1.1 1.1 1.2	1.74	2.4	1.0
Brazil (soldiers)	45	X̄	—	—	—	2.3-5.5-2.0	—	—	—
		SD				0.7 1.0 1.1			
Turkey (military recruits)	6000	X̄	18–20	—	—	1.7-4.8-2.6	42.58	—	—
		SD				0.7 1.2 1.1	1.78		
University student samples									
USA (Princeton) (1962)	101	X̄	19.7	179.1	73.5	3.4-4.6-2.9	42.89	4.4	
		SD	2.14	7.1	9.6	1.2 1.1 1.3	1.62	2.1	
USA (Monterey) (1972)	66	X̄	16–52	177.6	74.4	3.7-5.0-2.6	43.43	—	—
		SD	6.3	7.1	17.1	1.2 1.1 1.2	2.15		
USA (San Diego) (1973–75)	328	X̄	23.1	179.0	74.6	3.1-5.1-2.7	42.62	3.9	1.7
		SD	4.4	6.4	8.8	1.2 1.1 1.1	1.79	2.3	1.0
USA (Hawaii) (1955)	104	X̄	18–19	166.9	59.0	2.8-5.1-3.3	43.05	3.5	1.5
		SD	0.9	6.1	7.3	0.8 0.9 1.2	1.61	2.1	0.9
USA (Iowa) (1957)	50	X̄	19.1	170.0	67.0	3.1-4.1-2.9	43.41	—	1.4
		SD	1.4	6.2	7.1	0.8 0.8 1.2	1.67		0.8
USA (Missouri) (1980s)	161	X̄	19.9	178.7	75.0	4.0-4.4-2.6	42.38		
		SD	1.6	6.8	10.9	1.6 1.4 1.2			
USA (Berkeley) (1969)	133	X̄	22.0	177.3	75.6	4.0-5.1-2.3	—	—	—
		SD	3.1	7.2	11.0	1.6 1.3 1.1			
Canada (British Columbia) (1977)	153	X̄	21.3	178.6	72.5	2.8-4.9-2.8	42.91	3.7	1.6
		SD	2.9	7.1	8.6	1.2 1.1 1.1	1.49	2.5	1.0
England (London) (1950–55)	274	X̄	21.5	179.2	70.2	2.9-4.2-3.2	43.61	3.4	1.4
		SD	2.1	6.3	7.8	0.9 0.8 0.8	1.45	1.9	0.8

(continued)

Table 3.2. (*Continued*)

Sample/year	N	Statistic	Age (yr)	Height (cm)	Weight (kg)	Somatotype	HWR	SDM	SAM
Czechoslovakia (Prague) (1960s–1970s)	302	$\bar{X}$	20.1	178.5	72.5	3.4-4.3-3.0	43.40	—	—
		SD	1.4	6.0	8.4	1.2 0.9 1.1			
South Africa (Johannesburg) (1970s)	425	$\bar{X}$	18–25	176.5	69.8	3.2-4.7-3.0	42.87	—	—
		SD	9.6	6.5	9.5	0.7 0.9 1.1			
Hungary (Budapest) (1985)	569	$\bar{X}$	20	177.1	68.7	4.3-4.8-3.2	43.25[f]		
		SD		6.6	7.9	1.5 1.1 1.1	—		
South Korea (Pusan) (1984)	30	$\bar{X}$	21.2	170.2	61.9	3.2-4.4-2.9	43.03	3.0	1.5
		SD	1.5	4.5	5.0	0.8 0.9 1.0	—	1.7	0.7
Older men									
USA (San Diego businessmen) (1964)	56	$\bar{X}$	48.3	177.4	80.6	4.9-5.6-1.7	41.23	—	—
		SD	6.4	5.9	10.2	1.5 0.9 0.8	1.29		
USA (Boston veterans) (1963)	130	$\bar{X}$	81.1	169.3	69.8	4.6-4.7-2.1	41.31	4.0	1.8
		SD	3.5	5.7	10.7	1.4 0.8 1.2	1.94	2.5	1.0

[a]N = 207; [b]N = 213; [c]N = 203; [d]N = 231; [e]N = 187; [f]calculated from mean height and weight. $\bar{X}$ = mean.

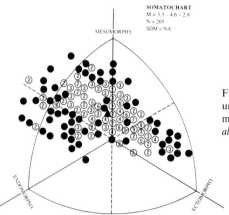

Fig. 3.6. Somatotype distribution of male urban Mexicans from Mexico City. ▲ = mean somatotype. (Data from De Garay *et al.*, 1974.)

in samples of predominantly European ancestry, are significantly less endomorphic and mesomorphic, and more ectomorphic than other samples and in this are closer to the Nilotes.

On the whole the Canadian, USA (FAA) and British distributions overlap a good deal. Yet, the number of extreme endo-mesomorphs (beyond the northwest somatochart boundary) among the British tunnel workers and the Canadians exceeds those in the USA and Mexican samples.

Ethnic samples

The ethnic samples include Wainwright Eskimos (N = 81), Caingang Indians of Brazil (N = 48), Manus islanders of Papua New Guinea (N = 100), Nilote villagers of the southern Sudan (N = 52), Spitians from India (N = 52), and seven subsamples from northwest India (N = 816). The somatotype distributions in samples of Eskimos (Fig. 3.7), Caingang

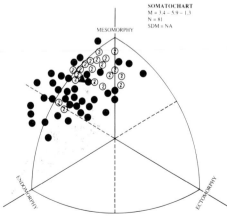

Fig. 3.7. Somatotype distribution of male Eskimos from the village of Wainwright, Alaska. ▲ = mean somatotype.

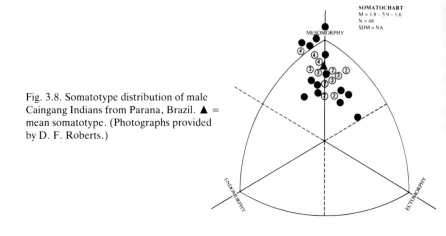

Fig. 3.8. Somatotype distribution of male Caingang Indians from Parana, Brazil. ▲ = mean somatotype. (Photographs provided by D. F. Roberts.)

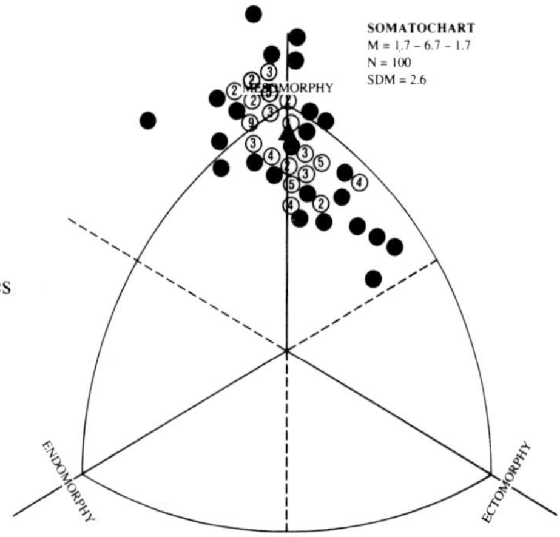

Fig. 3.9. Somatotype distribution of males from Pere village, Manus Island, Papua, New Guinea. ▲ = mean somatotype.

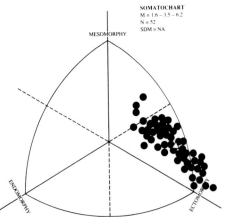

Fig. 3.10. Somatotype distribution of male Nilotes from southern Sudan. △ = mean somatotype. (Photographs provided by D. F. Roberts.)

(Fig. 3.8), Manus (Fig. 3.9) and Nilotes (Fig. 3.10) are markedly more restricted and dominated by extreme somatotypes than the distributions of the nationality samples. Although the restriction could be due to the sample sizes, it is more likely to be representative of the adult males in their respective villages or nearby territory. Even though the ages spanned several decades, it was impractical to divide the samples by age groups. The age differences in somatotype appeared to be less than in the 'Western' reference groups presented earlier.

These four samples fall into three somatotype groupings. The Eskimos are primarily endo-mesomorphic. The Manus and the Caingang are conspicuously mesomorphic, with many extreme mesomorphs. The Nilotes are strikingly ectomorphic, with many extreme ectomorphs. The limited range of values on a single component in each sample produces an elliptoid distribution. The Manus and Caingang overlap significantly, and to some degree the most mesomorphic of the Eskimos overlap with these two groups. The Nilotes overlap with none of the small ethnic samples, but do overlap somewhat with the Indians.

The Spitians, who are ethnically related to the Tibetans and the Mongolians and live in the Spiti Valley of the Himalayas, are separated from the other Indian samples. They are balanced mesomorphs, with a half-unit more endomorphy than ectomorphy. Their mean somatoplot lies one-third of the distance between the somatoplots of the Eskimos and Nilotes (Fig. 3.1).

The Indians consist of seven subsamples, including Gaddis (Rajput and Brahmin), Punjabis, Jat-Sikhs and Banias. All samples were collected in northwest India. Except for two groups of Rajput and Brahmin Gaddis, who are 25–45 years old, all are 18–25 years old. Figure 3.11 gives mean somatotypes, and Fig. 3.12(*a*) and (*b*) the distributions of Rajput and

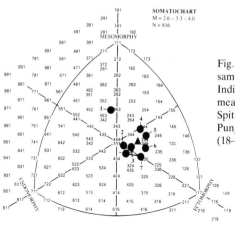

Fig. 3.11. Mean somatotypes of ethnic samples of male Indians from northwest India. The triangle represents the weighted mean of all samples (N = 816) except the Spitians. 1 = Spitians; 2 = Jat-Sikh; 3 = Punjabi; 4 = Gaddi (25–45 yr); 5 = Gaddi (18–20 yr); 6 = Jat-Sikh; 7 = Bania.

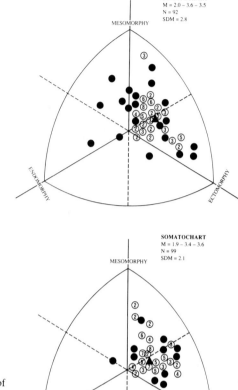

Fig. 3.12(*a*). Somatotype distribution of male Gaddi Rajput Indians. (*b*). Somatotype distribution of male Gaddi Brahmin Indians. ▲ = mean somatotype. (Redrawn from Singh, 1981.)

Brahmin Gaddis (from Singh, 1981). Almost all the Gaddis are ecto-mesomorphs, mesomorph–ectomorphs and meso-ectomorphs – that is, less endomorphic than the others. The older Gaddis are slightly more mesomor-phic than the 18–20-year-olds. One sample of Jat-Sikhs is more endomor-phic and less ectomorphic than the other. The means of the remaining four samples are near or on the ectomorphic axis. All subsamples lie about halfway between the Spitians and the Nilotes.

Military samples

The military groups from the United States, England, Belgium, Brazil and Turkey include two military academies that train officer cadets, three Army and two Navy samples (N = 7196). The mean somatoplots (Fig. 3.13) show that four of the groups have balanced mesomorphy somatotype means, with West Point, Sandhurst and Belgian samples close

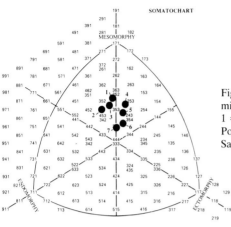

Fig. 3.13. Mean somatotypes of male military personnel from various countries. 1 = Brazil; 2 = USN–Hawaii; 3 = Portsmouth; 4 = Turkey; 5 = Belgium; 6 = Sandhurst, UK; 7 = West Point, USA.

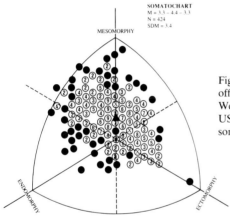

Fig. 3.14. Somatotype distribution of male officer cadets in their first year (1947) at West Point Military Academy, West Point, USA. X = 10 or more subjects; ▲ = mean somatotype.

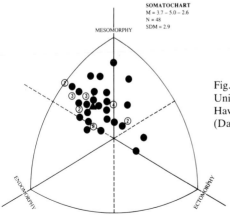

Fig. 3.15. Somatotype distribution of male United States Navy personnel stationed in Hawaii (1942). ▲ = mean somatotype. (Data from A. R. Behnke.)

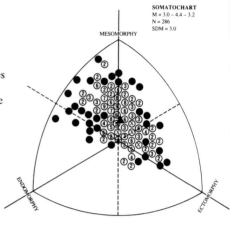

Fig. 3.16. Somatotype distribution of males entering the Royal Military Academy at Sandhurst, England, 1952. X = 10 or more subjects; ▲ = mean somatotype. (Photographs provided by J. M. Tanner.)

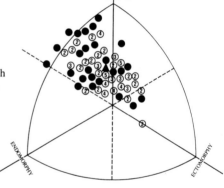

Fig. 3.17. Somatotype distribution of British Navy personnel at Portsmouth. ▲ = mean somatotype. (Photographs provided by D. F. Roberts.)

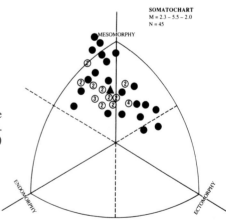

Fig. 3.18. Somatotype distribution of male Brazilian soldiers. ▲ = mean somatotype. (Photographs provided by D. F. Roberts.)

together. The Brazilians are more mesomorphic, with a somatotype near 2-5-2. The two Navy samples, Portsmouth and the US Hawaii, are somewhat more endo-mesomorphic. The US Hawaii sample is the most endo-mesomorphic and the oldest. The Turkish soldiers are the most ecto-mesomorphic. There are few extreme endo-mesomorphs in comparison with the nationality samples in the previous section. There are no somatotypes higher in endomorphy than mesomorphy except for a few in the Sandhurst and West Point samples. Except for a slightly greater range on all three components, the West Point sample closely resembles the Sandhurst sample (see Figs. 3.14–3.18).

University samples

There are 13 samples (N = 2696) of university students: seven from the United States, and one each from Canada, England, Czechoslovakia, South Africa, Hungary, and South Korea. Two groups are medical students from London and Johannesburg. One of the US groups consists of first- and second-generation Japanese students at the University of Hawaii. A large majority of the students are of Northwest European or Caucasian origin. Except for the medical students, the samples represent a variety of faculties. Because the students in Vancouver, San Diego and Iowa were enrolled in physical education classes at the time of the studies, they probably represent somewhat more physically active groups than the others.

The means (Fig. 3.19) closely cluster near and between the somatotypes 3-4-3 and 3-5-3. All fall within the category of balanced mesomorphy, except for the Hungary, Berkeley and Missouri means which are endomorph–mesomorphs. In several of the samples (Figs. 3.20–3.27) there are a few more endo-mesomorphs than ecto-mesomorphs (Princeton, San Diego, Monterey, Prague). The greatest number of extreme endo-mesomorphs are

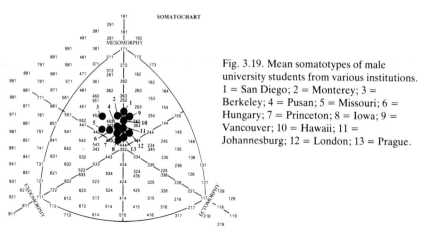

Fig. 3.19. Mean somatotypes of male university students from various institutions. 1 = San Diego; 2 = Monterey; 3 = Berkeley; 4 = Pusan; 5 = Missouri; 6 = Hungary; 7 = Princeton; 8 = Iowa; 9 = Vancouver; 10 = Hawaii; 11 = Johannesburg; 12 = London; 13 = Prague.

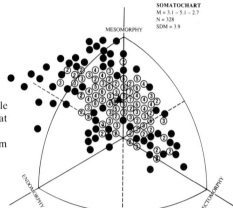

Fig. 3.20. Somatotype distribution of male students in physical conditioning classes at San Diego State University, San Diego, USA. ▲ = mean somatotype. (Data from Aubry, 1976, unpublished.)

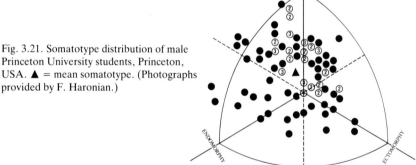

Fig. 3.21. Somatotype distribution of male Princeton University students, Princeton, USA. ▲ = mean somatotype. (Photographs provided by F. Haronian.)

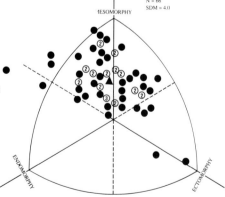

Fig. 3.22. Somatotype distribution of male students at Monterey Peninsula College, Monterey, USA. ▲ = mean somatotype.

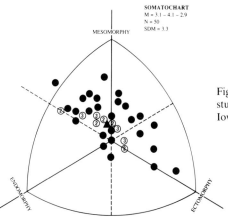

Fig. 3.23. Somatotype distribution of male students at the State University of Iowa, Iowa City, USA. ▲ = mean somatotype.

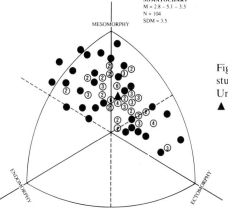

Fig. 3.24. Somatotype distribution of male students of Japanese ancestry at the University of Hawaii, Honolulu, USA. ▲ = mean somatotype.

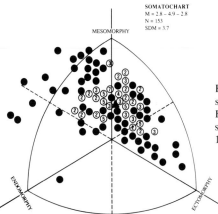

Fig. 3.25. Somatotype distribution of male students at three west coast universities in British Columbia, Canada. ▲ = mean somatotype. (Redrawn from Ross & Ward, 1982.)

Fig. 3.26. Somatotype distribution of male medical students at the University of London, England (1952–55). ▲ = mean somatotype. (Photographs provided by J. M. Tanner.)

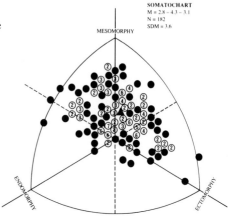

Fig. 3.27. Somatotype distribution of male students at Charles University, Prague, Czechoslovakia. (From Štěpnička *et al.*, 1979*a*.)

Fig. 3.28. Somatotype distribution of male business and professional men aged 35-61 years in a fitness program, San Diego, USA. ▲ = mean somatotype.

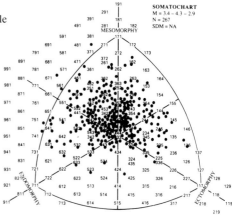

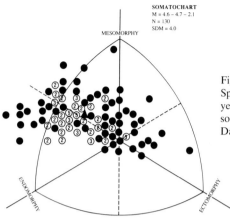

Fig. 3.29. Somatotype distribution of male Spanish–American war veterans aged 75–91 years in Boston, 1959. ▲ = mean somatotype. (Photographs provided by A. Damon.)

in the San Diego and Vancouver samples. Except for the Princeton and Prague samples there are few dominant endomorphs. There are some fairly ectomorphic students in each sample.

The Mexico City sample, described in the nationality samples above, could be classified with this group because a large number of the group were students. Their mean somatotype is near the centre of the means for the student samples.

Middle-aged and older men

There are two samples of middle-aged and older Americans. One is a sample of San Diego business and professional men in an adult fitness programme. The other is a group of 72–91-year-old male outpatients at a Veterans' Administration hospital in Boston. It is clear from Figs. 3.28 and 3.29 that they are predominantly endo-mesomorphs, endomorph–mesomorphs and meso-endomorphs. The San Diego sample means are close to the means for the 40+-year-old groups of USA (FAA) trainees, Canadians, and British tunnel workers. The Veterans group has a balanced endomorph–mesomorph mean of 4.6-4.7-2.1. The subjects are scattered evenly on either side of that category. Both samples include extreme meso-endomorphs and endo-mesomorphs.

Range of component ratings

Most of the samples (Table 3.3) include ratings in endomorphy as low as 1. A few had lowest ratings of $1\frac{1}{2}$ and 2. The upper end of the range for endomorphy in most samples varied between 5 and 8. The exceptions were the Nilotes with a high of $2\frac{1}{2}$, the Boston veterans with a high of $10\frac{1}{2}$ and USA (SEL) with $11\frac{1}{2}$.

Table 3.3. *Correlations between somatotype components, and range of component values for male samples*

Sample	Correlations			Ranges		
	endo with meso	endo with ecto	meso with ecto	endo-morphy	meso-morphy	ecto-morphy
Canada (YMCA-LIFE)	0.27	−0.53	−0.82	1.0–8.0	2.0–7.5	0.5–6.0
Canada (YMCA-LIFE) (20–29 yrs)	0.35	−0.65	−0.71	1.0–7.5	2.0–9.5	0.5–6.5
USA (FAA) (20–29 yrs)	0.03	−0.61	−0.65	1.0–7.5	2.0–8.0	0.5–6.0
USA (northeast and midwest) (18–29 yrs)	—	—	—	1.0–11.5	1.0–9.0	0.5–7.0
England (tunnel) (17–29 yrs)	0.32	−0.55	−0.76	1.0–6.5	3.0–8.0	0.5–6.0
Mexico (urban)	0.47	−0.69	−0.85	1.0–7.5	2.5–7.5	0.5–6.5
USA (Eskimo)	−0.33	−0.59	−0.23	2.0–6.5	4.5–6.0	0.5–3.0
Brazil (Caingang)	—	—	—	1.0–3.0	4.5–8.0	0.5–3.0
Papua New Guinea (Manus)	0.32	−0.33	−0.80	1.0–4.5	4.5–9.0	0.5–4.0
Sudan (Nilote)	0.45	−0.53	−0.86	1.0–2.5	2.0–5.5	4.0–9.0
USA (West Point)	−0.08	−0.42	−0.74	1.5–5.5	2.0–7.0	1.0–7.0
USA (Navy, Hawaii)	−0.07	−0.37	−0.76			
England (Sandhurst)	−0.07	−0.29	−0.78	1.5–5.5	2.5–7.0	1.0–5.5
England (Portsmouth)	—	—	—	1.5–5.5	3.0–7.0	0.5–5.0
Belgium (soldiers)	0.45	−0.70	−0.84	1.0–7.5	2.0–8.0	0.5–6.0
Brazil (soldiers)	—	—	—	1.0–4.0	4.0–8.0	0.5–4.5
USA (Princeton)	−0.27	−0.25	−0.72	1.0–7.5	2.0–7.0	0.5–6.0
USA (Monterey)	0.28	−0.54	−0.77	1.0–8.5	2.0–7.5	0.5–7.0
USA (San Diego)	0.39	−0.63	−0.59	1.5–8.5	2.0–8.0	0.5–6.0
USA (Hawaii)	0.18	−0.49	−0.89	1.0–5.0	3.0–7.0	1.0–6.0
USA (Iowa)	0.31	−0.62	−0.70	1.5–5.0	2.0–6.5	1.0–6.5
England (London)	−0.01	−0.41	−0.74	1.0–6.5	1.5–6.0	1.0–7.0
Czechoslovakia (Prague)	−0.23	−0.58	−0.52	1.0–6.5	2.0–7.5	0.5–6.0
South Africa (Johannesburg)	—	—	—	1.0–8.5	2.5–7.0	0.5–6.0
USA (San Diego businessmen)	—	—	—	2.5–8.0	4.0–8.0	0.5–3.5
USA (Boston veterans)	0.32	−0.62	−0.64	1.0–10.5	3.0–7.0	0.5–6.0

The lowest value in mesomorphy was 1.5 for a St. Thomas' (London) medical student. Several samples had lowest values of 2. Several samples had highest values between 8 and $9\frac{1}{2}$. A Canadian subject had a rating of 9.5, and several Manus males had ratings of 9.

The lowest ratings in ectomorphy in many samples were $\frac{1}{2}$ and 1 in others. For the Nilotes the lowest value was 4 and the highest 9. No rating exceeded 7 in the remaining samples.

Table 3.4. *Catalogue of characteristics of adult female samples*

Nationality samples

Canada	National survey
Sample	Participants in YMCA–LIFE programme, adult females. Collected under direction of Donald Bailey, 1976–78. Broad geographic and socio-economic spectrum. Probably relatively healthy sample. Height and weight similar to national Canadian and USA surveys
Age range	15–69 years
Ethnicity	Predominantly White
Number	4629
	Ages 15–19 = 235
	Ages 20–29 = 1752
	Ages 30–39 = 1201
	Ages 40–49 = 787
	Ages 50–59 = 498
	Ages 60+ = 156
Rating method	Heath–Carter anthropometric, by Bailey and colleagues
Reference	Bailey *et al.* (1982)
Mexico	Mexico City, DF
Sample	Urban university students, technical and clerical workers. Measured and photographed by J. Faulhaber and co-workers, as a reference group for athletes measured at 1968 Mexico City Olympic Games
Age range	12–29 years
Ethnicity	40 Caucasian, 46 Mestizo
Number	86
Rating method	Heath–Carter anthropometric
Reference	De Garay *et al.* (1974)
India	Punjab
Sample	Jat-Sikh (N = 502) and Bania (N = 510) females. Jat-Sikh are traditionally agricultural, mostly living in villages. Banias are traditionally traders, mostly living in cities. Both are endogamous at the caste level. Jat-Sikhs lead a strenuous life, Banias are relatively sedentary
Age range	20–80 years
Ethnicity	Jat-Sikh and Bania
Number	1012
	Ages 20–29 = 215
	Ages 30–39 = 218
	Ages 40–49 = 238
	Ages 50–59 = 171
	Ages 60–69 = 110
	Ages 70–80 = 60
Rating method	Heath–Carter anthropometric (with height correction for endomorphy)
Reference	Singal & Sidhu (1984)

Ethnic samples

Alaska	Wainwright, Alaska
Sample	Village with population of about 300 (approx. 40 households), about 145 km SW of Point Barrow. Photographs by C. Dotter of Univ. Oregon Medical School, 1958; anthropometry by Jamison & Zegura
Age range	16–75 years

(*continued*)

Table 3.4. (*Continued*)

Ethnicity	Eskimos and a few Eskimo hybrids
Number	76
Rating method	Some anthropometric and some photoscopic, by Heath
References	Heath (1973, 1977), Heath *et al.* (1968)
Papua New Guinea	Pere Village, Manus Province
Sample	Adult females measured and photographed by Theodore Schwartz & Barbara Heath, 1966
Age range	18–72 years
Ethnicity	Melanesian
Number	111
Rating method	Heath–Carter photoscopic, by Heath
References	Heath *et al.* (1968), Heath (1973)
India	Punjab
Sample	Jat-Sikh (N = 502) and Bania (N = 510) females. Jat-Sikhs are traditionally agricultural, mostly living in villages. Banias are traditionally traders, mostly living in cities. Both are endogamous at the caste level. Jat-Sikhs lead a strenuous life, Banias are relatively sedentary
Age range	20–80 years
Ethnicity	Jat-Sikh and Bania
Number	1012
Rating method	Heath–Carter anthropometric (with height correction for endomorphy)
Reference	Singal & Sidhu (1984)

University student samples

USA	Honolulu, Hawaii (University of Hawaii)
Sample	Hawaii-born females, both parents of Japanese ancestry. 50% first generation, 19% second generation, and 30% one parent born in Japan the other in Hawaii. Students at the University of Hawaii, volunteers from physical education classes. Data collected by Carey D. Miller in 1955
Age range	18–22 years
Ethnicity	Japanese
Number	104
Rating method	Heath photoscopic, by Heath
Reference	Heath *et al.* (1961)
England	London
Sample	Students enrolled in London colleges of home economics and music. Data and photographs collected by J. M. Tanner
Age range	17–32 years
Ethnicity	White
Numbers	Three subsamples: N = 90, N = 26, N = 116
Rating method	Heath–Carter photoscopic, by Heath
Reference	Heath *et al.* (1961)
Czechoslovakia	Prague
Sample	Students at Charles University, Prague. Data collected in late 1960s and early 1970s

(*continued*)

Table 3.4. (*Continued*)

Age range	17–30 years
Ethnicity	White
Number	286
Rating method	Heath–Carter photoscopic, by Štěpnička
References	Štěpnička *et al.* (1979*a*), J. Štěpnička, personal communication
USA	California (various colleges)
Sample	Volunteer students from various faculties in physical education classes at San Diego State University, California State College at San Luis Obispo, and Southwestern Community College in Chula Vista. Data collected 1968–72
Age range	16–25 years
Ethnicity	Mostly White or Mexican-American, a few Black and Oriental
Number	74
Rating method	Heath–Carter anthropometric and anthroposcopic plus photographs, by Carter
Reference	Carter *et al.* (1978)
USA	San Diego, California (San Diego State University)
Sample	Students from various faculties at San Diego State University, voluntarily enrolled in three types of conditioning classes, 1974–5
Age range	18–37 years (positively skewed: 90% 18–24 years)
Ethnicity	Over 90% White or Mexican-American, a few Black and Oriental
Number	143
Rating method	Heath–Carter anthropometric, by S. P. Aubry
Reference	S. P. Aubry (unpublished study, San Diego State University, 1975)
USA	Northeast and midwest states
Sample	Undergraduate students enrolled in routine physical education classes at five northeast and five midwest colleges and universities. Data collected and photographs taken by Heath and assistants, in a study on female somatotypes at Constitution Laboratory, Columbia University Medical Center, 1948–50
Age range	17–24 years (?)
Ethnicity	Majority White, a few Black and Oriental
Number	2434
Rating method	Rated (1948–50) by Sheldon photoscopic method; 2434 re-rated from a total of 3529 in about 1960 from data (without photographs) by Heath's modifications (1963), by Heath
Reference	Heath *et al.* (1961)
USA	Northeast and midwest colleges and universities. Subsamples of above study of college women
Sample	A random weighted subsample of the ten colleges (N = 207, approx. 6%). The subsample was used to represent the population for plotting and statistical purposes
Age range	17–24 years
Ethnicity	Majority White, a few Black and Oriental
Number	207
Rating method	Heath–Carter photoscopic, by Heath (1977)
Reference	Heath *et al.* (1961)

(*continued*)

Table 3.4. (*Continued*)

USA	Monterey, California (Monterey Peninsula College)
Sample	Community college students in a physical education 'figure control' class (N = 68) and students in a physical anthropology class (N = 28), 1967 and 1972
Age range	?
Ethnicity	Majority White, a few Black and Oriental
Number	96
Rating method	Heath–Carter anthropometric plus photographs, by Heath
Reference	Unpublished data
USA	Berkeley, California (University of California)
Sample	Volunteers attending University of California, Berkeley. Considered normal and healthy
Age range	Mean = 21.4 years
Ethnicity	Presumably the majority were White
Number	128
Rating method	Heath–Carter anthropometric, by Wilmore (with estimated medial calf skinfold)
Reference	Wilmore (1970)
USA	Urbana, Illinois (University of Illinois)
Sample	Healthy student volunteers enrolled in a weight control and conditioning class at University of Illinois, 1974
Age range	17–22 years
Ethnicity	White
Number	31
Rating method	Heath–Carter anthropometric, by Slaughter
Reference	Slaughter & Lohman (1976)
Canada	British Columbia (three west coast universities)
Sample	Students recruited from an exercise class, a campus residence, and a physical education class for elementary school teachers from Simon Fraser University, University of British Columbia and University of Victoria, 1976
Age range	17–31 years
Ethnicity	White and a few Oriental
Number	94
Rating method	Heath–Carter anthropometric, by Ross
References	Hebbelinck *et al.* (1980), Carter *et al.* (1982)
South Africa	Johannesburg (University of Witwatersrand)
Sample	Second-year medical students, University of Witwatersrand, 1975–8
Age range	18–25 years
Ethnicity	Caucasoid (White)
Number	46
Rating method	Heath–Carter anthropometric and photoscopic by author. Photoscopic proficiency obtained by studying criteria and a criterion set of photographs rated by Heath
Reference	Gordon (1984)
Japan	Tokyo (Ochanomizu University)

(*continued*)

Table 3.4. (*Continued*)

Sample	Students at Ochanomizu University, a women's university in Tokyo. Data and photographs provided by Yanagisawa, 1968
Age range	College age, young
Ethnicity	Japanese
Number	200
Rating method	Heath–Carter photoscopic, by Heath
Reference	Unpublished data
USA	Kirksville, Missouri (Northeast Missouri State University)
Sample	Students at Northeast Missouri State University
Age range	Mean age = 19.6 years
Ethnicity	White
Number	206
Rating method	Heath–Carter anthropometric method, by authors
Reference	Bale *et al.* (1985*a*)
Portugal	Lisbon (Technical University of Lisbon)
Sample	Freshman students in schools of economics, agronomics and architecture
Age range	18–19 years
Ethnicity	White
Number	41
Rating method	Heath–Carter anthropometric, by authors
Reference	Sobral *et al.* (1986)
Hungary	Several regions: Debrecen, Jaszbarény, Keiskemét, Sárospatak
Sample	Students at four teacher training colleges in 1973–4. One college was for kindergarten teachers and nurses
Age range	18–21 years
Ethnicity	White
Number	1069
Rating method	Heath–Carter anthropometric, by author
Reference	Papai (1980)

In other special samples (see Chapters 6 and 7) there are some values that exceed these ranges. Some athletes in weight lifting and body building have ratings of 10 to 12 in mesomorphy. The ranges in all known samples were from 1 to 15 in endomorphy, from 1 to 12 in mesomorphy, and from $\frac{1}{2}$ to 9 in ectomorphy. It is possible that higher values in any of the three components may be found, but low values are limited to a rating of $\frac{1}{2}$ by definition.

Correlations between somatotype components

Correlation coefficients between endomorphy and mesomorphy (Table 3.3) range from -0.33 to -0.47, with coefficients close to zero in several series. Endomorphy and ectomorphy are negatively correlated, with coefficients ranging from -0.25 to -0.70. Mesomorphy is also negatively correlated with ectomorphy, with a range of -0.23 to -0.89. The correlations vary with the sample and to some extent are probably due to the range of values of some components within the samples. The negative correlations of endomorphy and mesomorphy with ectomorphy are to be expected because endomorphy and mesomorphy describe relative mass of the subject and as mass increases ectomorphy decreases. In general, high values in endomorphy or mesomorphy go with lower values in ectomorphy. This is more pronounced when moderate to high endomorphy and mesomorphy are combined. The correlations between endomorphy and mesomorphy are quite varied and indicate little mutual predictive value. This means that some subjects can have high mesomorphy in the presence of widely variable endomorphy scores and vice versa.

Variation in somatotypes of adult women

There are 21 samples, and several subsamples, consisting of 11 189 subjects from various parts of the world, to illustrate the variations in somatotypes of women (Tables 3.4 and 3.5). The data are divided into groups representing nationality, ethnic and university student samples. There are no female military samples.

Nationality samples

The nationality samples are from a Canadian national survey (N = 4629), a study of urban Mexicans in Mexico City (N = 86), and a study of Punjabi women in India (N = 1012). The means (Fig. 3.30) show that the Indians are more endomorphic and less ectomorphic and mesomorphic than the Canadian and Mexican women. The Mexicans are slightly more endomorphic than the Canadians. The somatocharts of the Canadians by age group (Fig. 3.31) show that in the course of six decades distributions shift to the left towards meso-endomorphy and lower ectomorphy. The somatotype

Fig. 3.30. Mean somatotypes of female Canadians (1), Mexicans (2) and Indians (3).

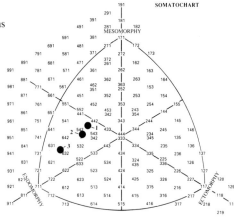

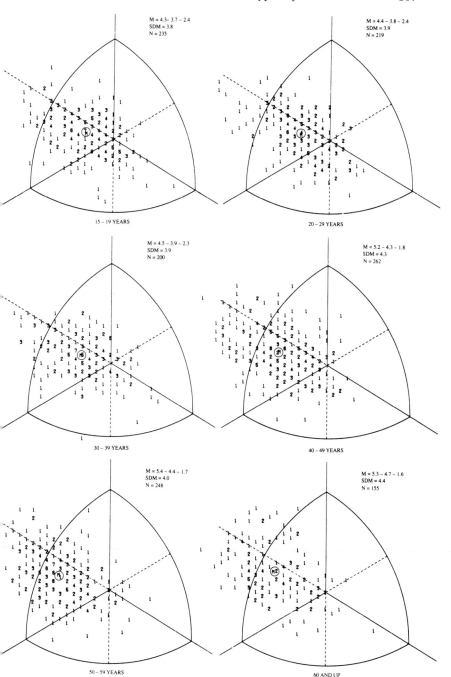

Fig. 3.31. Somatotype distributions of Canadian females by age groups.
M = mean; 1, 2, 3 = number of subjects at a point. (From Bailey *et al.*, 1982.)

Table 3.5. *Somatotypes of adult females*

Sample/year	N	Statistic	Age (yr)	Height (cm)	Weight (kg)	Somatotype	HWR	SDM	SAM
Nationality samples									
Mexico (urban) (1968)	86	X̄	19.0	158.0	53.8	5.2-3.9-2.3	42.06	4.6	2.0
(12–34 yrs)		SD	3.5	5.4	7.7	1.6 1.0 1.2	1.74	2.3	1.0
Canada (YMCA-LIFE)	4629	X̄	34.9	163.4	60.2	4.7-4.0-2.2	41.70	—	—
(1976–78)		SD	11.9	6.2	8.8	1.4 1.3 1.1			
15–19 yrs	235	X̄	18.9	164.6	59.6	4.3-3.7-2.4	42.14	3.8	1.7
		SD	0.9	5.8	8.6	1.2 1.1 1.1	—	2.2	0.9
20–29 yrs	1752	X̄	25.0	163.6	58.8	4.4-3.7-2.4	42.08	3.9	1.7
		SD	2.8	6.2	8.1	1.3 1.2 1.1	—	2.2	1.0
30–39 yrs	1201	X̄	34.4	163.6	59.7	4.6-3.9-2.3	41.87	3.9	1.8
		SD	2.9	6.0	9.0	1.4 1.2 1.1	—	2.3	1.0
40–49 yrs	787	X̄	44.6	163.0	62.6	5.1-4.4-1.8	41.05	4.3	1.9
		SD	3.0	6.2	9.1	1.4 1.3 1.0	—	2.3	1.0
50–59 yrs	498	X̄	54.2	162.4	62.4	5.4-4.5-1.8	40.95	4.0	1.8
		SD	2.8	6.1	9.1	1.4 1.3 1.0	—	2.2	1.0
60+ yrs	156	X̄	63.8	160.6	61.4	5.3-4.7-1.7	40.71	4.4	2.0
		SD	3.4	6.3	8.8	1.4 1.4 1.2	—	2.2	0.9
India (Punjabi) (1975–76)	1012	X̄	20–80	—	—	6.5-3.6-2.2	—	—	4.3
		SD				1.8 1.1 1.5			1.5
20–29 yrs	215	X̄		—	—	5.6-3.0-3.0	—	—	2.6
		SD				1.7 0.9 1.5			2.6
30–39 yrs	218	X̄		—	—	6.7-3.6-2.2	—	—	4.5
		SD				1.9 1.0 1.5			1.7
40–49 yrs	238	X̄		—	—	7.1-4.0-1.9	—	—	5.2
		SD				2.0 1.3 1.5			
50–59 yrs	171	X̄		—	—	6.8-3.0-2.0	—	—	4.8
		SD				1.8 1.1 1.5			1.0

Sample	n		Age						
60–69 yrs	110	X̄		—	—	—	—	6.8-3.9-1.9	4.9
		SD		—	—	—	—	1.8 1.1 1.4	0.9
70–80 yrs	60	X̄		—	—	—	—	5.5-3.6-2.3	3.2
		SD		—	—	—	—	1.8 0.8 1.4	0.6
Ethnic samples									
USA (Eskimo) (1968)	76	X̄	16–75	154.7	66.7	38.45	3.9	6.4-4.8-0.8	—
		SD		5.6	13.2	2.46	2.4	2.1 0.6 0.4	—
Papua New Guinea (Manus) (1966–68)	111	X̄	18–72	151.0	45.6	42.55	3.3	3.1-4.5-2.5	—
		SD		4.8	6.6	1.65	1.8	1.2 0.7 0.9	—
India (Jat-Sikh)	502	X̄	20–80	—	—	—	—	5.8-3.4-2.7	3.1
		SD						1.9 1.0 1.7	1.7
20–29 yrs	117	X̄						5.1-2.9-3.4	1.7
		SD						1.7 0.9 1.6	2.7
30–39 yrs	119	X̄						5.8-3.3-2.7	3.1
		SD						1.9 1.0 1.7	1.9
40–49 yrs	109	X̄						6.2-3.6-2.4	3.8
		SD						2.1 1.2 1.8	1.4
50–59 yrs	80	X̄						6.2-3.6-2.5	3.7
		SD						1.8 1.0 1.7	1.5
60–69 yrs	51	X̄						6.5-3.6-2.2	4.3
		SD						1.8 1.1 1.5	1.5
70–80 yrs	26	X̄						5.3-3.4-2.8	2.5
		SD						1.8 0.7 1.5	1.3
India (Bania)	510	X̄						7.2-3.9-1.8	5.4
		SD						1.9 1.1 1.3	1.2
20–29 yrs	98	X̄						6.3-3.1-2.5	3.8
		SD						1.7 0.8 1.3	2.6
30–39 yrs	99	X̄						7.8-4.0-1.6	6.2
		SD						1.9 1.1 1.3	1.4

(continued)

Table 3.5. (*Continued*)

Sample/year	N	Statistic	Age (yr)	Height (cm)	Weight (kg)	Somatotype	HWR	SDM	SAM
40–49 yrs	129	$\bar{X}$				7.8-4.4-1.4	—	—	6.4
		SD				2.0 1.4 1.3	—	—	0.4
50–59 yrs	91	$\bar{X}$				7.3-4.1-1.6	—	—	5.7
		SD				1.9 1.1 1.3	—	—	0.7
60–69 yrs	59	$\bar{X}$				7.0-4.1-1.7	—	—	5.3
		SD				1.8 1.2 1.3	—	—	0.5
70–80 yrs	34	$\bar{X}$				5.7-3.7-2.0	—	—	3.7
		SD				1.8 0.8 1.3	—	—	0.3
Mexico (urban) (White) (1968)	40	$\bar{X}$	19.3	161.4	53.5	4.6-3.5-2.9	42.98	—	—
		SD	3.2	5.9	7.1	1.4 0.7 1.0	1.32	—	
Mexico (urban) (Mestizo) (1968)	46	$\bar{X}$	18.7	155.0	54.0	5.7-4.3-1.8	41.26	4.4	2.0
		SD	3.7	5.0	8.2	1.7 1.0 1.2	2.05	2.5	1.1
University students									
USA (Hawaii) (1955)	104	$\bar{X}$	18–22	155.7	49.9	4.4-3.7-3.1	42.45	3.1	1.3
		SD		4.4	6.8	0.7 0.7 1.1	1.55	1.6	0.7
England (London) home economics (1952–55)	90	$\bar{X}$	17–23	—	—	4.1-3.5-2.9	—	3.3	1.4
		SD				0.8 0.7 1.2		1.7	0.7
music (1951–52)	26	$\bar{X}$	19–32	—	—	4.7-3.6-2.4	—	3.4	1.4
		SD				0.9 0.8 1.1		1.7	0.7
Czechoslovakia (Prague) (1967–69)	286	$\bar{X}$	17–30	164.4	60.3	4.8-3.7-2.2	41.94	—	—
		SD		5.1	6.6	1.2 0.5 1.0			
USA (California) (1970)	74	$\bar{X}$	19.9	165.3	58.0	4.6-3.6-2.8	42.85	3.6	—
		SD	1.6	6.1	7.1	1.0 0.9 1.1	1.49	1.9	
USA (San Diego) (1974)	143	$\bar{X}$	21.4	165.8	59.3	4.2-3.7-2.6	42.61	3.9	1.7
		SD	2.8	5.8	7.3	1.3 1.0 1.1	1.57	2.2	0.9

Population	n		17–24						
USA (northeast and midwest subsample) (1948–50)	207	X̄	17–24	—	—	4.9-3.2-2.9	—	3.9	1.6
		SD				1.2 0.8 1.2		2.1	0.9
USA (northeast and midwest) (1948–50)	2434	X̄	17–24	165.2	57.6	4.5-3.2-3.3	42.78	—	—
		SD		6.0	7.3	0.9 0.9 1.2	1.72		
USA (Monterey) (figure control, 1967)	68	X̄		164.1	55.8	4.3-3.9-3.0	42.95	3.5	1.5
		SD				1.1 0.6 1.2		2.0	0.8
USA (Monterey) (Physical Anthropology, 1972)	28	X̄				4.4-3.8-2.6		4.2	1.4
		SD				1.5 0.7 1.3		2.1	0.8
USA (Berkeley) (1969)	128	X̄	21.4	164.9	58.6	4.1-3.8-2.5	42.46	—	—
		SD	3.8	6.6	7.1	1.6 1.0 1.0	—		
USA (Illinois) (1974)	31	X̄	19.8	165.9	59.7	5.1-3.2-2.8	42.44	—	—
		SD	1.3	5.8	7.4	1.3 1.0 1.3	—		
Canada (British Columbia) (1976)	94	X̄	20.6	165.7	57.5	4.0-3.5-2.9	43.02	3.7	1.6
		SD	2.6	6.1	6.4	1.2 1.0 1.0	1.37	1.9	0.8
South Africa (Johannesburg) (1977–78)	46	X̄	18.9	164.2	56.2	4.3-3.5-2.9	42.87	—	—
		SD			6.0	7.0 1.1 0.8	1.1		
Japan (Tokyo) (1968)	200	X̄		—	—	4.1-4.4-2.2		2.9	1.2
		SD				0.8 0.5 1.0		2.3	1.0
USA (Missouri) (1984)	206	X̄	19.6	164.9	60.9	5.1-3.4-2.4	41.92[a]	—	—
		SD	1.2	6.1	10.2	1.5 1.5 1.3	—		
Portugal (Lisbon) (1985)	41	X̄	18?	159.2	53.7	4.1-3.6-2.3	42.20[a]	—	—
		SD		5.3	7.4	1.2 1.0 1.2	—		
Hungary[b] (1973–74)	1069	X̄	19.6	160.8	55.9	5.8-3.2-2.0	42.06[a]	—	—
		SD		5.7	6.8	1.0 1.1 1.0	—		

[a]Calculated from means; [b]Four regional groups (no differences); X̄ = mean.

Fig. 3.32. Somatotype distribution of female urban Mexicans from Mexico City. ▲ = mean somatotype. (Data from De Garay *et al.*, 1974.)

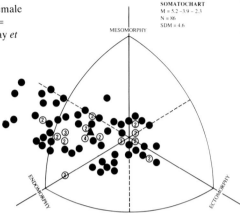

distribution of the urban Mexicans (Fig. 3.32), average age about 19 years, is similar to that of the 20–29-year-old Canadians.

The means for the Indian samples by age (Table 3.5, Fig. 3.33), calculated from Singal & Sidhu's (1984) data, show that the Indians, like the Canadians, become more meso-endomorphic from their twenties to their forties, but less so thereafter. The Indians show the largest age difference in somatotype between their twenties and thirties, while the Canadians show the largest differences between the thirties and forties.

Ethnic samples

The means on the somatoplots (Fig. 3.34) are derived from samples of Eskimos (Fig. 3.35), Manus villagers (Fig. 3.36) and subgroups from the Mexican (Fig. 3.37) and Indian (Fig. 3.38) nationality samples. The Mexi-

Fig. 3.33. Mean somatotypes of female Punjabi Indians by age groups. (From data in Singal & Sidhu, 1984.)

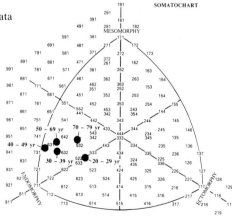

Fig. 3.34. Mean somatotypes of female
ethnic samples. 1 = Manus; 2 = Eskimos;
3 = Mexicans (Mestizos); 4 = Mexicans
(White); 5 = Indians (Bania); 6 = Indians
(Jat-Sikh).

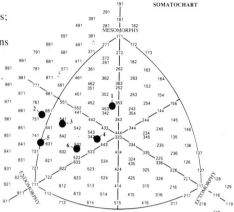

Fig. 3.35. Somatotype distribution of female
Eskimos from the village of Wainwright,
Alaska. ▲ = mean somatotype.

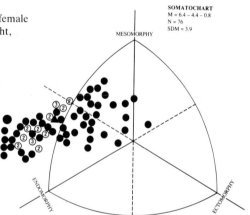

Fig. 3.36. Somatotype distribution of
females from Pere Village, Manus Island,
Papua, New Guinea. ▲ = mean
somatotype.

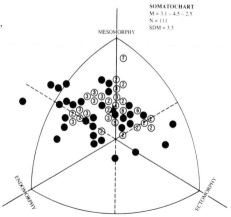

Fig. 3.37. Somatotype distribution of female
Mestizo urban Mexicans from Mexico City.
▲ = mean somatotype. (Data from De
Garay *et al.*, 1974.)

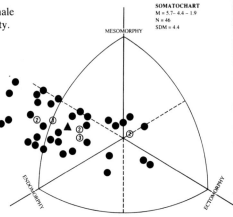

can subgroups are White (of European origin) and Mestizos (mixtures of native Indian and White). The Punjabi Indians are Jat-Sikhs and Banias.

The mean somatotypes of Eskimos, Banias and Manus are widely separated. The Eskimos and Banias (highest of all samples in endomorphy) are meso-endomorphic, with high endomorphy and mesomorphy. Because of small sample numbers, no age groupings were made for the Eskimos (ages 16–75) and the Manus (ages 18–72). Among the female nationality and ethnic groups, only in the Manus is mesomorphy greater than endomorphy alone; in a sample of 111, only five subjects have ratings in endomorphy 1 unit or more higher than in mesomorphy, and very few have ratings higher than 4 in endomorphy. The majority of Manus females are endo-mesomorphs, balanced mesomorphs and ecto-mesomorphs (Fig. 3.36).

The Mexican White and Mestizo subgroups differ in all three components. The Mestizos (Fig. 3.37) are more endomorphic and mesomorphic and less ectomorphic than the Whites (De Garay *et al.*, 1974).

Fig. 3.38. Mean somatotypes of female
Jat-Sikhs (circles) and Banias (squares) from
the Punjab, India, by decade. The numbers
refer to the age, e.g. 2 = twenties.
(Redrawn from Singal & Sidhu, 1984.)

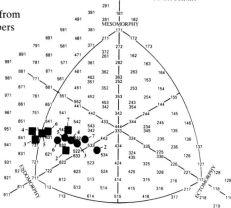

Comparing the Jat-Sikhs and Banias by age, the Banias are at all ages more endomorphic and less ectomorphic than the Jat-Sikhs. Both groups show a similar pattern of change. The means move to the west (greater meso-endomorphy) and return part way and to the east and slightly north (a decrease in meso-endomorphy). The Jat-Sikhs in the 20–29-year-old group are ecto-endomorphs, and the only age group whose mesomorphy is the lowest of the three components. Endomorphy and mesomorphy tend to increase up to the forties and then decrease. In both communities the oldest groups are the least endomorphic and are lower in mesomorphy. Singal & Sidhu (1984) suggested that the age changes may be associated with the climacteric in these females, and that the somatotype differences between Banias and Jat-Sikhs are probably due to different physical activity, dietary habits and genetic factors.

University samples

There are 16 samples (and some subgroups) of female university students (N = 5275) available for review. Nine of these are from North America, and there is one each from England, Hawaii, Japan, Czechoslovakia, Hungary, Portugal and South Africa. Means for the six US samples are plotted on Fig. 3.39, and group plots are on Figs. 3.40–3.43. Most of the students in the US samples were enrolled in physical education classes, some of them elective and some obligatory in order to meet graduation requirements. The students are from a variety of faculties and departments.

Means for the US samples (Fig. 3.39) are remarkably similar, concentrated in an area enclosed by the somatotypes 5-4-2, 4-3-3, 5-3-3 and 5-3-2. It is evident (Figs. 3.40–3.43) that the majority of the somatotypes are dominant in endomorphy. Many of them are meso-endomorphs, and next in

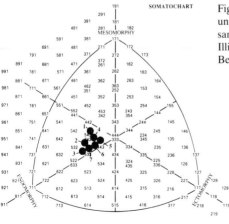

Fig. 3.39. Mean somatotypes of female university students from seven USA samples. 1 = California; 2 = Missouri; 3 = Illinois; 4 = San Diego; 5 = Monterey; 6 = Berkeley; 7 = USA (selected).

Fig. 3.40. Somatotype distribution of a random sample weighted by institution of approximately 6% (207/3529) of female students from 10 northeast and midwest universities, USA. ▲ = mean somatotype.

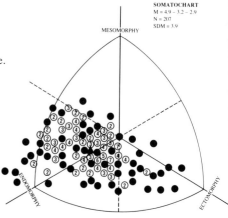

Fig. 3.41. Somatotype distribution of female students from three universities in California, USA. ▲ = mean somatotype. (Redrawn from Carter *et al.*, 1978.)

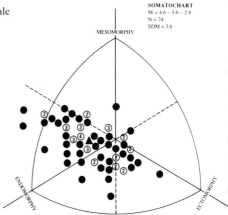

Fig. 3.42. Somatotype distribution of female students in physical conditioning classes at San Diego State University, San Diego, USA. ▲ = mean somatotype. (Redrawn from Aubry, 1976, unpublished.)

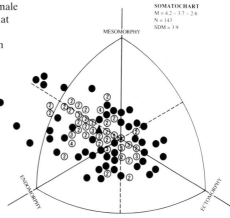

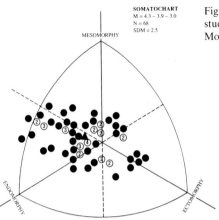

Fig. 3.43. Somatotype distribution of female students from Monterey Peninsula College, Monterey, USA. ▲ = mean somatotype.

frequency are balanced endomorphs and endomorph–mesomorphs, but all samples include some somatotypes to the east of the mesomorphic axis of the somatochart. Heath *et al.* (1961) reported a mean somatotype of 4.5-3.2-3.3 for the six series of US college females (N = 2434). The mean for the random sample (N = 3524) based on ten series (Table 3.5) is 4.9-3.2-2.9. A comparison with earlier ratings shows increased endomorphy (+0.4) and decreased ectomorphy (−0.4). The differences could be due to amendments of the method in the interim, e.g. to the somatotype–HWR table and to extension of the scales, particularly in endomorphy (Heath, 1963; Heath & Carter, 1967), but could also be due to sampling.

Somatotype distributions for five of the samples outside North America are shown in Figs. 3.44–3.49. The mean somatotypes (Fig. 3.44) for Hawaiian Japanese, western Canadians, English (home economics students), Portuguese and South African medical students are clustered together

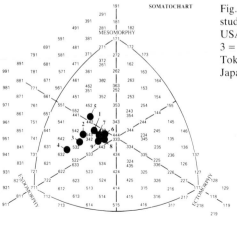

Fig. 3.44. Mean somatotypes of female students from universities other than in the USA. 1 = Portuguese; 2 = English; 3 = Czechoslovak; 4 = Hungarian; 5 = Tokyo Japanese; 6 = Canadian; 7 = Hawaii Japanese; 8 = English; 9 = South African.

Fig. 3.45. Somatotype distribution of female students at three west coast universities in British Columbia, Canada. (Redrawn from Ross & Ward, 1982.)

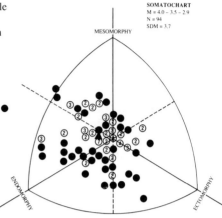

Fig. 3.46. Somatotype distribution of female students of Japanese ancestry at the University of Hawaii, Honolulu, USA. ▲ = mean somatotype.

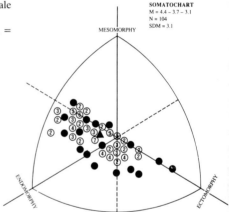

Fig. 3.47. Somatotype distribution of female students at Ochanomizu University, Tokyo, Japan (1968). X = 10; ▲ = mean somatotype. (Photographs provided by Professor Yanagisawa.)

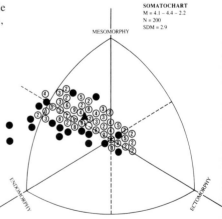

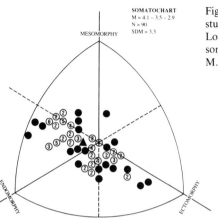

Fig. 3.48. Somatotype distribution of female students enrolled in home economics in London, England (1951–55). ▲ = mean somatotype. (Photographs provided by J. M. Tanner.)

between the somatotypes 5-4-3, 4-4-3 and 4-3-3. The English (music students) and Czechoslovakian sample means are further west, and the Hungarian students even more to the southwest, or more endomorphic, than the preceding four samples. The mean (4.1-4.4-2.2) for the Tokyo Japanese students is to the north, with almost no subjects less than 3 in mesomorphy. Their distribution (Fig. 3.47) is concentrated in a narrow band on either side of the ectomorphic axis from the central area to the edge of the endomorph–mesomorph arc. Comparison with the other distributions (Figs. 3.46–3.49) shows that the Tokyo students are the only sample in which mesomorphy exceeds endomorphy. In this respect they are similar to the Manus. Except for the Tokyo students, the somatotype distributions are generally similar to those of the US samples. The majority of somatotypes are meso-endomorphs. Their distributions are more or less ellipsoid, with the long

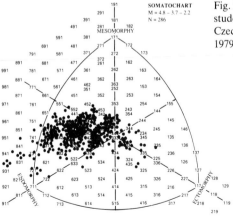

Fig. 3.49. Somatotype distribution of female students at Charles University, Prague, Czechoslovakia. (From Štěpnička *et al.*, 1979a.)

axis along or parallel to the ectomorphic axis. Only the distribution of somatotypes of the Canadian students is almost circular (Fig. 3.45).

Despite ethnic similarity, the Tokyo Japanese sample is less endomorphic and ectomorphic and more mesomorphic than the Hawaii Japanese sample (Fig. 3.46). Perhaps nutrition and other environmental factors account for the differences between these samples (Heath *et al.*, 1961). No intra-sample ethnic comparisons were made in the US samples, which were predominantly White with a small number of Blacks, Orientals and others. However, there are clear differences among the predominantly White samples, Tokyo Japanese, Mexican Mestizo, Indian, Eskimo and Manus samples. Although the western Canadian students are slightly less endomorphic and more ectomorphic than the 20–29-year-old Canadian national sample, their somatotype distributions are markedly similar. The somewhat greater physical activity inferred for the student sample (Hebbelinck *et al.*, 1980) may account for the differences.

Except for the Manus, the Tokyo Japanese and the Eskimos, high endomorphy with moderate to low mesomorphy and ectomorphy characterizes most available female distributions. There is substantial overlap of the distributions, which have a generalized mean of $4\frac{1}{2}$-$3\frac{1}{2}$-3. The reasons for the differences, apparently ethnic and environmental factors, will be profitable fields for research.

Range of component ratings

The range of component values in the female samples is from $1\frac{1}{2}$ to 10 in endomorphy, from $\frac{1}{2}$ to 6 in mesomorphy, and from $\frac{1}{2}$ to $6\frac{1}{2}$ in ectomorphy. Photoscopic ratings as high as 6 and 7 in mesomorphy are exceedingly rare in females. Such high ratings derived from anthropometry are usually spurious, partly because the anthropometric calculation of mesomorphy does not adequately correct for the fat in the limb girth and partly because adipose tissue around the joints in the obese may inflate bone diameters as well. Generally, in order to reconcile anthropometric and photoscopic ratings the mesomorphy rating must be reduced to agree with the distribution of somatotypes according to HWRs but, especially in athletes, there occur occasionally some mesomorphy ratings of $5\frac{1}{2}$ and 6 with low ratings for endomorphy. Because grossly obese females, particularly in physical education classes, are reluctant to volunteer as subjects, they are rare in these samples. Nevertheless there was high endomorphy in some Prague, USA, Eskimo, Mexican and Indian subjects. In the Canadian samples there were extreme mesomorphic endomorphs in all age groups.

In order to examine further the extremes in component ratings, subjects with high ratings in one component were selected from the total USA

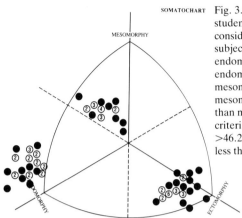

Fig. 3.50. Somatotype distribution of female students whose somatotypes were considered extreme in a sample of 3529 subjects. The criteria for extreme endomorphy were: HWR < 39.67; endomorphy > 7; mesomorphy <4. For mesomorphy, the criteria were: mesomorphy >4.4, with endomorphy lower than mesomorphy; and for ectomorphy the criteria were: ectomorphy >5.9; HWR >46.27, and endomorphy and mesomorphy less than ectomorphy.

student group (N = 3529). The criteria for high endomorphy were a HWR ratio less than 12.00 (39.67), an endomorphy rating greater than 7, and a mesomorphy rating of 4 or less. This selection yielded 27 subjects, or 0.8% of the total. The criteria for high mesomorphy were 4.5 or higher in mesomorphy, and endomorphy lower than mesomorphy; they were met by 24 subjects, or 0.7% of the total. The criteria for high ectomorphy were 6 or higher in ectomorphy, a HWR of 14.00 (46.28) or higher, and endomorphy and mesomorphy lower than ectomorphy, which 31 subjects, or 0.9% of the total, met. These selections yielded a range from $7\frac{1}{2}$ to 10 for endomorphy, from $4\frac{1}{2}$ to $5\frac{1}{2}$ for mesomorphy, and from 6 to 7 for ectomorphy. Figure 3.50 shows the somatoplots of the 82 (2.3%) extreme somatotypes selected according to the above criteria. As is shown in later chapters, more extreme endomorphy is found in selected samples of females and a wider range of dominant mesomorphy occurs among female athletes. There are no examples of female ectomorphy higher than 7. (Ectomorphy higher than 7 has been found only in the Nilote male sample. Having no sample of Nilote females, we do not know their range in ectomorphy.)

Sexual dimorphism

Comparisons of male and female somatotype means, as reported in Heath (1977) and Carter (1980*b*), clearly show sexual dimorphism. The number, mean somatotype, and somatotype attitudinal distance (SAD) for 15 male/female pairs of somatotype samples is shown in Table 3.6. These samples are from different countries, but have been measured and rated by the same methods and the same investigators for each pair. Eleven pairs from 10 countries or ethnic groups are plotted in Fig. 3.51. Urban Mexican Whites, the Vancouver students, the Monterey students and the University

Table 3.6. *Somatotype sexual dimorphism in different samples*

Samples		N	Somatotype	SAD
Papua New Guinea	Males	100	1.7-6.7-1.7	2.7
(Manus)	Females	111	3.1-4.5-2.5	
USA (Eskimo)	Males	81	3.4-5.9-1.3	3.2
	Females	76	6.4-4.8-0.8	
Canada (YMCA-LIFE)	Males	2259	3.6-5.0-2.2	1.5
(20–29 yrs)	Females	1752	4.4-3.7-2.4	
Canada (Vancouver,	Males	153	2.8-4.9-2.8	1.8
students)	Females	94	4.0-3.5-2.9	
Czechoslovakia	Males	302	3.4-4.3-3.0	1.7
(Prague, students)	Females	286	4.8-3.7-2.2	
Mexico (urban	Males	162	3.3-4.7-2.7	2.6
Mestizos)	Females	46	5.7-4.3-1.8	
Mexico (urban	Males	102	3.4-4.5-3.1	1.6
Whites)	Females	46	4.6-3.5-2.9	
USA (Hawaii,	Males	104	2.8-5.1-3.3	2.1
students)	Females	104	4.4-3.7-3.1	
England	Males	274	2.9-4.2-3.2	1.5
(students)	Females	116	4.2-3.5-2.8	
South Africa	Males	425	3.2-4.7-3.0	1.6
(Johannesburg,	Females	46	4.3-3.5-2.9	
medical students)				
USA (San Diego,	Males	328	3.1-5.1-2.7	1.8
students)	Females	143	4.2-3.7-2.6	
USA (Monterey,	Males	66	3.0-5.1-2.7	1.4
students)	Females	96	4.3-3.9-2.9	
USA (Berkeley,	Males	133	4.0-5.1-2.3	1.3
students)	Females	128	4.1-3.8-2.5	
India (Jat-Sikh,	Males	100	3.5-3.5-4.0	1.8
18–25 yrs)	Females	117	5.1-2.9-3.4	
India (Bania,	Males	71	3.0-3.0-4.0	3.7
20–29 yrs)	Females	98	6.3-3.1-2.5	

of California Berkeley students were omitted. The lines connecting the male and female means represent the two-dimensional distances between means. However, the SAD, the distance between the means in three dimensions, is a more accurate reflection of the true distances. The SAD between means for the White samples ranges from 1.3 to 1.8, and for the non-White samples from 1.8 to 3.7. The reason for the differences between White and non-White samples is not clear. The SAD for the Banias of India is the largest of all samples; the female and males are similar in mesomorphy; and there is a greater difference between female and male ectomorphy.

There are clear differences between male and female means in all

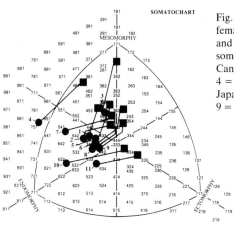

Fig. 3.51. Mean somatotypes of male ■ and female ● samples illustrating the magnitude and direction (in two dimensions) of somatotype sexual dimorphism. 1 = Canadian; 2 = Czechoslovak; 3 = Manus; 4 = Eskimo; 5 = USA; 6 = Hawaii Japanese; 7 = Mexican; 8 = South African; 9 = English; 10 = Jat-Sikh; 11 = Bania.

samples. The males are more mesomorphic and less endomorphic. Differences in ectomorphy are less in most samples. The dominance of endomorphy over mesomorphy is characteristic of females, except for the Manus and Tokyo Japanese students. In the majority of distributions there is an overlap of approximately one-third between the sexes, primarily in the endomorph–mesomorph region, or on either side of the ectomorphic axis from the central region to the northwest. There is almost no overlap between the male and female distributions for the Eskimos and the Manus. For most of the paired samples the connecting lines are across the ectomorphic axis and roughly parallel to each other. The Manus samples, with the males displaced northward into the mesomorphic sector, and the Banias, separated east to west, are exceptions. In general, the somatotype distributions and SADs of the White samples resemble each other more than they resemble those of the non-White samples.

Summary

There are many differences among samples of male and female Heath–Carter somatotypes: their distribution, the clear patterns of somatotype sexual dimorphism, and somatotype changes with growth and age. Presumably these are due in some measure to genetic factors, dietary patterns, physical activity, and other sociocultural or environmental factors, to which some of the within-population variation also is no doubt due.

The majority of male somatotypes are northeast of the ectomorphic axis, and the majority of the females are to the southwest of the ectomorphic axis. In some samples age, sex and ethnic characteristics limit the ratings in one or more components.

In general, in a given population, females are more endomorphic and less mesomorphic than males, and relatively close in ectomorphy. Distributions of males and females often overlap along the ectomorphic axis from the central to the endomorph–mesomorph sectors. It appears that there is greater somatotype variation in samples showing greater genetic heterogeneity. Although there is wide somatotype variation within individual samples of university students of northwest European descent, the means for each sex are close together. The means are approximately $4\frac{1}{2}$-$3\frac{1}{2}$-$2\frac{1}{2}$ for females and 3-$4\frac{1}{2}$-3 for males.

Somatotype frequency varies from population to population, and to some degree from sample to sample. Somatotypes close to $2\frac{1}{2}$-$6\frac{1}{2}$-$1\frac{1}{2}$ occur often in Manus and Caingang male distributions, but never in Nilote distributions. In general, somatotypes within plus or minus one rating unit (or 1.0 SAD) of the mean are common in any given population or sample.

PART II. VARIATION IN SOMATOTYPE USING OTHER METHODS

Data obtained from other somatotype systems are useful if the differences between methods and their limitations are remembered (see Chapter 2). There is a large body of data applying Sheldon's method to describe somatotypes in general and male somatotypes in particular, though there are some data inaccuracies, misleading presentations and interpretations.

Sheldon's somatotype distributions of men

Sheldon (1940) reported in *The Varieties of Human Physique* the frequencies of 76 somatotypes that he had identified in a male college sample of 4000 subjects (1940, Table 23A, p. 268), plotting them on a triangular diagram (1940, Fig. 10, p. 118). Then in *Varieties of Delinquent Youth* (1949) he introduced the 'clusterchart', the 'arc-sided' triangular shape well known in somatotype literature, which is the model for present-day somatocharts. Later in the book, figure 19 (p. 728) shows the frequencies of somatotypes in a male college population of 4000, in which 'each dot represents 20 cases'. Sheldon stated that this figure is based on the same statistical population as that represented in *The Varieties of Human Physique* (Table 23A, p. 268), and that the subjects were somatotyped by the complete somatotyping technique, which included measurements on the photographs and the 'anthroposcopic impressions' of the raters. The mean somatotype is given as 3.20-3.77-3.53, with standard deviations for each component of 1.2, 1.2 and 1.3 respectively. The subjects were 'drawn mainly from Harvard, the University of Chicago, and Oberlin, with smaller groups from three other colleges'.

In *Atlas of Men* (1954) Sheldon presented a 'new' somatotype frequency table (p. 30) said to represent 'the approximate incidence per 1000 for each of the 88 known somatotypes in an American population of 46 000'. But the corresponding somatochart (Fig. 1, p. 12) is the same that appears in *Varieties of Delinquent Youth*. It represents the frequencies in 4000 college males studied more than 14 years earlier, *not* the alleged frequencies in 46000 males.

Sheldon's stated somatotype frequencies and their plots are at variance, the discrepancies being shown in Figs. 3.52, 3.53 and 3.54. Figure 3.52 was plotted from the somatotype frequencies given in Table 23A (p. 268) in *The Varieties of Human Physique*, each dot representing 20 individuals. For the majority of somatotypes the dots were placed around the somatotype (e.g. there are eight dots around the somatotype 4-4-4, with an alleged incidence of about 16 per 4000). Some plot positions were interpolated to represent low frequency of given somatotypes.

Figure 3.53 (*Varieties of Delinquent Youth*, Figure 19, p. 728) that appeared in *Atlas of Men* as Figure 1 (p. 12) seems to be a freehand impression of the frequencies. The 196 dots – not the stated 200 – are apparently arbitrarily distributed, not clustered around the somatotypes they represent. Figure 3.52 here and Table 23A in *The Varieties of Human Physique* show there are only 15% more dominant mesomorphs than dominant ectomorphs, while Fig. 3.53 shows that there are 39% more mesomorphs than ectomorphs. (In this calculation the number of mesomorphs in the sector bounded by the line 5-5-1, 4-4-4 and 1-5-5 are included and those on the line are excluded.)

Figure 3.54, drawn from the frequencies in *Atlas of Men* (1954, Table 2, p. 30) and plotted in the same mannner as Fig. 3.53, is based on a 'sample

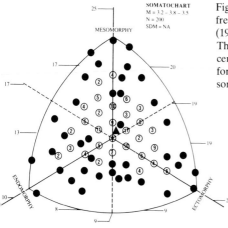

Fig. 3.52. Somatotype distribution based on frequencies for each somatotype in Sheldon (1940, pp. 268–9). There are 200 subjects. The numbers around the edge, and 8 at the centre, of the somatochart, are the counts for each somatotype category. ▲ = mean somatotype.

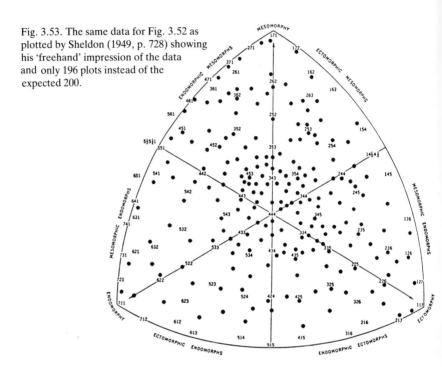

Fig. 3.53. The same data for Fig. 3.52 as plotted by Sheldon (1949, p. 728) showing his 'freehand' impression of the data and only 196 plots instead of the expected 200.

population [*sic*] of 46 000' representing the 88 somatotypes which Sheldon had identified by 1954 (an increase of 12 over the 76 reported in *The Varieties of Human Physique*). The mean somatotype for this distribution is 3.34-4.11-3.42, with standard deviations of 1.10, 1.03 and 1.18. Sheldon commented that 'our male population, in general, appears to be about one-eighth of a standard deviation more endomorphic than the college boys,

Fig. 3.54. Somatotype distribution based on frequencies for each somatotype in Sheldon (1954, p. 30). Plotted in the same manner as Fig. 3.52 with counts for each somatotype category. ▲ = mean somatotype.

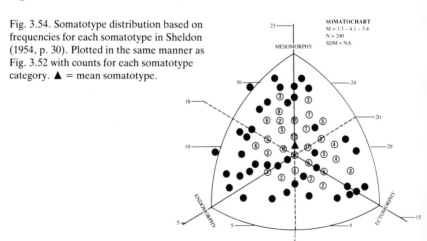

about one-third of a standard deviation more mesomorphic, and about one-tenth of a standard deviation less ectomorphic' (*Atlas of Men*, p. 32). However, the college men were included in these data.

Statistical tests applied to the somatotype frequency distributions plotted on Figs. 3.52, 3.53 and 3.54 showed:

1. There is a significant difference (chi squared = 66.13, df = 24) between the frequencies in somatotype categories in Figs. 3.53 and 3.54.

2. There is no difference (chi squared = 11.74) between the frequencies of somatotype categories in Figs. 3.52 and 3.54.

3. There is significantly higher ($t = 3.04$) mesomorphy in the general population (Fig. 3.54) than in the college students (Fig. 3.52). There is no difference between the two distributions on endomorphy and ectomorphy.

4. The SAD between the means of the general and student somatotypes is 0.4. This is a difference of slightly less than one-half unit, with the general mean tending towards slightly more endo-mesomorphy. (Both means are in the central category.)

This analysis indicates that the somatotype frequencies of Sheldon's college males and the general population are fairly similar, but there is a slight shift of approximately one-half unit in the direction of increased endo-mesomorphy in the latter. It also shows that the somatochart, drawn by Sheldon to represent college males, inaccurately represents the frequencies for this sample.

Sheldon's claim that *Atlas of Men* is based on an American male population of 46 000 is undocumented and is, at best, his own wishful thinking. Heath, who was closely involved with assembling the materials for *Atlas of Men* in the Constitution Laboratory, recalls that the actual data (i.e. somatotype photographs with age, height, weight and somatotype ratings) available for calculating frequencies and other statistics were substantially fewer than 46 000. She estimated that at the time the *Atlas* was assembled (1948–53) the samples of adult males totalled approximately 15 000. This allows for inclusion of the data from Army and Air Force studies by Sheldon, Dupertuis and Damon, plus a number of smaller samples. The military samples added up to a little over 9000 subjects (see Table 3.7). The total of 4000 students probably was exaggerated too. (The photographs and ratings of the Army sample of more than 35 000, collected under Hooton's aegis and rated by a different method, were not available to Sheldon.) Perhaps Sheldon estimated that 46 000 was about the number of subjects somatotyped, by any method, up to that time.

Sheldon's tables of alleged incidence per thousand of somatotypes in the

Table 3.7. *Somatotype ratings of males using methods other than that of Heath–Carter*

Sample/reference	N	Statistic	Age (yr)	Height (cm)	Weight (kg)	Somatotype	HWR	Rating method and rater
USA. Air Force Flight Personnel, 1967 (Laubach & Marshall, 1970)	2420	X̄	30.0	177.3	78.7	4.6-3.6-3.0	41.34	Parnell M4 by authors
		SD	6.3	6.2	9.7	0.9 1.0 1.0		
20–24 yrs	702	X̄		177.6	77.6	4.9-3.5-2.8	41.64	
25–29 yrs	679	X̄		177.3	78.2	4.5-3.6-2.9	41.46	
30–34 yrs	491	X̄		177.1	79.0	4.6-3.6-3.1	41.27	
35–39 yrs	329	X̄		177.7	80.4	4.3-3.8-3.2	41.17	
40–44 yrs	187	X̄		177.2	81.1	4.3-3.9-3.2	40.94	
45–49 yrs	31	X̄		174.0	80.5	4.1-4.4-2.9	40.30	
USA. Air Force Flight Personnel (Dupertuis, 1963) (From 1950)	3935	X̄	27.9	175.5	74.4	3.5-4.5-3.0	41.74	Sheldon by Dupertuis
		SD	4.2	6.2	9.5	0.9 0.8 1.1	1.57	
			18–54					
USA. Antioch College students (Laubach & McConville, 1966a)	63	X̄	19.0	174.9	68.4	3.9-4.3-2.8	42.77	Sheldon by Dupertuis
		SD	1.7	6.7	8.6	1.0 0.9 1.3	—	
			16–25					
USA. Antioch College students and Air Force personnel (Laubach & McConville, 1966b)	45	X̄	20.7	175.5	71.8	4.1-4.6-2.6	42.22	Sheldon by Dupertuis
		SD	3.5	8.5	9.3	0.9 0.9 1.4	—	
			17–35					
USA. Antioch College students and Air Force personnel (Laubach & McConville, 1969)	77	X̄	21.1	175.7	78.2	3.9-4.5-2.6	42.39	Sheldon by Dupertuis
		SD	4.5	6.5	9.9	0.8 0.8 1.1	—	

Sample	n		Age	Stature	Weight	Somatotype	Index	Method
USA. Army, White (Newman, 1952)	39321	$\overline{X}$ SD	23.9	—	—	4.3-3.9-4.3 1.0 0.9 1.0	—	Hooton
USA. Army, Negroid (Newman, 1952)	3218	$\overline{X}$ SD	23.2	—	—	3.9-3.8-4.2 0.9 0.8 1.0	—	Hooton
USA. Army, Mongoloid (Newman, 1952)	492	$\overline{X}$ SD	24.1	—	—	4.1-3.9-4.2 0.8 0.7 1.0	—	Hooton
USA. Yale University students (Haronian & Sugarman, 1965)	1004	$\overline{X}$ SD	—	—	—	3.7-4.3-3.3 1.0 0.9 1.2	—	Sheldon Tr. Index by authors
USA. Princeton University students (Haronian & Sugarman, 1965)	102	$\overline{X}$ SD	19.7 2.1	179.1 7.1	73.5 9.6	3.8-4.7-3.0 0.9 1.0 1.2	42.87 1.62	Sheldon Tr. Index by authors
USA. Princeton University students (Haronian & Sugarman, 1965)	1000	$\overline{X}$ SD	19.7 2.1	179.1 7.1	73.5 9.6	3.5-3.7-3.3 1.0 1.2 1.2	42.87 1.62	Parnell M.4 by authors
USA. Middle-aged white collar Vets and technical workers, Boston (Damon, 1965a)	114	$\overline{X}$ SD	mid-aged	175.8 7.2	77.5 11.2	3.4-4.3-3.1 1.0 0.7 1.0	41.23	Sheldon by Damon
USA. University of Iowa Freshmen (Sills & Mitchem, 1957)	433	$\overline{X}$ SD	18–20?	—	—	2.3-4.1-2.6 1.3 1.0 1.4	—	Sheldon by authors
USA. University of Iowa, High School (Everett & Sills, 1952)	400	$\overline{X}$ SD	17.9 2.3 14–29	174.9 7.8	69.0 12.0	1.7-3.7-3.3 1.1 1.1 1.4	42.64	Sheldon by authors
USA. Army Recruits, White (from Sheldon) (Damon et al., 1962)	400 (235)	$\overline{X}$ SD	22.3	175.3 6.6	73.5 10.9	3.0-4.4-3.0 1.2 1.0 1.1	41.85	Sheldon

(continued)

Table 3.7. (*Continued*)

Sample/reference	N	Statistic	Age (yr)	Height (cm)	Weight (kg)	Somatotype	HWR	Rating method and rater
USA. Air Force flight personnel (Dupertuis & Emanuel, 1956)	500	$\bar{\mathrm{X}}$ SD	18–42 25.8 4.1	175.7 7.0	73.9 10.7	3.3-4.4-3.0 1.3 1.2 1.5	42.05 2.28	Sheldon by Dupertuis
USA. Air Force flight personnel (Dupertuis & Emanuel, 1956)	500	$\bar{\mathrm{X}}$ SD	25.8 4.1	175.7 7.0	73.9 10.7	3.9-3.4-3.7	42.05 2.28	Hooton, by Hooton/Stagg
USA. Navy (including divers) (Dupertuis et al., 1951a)	81	$\bar{\mathrm{X}}$	26.6 18–46	176.8	74.8	3.4-4.6-2.7	41.96	Sheldon by Dupertuis
USA. University students (Sheldon, 1940)	4000	$\bar{\mathrm{X}}$ SD	—	—	—	3.2-3.8-3.5 1.2 1.2 1.3	—	Sheldon (from *Atlas of Men*, p. 13)
USA. Harvard students (Dupertuis in Sheldon, 1940)	1000	$\bar{\mathrm{X}}$ SD	—	—	—	3.2-3.7-3.6 1.2 1.2 1.3	—	Sheldon by Dupertuis
USA. General population (Sheldon, 1954)	15000[a]	$\bar{\mathrm{X}}$ SD	18–65	174.1	—	3.3-4.1-3.4 1.1 1.0 1.2	—	Sheldon (from *Atlas of Men*, p. 32)
USA. Naval Air cadets (McFarland & Franzen in Damon, 1955)	570	$\bar{\mathrm{X}}$ SD	—	—	—	3.1-4.4-3.2 0.9 0.8 1.0	—	Sheldon
USA. American Air Force cadets (Sheldon 1949; p. 797 in Damon, 1955)	2500 3000	$\bar{\mathrm{X}}$ SD	—	—	—	3.3-4.3-3.2 0.9 0.8 1.1	—	Sheldon

Group	n		Age	Height	Weight	Somatotype		Method
...cadets (Damon, 1955)	630	X̄	23.2	176.0	69.4	3.3-4.2-3.0	42.83	Sheldon by Damon anthroposcopic
		SD				1.1 1.1 1.1		
USA. Air Force Photo Reconnaissance Officers (Damon, 1955)	150	X̄	23.7	176.5	71.2	3.1-4.2-3.1	42.58	Sheldon by Damon anthroposcopic
		SD				1.0 1.0 1.1		
USA. Air Force combat pilots (Damon, 1955)	156	X̄	24.7	176.1	71.2	2.8-4.7-3.0	42.49	Sheldon by Damon anthroposcopic
		SD				1.1 0.6 1.3		
USA. Bus truck drivers (Damon & McFarland, 1955, some photos)	(269)	X̄	37	173.6	75.8	3.6 4.6 2.5		Sheldon by Damon anthroposcopic
	(252)	SD				1.3 0.9 1.2		
USA. Army recruits, Fort Dix (Damon & Polednak, 1967b)	1987	X̄	21.0	—	—	3.8-4.4-3.2	42.18	Sheldon by Damon & Sheldon
		SD	17–29			1.1 1.0 1.1	2.28	
USA. Italian/American, Boston (Damon & Polednak, 1967b)	141	X̄	42.6	—	—	4.1-4.2-3.2	40.23	Sheldon by Damon & Sheldon
		SD	22–61			1.0 0.8 0.9	1.79	
USA. Soldiers, Fort Devens MA, White (Damon et al., 1962)	199	X̄	23.7	176.3	73.9	3.9-4.3-2.9	42.01	Sheldon by Damon
		SD	4.7 17–46	7.1	11.8	1.1 0.8 1.1	—	
USA. Soldiers, Fort Devens MA, White (Damon et al., 1962)	170	X̄	24.2	175.5	73.0	3.9-4.1-2.9	41.99	Sheldon by Damon
		SD	5.3 18–50	6.9	10.9	1.1 0.9 1.1	—	
USA. Soldiers, Fort Devens MA, Black (Damon et al., 1962)	65	X̄	27.1	176.3	74.8	2.9-5.1-3.0	41.84	Sheldon by Damon
		SD	7.5	7.6	11.8	1.0 0.8 1.1	—	
USA. Army recruits, Black (Damon, 1957)	25	X̄	22.5	175.8	69.4	2.5-5.2-3.4	42.78	Sheldon
		SD		6.1	9.1	0.7 0.8 0.9		

(continued)

Table 3.7. (*Continued*)

Sample/reference	N	Statistic	Age (yr)	Height (cm)	Weight (kg)	Somatotype	HWR	Rating method and rater
Mexico. Medical students, Universidad Técnico de Nacional Autonoma de Mexico (Vargas *et al.*, 1975)	89	$\bar{X}$ SD	20.6 1.6	— 	— 	4.0-4.2-3.2		Parnell M.4 by authors
Mexico. Technical Instituto Técnico de la Procuraduría del Territorios Federales (Vargas *et al.*, 1975)	124	$\bar{X}$ SD	24.5 2.5	— 	— 	3.9-4.5-3.0		Parnell M.4 by authors
Mexico. University students, technical and office workers, Mexico City (Villanueva, 1976) 15–65 yrs	300	$\bar{X}$ SD	24.4 8.1	170.4 7.0	63.1 8.8	3.7-4.1-2.8 0.8 0.9 1.2	42.94 1.79	Sheldon Trunk Index by Villanueva
	265	$\bar{X}$ SD		170.4 7.0	63.1 8.8	3.8-4.3-3.5	42.94 1.79	Parnell M4 by Villanueva
Czechoslovakia. Charles University students Štěpnička, 1972)	302	$\bar{X}$ SD	20.1 1.4	178.1 6.0	72.0 8.0	3.4-4.0-3.3 1.1 1.0 1.2	43.14	Sheldon by Štěpnička
Greece. Military Personnel, NATO (Dupertuis, 1963)	113	$\bar{X}$ SD	22.4 18–32	171.1 5.7	67.0 7.0	3.7-4.1-2.7 0.8 0.7 1.0	42.13 1.50	Sheldon by Dupertuis
Italy. Military Personnel, NATO (Dupertuis, 1963)	135	$\bar{X}$ SD	25.3 19–55	171.5 6.6	71.6 9.4	3.9-4.4-2.4 0.9 0.7 1.0	41.30 1.57	Sheldon by Dupertuis

Group	n		Age	Height	Weight	Somatotype		Method
Turkey. Military Personnel, NATO (Dupertuis, 1963)	92	X̄ / SD	24.2 / 19–39	169.9 / 5.6	65.7 / 8.1	3.3-4.5-2.4 / 0.9 0.8 1.0	42.11 / 1.54	Sheldon by Dupertuis
Netherlands. PE and other students, Amsterdam (Brouwer, 1957)	91	X̄ / SD	—	—	—	3.2-3.9-3.4 / 0.7 0.9 0.7	—	Sheldon by Brouwer (ok'd by Tanner)
Britain. Students, Loughborough College of Technology (Jones et al., 1965)	169	X̄ / SD	19.7 / 1.1 / 18–22	177.4 / 6.3	68.6 / 8.5	2.5-4.4-3.4	43.44	Sheldon by Jones (ok'd by Tanner)
UK. Medical students (Tanner, 1951b)	46	X̄ / SD	22.1 / 4.1	179.4 / 4.7	69.9 / 7.7	3.4-3.7-3.7 / 1.1 1.0 1.1	43.48	Sheldon by Tanner
UK. Medical students (Tanner, 1954b)	101	X̄ / SD	21.0 / 19–28	179.3 / 6.4	70.4 / 7.8	3.2-4.2-3.7 / 0.9 1.0 1.3	43.42 / 1.57	Sheldon by Tanner
UK. Medical students (Tanner, 1954b)	162	X̄ / SD	—	—	—	2.9-4.2-3.8	—	Sheldon by Tanner
UK. Oxford students (Tanner, 1952a)	283	X̄ / SD	—	—	—	3.4-3.5-3.7 / 0.8 0.9 1.1	—	Sheldon by Tanner
UK. Oxford students (Tanner, 1954b)	171	X̄	—	—	—	3.3-3.6-3.0	—	Sheldon by Tanner
UK. Sandhurst cadets (Tanner, 1954)	287	X̄ / SD	18.5 / 0.6	—	—	3.1-4.5-3.7	43.14	Sheldon by Tanner
UK. Oxford students (Parnell, 1957a)	344	X̄ / SD	21.6	177.8	67.4	3.4-3.7-3.9 / 0.8 0.8 1.2	43.69	Parnell M.4 + photos
UK. Chelsea pensioners (Parnell, 1954b)	50	X̄ / SD	73 / 56–84	167.8 / 0.9	65.9 / 1.6	2.5-5.0-3.5	41.54	Parnell M.4
New Zealand. University students (Somerset, 1953)	39	X̄ / SD	21.3 / 18–27	—	—	3.4-4.3-2.8	—	Sheldon (metric) by Somerset

(continued)

Table 3.7. (*Continued*)

Sample/reference	N	Statistic	Age (yr)	Height (cm)	Weight (kg)	Somatotype	HWR	Rating method and rater
New Zealand. Otago University Liberal Arts students (Cockeram, 1975)	28	$\bar{X}$	22.2	177.1	70.3	3.7-4.6-3.5	42.91	Parnell M.4, by Cockeram
		SD	4.6	5.2	8.1	0.9 0.9 1.4		
New Zealand. Air Force (Carter & Rendle, 1965)								
16–24 yrs	307	$\bar{X}$		174.3	67.4	3.7-4.6-3.3	42.94	M.4, by authors
		SD		6.5	8.6	0.8 0.9 1.2	1.65	
25–34 yrs	77	$\bar{X}$		175.4	73.8	3.5-4.6-3.3	41.95	M.4, by authors
		SD		7.0	13.1	0.9 0.9 1.2	1.95	
35–44 yrs	49	$\bar{X}$		174.1	75.1	3.5-4.9-3.3	41.39	M.4, by authors
		SD		6.4	10.8	1.0 0.8 1.1	1.92	
45–52 yrs	21	$\bar{X}$		169.9	74.1	3.6-5.5-3.0	40.63	M.4, by authors
		SD		4.0	10.0	0.9 0.8 1.0	1.85	
Total	454	$\bar{X}$	16–52	174.2	69.6	3.6-4.7-3.3	42.48	M.4, by authors
		SD		6.4	9.6	0.8 0.9 1.1	1.72	
India. College/Tech students Chandigarh, North India (Berry, 1972)	1000	$\bar{X}$	21.4	168.7	54.1	2.9-3.0-4.6	44.63	Sheldon by Berry
India. Nagpur Medical and Science students Central India (Berry & Deshfukh, 1964)	877 (682)	$\bar{X}$ SD	21.4 18–26	166.8	51.6	3.0-3.3-4.6	44.86	Sheldon by Berry, corrected by Tanner
Japan. North Honshu office workers	544	$\bar{X}$	19–45	—	—	3.0-5.5-1.4	—	Sheldon modified
		SD				0.7 0.8 0.6		

			Age (yrs)	Stature (cm)	Weight (kg)	Somatotype		Method
19–25 yrs	182	X̄		—	—	3.3-5.4-1.5	—	Ditto above
26–35 yrs	232	X̄		—	—	3.0-5.6-1.4	—	above
36–45 yrs	125	X̄		—	—	2.9-5.4-1.5	—	
[NOTE: error in Nos. above, 5 too few]								
Taiwan. Students, medical students, technicians and faculty, Chinese (Chen et al., 1963)	31	X̄ SD	23.0 3.1 17–29	167.6 5.2	52.6 4.4	2.8-3.6-4.4 1.2 0.9 1.2	44.76 1.32	Sheldon by Damon anthropometric
Africa. Nilotes, Shilluk (Roberts & Bainbridge, 1963)	48	X̄ SD	18–45	178.7 7.2	58.0 7.4	2.1-2.5-5.6	46.28 1.79	Sheldon, mod. of Danby, by R & B
Africa. Nilotes, Dinka (Roberts & Bainbridge, 1963)	227	X̄ SD	18–45	181.3 6.4	58.0 5.9	1.6-2.3-6.1	46.95 1.36	Ditto above
Africa. Nilotes, Ageir, Dinka (Roberts & Bainbridge, 1963)	52	X̄ SD	18–45	182.6 4.9	59.0 6.0	1.7-2.4-6.2	46.95 1.45	Ditto above
Kenya. Tribal Kikuyu (1944) (Danby, 1953)	98	X̄ SD	18–25	163.2 5.9	51.2 5.3	2.1-3.5-3.7	43.95	Sheldon modified by Bullen & Hardy & Danby
Kenya. Prison Kikuyu (Danby, 1953)	68	X̄ SD	18–25	166.3 6.7	56.5 5.8	2.2-4.6-2.8	43.34	Sheldon modified by Bullen & Hardy & Danby
South Africa. Tonga, Bantu speaking (Tobias, 1972)	333	X̄	—	—	—	2.6-4.0-3.4	—	Sheldon modified by Morris/Jacobs by Tobias

[a]For an explanation of number of subjects, see p. 127. X̄ = mean.

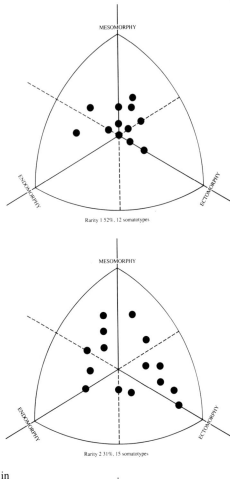

Rarity 1 52%, 12 somatotypes

Rarity 2 31%, 15 somatotypes

Fig. 3.55. Somatotypes plotted by Heath in 1952 to show the most common (left hand somatocharts) to the least common (right hand somatocharts) somatotypes according to the frequencies of Sheldon (1954, p. 30).

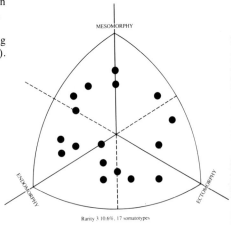

Rarity 3 10.6%, 17 somatotypes

Fig. 3.55 (*Continued*)

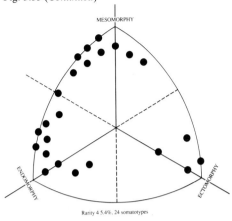

Rarity 4 5.4%, 24 somatotypes

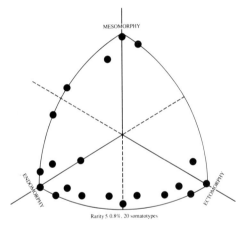

Rarity 5 0.8%, 20 somatotypes

American male population interested Heath in 1952 and she plotted the distributions shown in Fig. 3.55. Apparently 27 of the 88 known male somatotypes accounted for about 83% of the male population; about 4 out of 5 male somatotypes fell within the areas plotted on the first two somatocharts; and 61 somatotypes (almost 70%) accounted for a mere 17% of the population. Sheldon unaccountably disliked this graphic representation.

Sheldon's somatotype distributions of women

Sheldon published little on somatotypes in women. Table 22 (p. 127) in *The Varieties of Human Physique* shows the incidence of somatotype components for '2500 women' rated by photoscopy only, and there are highly misleading drawings of nine allegedly prototypical female somatotypes in the Appendix. He observed that 'the same 76 somatotypes that are found among men seem to occur among women, and no more, although the distribution of the population among the somatotypes is different' (p. 66). He says nothing about rating scales for women.

In *Varieties of Delinquent Youth* (1949) Sheldon included somatocharts of women with cancer of the breast, cancer of the uterus, gallbladder disease and duodenal ulcer. He speculated briefly on their somatotype distributions compared with the general population, but included no data for 'normal' female somatotype distribution, and no comments on methodology.

In *Atlas of Men* (1954) Sheldon stated that an *Atlas of Women* would be forthcoming in 'two or three years', and went on to discuss briefly the issues of rating scales and height–weight criteria for males and females. He said that in 1949, following controversial discussions, he and his colleagues had decided on a single somatotyping system for both sexes, and 'to let the female somatotypes fall where they would' (*Atlas of Men*, p. 13). He included a somatochart (Fig. 2, p. 13) representing 'the distribution of somatotypes for a female college population of 4000, each black dot representing 20 cases', or 200 subjects. He drew some comparisons between male and female distributions, but provided no frequency distribution data for the female samples. He commented that the commonest female somatotypes are close to 5-3-3 with a HWR of 12.85. In view of the discrepancies noted for Figure 1 (*Atlas of Men*, p. 12), and in the absence of data on frequency distributions, it is risky to assume that this female distribution accurately reflects the actual values. It probably does reflect the general trend of the distribution, just as Figure 1 does for the males. Also, due to the constraints of the 7-point scale, all Sheldon distributions to some degree differ from distributions of the same series rated by the Heath–Carter and other methods.

Heath recalls that by 1954 when the *Atlas* was published, there were almost 3000 college females photographed under her direction. She had also taken part in collecting data and photographs for approximately 3000 women of all ages in hospital and outpatient clinic populations. Although this was an adequate sample for its construction, the promised *Atlas of Women* was never published.

Samples of men rated by other methods

There are about 50 samples (Table 3.7) plus subsamples grouped as nationality and ethnic, military, and student samples, all rated by methods other than the Heath–Carter method. There is some overlap in student samples studied by different authors, and several separate studies of a single sample of military personnel. For comparison of methods see Chapter 2.

Table 3.8. *Somatotype ratings of females using methods other than that of Heath-Carter*

Sample/reference	N	Statistic	Age (yr)	Height (cm)	Weight (kg)	Somatotype	HWR	Method
USA. University students (Sheldon, 1940, p. 127)	2500	$\overline{X}$	—	—	—	3.6-3.2-3.6	42.45	Sheldon anthropo-
		SD				1.3 1.2 1.3		logical only
USA. University students (Bullen & Hardy, 1946)	176	$\overline{X}$	—	—	—	3.1-3.4-3.4 [estimated from graph]	—	Sheldon modified by Bullen/Hardy
USA. Presbyterian Hospital, White outpatients (Damon, 1960)	145	$\overline{X}$	46.2	159.0	63.7	4.8-3.5-2.7	40.10	Sheldon
		SD	14.4	7.2	12.3	0.7 0.7 1.0	2.35	by Damon
USA. Presbyterian Hospital, Black outpatients (Damon, 1960)	103	$\overline{X}$	37.0	159.3	64.8	4.7-4.1-2.8	40.00	Sheldon
		SD	12.2	6.5	14.7	1.0 0.6 1.0	2.68	by Damon
USA. Students, Washington State University (Brown, 1960)	58	$\overline{X}$	20.2	166.6	63.1	5.7-2.9-3.6	41.85	Sheldon? by
		SD	—	5.9	0.7	0.8 0.8 0.9	1.82	Sheldon
Netherlands. PE, other students, Amsterdam (Brouwer, 1957)	54	$\overline{X}$	—	—	—	4.7-3.0-2.0	—	Sheldon
		SD				0.6 0.6 0.7		by Brouwer (Tanner OK'd)
UK. Oxford students (Parnell, 1954, 1958)	671	$\overline{X}$	—	—	—	4.8-3.5-3.0	—	Parnell M.4
Taiwan. Chinese (Chen et al., 1963)	29	$\overline{X}$	23.3	154.5	49.3	4.4-3.4-3.7	42.32	Sheldon
		SD	4.7	5.4	6.9	1.0 0.8 1.3	1.75	by Damon

$\overline{X}$ = mean.

Sheldon himself rated a number of the samples; Dupertuis and Damon rated some others. All methods used a 7-point rating scale, except for the Roberts & Bainbridge study of the African Nilotes, for which they extended the rating scale for ectomorphy. In some studies, where there are sub-samples by age, it is difficult to judge the significance and validity of the ratings because they used age-adjusted scales, which assumed that the somatotype did not change over age. Also the same ratings at different ages were given to conspicuously different looking somatotypes. This is true for the Sheldon and Parnell systems in particular.

Samples of women rated by other methods

The small number of samples rated by other methods (Table 3.8) may reflect poor cooperation of females, or the lack of female military groups. Nonetheless, the differences between male and female mean somatotypes show the same pattern of sexual dimorphism seen in the Heath–Carter data. The females are more endomorphic and less mesomorphic than the males, and are similar in ectomorphy.

Summary

In Part I, in a wide variety of male and female samples, the Heath–Carter somatotypes showed greater variation than those rated by other methods. It is likely that genetics, sex, physical activity, diet and age are major influences on the variation. It is clear that extended rating scales are valuable in discriminating between physiques and phenotypic ratings, and in showing age changes.

Part II raised questions about Sheldon's somatotype distributions and the alleged numbers of subjects in Sheldon's studies. Tabulated lists of male and female somatotypes in samples rated by other methods show the variations obtained by these methods and provide a historical record.

4 Growth and aging

Introduction: the phenotypical approach

Heath–Carter phenotypic somatotype ratings, covering as they do wide variations in shape, absolute and relative size, and body composition, are well suited for analysing the widely recognized changes in human beings during growth, maturation and the processes of aging. The Sheldon method with its assumption of lifelong, stable, genetically determined somatotypes, does not.

Although at present information is incomplete, the several longitudinal and cross-sectional studies reviewed here offer some general answers to the question of somatotype permanence. There is more information on boys and girls aged 6–12 years and on boys aged 12–18 years than on children younger than 6 years and on adolescent girls. There is relatively little longitudinal information on adults.

Somatotyping young children

With minor limitations, the method of Heath & Carter (1967) can be applied to both sexes at all ages. From photographs and anthropometric data experienced raters can make reliable and valid somatotype ratings. By itself an anthropometric somatotype is an estimate of the criterion rating, but it may be less valid for young children than for older children and adults. In early studies anthropometric ratings appeared to be reliable for children aged 10 years and older. Several investigators (e.g. Duquet *et al.*, 1975; Štěpnička, 1976*b*; Slaughter *et al.*, 1977*b*, 1980; Singh & Sidhu, 1980; Holopainen *et al.*, 1984; Eiben, 1985) have used anthropometric somatotypes from ages 6–10 years, and Pařízková *et al.* (1984) used this method on 3½–6-year-old children, and Amador *et al.* (1983) on 1½–5½-year-olds.

Experience with rating young Manus children and children at the Gesell Institute (New Haven, Connecticut) led Heath to believe that somatotype photographs were essential for rating small children, and that it was difficult to make fine discriminations at ages less than 6 years. Obviously, lack of illustrations and reference ratings is a problem.

To allow rating of children Heath further modified (Table 4.1) the revised

Table 4.1. *Somatotypes and HWRs (Adapted by Heath for somatotype ratings of children.)*

HWR	Adult somatotypes	Children's somatotypes
14.40	2-2-6	
14.30	1-3-6, 3-1-6, 2-3-7, 3-2-7	1-2-5
14.20	1-3-5, 3-1-5, 2-2-5, 1-4-6, 2-3-6, 3-2-6	
14.10		2-2-5, 1-3-5
14.00	1-4-5, 2-3-5, 3-2-5, 2-4-6, 4-2-6, 3-3-6	2-3-5, 1-3-4
13.90		2-2-4, 1-3-4
13.80	1-4-4, 4-2-5, 2-4-5	
13.70	3-3-5	2-4-5, 2-3-4, 3-2-4, 1-3-3
13.60	1-5-4, 2-4-4, 4-3-5, 5-2-5, 3-4-5	3-3-4
13.50	2-5-4	1-3-2, 2-3-3

13.40								
13.30		1-5-3	3-4-4 5-2-4			2-3-2	3-3-3	4-3-4
13.20		1-6-3 2-5-3 3-4-3	3-5-4 5-3-4 4-4-4					
13.10		1-6-2 2-5-2 5-2-2			2-4-2	3-3-2 4-2-2		4-3-3
13.00		2-6-3 3-5-3 4-4-3	6-2-3 5-3-3					
12.90		1-7-2 5-3-2 4-4-2		2-4-1	3-3-1		4-3-2
12.80		2-6-2 6-2-2 3-5-2	3-6-3 4-5-3 6-3-3	5-4-3			
12.70						4-3-1	3-4-1
12.60	3-5-1		4-5-2 2-7-2 3-6-2				
12.50	1-7-1 7-1-1 · 2-6-1 6-2-1		5-4-2 7-2-2 6-3-2			5-3-1	4-4-1
12.40	1-8-1 2-7-1 · 7-2-1 3-6-1 · 6-4-1 5-5-1		3-7-2 4-6-2 5-5-2				
12.30			7-3-2 6-4-2			6-3-1	5-4-1
12.20							
12.10						7-3-1	6-4-1

Table 4.2. *Somatotypes not in the Heath HWR table.*
Frequencies of occurrence by sex in Belgian children

Somatotype	Female	Male	Total
1-2-4		1	1
1-2-5	4	7	11
1-3-3	1	1	2
1-3-4	19	50	69
1-3-5	40	13	53
1-4-2		1	1
1-4-3	6	17	23
1-4-4	31	13	44
1-5-2	1	3	4
2-3-3	2	1	3
2-3-4	26	35	61
2-4-2	1	3	4
2-4-3	36	44	80
3-3-3	5	5	10
3-4-2	6	7	13
Subtotal	178	201	379
Total (all children)	351	277	628

(From Duquet, 1980).

somatotype/height–weight ratio (HWR) scale (Heath, 1963; Heath & Carter, 1967). Duquet (1980) developed a similar scale (Table 4.2) during his study of young Belgian children. These investigators recognized the need to establish for children somatotype ratings with values lower in all components than those for adults with the same HWRs. For example, if the HWR is 13.20 for adult somatotypes 1-5-4, 1-6-3, 2-5-3, 3-5-4, 3-4-3, 4-3-3, 4-4-4 and 5-2-3, it follows that 13.20 is the logical HWR for the somatotypes 2-4-1, 3-4-2, 3-3-1 and 4-2-1. Moreover a growing child at various ages may have, for example, the same 13.20 HWR and be rated 3-3-1, 3-4-2 or 2-5-3.

The rating forms (Heath & Carter, 1967; Carter, 1980*a*) for anthropometric somatotypes provide mesomorphy and ectomorphy scales adjusted for stature, but no similar adjustment for endomorphy and the sum of three skinfolds. On the assumption that skinfolds diminish during child growth in proportion to increase in stature, Hebbelinck *et al.* (1973) suggested multiplying the sum of three skinfolds by (170.18/height in cm) before rating endomorphy, accepting a reference height of 170.18 cm (Ross & Wilson, 1974). Application of this procedure in several groups of children's data indicates that, despite its statistical limitations, the correction is basically sound. In a study of a group of 70 boys age 11, J. P.Clarys and J. E. L. Carter (unpublished) found that the anthropometric endomorphy ratings average one-half unit lower than the photoscopic ratings. Ross *et al.* (1978) obtained similar results in a comparison of photoscopic and anthropometric ratings of endomorphy for 8–14-year-old boys (N = 26) and girls (N = 14). The uncorrected anthropometric ratings were 1 unit lower for boys and $1\frac{1}{2}$ units lower for girls. After height corrections the ratings were $\frac{1}{2}$ unit and 1 unit lower. In these samples the correction was in the right direction, but insufficient to match the photoscopic rating. It appears that when anthropometric ratings alone are made the correction should be used for endomorphy in children and for subjects greatly deviant from the reference height of 170.18 cm. In the case of anthropometric plus photoscopic ratings, inspection of the photograph and somatotypes appropriate to the given HWR indicates the logical correction to the final somatotype.

Cross-sectional and longitudinal studies

Longitudinal studies are required for observation of somatotype development and change, but they suffer from inevitable attrition and they demand exceptionally dedicated investigators and cooperative subjects. Such studies can provide valuable histories of growth patterns which, when nutrition and exercise as well as somatotype data are incorporated, shed light on somatotype stability and somatotype change.

However, comparative cross-sectional studies suggest valuable inferences. The differences between age groups that they describe may in some circumstances represent individual age changes,, if they are not due to differences in other causal factors.

Usually analysis of the somatotype as a whole (see Appendix II) is more useful than separate component analysis in both longitudinal and cross-sectional studies. However, the two analyses can support one another.

Children and adolescents aged 2–18 years
Longitudinal studies

Zuk (1958) was the first to use Heath ratings of longitudinal data. He reported the somatotypes of 74 males and 78 females at ages 12 and 17, and 38 males and 39 females at age 33, in the Berkeley Growth Study, directed by Harold E. Jones at the Institute of Child Welfare, University of California, Berkeley. The means given in Table 4.3 and the plot of mean somatotypes (Fig. 4.1) show, as expected, that with age the males tended to become more mesomorphic and the females more endomorphic.

The Medford Growth Study (1956–68), directed by H. Harrison Clarke of the University of Oregon, was a large mixed longitudinal and cross-sectional study of boys aged 7 to 18 (Clarke, 1971) that gave rise to a number of publications. The data included anthropometry, somatotypes and other tests and measurements. When Heath rated the series by her modified

Table 4.3. *Mean somatotypes by age in longitudinal studies using the Heath–Carter methods*

Sample, reference and method	N	Age (yr)	Somatotype
Males			
Oakland, California (Zuk, 1958)	74	12	3.2-3.3-3.9
photoscopic by Heath	74	17	2.9-4.1-4.3
	38	33	3.0-4.7-3.9
Medford, Oregon (Clarke, 1971),	106	9	3.4-4.2-2.9
photoscopic by Heath	106	10	3.6-4.2-2.9
	106	11	3.4-4.2-3.2
	106	12	3.5-4.1-3.2
	106	13	3.5-4.0-3.2
	106	14	3.3-4.1-3.4
	106	15	3.3-4.1-3.5
	106	16	3.4-4.2-3.4
Manus, Admiralty Islands	15	2	3.9-5.0-1.0
(Heath & Carter, 1971)	12	4	3.6-5.0-1.2
	15	6	2.1-5.0-1.4
	25	8	1.5-5.1-2.1
	24	10	1.9-5.1-2.3
	20	12	1.7-5.0-2.8
	16	14	1.6-4.8-3.1
	11	16 + 17	2.4-6.1-2.2
	9	20	2.1-6.1-1.6
Prague infants, Czechoslovakia	28	3.5	3.3-5.7-0.9
(Pařízková *et al.*, 1984),	28	4.0	3.2-5.6-0.9
anthropometric $(En_H)^a$	28	4.5	3.0-5.4-1.2
	28	5.0	2.6-5.2-1.7
	28	6.0	2.6-5.0-2.2
Prague boys, Czechoslovakia (Pařízková	39	10.7	2.5-4.1-3.4
& Carter, 1976; Carter & Pařízková,	39	11.7	2.4-3.9-3.6
1978), anthropometric (No En_H)	39	12.7	2.1-4.0-3.6
	39	13.7	2.4-4.0-3.8
	39	14.6	1.5-3.8-3.8
	39	15.7	1.6-3.8-3.7
	39	16.8	1.7-3.8-3.5
	39	17.8	2.2-4.3-3.3
	14	24	2.7-5.4-2.5
Prague, Czechoslovakia (Bok &	11	15	2.3-4.3-3.4
Tlapáková, 1982), anthropometric	11	16	2.2-4.2-3.5
(No En_H)	11	17	1.8-4.6-3.4
Belgium (Claessens, 1981),	210	13	2.2-4.1-3.6
anthropometric	210	14	2.3-4.1-3.7
	210	15	2.6-3.8-3.7
	210	16	2.7-4.1-3.8
	210	17	2.6-4.0-3.6
	210	18	2.6-3.9-3.4
Females			
Oakland, California (Zuk, 1958),	78	12	3.9-3.3-3.5
photoscopic by Heath	78	17	4.3-3.5-3.3
	39	33	4.7-3.9-2.8

Table 4.3. (*Continued*)

Sample, reference and method	N	Age (yr)	Somatotype
Manus, Admiralty Islands (Heath &	11	2	4.3-4.5-0.9
Carter, 1971), photoscopic by Heath	13	4 + 5	2.8-4.5-1.5
	12	6	2.5-4.4-1.5
	31	8	2.0-4.5-2.3
	21	10	2.1-4.3-2.9
	13	12	2.5-4.3-3.1
	12	14	2.5-4.2-3.2
	16	16	4.0-4.7-2.6
	10	22	4.5-4.4-2.2
Prague infants, Czechoslovakia	28	3.5	3.7-5.2-0.7
(Pařízková *et al.*, 1984),	28	4.0	3.4-5.4-0.9
anthropometric (En$_H$)[a]	28	4.5	3.4-5.4-1.1
	28	5.0	3.4-5.1-1.3
	28	6.0	3.2-4.5-2.0
Prague, Czechoslovakia (Bok &	21	15	3.9-3.6-2.9
Tlapáková, 1982), Heath–Carter	21	16	4.0-3.8-2.6
anthropometric (No En$_H$)	21	17	3.9-4.0-2.7

[a] (En$_H$) = height-corrected endomorphy

method (1963), she showed that with age there was an increase in mean ectomorphy while endomorphy remained fairly constant and mesomorphy showed a diminution into adolescence followed by a consistent rise.

Morton's (1967) report on the Medford boys at ages 9 to 16 showed that the mean somatotypes clustered about 3-4-3 in the upper central somato-chart, with a moderate southeasterly tendency toward increasing ectomor-phy. From ages 15 to 18, Kurimoto (1963) found that the average mesomor-phy increased from 4.1 to 4.7 while the average ectomorphy decreased from 3.7 to 3.2.

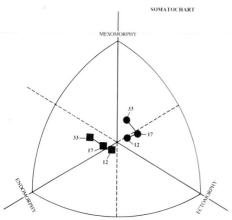

SOMATOCHART

MESOMORPHY

ENDOMORPHY

ECTOMORPHY

Fig. 4.1. Mean somatotypes of Berkeley males ● and females ■ at 12, 17 and 33 years. (From data in Zuk, 1958.)

Fig. 4.2. Somatotypes of two boys from
Medford followed from ages 7 to 17 years.
(Data from H. H. Clarke.)

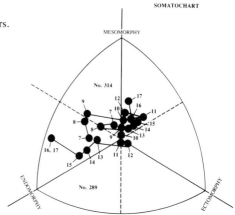

Sinclair (1966, 1969) in studies of longitudinal differences in somatotypes of boys aged 9–12 and 12–17, found little change in the younger group, but the significant differences he found in the 12–17-year group suggested accelerated rates of change as the boys entered maturity. There were high inter-age correlations (0.8–0.9) of somatotype components for adjacent years, and much lower correlations (0.50, 0.60 and 0.67) between ages 2–5 years apart.

Disregarding changes of one-half unit in component ratings as possible rating errors, Sinclair found that 45% of the endomorphy ratings changed by one unit or more; that 20% of the mesomorphy ratings changed by one unit or more; and 44% of the ectomorphy ratings changed by one unit or more. The greatest difference between any two ratings during the period of the study ranged up to $2\frac{1}{2}$ units for each component within the age spans. Many boys showed marked changes in somatotype during the years of the study; and a few showed no major changes. The somatoplots (Fig. 4.2) of subjects No. 289 and No. 314 show the courses of somatotypic change for two boys from age 7 to 17. Although the two boys started with almost the same somatotypes, they followed markedly different paths and had widely separated somatotypes at age 17. (See Fig. 4.3 for their somatotype photographs.)

A 30-year mixed longitudinal somatotype study, initiated by Margaret Mead in 1953 in Pere Village, Manus Island, (now also called Manus Province), Territory of Papua New Guinea (administered by Australia until the independence of Papua New Guinea in 1975) is probably the longest and most extensive study carried out on an intact group (a village with a population averaging about 500). Heath & Carter (1971) described the somatotypes of children and adolescents ranging from 2 to 22 years, based on anthropometry and somatotype photographs of 223 boys and 215 girls collected in 1953, 1966 and 1968 (see Table 4.3). Because of small sample

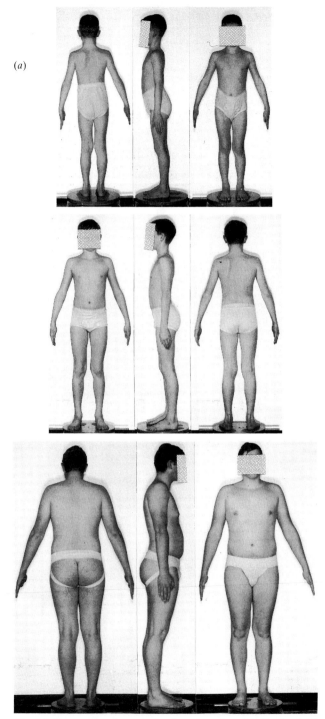

(a)

Fig. 4.3. Somatotype photographs of two boys from Medford at 8, 12 and 17 years. (*a*) For subject 289 the somatotypes are 4-4-2½, 3-3-3½, 6-3½-1½; and (*b*) for subject 314 the somatotypes are 4-4-1½, 2½-4-3, 2½-5-3. (Photos from H. H. Clarke.)

149

(b)

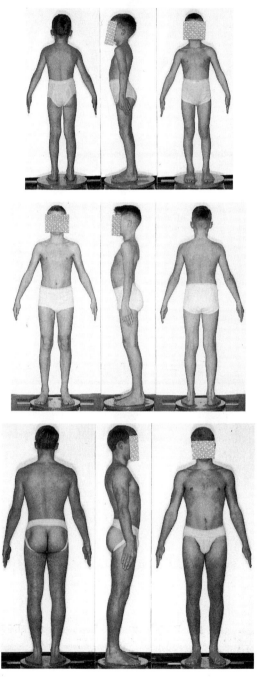

Fig. 4.3(b).

150

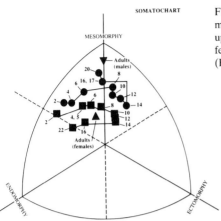

Fig. 4.4. Mean somatotypes by age of Pere males ● and females ■ from two years and upward. Means for adult males ▼ and females ▲ from Pere are also plotted. (From data in Heath & Carter, 1971.)

numbers, some yearly age groups were combined (see Figure 4.4 for mean somatotypes at 2-year intervals), for boys at 16 and 17 years and for girls at 4 and 5 years. The mean somatotypes for boys and girls at ages 2–6 years are close and their distributions overlap considerably. From age 8 male and female somatotypes diverge increasingly, with marked divergence from age 16. From age 2 to 8 male endomorphy decreases and ectomorphy increases. From age 8 to 14 male mesomorphy decreases slightly and ectomorphy continues to increase. From age 14 mesomorphy increases dramatically, with a mean somatotype of 2-6-1½ at age 20. The mean adult male somatotype is 1.7-6.7-1.7.

The somatotype pattern for girls at age 2–14 is similar to that of the boys, with lower mesomorphy but slightly higher endomorphy. Between age 16 and 22 the girls become markedly more endomorphic and less ectomorphic. At age 16 the girls' somatotypes are close to the mean for their mothers and grandmothers (3-4½-2½) but in their early twenties their endomorphy is significantly higher. Follow-up studies on the Manus population have confirmed and expanded upon these findings (R. F. Shoup, 1987).

In a Czechoslovakian (Prague) study of 39 boys measured annually from age 11 to 17 Pařízková & Carter (1976) found substantial individual somatotype changes, although the somatoplots for the two groups were similar. Sixty-seven per cent of the boys changed component dominance, and all the boys changed somatotype one or more times (Fig. 4.5). Fourteen of the boys when re-somatotyped at age 24 (Carter & Pařízková, 1978) showed that mesomorphy had increased significantly, endomorphy had increased slightly, and ectomorphy had decreased slightly. Inter-age correlations indicate that prediction between years is generally poor, and becomes increasingly poor with greater intervals between years.

A somatoplot (Fig. 4.6) shows the migratory distances, (MD), or changes

Fig. 4.5. Mean somatotypes by age of Prague males followed from 11 to 14 years. (Redrawn from Carter & Pařízková, 1978.)

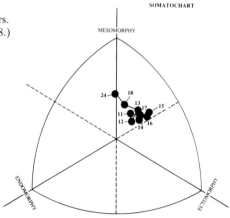

Fig. 4.6. Somatotypes and migratory distances (MD) of four boys followed from ages 11 to 18. (Redrawn from Pařízková & Carter, 1976.)

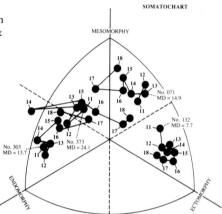

Fig. 4.7. Mean somatotypes of Prague children aged 3½ to 6 years (males = ●; females = ■), from Pařízková *et al.* (1984). Mean somatotypes for male ▼ and female ▲ adolescents aged 15 to 17 years are taken from Bok & Tlapáková (1982).

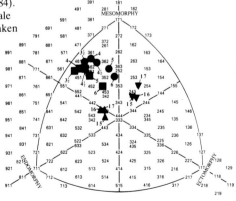

between successive years, for four boys. The MD is the sum of somatotype dispersion distances (SDD) between successive somatoplots. Over the 7 years, the MDs range from a low of 7.7 to a high of 24.1, with a mean of 15.1. As the figure shows, No. 132 was the most stable and No. 373 the least. Two subjects in the Saskatchewan Growth Study of boys aged 10–16 showed similar patterns of change (Weese *et al.*, 1975).

The Heath–Carter anthropometric somatotype method was used in a study of 58 boys and girls, aged $3\frac{1}{2}$, $4\frac{1}{2}$, 5 and 6 years, in two Prague kindergartens (Pařízková *et al.*, 1984). Mean somatotypes were recalculated from the authors' mean anthropometric values (Table 4.3, Fig. 4.7), because there were no height corrections for endomorphy estimated from skinfolds. At the earliest ages both sexes were endo-mesomorphs, but there were increasing differences in the fifth and sixth years. The means for the Prague and Manus children were close, despite slight differences in somatotype method.

In another study in Prague, Bok & Tlapáková (1982) somatotyped 11 boys and 21 girls aged 15, 16 and 17. The girls, with a mean of $4\text{-}3\frac{1}{2}\text{-}3$ at 15 and of $4\text{-}4\text{-}2\frac{1}{2}$ at 16 and 17 (Fig. 4.7), on either side of the ectomorphy axis, tended towards slightly increased mesomorphy and endomorphy and decreased ectomorphy. The boys, with means near $2\text{-}4\text{-}3\frac{1}{2}$ at age 15 and 16, and $2\text{-}4\frac{1}{2}\text{-}3\frac{1}{2}$ at 17, are northeast of the girls. Whole somatotypes did not change greatly; for both boys and girls the differences in means were approximately one-half unit (SAD = 0.53), but there were individual changes greater than one unit in a given component.

In a longitudinal study of 210 Belgian boys at ages 13–18 years, Claessens (1981) reported on the stability of anthropometric somatotypes. For endomorphy there were high between-years correlations ($r = 0.86$–0.90) and a lower correlation ($r = 0.79$) for the 5-year interval. Most of the differences in component means were approximately one-half unit. The mean somatotypes for the six age groups are close together on the somatochart (Fig. 4.8); i.e. at age 13 the mean was $2\text{-}4\text{-}3\frac{1}{2}$ and at 18 it was $2\frac{1}{2}\text{-}4\text{-}3\frac{1}{2}$.

The longitudinal Harpenden Growth Study established by Tanner & Whitehouse in 1948 followed the same children from 1949 to 1970. The subjects were healthy, well-nourished boys and girls attending the 'usual schools'. Interested in finding out how the great variations in size, shape and physique of young adult men and women developed in childhood, the authors examined the children every 6 months (every 3 months during puberty).

In *Atlas of Children's Growth* (Tanner & Whitehouse, 1982) the authors assigned Sheldon (1940, 1954) somatotypes to the normal subjects when they reached young adulthood. They presented the photographs of the children at various ages, with somatotype ratings of the last one only. They

Fig. 4.8. Mean somatotypes of male
Belgians followed from 13–18 years. (From
data in Claessens, 1981.) The mean
somatotype ▲ is for Belgian military
servicemen aged 20.3 years.
(Data provided by J. Clarys.)

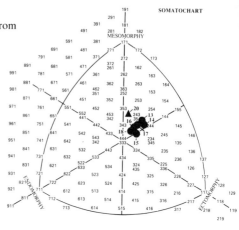

said that it was difficult to make correct ratings at age 6, and suggested that
the difficulty was due to real changes in shape between age 6 and 18, or to the
observer's lack of sensitivity to small differences. They noted that under-
nutrition and other environmental pressures influence size, but not shape:
'The genotype has a stronger control over shape than over size' (p. 1). They
also commented that the Sheldon system ignores size but considers body
shape and tissue composition as revealed in the photographs. It is clear that
they rated the young adult, irrespective of changes in shape shown in the
photographs and anthropometry. Tanner & Whitehouse's evaluation of
predictability of shape related to somatotype, graphs and interpretations are
confusing, because they used age-adjusted standard deviation scores for
shape variables but somatotypes at age 18 only.

Heath assigned Heath–Carter photoscopic ratings to all photographs in
Atlas of Children's Growth. (The ratings by subject are presented in
Appendix III. For the photographs see the *Atlas*.) The somatotypes, plotted
at 2-year intervals, show changes in three boys and three girls (Fig. 4.9*a,b*).
Subjects 4, 21 and 41 become less endomorphic with age; subject 15 more
ectomorphic; and subjects 25 and 46 first less endomorphic and later more
endomorphic. There is no pattern consistent for all children.

A study of 7-year-old Bulgarian boys and girls, who were followed for one
year, showed that although there were many changes by component, the
mean somatotype did not change much (Toteva, 1986). Mesomorphy
increased and ectomorphy decreased slightly in both sexes. The mean
somatotype for the boys was $2\frac{1}{2}$-$4\frac{1}{2}$-3, and for the girls 3-4-3.

Longitudinal studies show that both individual and group somatotypes
change with age. Individual patterns of change are particularly important,
because group means are likely to mask variability.

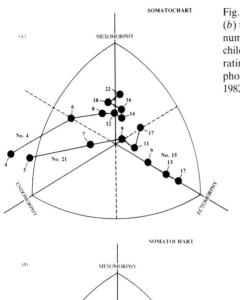

Fig. 4.9. Somatotypes of (*a*) three boys and (*b*) three girls followed longitudinally. The numbers near the circles are the ages of the children when rated. Based on photoscopic ratings by Heath. (See Appendix III. From photographs in Tanner & Whitehouse, 1982.)

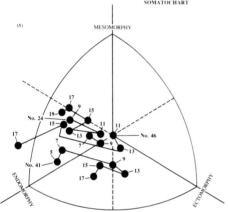

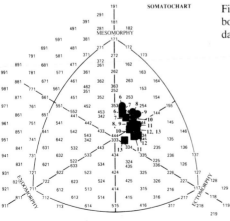

Fig. 4.10. Mean somatotypes of Belgian boys ● and girls ■ at 6–13 years. (From data in Duquet, 1980.)

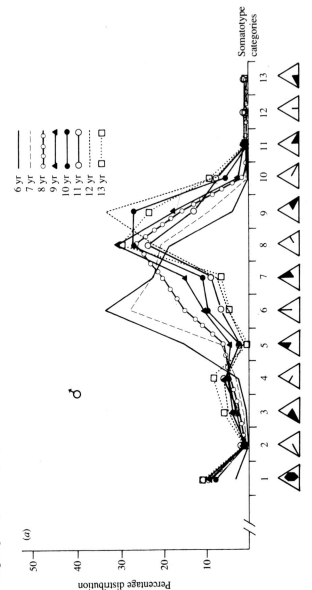

Fig. 4.11. Percentage of somatotypes by category for (*a*) Belgian boys and (*b*) Belgian girls. (From Duquet, 1980.)

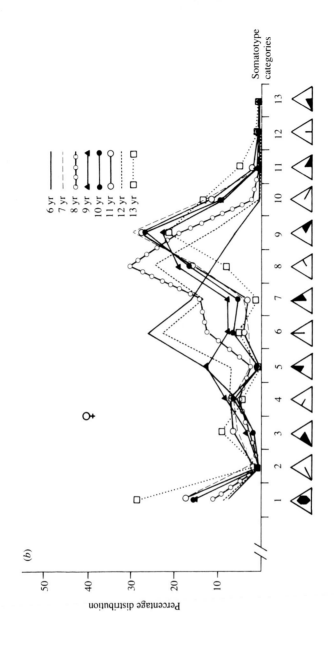

(b)

Fig. 4.12. Mean somatotypes for
Czechoslovak boys ● and girls ■ at 8–14
years. (From data in Štěpnička, 1976*b*.
Endomorphy is height corrected.)
Twenty-year-old male and female
university students in Czechoslovakia
are also shown (▼ and ▲ respectively)
(Štěpnička *et al.*, 1979*a*).

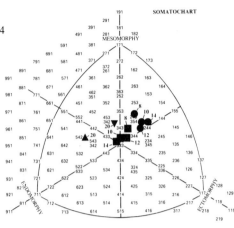

Cross-sectional studies

Several large cross-sectional somatotype studies in Belgium, Czechoslovakia, Finland, Hungary, India and Venezuela provide substantial
information from ages 5 to 18 years (see Table 4.4 for summaries).

In a study of 8554 Belgian primary school boys and girls from the
Performance and Talent Project, Duquet *et al.* (1975) analysed anthropometric somatotypes by age. Subsequently, Duquet (1980) made photoscopic
ratings. Values in Table 4.4 and Figs. 4.10 and 4.11(*a*,*b*) are from Duquet
(1980). At all ages (6 to 13 years) boys and girls increased in endomorphy,
with greater increases for the girls. Both groups decreased in mesomorphy
and increased in ectomorphy. The trend toward increasing ectomorphy was
more consistent for the boys. The girls had significantly higher endomorphy
than boys at all ages, and higher ectomorphy at all ages except 13 years.

Fig. 4.13. Somatotype distribution of 12-
year-old boys from Czechoslovakia. ▲ =
mean somatotype. (Redrawn from
Stěpnička, 1976*b*.)

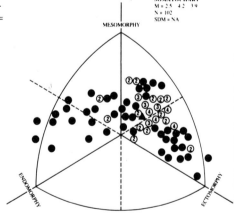

There were no differences in mesomorphy. Mean somatotypes proceeded from balanced mesomorphy ($2\frac{1}{2}$-4-$2\frac{1}{2}$) toward mesomorphy-ectomorphy ($2\frac{1}{2}$-$3\frac{1}{2}$-$3\frac{1}{2}$) in boys, and toward a more central somatotype (3-$3\frac{1}{2}$-$3\frac{1}{2}$) in girls. The girls tended toward greater dispersions, with increased endomorphic and ectomorphic dominance. Distribution by somatotype categories (Fig. 4.11*a,b*) shows increasing heterogeneity within sexes, especially for girls.

In a study of 106 school boys at ages 10 to 12 years in Mechelen, Belgium, Clarys *et al.* (1970) found a mean anthropometric plus photoscopic somatotype of 2-$4\frac{1}{2}$-$3\frac{1}{2}$. This mean, which is slightly more ecto-mesomorphic than that for the 11-year-olds in the study above, may be due to ratings based on both anthropometry and photographs.

Štěpnička (1976*b,c*) studied 403 boys and 400 girls, aged 8, 10, 12 and 14, in the second, fourth, sixth and eighth classes in Bohemian and Moravian schools. The mean somatotypes of the boys shifted from ecto-mesomorphy to balanced ectomorph-mesomorphy. The mean somatotypes of the girls shifted from ecto-mesomorphy to the center of the somatochart. Both distributions of 12-year-olds display examples of dominant mesomorphy, but the boys are higher and tend toward the northeast. (See Figs. 4.12 [from Table 4.4], 4.13, 4.14).

A comparison of Czech children (Figs. 4.7 and 4.12) with Manus children (Fig. 4.4) shows similar somatotype pathways, with Czech means, especially in the latter ages, slightly lower in mesomorphy.

Four studies of boys and girls in Hungary show a wide range of somatotypes by age. Szmodis (1977) obtained anthropometric somatotypes on about 1400 children of both sexes aged between 5 and 17. In addition to ordinary schooling, these children participated in sports training. The mean

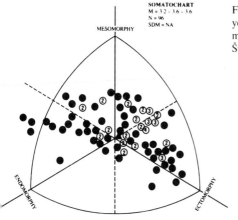

Fig. 4.14. Somatotype distribution of 12-year-old girls from Czechoslovakia. ▲ = mean somatotype. (Redrawn from Štěpnička, 1976*b*.)

Table 4.4. *Mean somatotypes by age in cross-sectional studies using the Heath–Carter methods*

Sample, reference and method	N	Age (yr)	Somatotype
Males			
Belgium (Duquet, 1980), Heath–Carter	149	6	2.0-4.2-2.4
anthropometric and photoscopic	409	7	2.0-4.0-2.4
	469	8	1.9-3.8-3.1
	536	9	2.0-3.7-3.3
	536	10	2.1-3.6-3.5
	871	11	2.1-3.5-3.7
	721	12	2.2-3.4-3.7
	123	13	2.3-3.5-3.6
	3814		
Bohemia & Moravia (Štěpnička, 1976*b*),	101	8	2.3-4.5-3.2
Heath–Carter anthropometric $(En_H)^a$	101	10	2.3-4.2-3.8
	102	12	2.5-4.2-3.9
	99	14	2.1-4.3-3.9
	403		
Hungary (Farmosi, 1982), Heath–Carter	20	9	2.7-4.6-3.0
anthropometric (No En_H)	28	10	2.3-4.4-3.4
	39	11	2.6-4.3-3.8
	40	12	2.8-4.5-3.5
	31	13	3.0-4.4-3.6
	32	14	3.1-4.3-3.7
	57	15	3.0-4.1-3.8
	95	16	3.2-4.6-3.2
	71	17	3.4-5.0-2.6
	81	18	3.4-5.4-2.8
	63	24–25	3.4-5.4-2.2
	557		
Kormend, Hungary (Eiben, 1985), Heath–	41	68-6^b	2.5-4.1-2.3
Carter anthropometric (En_H)	75	78-6	3.3-4.1-2.9
	53	68-7	2.1-3.9-3.1
	78	78-7	2.8-4.1-2.8
	53	68-8	2.8-3.8-3.3
	80	78-8	3.3-4.0-3.0
	67	68-9	2.1-3.7-3.8
	94	78-9	3.4-3.9-3.4
	51	68-10	2.1-3.6-4.0
	59	78-10	3.7-3.8-3.5
	60	68-11	2.5-3.6-3.9
	94	78-11	3.7-3.8-3.7
	57	68-12	2.1-3.7-2.9
	117	78-12	3.6-3.7-3.6
	84	68-13	2.2-3.5-4.2
	103	78-13	3.5-3.6-3.9
	85	68-14	2.1-3.9-3.9
	82	78-14	3.3-3.7-3.8

(*continued*)

Table 4.4. (*Continued*)

Sample, reference and method	N	Age (yr)	Somatotype
	140	68-15	2.0-3.6-3.9
	110	78-15	3.1-3.6-3.8
	109	68-16	2.1-3.8-3.6
	90	78-16	3.1-3.6-3.5
	89	68-17	2.2-3.7-3.5
	88	78-17	3.0-3.3-3.8
	25	68-18	2.0-3.6-3.5
	71	78-18	3.4-3.8-3.2
	2055		
Central Finland (Holopainen *et al.*, 1984;	18	7	2.6-4.3-2.7
and personal communication, 1985),	46	8	2.5-4.2-2.9
Heath–Carter anthropometric (En_H)	44	9	2.7-3.9-3.3
	48	10	3.2-4.0-3.1
	42	11	3.4-4.2-3.0
	38	12	3.4-3.8-3.4
	45	13	3.0-3.8-3.6
	62	14	2.7-3.6-3.9
	57	15	2.5-3.4-4.0
	30	16	2.9-3.8-3.8
	430		
Gaddi, Rajput, India (Singh & Sidhu, 1980),	29	4	3.0-4.7-0.9
Heath–Carter anthropometric (En_H)	42	6	2.4-4.4-1.6
	56	8	1.9-3.7-3.2
	63	10	1.7-3.4-3.7
	59	12	1.4-3.1-4.8
	41	14	1.6-3.1-4.8
	56	16	1.6-3.3-4.5
	30	18	1.9-3.4-4.1
	31	20	1.8-3.2-3.9
	407		
Jat Sikh Punjab, India (Kansal, 1981),	43	10	1.8-3.4-4.4
Heath–Carter anthropometric (En_H)	97	12	1.7-3.1-5.3
	83	14	1.7-3.1-5.3
	83	16	1.8-2.9-5.2
	43	18	2.1-2.9-4.8
	30	20	2.3-3.0-4.4
	379		
Bania, Punjab, India (Kansal, 1981),	30	10	2.3-3.5-4.1
Heath–Carter anthropometric (En_H)	37	12	2.5-3.1-4.6
	31	14	2.0-2.9-5.2
	35	16	2.1-2.9-5.0
	43	18	2.7-3.0-4.8
	38	20	3.1-3.1-4.3
	214		

(*continued*)

Table 4.4. (*Continued*)

Sample, reference and method	N	Age (yr)	Somatotype
Bangalore, India (Rangan, 1982), Heath– Carter anthropometric (En$_H$)	60	13	3.4-2.6-4.9
	60	14	3.2-2.6-5.0
	60	15	3.2-6.4-4.8
	180		
Oamaru, New Zealand (Carter, unpubl.), Heath–Carter photoscopic, by Carter	103	12–17	3.4-4.5-2.7
Harpenden, England (Tanner & Whitehouse, 1982), Carter & Heath (1986), Heath–Carter photoscopic, by Heath	23	19.1 17–22	2.9-4.0-3.4
Chula Vista, California (Haley, 1974), Heath–Carter anthropometric (En$_H$)	34	15	2.9-4.2-3.6
Illinois, USA (Slaughter *et al.*, 1977*b*), Heath–Carter anthropometric (No En$_H$)	68	7–12	2.6-4.1-3.3
Caracas, Venezuela (Pérez *et al.*, 1985), Heath–Carter anthropometric (En$_H$)	40	8.4–11.3	2.5-4.9-2.4
	127	11.4–15.3	2.7-4.4-3.2
	151	15.3–21.3	2.7-4.4-3.0
	318		
Ile-Ife, Nigeria (Toriola & Igbokwe, 1985; Toriola, personal communication), Heath–Carter anthropometric (En$_H$)	25	10	2.5-3.7-2.9
	32	11	2.8-3.7-2.7
	30	12	2.7-3.8-2.6
	50	13	2.7-3.9-2.5
	42	14	2.9-4.0-2.3
	60	15	2.9-4.4-2.4
	38	16	3.3-4.9-2.2
	24	17	3.7-4.9-2.3
	24	18	3.8-5.0-2.4
	325		
Londrina, Brazil (Guedes, 1983), Heath– Carter anthropometric	30	11.5	2.6-4.0-3.5
	30	12.6	2.4-4.2-3.6
	30	13.7	2.8-3.8-3.5
	30	14.4	2.6-3.9-3.4
	30	15.5	2.1-3.7-3.5
	30	16.4	2.2-4.0-3.3
	180		
Females			
Belgium (Duquet, 1980), Heath–Carter anthropometric and photoscopic	210	6	2.3-4.1-2.8
	801	7	2.3-4.0-3.0
	734	8	2.2-3.8-3.4
	656	9	2.4-3.7-3.6
	795	10	2.5-3.6-3.8
	862	11	2.7-3.5-3.9
	614	12	2.7-3.5-4.0
	68	13	3.1-3.3-3.7
	4740		

(*continued*)

Table 4.4. (*Continued*)

Sample, reference and method	N	Age (yr)	Somatotype
Bohemia & Moravia (Štěpnička, 1976*b*),	102	8	2.9-3.9-3.3
Heath–Carter anthropometric (En$_H$)	100	10	3.4-3.7-3.6
	96	12	3.2-3.6-3.6
	102	14	3.6-3.5-3.2
	400		
Bakony, Hungary (Bodzsar, 1982), Heath–	258	10	3.9-3.6-3.6
Carter anthropometric (No En$_H$)	273	11	3.4-3.6-3.9
	272	12	3.3-3.4-4.1
	249	13	4.4-3.4-3.7
	261	14	3.9-3.2-3.9
	1313		
Kormend, Hungary (Eiben, 1985), Heath–	26	68-6[b]	2.9-3.9-2.4
Carter anthropometric (En$_H$)	71	78-6	4.2-3.9-2.8
	43	68-7	2.7-3.5-3.3
	79	78-7	4.1-3.8-2.9
	39	68-8	3.1-3.5-3.5
	74	78-8	3.8-3.6-3.3
	52	68-9	3.2-3.5-3.6
	91	78-9	4.4-3.6-3.3
	46	68-10	3.5-3.4-3.3
	65	78-10	4.2-3.4-3.6
	48	68-11	3.4-3.3-3.6
	67	78-11	4.1-3.3-2.9
	43	68-12	3.1-3.2-3.7
	75	78-12	4.5-3.4-3.3
	72	68-13	3.5-2.9-3.6
	90	78-13	4.5-2.9-3.5
	81	68-14	4.0-3.2-3.2
	107	78-14	4.8-3.0-3.4
	73	68-15	5.0-3.4-2.7
	58	78-15	4.9-3.1-3.3
	45	68-16	4.8-3.2-2.8
	60	78-16	5.4-3.4-3.1
	65	68-17	4.7-3.2-2.6
	75	78-17	5.7-3.5-2.6
	34	68-18	5.3-3.6-2.1
	42	78-18	5.2-3.2-2.9
	1612		
Central Finland (Holopainen *et al.*, 1984;	26	7	3.2-3.7-2.7
and personal communication, 1985),	39	8	3.6-3.6-3.0
Heath–Carter anthropometric (En$_H$)	48	9	3.6-3.8-3.1
	43	10	3.8-3.4-3.2
	48	11	3.4-3.2-3.6
	43	12	3.7-3.2-3.5
	46	13	3.6-2.7-3.8
	67	14	4.0-2.9-3.3
	56	15	4.0-2.2-3.3
	27	16	4.6-3.1-2.9
	443		

(*continued*)

Table 4.4. (*Continued*)

Sample, reference and method	N	Age (yr)	Somatotype
Punjab, India (Sidhu *et al.*, 1982), Heath–Carter anthropometric (En$_H$)			
Jat-Sikh	150	19.3	4.1-2.4-3.6
Bania	161	19.1	4.2-2.3-4.0
Harpenden, England (Tanner & Whitehouse, 1982), Carter & Heath (1986), Heath–Carter photoscopic by Heath	32	18.2 (14–20)	4.6-3.6-2.7
San Diego, USA (Fisher, 1975), Heath–Carter anthropometric (En$_H$)	30	15	3.7-3.3-3.2
Illinois, USA (Slaughter *et al.*, 1980) Heath–Carter anthropometric (No En$_H$)	50	9.9	3.2-3.3-3.5
Caracas, Venezuela (Pérez *et al.*, 1985), Heath–Carter anthropometric (En$_H$)	42	8.4–11.3	3.0-4.3-2.4
	145	11.4–15.3	3.9-3.7-2.6
	151	15.3–21.3	4.5-4.0-1.9
	338		
Ile-Ife, Nigeria (Toriola & Igbokwe, 1985; Toriola, personal communication), Heath–Carter anthropometric (En$_H$)	21	10	3.2-3.8-2.6
	29	11	3.3-3.7-2.4
	32	12	3.6-3.8-2.6
	44	13	3.7-3.8-2.5
	41	14	4.6-3.8-2.4
	52	15	4.9-3.8-2.6
	33	16	4.9-3.8-2.4
	28	17	4.8-4.0-2.3
	20	18	5.0-3.9-2.1
	300		
Londrina, Brazil (Guedes, 1983), Heath–Carter anthropometric	30	11.3	3.1-3.6-3.2
	30	12.6	3.2-3.4-3.2
	30	13.4	3.7-3.4-3.0
	30	14.5	3.8-3.4-3.0
	30	15.5	4.0-3.8-2.3
	30	16.4	4.0-3.4-2.6
	180		

[a] En$_H$ = height-corrected endomorphy.
[b] The first number is the year, the second number the age.

somatoplots (Fig. 4.15) show that from age 5 to 13 the boys tend toward higher ectomorphy, and from age 14 to 17 slightly more toward balanced ectomorphy-mesomorphy. The distribution of the girls' somatotypes is almost wholly confined to central somatotypes, and with decreasing mesomorphy tend toward the lower right in the central somatotype area. (Mean somatotypes were not provided, but *x*,*y* coordinates were given.)

Between 1977 and 1979, Farmosi (1982) conducted a somatotype study of 494 boys between ages 9 and 18 and of smaller groups of boys aged up to 26.

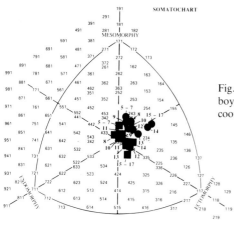

Fig.4.15. Mean somatotypes of Hungarian boys ● and girls ■ at 5–17 years. (From x, y coordinates in Szmodis, 1977.)

The subjects in the large sample were training for competition in eleven sports. Some subjects in the smaller groups were students at a college of physical education. The mean somatoplots (Fig. 4.16) show a decrease in mesomorphy to age 15. At age 16 endomorphy and mesomorphy shift upward dramatically to a mean of 3-4½-3 (balanced mesomorphy). The shift toward endo-mesomorphy continues through ages 17 and 18. Although at all ages the Farmosi means are slightly more endo-mesomorphic than those of Szmodis, in the 9–15 age range the patterns are similar. The slight divergence may be due either to differences in sport groups or to small differences in methodology.

In a somatotype study of 1313 girls, aged 10 to 14 years, from the Bakony Hills of Hungary, Bodzsar (1982) grouped the subjects according to occupation of their fathers, and according to size of family, number of siblings, and earnings of fathers. Intellectuals and physical workers formed the two

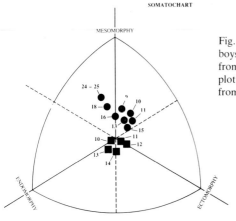

Fig. 4.16. Mean somatotypes for Hungarian boys ● at 9–18 years, and at 24 and 25 years from data in Farmosi (1982). Means are also plotted for Hungarian girls ■ at 10–14 years from data in Bodzsar (1982).

largest categories. The mean somatotypes (Fig. 4.16) of the total sample are central, and tend toward increasing endomorphy and decreasing mesomorphy with age. Offspring of intellectuals tend toward lower endomorphy than offspring of physical workers. Endomorphy also decreased with number of siblings or number of family members. Somatotypic differences according to per capita earnings are equivocal. Without further investigation of the different nutritional habits of the physical workers and intellectuals it is impossible to explain inconsistencies. At most ages the girls in Bodzsar's study were slightly less mesomorphic and more endomorphic than those in Szmodis' study.

An extensive growth study, the fourth from Hungary, was conducted in Kormend, a small town in Western Hungary. In 1958, 1968, and 1978, almost every healthy child aged between 3 and 18 was measured. In 1968, Eiben (1985) made anthropometric somatotype assessments on 914 boys and 667 girls from ages 6 to 18, and in 1978 on 1141 boys and 954 girls. Between 1958 and 1978, Kormend had changed from an agricultural village to an industrialized and urbanized town with a greatly increased population. The immigration brought changes in social structure, medical attention, nutrition, patterns of physical activity, and genetic balance. These factors may have influenced the growth patterns and secular trends in the children. This study shows significant differences in somatotype distributions with age and sex as well as a dramatic secular trend in somatotypes in the decade between 1968 and 1978. (See Table 4.4 and Fig. 4.17a,b.) The means for endomorphy in Table 4.4 have been corrected, using the height for each age from Eiben (1982), because endomorphy was not height-corrected in Eiben's 1985 report.

The somatocharts show that the Kormend children became more endomorphic between 1968 and 1978. This is true at all ages, except for the oldest girls. In general, in 1978 endomorphy was from 0.6 to 1.5 units higher for both boys and girls, and mesomorphy and ectomorphy differed by 0.5 or less across the ages, with a slight tendency towards increased mesomorphy and decreased ectomorphy (except at the oldest ages). The dispersion of the somatotypes about their means (SDM) were higher at all ages and in both sexes in 1978. Eiben suggested that this was due to earlier onset of puberty and increased genetic variability due to migration. For the boys the mean somatoplots by age in 1968 and 1978 show the same patterns of change. At age 6 the means are near the 3-4-3 somatotype, move towards greater ectomorphy by the early teens, and towards greater mesomorphy by age 18. At age six the means for the girls begin to the northeast of the somatochart center, move southeast through the central region, and west or southwest during the late teens.

The findings of Szmodis, Farmosi, and Bodzar are similar to those of

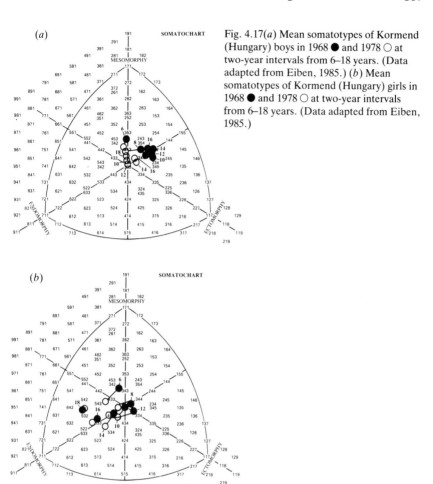

Fig. 4.17(*a*) Mean somatotypes of Kormend (Hungary) boys in 1968 ● and 1978 ○ at two-year intervals from 6–18 years. (Data adapted from Eiben, 1985.) (*b*) Mean somatotypes of Kormend (Hungary) girls in 1968 ● and 1978 ○ at two-year intervals from 6–18 years. (Data adapted from Eiben, 1985.)

Eiben. The slightly higher mesomorphy and lower endomorphy of the sport oriented children in the first two studies could be expected. However, the differences would have been smaller if height corrections for endomorphy had been made in the former three studies.

Holopainen *et al.* (1984) used the Heath–Carter anthropometric method in a somatotype study of 452 boys and 462 girls, aged 7 to 16, from first to ninth comprehensive school grades in Central Finland. The means for the girls (Fig. 4.18*a,b*), in or near the central category, show decreasing meso-morphy from age 7 to 13, followed by slight increases in endomorphy from 14 to 16. At the outset the boys show a mean balanced mesomorphy, followed by a tendency toward increasing ectomorphy, and balanced ecto-mesomorphy close to 3-4-4 at 14 to 16. For both boys and girls there were significant differences between somatotype categories by age. In per cent

Fig. 4.18. Mean somatotypes for (*a*) Finnish
girls and (*b*) Finnish boys, at 7–16 years.
(Redrawn from Holopainen *et al.*, 1984.)

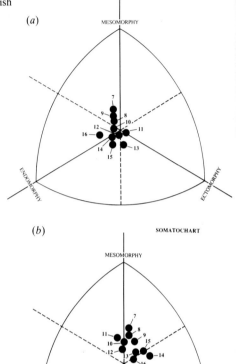

Fig. 4.19. Mean somatotypes of Gaddi
Rajput boys from 4 to 20 years of age.
(Redrawn from Singh & Sidhu, 1980.)

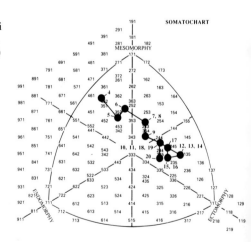

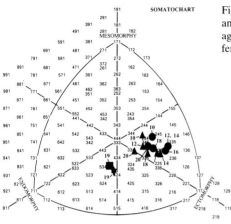

Fig. 4.20. Mean somatotypes of Jat-Sikh ● and Bania ▲ males from 10 to 20 years of age (Kansal, 1981); Jat-Sikh ■ and Bania ★ females at age 19 (Sidhu *et al.*, 1982).

frequency by categories the boys showed a dramatic increase in the number of ecto-mesomorphs, with a decrease in endo-mesomorphs. From the 9–10 year group onward the girls showed a steady increase in endomorphy with a decrease in the number of central somatotypes.

Several somatotype studies of young Indians yielded data significantly different from the foregoing. These are cross-sectional studies of boys aged 4 to 21 years, with a combined total of 2246 subjects from different ethnic, economic, dietary, and regional samples. (See Table 4.4, Figs. 4.19, 4.20, 4.21.)

Singh (1978) and Singh & Sidhu (1980) somatotyped 786 Gaddi Rajput males aged 4 to 20 years from the Chamba district of Hamachal Pradesh (Northwest India) in 1974–1975. Sample sizes ranged from 29, at age 4, to 71, at age 9. The Gaddis are semi-agricultural and semi-nomadic people living at altitudes of 1500 to 2500 m. (See Table 4.4 for mean somatotypes at

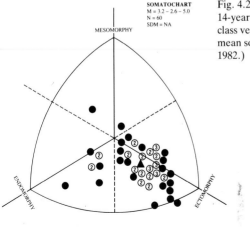

Fig. 4.21. Somatotype distribution of Indian 14-year-old boys who were from middle class vegetarian families in Bangalore. ▲ = mean somatotype. (Redrawn from Rangan, 1982.)

2-year intervals, and Fig. 4.19 for mean somatoplots at each age.) The somatotype distributions and means were endo-mesomorphic at the youngest ages, ecto-mesomorphic from 7 to 9 years, and meso-ectomorphic from 10 to 20 years. The direction of the shift in means was to the southeast and parallel to the ectomorphic axis until age 14 (ectomorphy = 4.8), when the trend turned back towards slightly higher mesomorphy. The migratory distance (MD) from 4 to 20 years was 16.4.

In 1976–1977, in the Patiala and Sangar districts and in the city of Chandigarh in Punjab State, India, Kansal (1981) collected data on 799 Jat-Sikh and 421 Bania males aged 10 to 21 years. The life style of Jat-Sikhs is traditionally agricultural, usually requiring strenuous physical work, while the Banias mostly engage in business and commerce, and have a more sedentary way of life. The means at two-year intervals (Table 4.4, Fig. 4.20) show that the Jat-Sikhs are more ectomorphic than the Banias. They are less endomorphic and more ectomorphic, but are similar in mesomorphy. The greatest linearity is between 12 and 16 years of age for the Jat-Sikh (ectomorphy = 5.25) and at 16 years for the Bania (ectomorphy = 5.27).

In a somatotype study of 420 secondary school boys, at ages 13 to 15, from Bangalore City District, Rangan (1982) separated the subjects into age, socio-economic, vegetarian and non-vegetarian groups. From the total sample he drew at random a sample of 60 boys from each group. He found no inter-age differences in somatotype components or in per cent frequencies of somatotypes by category (Fig. 4.21). More than two-thirds of the boys were dominant in ectomorphy, and more than one-half were low in mesomorphy (3 or less). There were no somatotype differences between vegetarian and non-vegetarian boys. The average somatotype for these groups was 3-2½-5. There were almost no dominant mesomorphs and few dominant endomorphs. In this sample mesomorphy is markedly lower and ectomorphy markedly higher than in other studies of boys' somatotypes.

In Rangan's study, from an essentially urban population, boys from middle and lower classes were more ectomorphic than those from the upper class, who had more central somatotypes and higher endomorphy. The mean somatotypes for the lower and higher classes were 2.7-2.5-5.4 and 3.3-2.5-4.6. Those in the lower class were 8.4 cm shorter and 8.8 kg lighter than the higher class. This suggests that the increases in weight and differences in somatotypes are due to increased endomorphy in the higher class boys. Rangan compared his findings with those of Kansal's (1981) Punjabi Jat-Sikhs (agricultural) and Banias (trading), and Singh's (1978) Rajput Gaddis (tribal) (see Table 4.4 and Fig. 4.22). In both groups mean endomorphy is approximately one unit lower than in the Bangalore sample. The Punjabi boys were slightly higher in ectomorphy, except for the 12-year-old Banias, who were similar to their age mates in the other studies. Rangan speculates that low mesomorphy and high ectomorphy for his subjects may

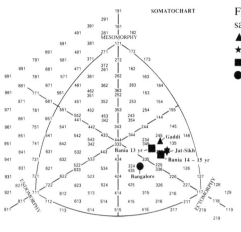

Fig. 4.22. Mean somatotypes of four samples of Indian boys aged 13–15 years. ▲ = Rajput Gaddi (Singh, 1978); ★ = Punjabi Jat-Sikh (Kansal, 1981); ■ = Punjabi Bania (Kansal, 1981); and ● = Bangalore (Rangan, 1982).

be in part genetic and in part due to sub-nutrition. Berry & Deshfukh (1964) made similar inferences in their study of male college students in Nagpur.

These studies of Indian boys show that they are more ectomorphic and less endomorphic and mesomorphic than boys of European descent, particularly at older ages. There are also differences by ethnicity and socio-economic status. However, the pattern of somatotype changes with age is similar to those of other samples.

In the only study of Indian girls, Sidhu *et al.* (1982) reported the somatotypes of 150 Jat-Sikh and 161 Bania girl students in the Punjab. Their ages ranged from 16 to 23 years, with an average of 19.2 years, which is older than most of the boys studied. The mean somatotypes were 4.1-2.4-3.6 (Jat-Sikh) and 4.2-2.3-4.0 (Bania), with no differences between component means. The means are in the endomorph-ectomorph category and may be compared with the older boys from Kansal's (1981) study. Figure 4.20 shows that there is typical sexual dimorphism, with the female means southwest of the male means.

Haley (1974) and Fisher (1975) conducted somatotype studies of 15-year-old boys and girls in two San Diego County junior high schools. The mean somatotype (Fig. 4.23) for the 34 boys was 2.9-4.2-3.6 and for the 30 girls it was 3.6-3.3-3.2 (Fig. 4.24). In a study of 68 7–12-year-olds (mean = 10.0 yr) in the University of Illinois Sport Fitness Program in summers of 1974 and 1975, Slaughter *et al.* (1977*b*, 1980) found an average somatotype of 2.6-4.1-3.3 for boys, and 3.2-3.3-3.5 for girls (means for separate ages were not given.)

In 1978 in Caracas, Venezuela, Pérez *et al.* (1985) obtained anthropometric somatotypes on 318 boys and 337 girls (ages 8.4 to 21.3 years) of differing socio-economic status. The subjects were Latin American Mestizos, classified by age into pre-adolescents (N = 82) 8.4 to 11.3 years (mean =

Fig. 4.23. Somatotype distribution of Chula Vista 15-year-old boys. ▲ = mean somatotype. (Redrawn from Haley, 1974.)

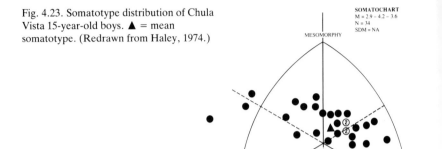

10.3), adolescents (N = 272) 11.4 to 15.3 years (mean = 13.5), and young adults (N = 301) 15.4 to 21.3 years (mean = 17.2). The boys decreased in mesomorphy and increased in ectomorphy between the two youngest age groups, and reversed these trends slightly in the oldest group (see Fig. 4.25). The percentage of dominant ectomorphs in the boys' group increased from 5% in the youngest to 22.1% in the middle group, and decreased to 15.9% in the oldest group. The girls showed a large increase in endomorphy (1½ units) from the youngest to the oldest groups, together with slight decreases in mesomorphy and ectomorphy. The percentage of girls dominant in endomorphy increased from 16.7% in the youngest age group to 25.5% and 39.3% in the middle and oldest age groups respectively. These patterns of change among means are similar to those observed in other samples in Table 4.4.

In 1954, Carter photographed and measured 103 boys at Waitaki Boys High School in Oamaru, New Zealand. The boys were 12–17 years of age. In

Fig. 4.24. Somatotype distribution of Point Loma 15-year-old girls. ▲ = mean somatotype. (Redrawn from Fisher, 1975.)

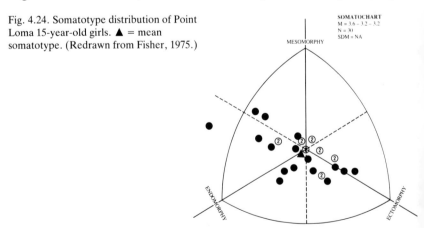

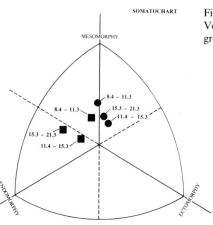

Fig. 4.25. Mean somatotypes for Venezuelan boys ● and girls ■ for three age groups. (Pérez *et al.*, 1985.)

1970 Carter (unpublished) somatotyped the subjects by the Heath–Carter photoscopic method. The mean somatotype was 3.4-4.5-2.7. The greatest percentages were endo-mesomorphs (26.2%) and balanced mesomorphs (17.5%). (See Fig. 4.26.) These subjects are more endo-mesomorphic than those in samples of similar age groups. Their mean is closest to the Hungarian 16–18-year-olds (Farmosi, 1982) and the Medford 12–16-year-olds (Clarke, 1971). Many of the Waitaki boys were boarders from farm families on the South Island. Perhaps the physiques of these boys reflect some of the endo-mesomorphic ruggedness that can be seen in their parents, many of whom are descended from mid-nineteenth century pioneer settlers.

Heath re-rated photoscopically (Carter & Heath, 1986) the 23 boys, mean age 19.1, in *Atlas of Children's Growth* (Tanner & Whitehouse, 1982). The mean Heath somatotype of the boys was ecto-mesomorphic (2.9-4.0-3.4) (Fig. 4.27(*a*)). The mean was similar to the means for comparable age groups in the studies of Pařízková & Carter (1976), Bok & Tlapáková (1982)

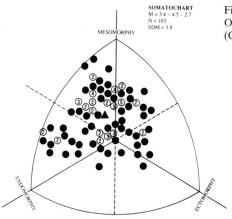

Fig. 4.26. Somatotype distribution of Oamaru (New Zealand) high school boys (Carter, unpubl.). ▲ = mean somatotype.

Fig. 4.27. Somatotype distributions of (*a*)
young males and (*b*) young females in the
Harpenden growth study. ▲ = mean
somatotype. Rated by Heath from data in
Tanner & Whitehouse (1982).

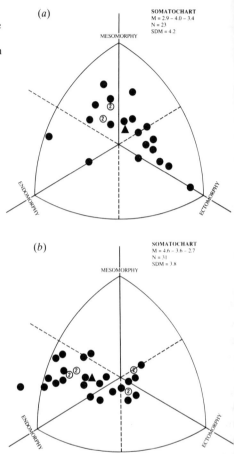

and Claessens (1981); but different from the means of Zuk (1958), Heath &
Carter (1971) and Farmosi (1982).

For the 32 girls, mean age 18.2, the mean Heath somatotype was
4.6-3.6-2.7 (Fig. 4.27(*b*)). The girls were meso-endomorphs, with means
similar to those of Zuk (1958) and Bok & Tlapáková (1982) at age 16,
different from those of Heath & Carter (1971) and Bok & Tlapáková (1982)
at age 17.

In a study of Iowa boys (N = 123) aged 12 to 17 years (Carter, 1958), the
mean somatotypes changed from 3½-3½-2½ at 12–13 years to 2½-4-3 at 14–15
years, and to 2½-4½-3 at 16–17 years. Although the ratings (made by Sills and
Carter) were phenotypical, they were not made by the Heath–Carter
method. Nevertheless, the results are similar to Heath ratings of Oregon
boys at the same ages (Clarke, 1971).

Amador *et al.* (1983) calculated anthropometric somatotypes of 362
children between 1.5 and 5.5 years (190 boys and 172 girls) in Havana, Cuba.

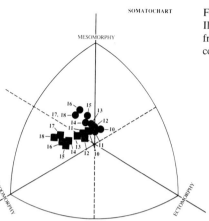

Fig. 4.28. Mean somatotypes by age of Ile-Ife (Nigeria) boys ● and girls ■. Plotted from data provided by A. Toriola (personal communication).

They demonstrated sexual dimorphism at all ages, with endomorphy decreasing and ectomorphy increasing with age. Average mesomorphy was higher in these children than in Manus (Heath & Carter, 1971) or Prague children (Pařízková *et al.*, 1984). The authors pointed out that the relatively wide humeral and femoral epicondyles contribute disproportionately to the high mesomorphy ratings. This is an example of the questionable validity of anthropometric somatotype rating of young children.

Guedes (1983) somatotyped 360 school children (180 boys and 180 girls) aged 11 to 16 years in Londrina, Brazil (Table 4.4). The boys' somatotypes, which clustered around 2½-4-3½, became slightly less endomorphic with increasing age. The girls became much more endomorphic and less ectomorphic, with a mean of 3-3½-3 at 11 years and a mean of 4-3½-2½ at age 16. Sexual dimorphism was greater from ages 13 to 16 than at ages 11 and 12.

Toriola & Igbokwe's (1985) study of 625 students (325 boys, 300 girls) in secondary schools in Ile-Ife, Nigeria, provides the only available somatotype data on African children. They were Yoruba speaking children aged 10 to 18 years, of heterogeneous socio-economic status (Toriola, personal communication). From means close to 3-4-3 for both sexes at age 10 (Table 4.4, Fig. 4.28) the boys' somatotypes moved in a northwesterly direction and those for the girls in a southwesterly direction. Thus, sexual dimorphism increased with age. The Nigerian adolescents are slightly more endomorphic than most samples of the same ages presented in this chapter.

Instability of somatotypes in children

The foregoing data show that somatotypes of individual children are subject to significant changes during childhood and adolescence. It also appears that the physiques of some children are fairly stable over some

Fig. 4.29. Schematic representation of the general pathway of children's somatotypes from infancy to adolescence and into adulthood.

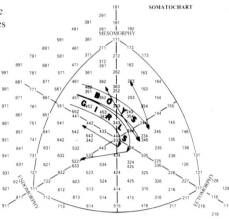

periods in their growth. Group means and distributions show that in general the somatotypes of children 2 to 6 years of age progress from endo-mesomorphy toward balanced mesomorphy for boys, and toward central somatotypes for girls (Fig. 4.29). Thereafter, the boys tend to decrease in mesomorphy and increase slightly in ectomorphy into mid-adolescence, when there is a dramatic reversal toward ecto-mesomorphy, balanced mesomorphy or endo-mesomorphy. In adolescence the girls tend toward decreasing mesomorphy, followed by increasing endomorphy, with means moving toward the central somatotype region and settling in the southwest and west of endo-mesomorphy and balanced endomorphy-mesomorphy. Individuals, of course, may differ greatly from the general trends. Moreover, the 18–21 year somatotype of a boy who increases from 4 to 5 in mesomorphy in late adolescence depends greatly on whether he increases in stature, and maintains, increases or decreases his endomorphy. For example, he may be a 2-4-4 who becomes a 3-5-2, 3-5-3, 4-5-2, 4-5-3, 2-5-2 or 2-5-3.

Although there are probably genetic factors in the development of endomorphy and mesomorphy, nutrition and exercise play important roles. In longitudinal studies, although somatotypic differences in adjacent years are often small, differences are evident at intervals of two years or more. Possibly the one-half unit rating scale is not fine enough to register subtle changes in childhood physiques. It may be that decimalized ratings used for anthropometric somatotypes reflect component trends and changes more smoothly.

Ethnic differences

Children in Manus are more mesomorphic than children in samples of Northern European populations. Boys in India are less mesomorphic and more ectomorphic than other known groups. There appear to be some ethnic differences within samples from European populations. Also there are small discrepancies which could be due to methodological differences, e.g. photoscopic versus anthropometric ratings, differences in anthropometric technique and use of a height correction for endomorphy with the anthropometric method.

Sexual dimorphism

There is obvious sexual dimorphism in somatotype distributions within studies of a single ethnic group. Although differences at ages 2 to 6 years are somewhat less than at ensuing ages, they are apparent from the earliest years. In general, it is evident that boys are more mesomorphic and less endomorphic than girls; these differences increase after adolescence. This form of sexual dimorphism is well known in adults. It appears that the somatotype reflects the general biological differences in shape and composition of boys and girls, as reported in Hall (1982).

Maturation and somatotype

Although there are a number of studies, there are no definitive data on early and late maturing somatotypes, fast and slow maturing somatotypes, or variability of individual somatotypes. Unfortunately, studies of these questions, confined almost exclusively to boys, have used different indicators of maturity, different somatotype methods and different kinds of analyses (Carter, 1980*b*). In this chapter, studies using the Heath–Carter method show that the somatotypes of many children change during adolescence, and often reverse component dominance more than once. It appears that the conclusions of other studies on this point are questionable, or, at best, equivocal.

It has been observed that in general somatotypes of boys increase in mesomorphy in adolescence (which is as expected from the growth in muscle that then occurs), while girls grow less mesomorphic and more endomorphic. This process extends over three or four years, probably longer in some individuals. Coaches and physical education teachers are aware that early maturing boys commonly out-perform their later-maturing peers in athletic ability and physical performance tests (Clarke, 1971).

Zuk (1958) found different maturation patterns for boys and girls at ages 12 and 17. Early maturing boys were slightly more mesomorphic than their peers. Early maturing girls were more endomorphic and less ectomorphic than their peers. But a follow-up at age 33 showed no somatotype differences between early and late maturers. In a summary of several samples of boys aged 9–16 Clarke (1971) showed that for ages 9 to 11 there were low positive correlations (high 0.20s and low 0.30s) of endomorphy with skeletal age; and that at ages 9 to 11 and at age 16 there were low negative correlations of ectomorphy with skeletal age. Except at age 9, mesomorphy correlated significantly (0.27 to 0.45) with skeletal age. He also found that when he separated the sample at age 15 into normal, retarded and advanced maturation groups, mesomorphy in the advanced group was higher than in the other two, with respective means of 4.6, 3.8 and 3.9. In addition, at age 17 the mean mesomorphy of the advanced group was significantly higher than that of the retarded group. The mean ectomorphy of the retarded maturity group at age 17 was significantly higher than that of the advanced group, with means of 3.7 and 2.9 respectively.

In a study of somatotype and skeletal maturity of 12-year-old boys, Borms *et al.* (1972) showed that compared to other somatotype groups a slightly higher percentage of dominant endomorphs matured early. In a more comprehensive study of Belgian children aged 6 to 13 years, Borms *et al.* (1977) found that although early maturers tended to higher endomorphy, it was not significant. They found that children of advanced and retarded skeletal ages did not differ in somatotype until age 11. From that time onward there was a tendency toward higher endomorphy in the skeletally advanced group.

In studies using somatotype methods other than that of Heath–Carter some investigators report that mesomorphs tend to mature earlier than others. Others report that endomorphs and endo-mesomorphs tend to mature early. However, irrespective of method, most studies find that late maturation occurs with high ectomorphy or meso-ectomorphy (Acheson & Dupertuis, 1957; Beunen, 1973–4; Beunen *et al.*, 1981*a*; Dupertuis & Michael, 1953; Hunt *et al.*, 1958; Livson & McNeill, 1962; Tanner & Whitehouse, 1982).

It is believed that hormonal control of growth is one mechanism associated with onset of adolescent maturation and changes in somatotype. In boys, increased secretion of androgens stimulates musculoskeletal development and increasing mesomorphy. It is also believed that increases in fatness or adiposity triggers early maturation. It is possible that both these mechanisms may be active concomitantly, which could lead to some confusion over whether endomorphy or mesomorphy is more closely related to advanced maturation.

Changes in adult somatotypes

Somatotypes of adults followed from adolescence into adult ages continue to change, as established in studies of Zuk (1958), Heath & Carter (1971), and Carter & Pařízková (1978). Men tend toward increases in mesomorphy and endomorphy. Women become more endomorphic. The male tendency toward increasing endo-mesomorphy is supported by the cross-sectional studies of Canadians (Bailey *et al.*, 1982), studies of British tunnel workers (King, personal communication), and studies of United States Federal Aviation Administration Trainees (see Chapter 3). In all these samples endomorphy increases by decade and mesomorphy increases or remains the same in adjacent decades. Newman (1952) found the same tendencies for endomorphy and mesomorphy in a somatotype study of approximately 40 000 individuals, photographed by the US Army and rated by Hooton and associates at Harvard University. Average endomorphy increased from about 4.1 to 4.8, mesomorphy increased from 3.6 to 4.1, and ectomorphy decreased from 4.6 to 3.8. Although the magnitude of the Hooton component ratings is likely to be different from Heath–Carter ratings, the pattern of values by age is similar. As Damon (1965*b*) pointed out, the principle of selective survival may well apply to cross-sectional samples. It may be that certain somatotypes are better adapted to survival. Or, it may be that in general the somatotypes found in the series of British tunnel workers, aviation trainees and the U.S. army are better suited to their occupations. Although at present there are no supportive data, occupational selection may be a factor in the Canadian sample.

Diet, physical activity and disease may be other factors affecting adult somatotypes. In studies of young and older groups of males undergoing selective physical training Carter & Phillips (1969) and Carter & Rahe (1975) reported decreases in endomorphy and increases in mesomorphy. Exercise and diet can also be used for maintaining stable somatotypes as well as for changing them. Studies of body builders demonstrate that markedly high levels of mesomorphy can be reached and sustained at least to the late thirties. Some people, especially those who are conscious of health and fitness, maintain the same somatotype with minor fluctuations for several decades through appropriate diet and exercise.

The studies discussed above show that prediction of adult somatotypes from childhood somatotypes is uncertain, and that there can be significant variation in adult somatotypes. Obviously prediction of a single variable like adult weight from a child's weight at a given age is risky. Foreseeing an adult somatotype from a given point in childhood is enormously more complicated. It would require, for example, accurate anticipation of the intensity and duration of the adolescent growth spurt, accompanying maturational changes in musculoskeletal structure and composition, and changes in fat deposition. At present we recognize, but cannot evaluate, genetic factors that influence growth patterns and the evolution of somatotypes. We know that environmental factors (including disease) may alter the somatotype. We also know that if environmental conditions are held more or less constant the adult somatotype may remain almost unchanged into old age, when degeneration of tissues alters shape and composition. Perhaps future research will identify some of the factors that cause changes in somatotypes.

Other studies of children's somatotypes

The choice of somatotype method greatly influences the findings in both longitudinal and cross-sectional studies of growth and maturation. The somatotype methods (described in Chapter 2) that allow for variation over time consistently support the concept of change in shape and/or composition (somatotype). The studies of Bodel (1950), Barton & Hunt (1962), Hunt & Barton (1959), and Claessens (1981) also support this model.

If either the Sheldon photoscopic or the Sheldon Trunk Index method is used, the stability of the somatotype cannot be questioned. As Walker (1978, p. 113) states:

> A person's somatotype never changes. One can make such a flat statement with confidence because it requires no empirical demonstration; it is built into the definition of somatotyping. (In just the same way a person's maximal height never changes. A value assigned as a maximum may later be superseded. But the earlier value is not then considered one of several alternative maximums; it is treated as an error of assignment.)

In practice, studies of the stability of somatotype, or prediction of adult somatotype, using Sheldon's methods, become exercises in the error of the method rather than tests of stability. Frequently the prediction error is similar to error between methods and within the same method. For example, Singh (1976) and Walker (1979) obtained different results with the trunk index method. Examples of studies which used a single adult somatotype (the genotype model) are: Parnell (1958), Livson & McNeill (1962), Petersen (1967), Walker (1978), Walker & Tanner (1980), and Tanner & Whitehouse (1982).

Summary

A synthesis of longitudinal and cross-sectional studies shows:

1. There are significant differences in mean somatotypes between age groups, increasing with length of intervals.
2. In general, boys at young ages move from endo-mesomorphy to ecto-mesomorphy and balanced ectomorphy-mesomorphy. During adolescence, with increased muscle mass and complete ossification, mesomorphy increases and ectomorphy decreases.
3. In general, girls, like boys, move from endo-mesomorphy and balanced endomorphy-mesomorphy toward central somatotypes. In adolescence and early maturity they move toward balanced endomorphy-mesomorphy and meso-endomorphy.

4. Often small between-age comparisons of mean somatoplots mask significant between-age changes in individuals. In studies which take into account individual changes (Sinclair, 1966, 1969; Pařízková & Carter, 1976; Carter & Pařízková, 1978; Heath & Carter, 1971; Kurimoto, 1963; Morton, 1967; and Bok & Tlapáková, 1982) large numbers of subjects changed somatotypes from year to year, and changes were greater at intervals of two years and more. Because of differences in direction, changes in some subjects cancel changes in others, resulting in an apparent stability of means.

5. Although we know that many children change somatotype, and that some children have relatively stable somatotypes, we cannot predict which subjects' somatotypes are stable or why. However, it appears that with relatively unchanging patterns of diet and exercise the more ectomorphic somatotypes are the most stable.

5   Genetics and somatotype

Introduction

Sheldon and his associates maintained that the somatotype was fixed for life, was a genetically determined entity, and saw their somatotype rating as an estimate of the 'morphogenotype' which was an extrapolated, unalterable concept of component ratings that had been fixed by the genes (Sheldon, 1949). This view fitted well with the long series of constitutional typologies, of which, as Hunt (1981) pointed out, Sheldon's contribution was the last major system.

Sheldon's somatotype method, his Trunk Index method, Parnell's M.4 method, and Petersen's method for children, are all based upon rating scales and concepts dependent on the assumption that the somatotype is fixed. In practice these methods concede that the physique changes its appearance, but insist that the same somatotype rating be given at all ages and irrespective of measurements and appearance. On the other hand, the regression equation approaches of Damon *et al.* (1962) and Munroe *et al.* (1969), as well as those of Heath (1963) and of the Heath–Carter adaptations (1966, 1967), and the modifications of Cureton (1947), Bullen & Hardy (1946), Danby (1953), Kraus (1951), Roberts & Bainbridge (1963) and Hooton (1959), all concentrate more on rating the physique as it is than on assuming genetic predetermination. Hunt (1952) maintained that it was safer to assume that 'body build photographs' depicted a transitory episode in an individual's somatic biography than to try to find body build ratings that are constant throughout life. Obviously any investigation of somatotype genetics will need careful attention to the method used.

There have been many opinions expressed but few investigations. The relationship between the genotype and the phenotype is a dynamic one (Dobzhansky, 1941), so that a phenotype must be the consequence of interactions between a certain genotype and a certain environment.

Hulse (1981) stressed the importance of biological plasticity during development and noted (1971) that present diversity in the human species is due partly to evolution, which has led to varying gene frequencies in different breeding populations, and partly to the effect of environment on the developing organism, so that gene action will lead to different consequences under different circumstances. For example, although the tendency

182

toward obesity may be genetically determined, genes alone do not make a person obese; he must also eat sufficient so that his energy intake exceeds his energy expenditure. One question to be asked then is 'In a given population in a given environment how much physique variation is attributable to genes, how much to environment and how much to interaction between them?'

At the next level of complexity, Dobzhansky (1967) asked whether human populations are polymorphic for genes or linked groups of genes that determine complexes of traits inherited as units. If Sheldon and his followers were correct that these complexes consist not only of morphological traits, but also of conjoined psychic traits, 'one could then imagine that there exist three gene complexes, forming multiple allelic series which determine respectively the so-called endomorphic, mesomorphic, and ectomorphic components.' (p. 15). Differences between populations would then derive to some extent from differences in frequency of such gene complexes. The second question to be asked then is 'To what extent are differences between populations in physique distribution due to gene frequency differences?' Little attempt has been made to answer this.

Heritability and somatotypes

To answer the first question, there are of course the well-known demonstrations of genetic control from pathology, in which simply inherited and chromosomal syndromes show characteristic physiques as in the Laurence–Moon–Bardet–Biedl syndrome with its obesity, or in Down's syndrome (Buday & Eiben, 1982). But these are rare, and most physique variation is continuous, so that methods of quantitative genetics have to be applied. Although available data are limited, there are a few studies estimating genetic influence upon physiques, in various stages of growth and aging, that contain useful information on heritability and somatotype.

Everybody knows that biologically related persons often have in common similar and measurable phenotypic characteristics. This is an expression of Fisher's basic genetic theorem that the degree of recognizable similarity between individuals is proportional to the number of genes they have in common. To measure the genetic contribution to such characteristics in behavioural and biological sciences, twin, sibling and parent–offspring studies are used to study these relationships. The first of these, of course, necessitates efficient zygosity diagnosis by adequate serogenetic investigation which, regrettably, has not always been done.

According to Bouchard *et al.* (1980*b*) a character that shows continuous variation is influenced by several relatively small sources of variation. Observed phenotypic variance is generally believed to consist of genetic and environmental factors and interaction and covariation between some of them; and all contribute to the observed phenotypic variance. Usually a general linear model is used to quantify the total observed phenotypic variance in a sample. This is partitioned as follows:

$$V_P = V_G + V_E + V_{G \times E} + 2 \operatorname{cov} GE + e.$$

In this equation,

V_P = the total phenotypic variance
V_G = the genetic contribution to the variance
V_E = the environmental component of the variance
$V_{G \times E}$ = the interaction effect between genotypes and environment
$2 \operatorname{cov} GE$ = correlated variations between phenotypes and environments
e = the random error component.

In practice it is usually assumed that the effects of the covariance and interaction of terms are negligible and that the error term is a random source uncorrelated with the parameters of the full model. However, it usually increases the size of V_P, and unless corrected reduces the size of genetic estimates accordingly. Heritability is estimated as V_G/V_P. In twin studies heritability is estimated in several ways – comparison of concordance in monozygotic and dizygotic twins, comparison of their intra-pair variances or of their correlations.

Twin studies

Osborne & De George (1959), in the first and probably classic somatotype study of twins, included 59 pairs of identical and 53 pairs of fraternal twins, all American White, on whom Sheldon made the somatotype ratings. They compared the variances in monozygotic and dizygotic pairs and controls for each component and for overall somatotype. Though the intra-pair variances for dizygotic are consistently greater than monozygotic twins in each case, they only are significantly so in ectomorphy in

Table 5.1. *Heritability estimates based on variance (Hv) for somatotype components and ponderal index (HWR) as calculated by Bouchard (1977)*

Sex	Coefficient	Endo- morphy	Meso- morphy	Ecto- morphy	Somato- type	HWR
M	Hv	0.45	0.33	0.18	0.36	0.33
F	Hv	0.44	0.24	0.74	0.61	0.55

From Osborne & De George, 1959.

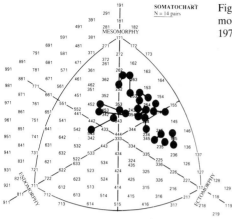

Fig. 5.1. Somatotypes of Prague monozygotic twins. (Redrawn from Kovář, 1977.)

females and total somatotype in females. The authors found no genetic basis to specific combinations of components, although gross combinations of different somatotypes showed significant differences between fraternal and identical twin pairs. Converted to heritabilities by Bouchard (1977), the estimates in the HWR and total somatotype are almost twice as high for females as for males and considerably more in ectomorphy (see Table 5.1 for the obtained heritability coefficients).

In Prague (Czechoslovakia), Kovář (1977) analysed the somatotypes (Heath–Carter anthropometric method) of fourteen pairs of identical and ten pairs of fraternal twin boys at ages 11 to 23. The somatocharts (Figs. 5.1, 5.2) show that the distances between the pairs are smaller for the monozygotic than for the dizygotic twins. The average difference between components for the identical twins was 0.21 for endomorphy, 0.25 for mesomorphy, and 0.28 for ectomorphy. For the fraternal twins the differences were 0.75 for

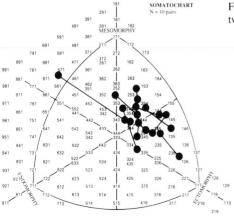

Fig. 5.2. Somatotypes of Prague dizygotic twins. (Redrawn from Kovář, 1977.)

Table 5.2. *Heritability index (H) for somatotype components using coefficients of intraclass correlation and interpair variance*

	Somatotype Component		
Coefficient	Endomorphy	Mesomorphy	Ectomorphy
H (correlation)	0.69	0.88	0.87
H (variance)	0.77	0.90	0.89

From Kovář, 1977.

endomorphy, 0.90 for mesomorphy, and 1.1 for ectomorphy. Note in Fig. 5.2 that one pair appears to influence these mean differences more than the others. Without this pair the mean differences (N = 9) for the three components are 0.56, 0.67, and 0.94. The component correlations between the identical twins are 0.83, 0.90, and 0.90. For the fraternal twins the correlations are much lower: 0.44, 0.15, and 0.22. Intra-pair variances differed significantly between the two twin groups for all three components. The heritability ranged from 0.69 to 0.90 (Table 5.2), relatively high by both the correlation and the variance methods. According to Table 5.2 heredity plays a greater part in ectomorphy and mesomorphy than in endomorphy.

In a review of two Polish somatotype studies of twins, Chovanová *et al.* (1981, 1982a) reported that Orczykowska-Świątkowska *et al.* (1978), using Welon's modification of Parnell's method, found the highest intra-pair correlations for endomorphy for both boys and girls. In a study of Sheldon somatotypes of boys and girls, Skibińska & Sklad (1979) found that mesomorphy, ectomorphy and endomorphy were influenced in descending order by heredity. Endomorphy was more strongly influenced by environment in girls than in boys, and mesomorphy and ectomorphy were more influenced by heredity in girls than in boys. They found that the somatotype as a whole was genetically determined to a greater degree in girls than in boys.

Chovanová and colleagues designed their own study of monozygotic and dizygotic twins of both sexes to investigate the relative influence of heredity and environment on the variability of components and whole somatotypes. They used the Heath–Carter anthropometric method, with a height correction for endomorphy. This was the first twin study to use somatotype dispersion distance (SDD) to evaluate the difference between somatotypes as a whole (Table 5.3).

The monozygotic group included 29 male and 24 female pairs; and the dizygotic group included 28 male and 15 female pairs. The ages ranged from 11.7 to 18.8 years, with most around 15 to 16 years. The male monozygotes were more similar in somatotype than the dizygotes. Intra-pair correlations were high (r = 0.86 to 0.89) for monozygotes and low for dizygotes (r = 0.12

Table 5.3. *Comparison of intra-pair correlations for somatotype components of dizygotic (DZ) and monozygotic (MZ) twins, boys*

Authors	Endomorphy			Mesomorphy			Ectomorphy		
	Kovář 1977	Orczykowska et al. 1978	Chovanová et al. 1982a	Kovář 1977	Orczykowska et al. 1978	Chovanová et al. 1982a	Kovář 1977	Orczykowska et al. 1978	Chovanová et al. 1982a
rMZ	0.83	0.85	0.86	0.90	0.83	0.89	0.90	0.83	0.88
rDZ	0.44	0.29	0.38	0.15	0.49	0.12	0.22	0.44	0.14

From Chovanová *et al.*, 1982a.

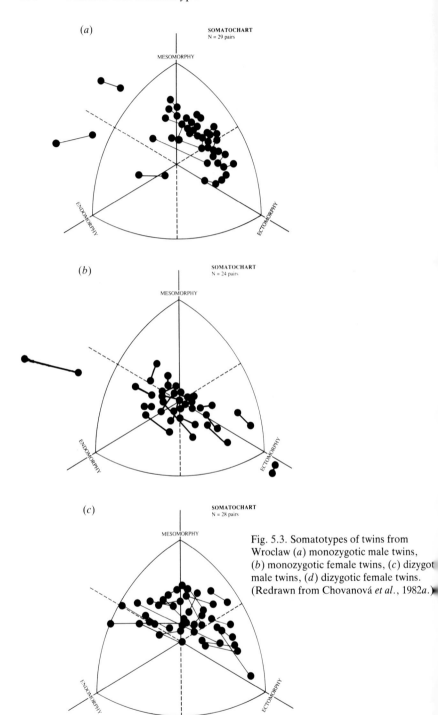

Fig. 5.3. Somatotypes of twins from Wroclaw (*a*) monozygotic male twins, (*b*) monozygotic female twins, (*c*) dizygotic male twins, (*d*) dizygotic female twins. (Redrawn from Chovanová *et al*., 1982*a*.)

Fig. 5.3 (*Continued*)   (*d*)

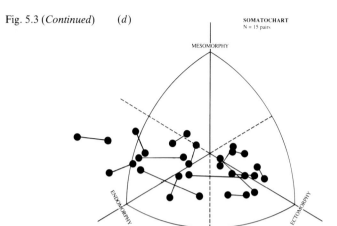

to 0.38). On transformed Z correlations there were significant differences between monozygotic and dizygotic pairs (Figs. 5.3, 5.4). There were greater intra-pair variances for somatotypes and SDDs. According to the heritability estimates male mesomorphy and ectomorphy were most influenced genetically, and endomorphy the least influenced. The variance heritability index was Hv = 0.78 for the SDD.

Female monozygotes were more alike somatotypically than dizygotes.

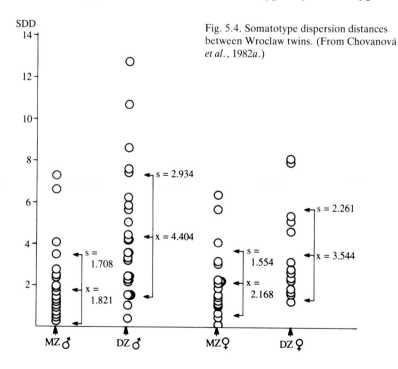

Fig. 5.4. Somatotype dispersion distances between Wroclaw twins. (From Chovanová *et al.*, 1982*a*.)

Intra-pair correlations were high (r = 0.77 to 0.99 for monozygotes, and r = 0.77 to 0.83 for dizygotes). When transformed correlations were compared, only ectomorphy differed. There were no differences in intra-pair variances. For the three indices of heritability ectomorphy showed the greatest genetic influence; endomorphy showed slight influence and mesomorphy showed none. For the somatotype as a whole (SDD), Hv = 0.60.

In general, male twins had greater heritability indices than female twins. The authors found intra-pair correlations for males in their study were similar to those of Kovář (1977) and Orczykowska-Światkowska *et al.* (1978). They found intra-pair correlations for female monozygotes similar to those in the Orczykowska-Światkowska study, and intra-pair correlations for the female dizygotes approximately 50% higher. The indices of heritability were similar for boys across studies. For girls the indices of heritability were more variable than those of Orczykowska-Światkowska *et al.* (1978) and of Skibińska & Sklad (1979). The Kovář and the Chovanová studies alone used the same somatotype method.

Family studies – parents, offspring, and siblings

Parnell (1958) used his M.4 method, with 11 year old standards for children and M.4 adult standards for adults, in a somatotype study of 45 families with 121 children (N = 63 boys, N = 58 girls). He gave no ages for parents and children. He drew a line between the somatotypes of the parents and a perpendicular from this line to the somatotype of the child (the 'parental line principle') to measure the similarity between parent and child somatotypes. He found that between 60% and 75% of the children (depending upon the strictness of the criterion) were close to the line between the parents. This indicates somewhat better than chance relationships between a child's somatotype and that of his parents. These relationships may not hold at other ages, because there was a different rating scale for children and age-adjusted scales for parents.

Withers (1964) applied Parnell somatotype ratings to 125 parent–child pairs, and reported correlations between father–son pairs (N = 10) for mesomorphy, and correlations between mother–son pairs for endomorphy. The author suggested, despite the small samples, that the father influenced his daughter's mesomorphy but not his son's ectomorphy. He also concluded that the mother contributed to her son's endomorphy.

Fisher (1975) investigated the differences in selected physical and psychological variables between 30 pairs of grandmothers (58 to 80 years) and their granddaughters at age 15. The mean Heath–Carter anthropometric somatotypes were 5.5-5.2-2.1 for the grandmothers and 3.6-3.3-3.2 for the granddaughters. Although the study did not assess heritability, it is obvious that the means were far apart.

Bouchard (1977) and Bouchard *et al.* (1980*a*) selected 208 sibling pairs from a sample of 239 Montreal French Canadian families with 415 children in a study of heritability of somatotype components. They used Heath–Carter anthropometric criteria for the children's somatotypes and calculated the ectomorphy of the parents from their HWRs. The average age of the parents was 44 years and the selected children were 10 years old. The mean somatotype for the children (N = 415) was 2.1-4.6-3.3, with standard deviations of 0.99, 0.92, and 1.07. By partial regression the authors examined seven socioeconomic indicators to account for the effects of environment common to members of the family. The overall sibling correlations were 0.40 for endomorphy, 0.30 for mesomorphy, and 0.38 for ectomorphy, but when the seven socioeconomic indicators were taken into account the correlations were reduced to 0.25, 0.21, and 0.27. Residual sibling correlations yielded broad heritability estimates (H_B) of 0.50 for endomorphy,

(*a*)

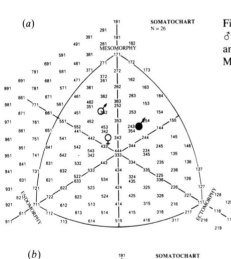

Fig. 5.5. Mean somatotypes for (*a*) parents ♂ and ♀ and sons, ♂, and (*b*) parents ♂ and ♀ and daughters ♀. (Redrawn from Medeková & Havlíček, 1982.)

(*b*)

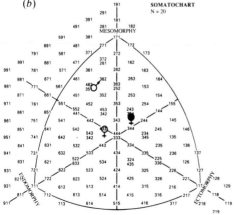

0.42 for mesomorphy and 0.54 for ectomorphy. Narrow heritability (H_N) for ectomorphy, controlling for familial indicators, was approximately 0.36, including a positive contribution from assortative mating. The authors pointed out that the much higher heritability estimates (Hv) reported by Kovář (1977) were probably inflated, due in part to the small sample size. Bouchard *et al.* (1980*a*) concluded that somatotype components are potentially influenced by non-genetic factors, and that the genetic contribution to individual physique components is only moderate, since H_B estimates tend to cluster round r = 0.50 when the sample size is adequate and there is some control over environmental influences in the analysis.

Medeková & Havlíček (1982) used Heath–Carter anthropometric somatotypes in a study of 29 Slovak families with 22 female and 27 male children. The average age of the parents was early thirties, and of the children, 7.3 years. It appears that the mothers were somatotypically closer to their children than the fathers, and closer to their daughters than to their sons. Correlations for ectomorphy were r = 0.47 between fathers and sons and r = 0.51 for mothers and daughters, the only significant correlations. (See Figs. 5.5 and 5.6 for SDD of children from their parents, and Fig. 5.7 for parent–child correlations.)

Bok (1981), in a study of the Heath–Carter anthropometric somatotypes of parents (N = 117) and groups of their children from ages 15 to 17, found smaller differences between the somatotypes of parents than between

Fig. 5.6. Somatotype dispersion distances (SDDs) between parents and children. (From Medeková & Havlíček, 1982.)

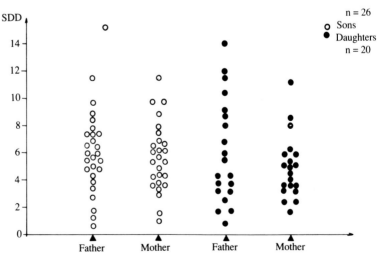

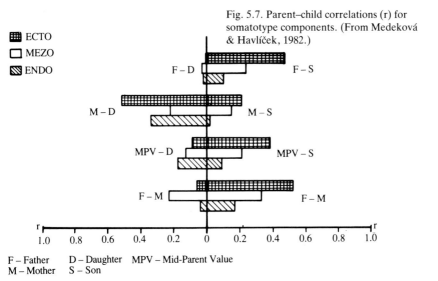

Fig. 5.7. Parent–child correlations (r) for somatotype components. (From Medeková & Havlíček, 1982.)

ECTO
MEZO
ENDO

F – Father   D – Daughter   MPV – Mid-Parent Value
M – Mother   S – Son

parents and children. He concluded that children's somatotypes could not be predicted from their parents' somatotypes.

Longitudinal studies

Useful information on the extent of genetic influence comes from investigation of environmental effects. There are some longitudinal studies of the influences of growth, diet and exercise on the stability of somatotype, but no longitudinal somatotype studies of families for comparisons.

Studies of longitudinal stability of children's somatotypes, using Heath–Carter phenotypical ratings, were carried out by Zuk (1958), Kurimoto (1963), Sinclair (1966), Heath & Carter (1971), Pařízková & Carter (1976) and Carter & Pařízková (1978). In general these studies showed only moderate stability for somatotype components during adolescence and slightly greater stability before adolescence. There were changes in ratings and component dominance for many subjects, and fairly stable ratings for a few subjects. As might be anticipated, correlations between adjacent years were higher for each component and lower when four or more years apart. Bodel (1950), and Hunt & Barton (1959), using different methods (Hooton's and Hunt's methods respectively), reported greater changes than these studies. (See Chapter 4 for greater details on these and other studies.)

But in children change with growth has to be discounted, for during growth we expect changes in body shape and composition and corresponding changes in the somatotype. It is likely that some difference in degree of change is due to variations in nutrition and physical activity. Clarke (1971) and Ross & Day (1972) concluded that exercise played a major role in the development of somatotypic differences in the success of young athletes. On the other hand, Pařízková & Carter (1976) found no clear evidence of the relationship of somatotype and physical activity levels of adolescent boys, but noted the problem of quantification and control of activity categories.

The confounding effect of growth is avoided in adult studies of environmental change provided that these are carried out over a short period at ages where any aging effect is minimal. Carter & Rahe (1975) demonstrated adult changes in somatotype through exercise in a group of young men, and Carter & Phillips (1969) also found changes in middle-aged men. As anticipated, the changes were toward lower endomorphy and higher mesomorphy for the younger men, and lower endomorphy only for the older men.

The effects of drastic nutritional deprivation on somatotype were investigated as part of the Minnesota starvation experiment begun in 1944 under the direction of Ancel Keys. Lasker (1947) examined the effect of partial starvation on somatotypes of 34 male volunteer conscientious objectors, who were subjected to a 'European type' of famine diet for 24 weeks – a diet designed to simulate that of prison camps in Europe during World War II. Somatotype photographs and measurements were taken at the beginning and end of the experiment. The photographs were rated by several somatotype methods. Two groups of measurements, taken from the photographs, used the methods described by Sheldon (1940), and the methods of Bullen and Hardy were also used. Also, James Andrews with several co-workers applied the observational criteria method of Hooton and averaged the ratings, and Lasker too applied Hooton's observational criteria (Hooton, 1946).

Body weight of the 34 subjects decreased by 24%. Two photometric somatotype ratings showed that endomorphy decreased approximately by 48%, mesomorphy decreased approximately by 43% and ectomorphy increased approximately by 78%. The two observational rating methods showed average decreases of 48% in endomorphy, 26% in mesomorphy, and 65% increases in ectomorphy. Averaging all four methods, the somatotype before diet was 3.4-3.8-3.3, and after the diet 1.8-2.5-5.7. Lasker observed that there was considerable individual variation, and that the somatotypes of all subjects changed markedly. The most endomorphic subjects' losses were greatest in endomorphy; the most mesomorphic subjects' losses were greatest in mesomorphy; and the most ectomorphic subjects seemed to change the least. The photometric observations showed the least change in measurements of wrists, hands and feet (areas of the least soft tissue). Lasker (1947) commented that 'somatotyping serves better as a measure of nutritional status than as a measure of inherent tendencies to specific constitutional types'. It is reasonable to infer that the observed changes would be reversed by return to the pre-experiment diet, and probably with specified exercise to assure return to normal muscle mass.

Heritability of tissue

It is generally inferred from long observation of close relatives that many aspects of body form are heritable, although environmentally modifiable. General patterns like leg and hand structure, tooth structure and many other body characteristics are recognizably similar in close relatives, although not predictable and fixed in the same way as single gene characters. There have been extensive studies of anthropometric characteristics, especially of the bony dimensions of the head and face; and the studies of heritability in major body tissues, like fat or adipose tissue, muscle and bone, as well as height and weight, are of relevance to that of the somatotype. For the somatotype is a *gestalt* assessment of body shape and form, which in turn depends on the tissues or compartments mentioned. If heritability for some tissues is higher than for others we would expect some aspects of the somatotype to be more stable than others.

In a summary of findings in twin studies, Kovář (1977) reported that on the average somatic features are more heritable than the majority of functional characteristics of the organism. He found that the established indices for heritability ranged from 0.73 to 0.98 for body height; 0.54 to 0.94 for body weight; 0.59 to 0.89 for trunk length; 0.57 to 0.86 for limb length; 0.53 to 0.93 for circumferences and breadths; 0.54 to 0.88 for skinfold thicknesses. The studies also indicated that hereditary influences are disproportionately higher for the fat-free component of body composition than for the fat component.

When he reviewed studies of the genetics of skeletal lengths, breadths, muscularity and fatness, Bouchard (1977) found that in twin studies broad heritability estimates of skeletal lengths ranged from 0.60 to 0.97, with a mean of 0.80. The composite indicators of length tended to be higher than segmental lengths. In most parent–child and full sib studies for body and segmental lengths average correlations were 0.30 to 0.50. In twin studies of skeletal breadths heritability coefficients for biacromial and biiliac breadths were approximately 0.64 and 0.60; coefficients for wrist and ankle breadth were approximately 0.80; and bicondylar humerus and femur breadths were 0.71 and 0.60. In several parent–child and full sib studies for skeletal breadths (biacromial, biiliac, biepicondylar humerus, biepicondylar femur) correlations averaged from 0.24 to 0.49 for various combinations.

Bouchard found none of the data amenable for analysis in tests for age-associated effects on skeletal lengths and breadths. He found that in samples of subjects living in relatively favourable socioeconomic conditions genetic effects in skeletal breadths were less than those in skeletal lengths.

Bouchard (1977) found that the few studies reporting on the genetics of human muscle tissue suggested that some of the variations in muscularity are genetic. Available studies consisted of small numbers over a wide age range with no control over environmental conditions. Some studies examined gross muscular size examined through cross sections or diameters; others examined ultra-structural muscle components. Studies using anthropometric indicators of muscle size appeared to identify a substantial non-genetic contribution to lean tissues. Broad heritability estimates for selected skinfolds in twins ranged from 0.30 to 0.77, with a mean of 0.55. The coefficients for parent–child and full sib studies averaged from 0.11 to 0.45.

In addition to strong environmental effects, there is apparent some genetic control of fat patterning and some differences in fat deposition. Bouchard (1977) found that in studies of breadths and lengths of bones, of skinfold thicknesses, and of height and weight, in Montreal families, heritability estimates for parent–child and reduced full sib correlations were lower when socioeconomic indicators were controlled for. Orvanová's (1984) conclusions resembled Bouchard's (1977) when she reviewed the heritability of anthropometric variables and somatotype from a wide range of studies. In general, heritability was higher for heights and lengths than for measures of widths and girths, and in most cases higher than fat tissue or skinfolds. The heritability of somatotype components varied somewhat between sexes. In general ectomorphy was more heritable than endomorphy and mesomorphy. Orvanová concluded that although widths and girths make some contribution to heritability, environment affects mesomorphy to a minor degree but affects endomorphy significantly.

Difference in somatotype methods and sizes and ages of samples limited the inferences to be drawn from studies of relationships between somatotypes of parents and children, siblings and twins. No studies but Bouchard's accounted for socioeconomic differences within samples.

Genotype and phenotype

Sheldon's rigid adherence to his concept of the unchangeable genetically determined somatotype discouraged studies designed to scrutinize the heritability of somatotype.

The discussions following Sheldon's (1951) paper, 'The somatotype, the morphophenotype, and the morphogenotype', at the Cold Springs Harbor Symposium on Quantitative Biology emphasized the problems raised by Sheldon's fiat. Eminent biologists and anthropologists (e.g., Boyd, Buzati, Dunne, Hunt, King, Montagu, Traverso, Warren and Washburn) seriously questioned the legitimacy of the concept of the morphogenotype. Twin studies were among the approaches suggested for clearing up the issue of somatotype heritability. It was also proposed that somatotype variations be observed among persons with recognizable single gene characters. The consensus among the discussants was that Sheldon begged the question of evidence for his claim of a genetic basis for somatotypes. Near the end of the discussion Sheldon made this slight concession: 'Causes and events – genotypes and phenotypes – lie in a continuum, and I'm afraid that there may be no way of cogently studying the genesis of the genotype without at the same time systematically studying the phenotype' (Sheldon, p. 381, 1951). Nonetheless he neither attempted a systematic study of the 'genesis of the genotype' nor encouraged others to do so.

Summary

It is apparent that the somatotype is the outcome of interaction between genotype and phenotype, but there has been relatively little investigation of the extent of genetic influence on normal somatotype variation within a population. Estimates from the several twin and family investigations that have been made vary from one study to another but cluster about moderate heritability levels, while experimental and comparative studies indicate the type of environmental variable that contributes to non-genetic variation.

6   *Sport and physical performance*

Introduction

The characteristics of physique apparently associated with success in sports and other forms of physical performance have always greatly interested scientists, artists, writers and lay persons. They have found that what is easily observed is difficult to quantify. As a result there are few theoretical or experimental studies relating physical characteristics and performance.

The number of studies using measurements or morphology of athletes has accelerated rapidly in the past century. These have focused on descriptions of athletes, comparisons of athletes between and within sports, relationships of physique to physiology and sport performance, or 'sport anthropometry'. Tittel & Wutscherk (1972) cited over one hundred such studies. In the past fifteen years there have been at least a hundred more. There have been fourteen major books and monographs reporting studies of Olympic athletes. The earliest is Knoll (1928), and the most recent is Carter (1984*a*). Borms & Hebbelinck (1984) published a comprehensive review of these.

Sargent (1887) used extensive anthropometry and functional tests on large numbers of Harvard University students. He observed that development of athletes was 'governed largely by the constitutional bias of the individual, the sport in which he is engaged, and the time devoted to it.' (p. 541). Although the calibre of athletic performance is higher today, Sargent asked many of the questions we are now asking. Can we predict outstanding athletic ability from body structure? Has a given athlete the physique best suited to his or her sport? Can special training alter the physique, or somatotype? Can we predict the adult somatotype from measurements and tests on the pre-adult athlete? Are the somatotypes of young athletes similar to successful Olympic athletes in given sports? What specific characteristics of physique are important to success in various kinds of physical performance? Although Sargent's descriptions of some physical characteristics, made a hundred years ago, are remarkably similar to contemporary findings, the early studies were largely descriptive, without substantive statistical comparisons.

Direct experimental study relating physical structure to level of performance is discouraged because it is difficult, if not impossible, to avoid

198

interfering with the development and training of the athlete. It appears that anatomical and biomechanical modelling of physique changes and performance may be more appropriate, and that future training may attempt to match the model.

Up to the 1940s most studies of physique and physical performance were based on selected anthropometric dimensions and ratios. Somatotyping introduced a promising method for investigating relationships between physical structure as a whole and performance in sports and tests of strength, speed and endurance. The Heath–Carter somatotype method shifted emphasis from the static or typological model of physique to the dynamic, which allowed research to take advantage of anatomical and biomechanical models of physique changes and performance. Moreover, the somatotype provides a physique summary that is more useful than lists of separate measurements or multivariate equations.

Somatotype and success in sport

The following pages present somatotype results derived from analysis of data collected in a large number of sports. The studies reviewed here include several levels of athletic competition and show that the nature and level of performance influence the degrees of association with somatotypes. As expected, studies of high-level performers in national and international competition yield the clearest information. In theory we expect to find that the most successful athletes have physical structures best suited to their particular sport, and that differences in physique will emphasize the importance of aspects of physique such as somatotype. The general hypothesis is, as Tanner (1964) said, that without the required physique an athlete is unlikely to reach a high level of success.

Anthropometry of growing children, which has been traditional for monitoring normal and abnormal growth patterns, has been extended to include young athletes, whose numbers are greatly increased with the growing emphasis on youth sports in many countries. The precocity of many athletes in sports like female gymnastics and swimming raises questions about the effects of training on growth and maturation. There are data suggesting there are different rates of maturation for different somatotypes and the somatotype changes through childhood and adolescence. Comparison of somatotypes of young athletes with those of Olympic athletes in the same sport helps to define the patterns of change.

Data for athletes shed light on limitations to human performance due to morphological characteristics like absolute and proportional size, body composition, maturation and somatotype. Somatotype quantification may shed light on performance and improve understanding of biomechanical limitations and the physiology of performance. Physical educators and coaches may find these insights helpful in making effective choices for their students and young athletes. It is important to extend the options to the sport that most closely fits the somatotype and to the training methods appropriate for a given somatotype in a given sport.

In a review of the somatotypes of athletes, Carter (1970) hypothesized that somatotype was an important selective factor for success in championship performance. He concluded that the higher the level of competition the narrower the distribution of somatotypes within a given sport, that there are clear somatotypic differences between some sports, and similarities between others. Since then a variety of studies have supported and amplified these conclusions.

It is appropriate to give an overview of the somatotype distributions in a variety of national and international sports competitions before reviewing the present knowledge of somatotypes associated with each sport. The following material reviews the studies from Olympic Games and other international competitions, from national studies, and comments on studies of adolescent and child athletes. Unless otherwise stated, the Heath–Carter somatotype method was used in all studies.

Olympic Games and other international studies

In the first somatotype study of Olympic athletes, Cureton (1951) (using his own method) studied swimmers and track and field athletes at the 1948 London Olympics. Tanner (1964) used Sheldon's (1954) method to somatotype athletes from the 1958 British Empire Games in Cardiff, Wales and from the 1960 Rome Olympics. The studies of De Garay *et al.* (1974) at the 1968 Mexico Olympics, and Carter *et al.* (1982) at the Montreal Olympics were the two most extensive. These studies included samples of both male and female athletes, medallists, finalists and less successful competitors from 23 sports. In Carter's (1984*b*) summary of data on Olympic athletes from 1948 to 1976, Heath–Carter ratings were applied to Cureton's and Tanner's studies to make them comparable with the 1968 and 1976 studies.

The somatotype distributions (Fig. 6.1) of athletes for all sports in the 1968 and 1976 Olympics show that the male somatotypes are concentrated around 2-5-2½ and the females around 3-4-3. Fifty per cent of both sexes are near these means. The whole distribution forms an ellipse, running a NW–SE direction. The majority of the males are dominantly mesomorphic and more mesomorphic and less ectomorphic than the females. Comparisons of athletes of both sexes in each Olympiad with reference groups (see Chapter 3, Figs. 3.6, 3.25, 3.32, 3.45) showed that the athletes were more

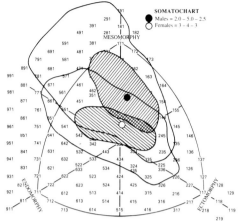

Fig. 6.1. Somatotype distribution of male (upper areas) and female (lower areas) athletes from combined 1968 and 1976 Olympic samples. The shaded areas enclose approximately 50% of somatotypes closest to their respective means. ● = males; ○ = females. (From Carter, 1984*b*.)

mesomorphic and less endomorphic (De Garay *et al.*, 1974; Carter *et al.*, 1982).

Means for male Olympians show greater somatotypic variation among sports than those of females, who do not participate in several of the sports (Figs. 6.2, 6.3). Although the means for male athletes are near 2-5-2$\frac{1}{2}$, mesomorphy is exceptionally high in upper weight classes in judo, weightlifting and wrestling, and comparatively lower in sports like basketball and fencing. Because of large event differences, track and field athletes are not included on the charts noted above.

In her study of athletes in the IXth Bolivarian Youth Games in Barquisimeto, Venezuela, in December 1981, Brief (1986) somatotyped 139

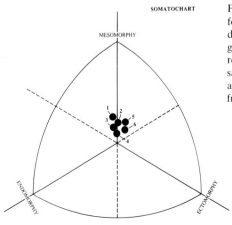

Fig. 6.2. Mean somatotypes for selected female Olympic sports. 1 = canoeing; 2 = diving; 3 = rowing; 4 = swimming; 5 = gymnastics; 6 = track and field. Plots represent the combined 1968 and 1976 samples. Note the concentration of means around the 3-4-3 somatotype. (Redrawn from Carter, 1984*b*.)

Fig. 6.3. Mean somatotypes for selected male Olympic sports. 1 = weight lifting; 2 = judo; 3 = wrestling; 4 = modern pentathlon; 5 = rowing; 6 = waterpolo; 7 = field hockey; 8 = fencing; 9 = gymnastics; 10 = canoeing; 11 = diving; 12 = boxing; 13 = swimming; 14 = cycling; 15 = basketball. Plots represent the combined 1968 and 1976 samples. Note the concentration of means around the 2-5-2½ somatotype. (Redrawn from Carter, 1984*b*.)

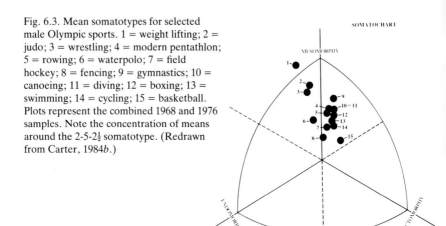

males and 60 females from Bolivia, Columbia, Ecuador, Panama, Peru and Venezuela (Fig. 6.4). The overall mean for males in 10 sports was 2-5-2½, the same as for Olympic athletes; and the means for individual sports were also close to the corresponding Olympic means. The overall mean for the six female sports was 3½-4-2½, with a somatochart distribution to the left of centre. The female means for four of the six sports centered around 4-4-2; and the means for the other two (track and gymnastics) were more ectomorphic and less endomorphic. There were significant differences between sports for both sexes. The athletes also differed from the urban Mexicans in De Garay *et al.* (1974).

The above studies are from the 'summer' Olympics. Chovanová (1976*b*) made the only study of international Alpine winter sportsmen. She included the somatotypes of 143 European athletes competing in downhill, cross-country, combined and jumping events in Czechoslovakia during 1970–71.

Fig. 6.4. Mean somatotypes of male and female athletes competing in the 1981 Bolivarian Youth Games. m = male athletes. m1 = boxing; m2 = judo; m3 = basketball; m4 = cycling; m5 = fencing; m6 = football; m7 = swimming; m8 = track and field; m9 = volleyball. f = female athletes. f1 = basketball; f2 = fencing; f3 = gymnastics; f4 = judo; f5 = swimming; f6 = track and field. (Redrawn from Brief, 1986.)

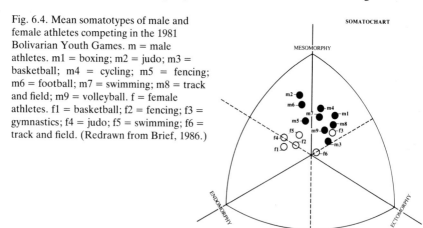

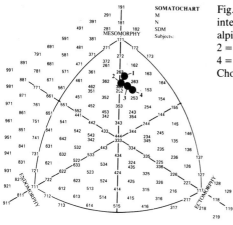

Fig. 6.5. Mean somatotypes of an international sample of male athletes in four alpine winter sports. 1 = ski jumpers; 2 = downhill; 3 = nordic combined; 4 = cross country. (Redrawn from Chovanová, 1976*b*.)

The means by event (Fig. 6.5) cluster around 2-5$\frac{1}{2}$-2$\frac{1}{2}$, which is slightly higher than the means for many other Olympic sports. Their distributions by event or sport are within the areas for summer Olympians, a combination of endo-mesomorphs, ecto-mesomorphs and balanced mesomorphs. Four of the athletes had ectomorphy higher than mesomorphy; all were low in endo-morphy, averaging near 2.

All of the above studies of international athletes show somatotype differences among sports for both sexes, despite mixtures in race/ethnicity and selection factors.

National studies

The frequent national differences in measurement techniques, race/ethnicity of samples, socio-economic status, selection and training methods present problems in comparing somatotypes by sport. The Olympic samples in particular consist of mixtures of racial and ethnic groups. Consideration of a variety of selection methods, involving complicated socio-economic and political criteria, weigh heavily in deciding which athletes to measure (de Garay *et al.*, 1974; Mechikoff & Francis, 1984).

Within a given country a single investigator or team usually has reasonable control over these factors. Within some countries athletes are likely to be racio-ethnically fairly homogeneous, although some national samples are heterogeneous. As in Olympic samples, in national samples there are consistent somatotypic differences between sports and consistent similarities with many levels in some sports. Some investigators have inferred that between-nation somatotypic differences within sports may be due in part to levels of development and even availability of appropriate somatotypes in some sports in a country. Empirically, it appears that there are few

Fig. 6.6. Mean somatotypes for male ● and female ■ Czechoslovak national level athletes. Males: 1 = gymnastics; 2 = weight lifting; 3 = bodybuilding; 4 = skiing, cross-country; 5 = skiing, downhill; 6 = football (soccer); 7 = volleyball; 8 = basketball; 9 = handball; 10 = ice hockey; 11 = track and field, sprinting; 12 = track and field, 400 m; 13 = 800 m, 1500 m; 14 = high jump; 15 = long jump; 16 = shot put; 17 = hammer throw; 18 = canoeing, 19 = wrestling; 20 = cycling. Females: 21 = modern gymnastics; 22 = handball (1967 and 1977); 23 = sprinting; 24 = hurdling, 100 m; 25 = orienteering; 26 = skiing, downhill. (Male means from Štěpnička, 1977; female means from Štěpnička *et al.*, 1979a.)

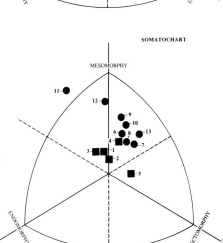

Fig. 6.7. Mean somatotypes for male and female Venezuelan national level athletes. Females: 1 = swimming; 2 = basketball; 3 = volleyball; 4 = gymnastics; 5 = track sprinting. Males: 6 = swimming; 7 = basketball; 8 = volleyball; 9 = gymnastics; 10 = track sprinting; 11 = weight lifting – light; 12 = weight lifting – heavy; 13 = track, 400 m. ● = male and ■ = female. (Redrawn from Pérez, 1981.)

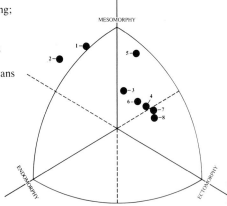

Fig. 6.8. Mean somatotypes for Chinese national level sportsmen. 1 = weight lifting; 2 = shot, discus, hammer throw; 3 = swimming; 4 = sprinting; 5 = gymnastics; 6 = long, triple jumping; 7 = middle, long distance; 8 = high jumping, pole vaulting. (Plotted from means in Zeng, 1985.)

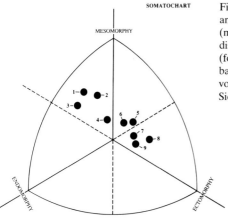

Fig. 6.9. Mean somatotypes for Indian state and national level sportsmen. 1 = wrestling (middle); 2 = weight lifting; 3 = shot, discus, hammer throw; 4 = field hockey (forward); 5 = cycling; 6 = sprinting; 7 = basketball; 8 = 5, 10 km running; 9 = volleyball. (Plotted from means in Sodhi & Sidhu, 1984.)

significant somatotypic differences between countries when the level, duration and intensity of training and appropriate somatotypes are similar. That is, it appears that biomechanical, physiological and technique demands of a particular sport limit the range of physiques that can satisfy the demands.

The national somatotypic similarities and differences are shown in the following studies: for Czechoslovakia, Štěpnička (1974a) and Štěpnička *et al.* (1979a); for Venezuela, Pérez (1981); for Brazil, Guimarães & De Rose (1980); for India, Sodhi & Sidhu (1984); for the USA, Carter (1970, 1971) and Thorland *et al.* (1981); for Hungary, Farmosi (1980b, 1982); for England, Bale (1983, 1986); for Nigeria, Toriola *et al.* (1985); for Australia, Withers *et al.* (1986, 1987); for the USSR, Heath (in Carter, 1970); for France, Boennec *et al.* (1980); for Korea, Shin (1985); for China, Zeng (1985) and for Cuba, Rodríguez *et al.* (1986).

Somatocharts (Figs. 6.6 to 6.10) show the means for Czechoslovak master

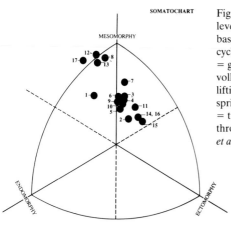

Fig. 6.10. Mean somatotypes for national level Cuban sportsmen, 1976–80. 1 = baseball; 2 = basketball; 3 = boxing; 4 = cycling; 5 = fencing; 6 = football (soccer); 7 = gymnastics; 8 = judo; 9 = rowing; 10 = volleyball; 11 = water polo; 12 = weight lifting; 13 = wrestling; 14 = track – sprinting; 15 = track – middle distance; 16 = track – long distance; 17 = field – throwing. (Plotted from means in Rodríguez *et al.*, 1986.)

class sportsmen and women (Štěpnička, 1974a, 1986; Štěpnička et al., 1979a), for Venezuelan national team male and female athletes (Pérez, 1981), for Chinese national male athletes (Zeng, 1985), for Indian state or national male athletes (Sodhi & Sidhu, 1984), and for national level male Cuban athletes (Rodríguez et al., 1986). These studies show that between-sport patterns of difference within countries are similar to each other and to Olympic means. However, in selected sports there are some between-countries differences in magnitude. Overall the male Czechoslovak athletes are the most mesomorphic and the Indians the least. It is likely that relatively low mesomorphy and high ectomorphy of the Indian athletes reflects the population from which the athletes are drawn (see Chapter 3). Nonetheless, Indian athletes are more mesomorphic than non-athletes. Both male and female Czechoslovak athletes are more mesomorphic and less endomorphic than their reference groups (see Chapter 3). Pérez found Venezuelan and Olympic athletes in the same sports were somatotypically similar, and that in general the Venezuelans were younger, shorter, lighter in weight, and smaller in biacromial and biiliac breadths. In most sports the ethnically mixed Cuban and Olympic means were quite similar. The Cubans ranged from approximately 20% Blacks in waterpolo to 80% in sprinting and boxing.

Studies such as those of Guimarães & De Rose (1980) and Thorland et al. (1981) included primarily young athletes. Alonso (1986) reported the somatotypes of 396 twelve year old boys and 196 girls in EIDE (Initial sports school for students) in Occidental Province. Carter (1988) summarized the findings of a large number of studies of somatotypes of children in sports. There are specific comparisons by sport later in this chapter. In general, there are somatotype differences between sports for both young male and female athletes, but less pronounced differences than among adult athletes. Sexual dimorphism also is less in youths than in adults.

As seen in the above studies, at the highest levels of sport there are apparent somatotypic differences between sports. There are also differences due to sex, level of competition, race-ethnicity, and age of the competitors. Further, there is evidence that in several sports the most successful somato-types have changed over time (Carter, 1984b; Mészáros & Mohácsi, 1982b; Štěpnička, 1986). These differences should be kept in mind as data for each sport are reviewed.

During the past 20 years there have been extensive and intensive studies of athletes. As a group, these studies have been well documented, have used fairly consistent methodology and analysis, and have been reviewed in depth elsewhere (Bale, 1983; Carter, 1970, 1971, 1978, 1981, 1984b, 1988; Sodhi & Sidhu, 1984; Štěpnička, 1972, 1974a, 1977, 1986). The following summaries by sport are presented in tabular form, with a minimum of commentary. The references given in the tables are not always repeated in the text. They are presented by sex, alphabetically by sport, and by event or position with the sport where appropriate. The sports for which there is little information are grouped with 'miscellaneous' at the end of the male and female sport sections.

Detailed analyses of each sport cannot be presented here because of limitations of space. The reader may refer to the original or review articles for more detail and for somatocharts. In the following sections mean somatoplots only are presented. Keep in mind possible between-sample differences in age of subjects, level of competition, training status, year of study, methods of measurement and analysis, race-ethnicity of subjects, and playing positions in team sports.

Male sports

Basketball (Table 6.1, Figure 6.11)

Analysis of the Olympic samples showed that the majority of basketball players were ecto-mesomorphs, ectomorph–mesomorphs, and meso-ectomorphs. Few players were higher in endomorphy than ectomorphy. In the 1968 sample de Garay *et al.* (1974) found no somatotypic differences between Black and White players. Empirical observation indicated that in general Olympic and college guards were highest in mesomorphy (some were as high as 6 and 7). Centres and forwards were highest in ectomorphy (some as high as 5 and 6).

It is not surprising that basketball players, who are relatively light for their height, are high in ectomorphy. However, in other sports there are tall athletes who are not markedly ectomorphic. Rowers, track and field

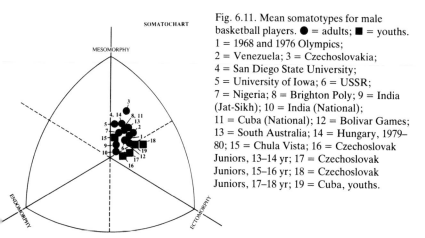

Fig. 6.11. Mean somatotypes for male basketball players. ● = adults; ■ = youths. 1 = 1968 and 1976 Olympics; 2 = Venezuela; 3 = Czechoslovakia; 4 = San Diego State University; 5 = University of Iowa; 6 = USSR; 7 = Nigeria; 8 = Brighton Poly; 9 = India (Jat-Sikh); 10 = India (National); 11 = Cuba (National); 12 = Bolivar Games; 13 = South Australia; 14 = Hungary, 1979–80; 15 = Chula Vista; 16 = Czechoslovak Juniors, 13–14 yr; 17 = Czechoslovak Juniors, 15–16 yr; 18 = Czechoslovak Juniors, 17–18 yr; 19 = Cuba, youths.

Table 6.1. *Basketball – male*

Sample, reference	N	Statistic	Age (yr)	Height (cm)	Weight (kg)	HWR	Somatotype
(a) *Adults*							
Mexico City Olympics	63	X̄	24.0	189.1	79.7	43.8	2.0-4.3-3.5
(De Garay *et al.*,		SD	4.1	8.5	9.9	1.2	0.7 0.9 0.9
1974)							
Mexico City +	68	X̄	—	—	—	43.9	2.0-4.2-3.5
Montreal		SD				1.2	0.7 0.9 0.9
(Carter, 1984*b*)							
Venezuela	21	X̄	20.0	186.1	76.1	43.9	1.9-4.4-3.7
(Pérez, 1981)		SD	3.5	7.4	8.7	1.7	0.5 1.0 1.0
Czechoslovakia	31	X̄	—	190.4	85.1	43.3[a]	2.0-5.5-3.1
(Štěpnička, 1974*a*,		SD		9.4	10.0		0.7 0.8 1.1
1986)							
San Diego State	10	X̄	20.6	190.0	83.4	43.5[a]	2.4-4.9-3.3
University		SD	1.2	8.1	9.0		0.6 0.6 0.7
(Carter, 1970)							
University of Iowa	10	X̄	19.6	183.9	79.4	43.5[a]	2.7-4.9-3.0
(Carter, 1970)		SD	1.0	9.7	5.6		0.6 0.8 1.4
USSR (Heath, in	8	X̄	—	192.5	87.5	43.4[a]	2.9-4.6-4.1
Carter, 1970)		SD		4.3	3.8		0.5 0.4 0.6
Nigeria (Toriola	12	X̄	26.8	178.3	65.4	44.3[a]	2.9-4.9-3.7
et al., 1985)		SD	4.2	6.1	8.7		0.4 0.4 0.6
Brighton Polytechnic	5	X̄	—	184.2	76.9	43.3	2.5-4.7-3.6
(Bale, 1986)		SD		7.4	7.0	1.0	0.8 0.4 0.8
India (Jat-Sikh) (Sidhu	23	X̄	22.2	—	—	—	3.0-4.0-3.5
& Wadhan, 1975)		SD					0.9 0.8 0.9
State	15	X̄	23.7	178.7	65.0	44.6[a]	2.5-3.4-4.0
		SD	2.7	6.5	9.5		1.0 0.7 1.0
District	13	X̄	—	173.2	59.9	44.3[a]	2.7-3.6-3.9
		SD		5.5	6.0		1.0 0.7 0.9
National (Sodhi,	12	X̄	—	185.6	76.7	43.7[a]	3.0-3.7-3.5
1980; Sodhi &		SD		—	—	—	—
Sidhu, 1984)							
Cuba, 1976–80	18	X̄	24.0	192.3	86.9	43.4[a]	2.2-4.4-3.2
(Rodríguez *et al.*,		SD	2.2	7.4	9.5		0.7 1.1 0.8
1986)							
Bolivar Games, 1981	12	X̄	21.5	188.2	79.9	43.8	2.0-3.7-3.5
(Brief, 1986)		SD	2.5	7.1	8.3	0.8	0.5 0.8 0.5
South Australia	11	X̄	25.7	190.3	82.1	43.8	2.1-4.5-3.5
(Withers *et al.*,		SD	3.1	11.7	10.9	1.2	0.5 0.8 0.9
1986)							
Hungary							
1972–75	36	X̄	—	186.6	76.4	44.0[a]	2.3-3.9-4.0
1979–80	22	X̄	—	186.2	76.5	43.9[a]	2.4 4.8 3.4
(Mészáros &							
Mohácsi, 1982*b*)							

Table 6.1. (*Continued*)

Sample, reference	N	Statistic	Age (yr)	Height (cm)	Weight (kg)	HWR	Somatotype
New Zealand (Lewis, 1966)	100	X̄	24.0	181.6	76.2	42.8[a]	3.0-5.0-2.8[b]
		SD	4.8	5.8	8.1		—
(b) *Youths*							
Chula Vista, Calif.	10	X̄	14.9	176.9	62.5	44.6	2.2-4.4-4.1
(Haley, 1974)		SD	0.3	3.5	7.1	1.1	0.5 0.7 0.9
Czechoslovak Juniors	29	X̄	13–14	—	—	—	3.0-3.5-4.0
Chovanová &	54	X̄	15–16	—	—	—	2.5-3.5-4.0
Zapletalová, 1980)	37	X̄	17–18	—	—	—	2.0-4.0-4.5
Cuba (Alonso, 1986)	25	X̄	12.5	163.3	49.3	44.5[a]	2.3-4.0-4.1

[a] Calculated from mean height and weight.
[b] Estimated Heath–Carter anthropometric somatotype.
X̄ = mean.

throwers, and American football linemen, who are as tall as many basketball players, are often 10 to 40 kg heavier and far less ectomorphic. The mean somatotype of 2.0-4.2- 3.5 of Olympic basketball players is closest to that of the markedly shorter long distance runners.

With the exception of the Indian and New Zealand samples, the somatotypes of basketball players fall within the Olympic distribution. The Czechoslovaks were one-half unit higher in mesomorphy than the others. The majority of the Venezuelans were ecto-mesomorphs, but 19% were dominant in ectomorphy. The Nigerian college players were as mesomorphic as the San Diego players, and slightly more ectomorphic. The Brighton sport science students, specializing in basketball, were somatotypically closest to the Nigerians.

The scatter of somatotypes of the Iowa, San Diego and Russian samples is similar. Endomorphy is moderately low (1.9 to 3) and consistent in all the above samples. Mean somatotypes of the Russian, Venezuelan and Nigerian players are close to that of the Olympians. Mean somatotypes of the Czechoslovaks differ most from the others.

Sodhi & Sidhu (1984) reported that Indian players were less endomorphic and more ectomorphic than the controls. The state and district level players were more meso-ectomorphic than the national ecto-mesomorphic players. At the lower playing levels somatotypes were more widely distributed. Indian players were less ecto-mesomorphic than Olympic players. The findings of Sidhu & Wadhan (1975) for Jat-Sikh (Punjabi) players and university students were similar.

Lewis (1966) used Parnell's M.4 method to somatotype three levels of 100

New Zealand players, including the national team. Except for lower endomorphy at the higher levels, body size and somatotypes were similar. The mean M.4 somatotype was $3\frac{1}{2}$- $4\frac{1}{2}$-$3\frac{1}{2}$, which would be 3-5-$2\frac{1}{2}$ if it were transformed to an approximate Heath–Carter rating. The New Zealand players appeared to be more endo-mesomorphic than the other samples, and most similar to the Iowa players. Possibly there are relatively few typical basketball physiques in New Zealand, or perhaps basketball does not appeal to those with appropriate physiques.

Haley (1974) somatotyped ten 15 year old basketball players at Chula Vista Junior High School, and Chovanová & Zapletalová (1980) somatotyped a sample of 120 of the best young Czechoslovak players at the 1977–78 National Championships.

The means at different ages in the Czechoslovak study showed that the boys became more mesomorphic, more ectomorphic and less endomorphic with increasing age. Guards were the most mesomorphic and centres the most ectomorphic. The Chula Vista team was slightly more mesomorphic than the 15–16 year old Czechoslovaks and were closer to the mean of the Olympic basketball players. However, the 17 and 18 year old Czechoslovaks were markedly less ecto-mesomorphic than their adult counterparts ($\overline{S}$ = 2.0-5.5-3.1).

Other studies that provide some data on basketball players are those of Bláha & Seifertová (1981), Caldeira, Vívolo & Matsudo (1986c), Boennec *et al.* (1980), and Muthiah & Sodhi (1980).

Body building (Table 6.2, Fig. 6.12(*a*), (*b*), (*c*))

Body builders, who use weight lifting and other specialized training to develop extraordinary muscle mass, body shape, definition, and aesthetic

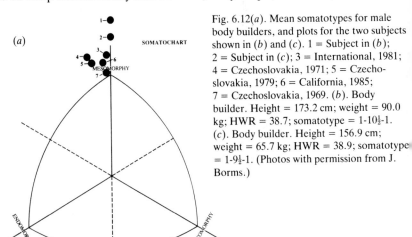

Fig. 6.12(*a*). Mean somatotypes for male body builders, and plots for the two subjects shown in (*b*) and (*c*). 1 = Subject in (*b*); 2 = Subject in (*c*); 3 = International, 1981; 4 = Czechoslovakia, 1971; 5 = Czechoslovakia, 1979; 6 = California, 1985; 7 = Czechoslovakia, 1969. (*b*). Body builder. Height = 173.2 cm; weight = 90.0 kg; HWR = 38.7; somatotype = 1-$10\frac{1}{2}$-1. (*c*). Body builder. Height = 156.9 cm; weight = 65.7 kg; HWR = 38.9; somatotype = 1-$9\frac{1}{2}$-1. (Photos with permission from J. Borms.)

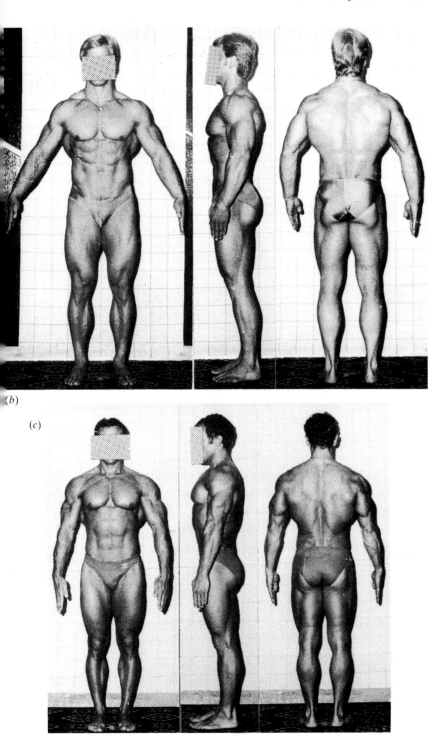

(b)

(c)

Table 6.2. *Body building – male*

Sample, reference	N	Statistic	Age (yr)	Height (cm)	Weight (kg)	HWR	Somatotype
International, Cairo, 1981 (Borms *et al.*, 1986)	66	X̄ SD	28.7 5.3	171.4 3.9	80.2 4.6	39.9 1.0	1.6-8.7-1.2 0.4 1.0 0.4
Czechoslovakia, 1969 (Štěpnička, 1974*a*, 1986)	102	X̄ SD	—	175.4 6.5	78.9 7.3	40.9[a]	1.8-7.9-1.4 0.7 0.9 0.6
Czechoslovakia, 1975 (Štěpnička *et al.*, 1979*a*)	32	X̄ SD	—	172.6 6.4	85.1 8.7	39.2[a]	2.0-8.2-0.7 0.5 0.7 0.3
Czechoslovakia, 1971 (Zrubák & Hrčka, 1976)	19	X̄	22.8	172.4	84.3	39.3[a]	2.5-9.0-1.0

[a]Calculated from mean height and weight.
X̄ = mean.

appeal, achieve exceedingly high ratings in mesomorphy and reduce subcutaneous fat to a minimum. The data for the World Amateur Body Building Championships in Cairo, Egypt, showed that the four weight classes differed in height and weight but not in somatotype. Although the range in mesomorphy was from $6\frac{1}{2}$ to 11, there appeared to be little chance of success for anyone lower than 8 in any class. Because *ipso facto* the lower the endomorphy the better the muscle relief and definition, a body builder with a rating of 2 or more in endomorphy could not expect a high rating in competition.

Studies of other body builders showed physiques similar to those in the championship sample. Štěpnička (1977) found 2-9-$1\frac{1}{2}$ somatotypes for the best competitors. The 1975 Professional Mr USA had a $1\frac{1}{2}$-$9\frac{1}{2}$-1 somatotype (Carter, unpublished). The mean somatotype for 5 competitors in the 1985 California Championships in San Diego was 1.4-7.9-0.9 (Grobl & Carter, unpublished). Carter (1978) observed that championship body builders were more mesomorphic than younger (novice) and older (master) competitors. Parnell (1957, 1958), using his M.4 method, and Štěpnička (1972), using Sheldon's method, found near maximum mesomorphy (means near 6.8) and showed all physiques clustered in the top angle of the somatochart.

Boxing (Table 6.3, Fig. 6.13)

From photographs and data supplied by Tanner, Heath re-rated the somatotypes of the British Empire Games boxers, some of whom were Olympic competitors. (Earlier ratings by Heath, published in Carter, 1970,

Table 6.3. *Boxing – male*

Sample, reference	N	Statistic	Age (yr)	Height (cm)	Weight (kg)	HWR	Somatotype
a) Adults							
British Empire Games	39	$\overline{X}$	21.5	171.5	65.4	42.7	2.0-5.7-2.3
Cardiff, 1958 (Heath unpubl.)		SD	2.7	7.5	10.2	1.1	0.6 0.7 0.9
Mexico City Olympics,	142	$\overline{X}$	22.9	169.3	63.1	42.7	1.9-5.3-2.6
1968 (De Garay et al., 1974)		SD	3.4	7.4	9.5	1.3	0.6 0.6 0.9
Montreal Olympics,	22	$\overline{X}$	23.9	172.6	66.3	42.8	1.7-5.1-2.7
1976 (Carter et al., 1982)		SD	3.6	10.4	10.9	1.0	0.4 0.8 0.7
Combined Olympics							
<60 kg	67	$\overline{X}$	—	—	—	43.2	1.6-4.9-3.0
		SD				1.0	0.4 0.7 0.7
60–79.9 kg	85	$\overline{X}$	—	—	—	42.4	2.0-5.5-2.5
		SD				1.1	0.5 0.8 0.8
80–89.9 kg	11	$\overline{X}$	—	—	—	41.5	2.6-6.1-1.8
Carter, 1984b)		SD				1.4	1.1 0.9 0.8
Hungary							
<63.5 kg	20	$\overline{X}$	21.0	165.2	56.8	43.0^a	3.0-5.0-2.8
		SD	2.6	7.0	4.9		0.7 1.0 1.1
63.6–75 kg	20	$\overline{X}$	20.0	173.7	68.6	42.4	2.6-5.3-2.5
		SD	2.5	5.7	3.9		0.5 1.0 0.7
>75 kg	18	$\overline{X}$	21.4	182.0	85.0	41.4	4.3-6.0-1.8
Farmosi et al., 1985)		SD	3.1	4.5	6.3		0.9 1.2 0.9
Cuba, 1976–80							
48–57 kg	18	$\overline{X}$	22.01	165.2	56.7	43.0^a	1.9-5.1-2.9
		SD	3.5	5.6	3.6		0.6 0.7 0.9
60–75 kg	12	$\overline{X}$	23.4	172.4	68.4	42.2^a	2.0-5.6-2.4
		SD	3.4	4.0	4.1		0.4 0.7 0.9
>75 kg	9	$\overline{X}$	22.0	184.0	85.3	41.8^a	2.2-6.1-2.1
Rodríguez et al., 1986)		SD	2.3	8.6	9.3		0.4 0.5 0.7
Bolivar Games, 1981	16	$\overline{X}$	20.4	170.4	62.1	43.2	1.4-4.8-3.0
(Brief, 1986)		SD	2.2	8.6	9.4	1.2	0.3 0.8 0.9
b) Youths							
Cuba (Alonso, 1986)	7	$\overline{X}$	12.5	148.4	40.7	43.1^a	2.5-4.4-2.6

Calculated from mean height and weight.
= mean.

preceded the completed modifications of Heath's method. The ratings presented here reflect the criteria of subsequent modifications.)

In studies of boxers means for total samples are useful as general indicators for between-sports comparisons, despite the somatotypic differ-

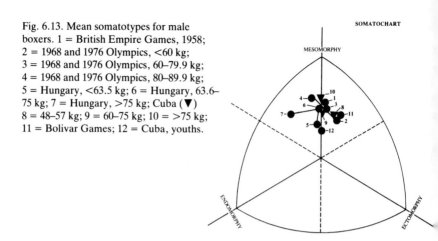

Fig. 6.13. Mean somatotypes for male boxers. 1 = British Empire Games, 1958; 2 = 1968 and 1976 Olympics, <60 kg; 3 = 1968 and 1976 Olympics, 60–79.9 kg; 4 = 1968 and 1976 Olympics, 80–89.9 kg; 5 = Hungary, <63.5 kg; 6 = Hungary, 63.6–75 kg; 7 = Hungary, >75 kg; Cuba (▼) 8 = 48–57 kg; 9 = 60–75 kg; 10 = >75 kg; 11 = Bolivar Games; 12 = Cuba, youths.

ences between weight classes or groups. In samples of Olympic boxers, endomorphy and mesomorphy increased and ectomorphy decreased with ascending order of weight group. There were similar patterns of difference in the 1968 and 1976 Olympic samples. Carter (1984*b*) summarized the findings by combining the two samples in three weight groups. In the Mexico City sample somatotype component comparisons among Whites, Blacks, Mestizos, and Orientals showed no differences in the light and middle weight groups. In the British Empire Games sample the mean weight of the five lightest boxers was 51.7 kg and the mean somatotype was ecto-mesomorphic (1.7-5.3-2.5); the mean weight of the five heaviest boxers was 83.5 kg and the mean somatotype was endo-mesomorphic (2.8-6.6-1.5).

Fig. 6.14. Mean somatotypes for male canoeist. 1 = Czechoslovakia; 2 = Olympics, 1968 and 1976; 3 = Hungary; 4 = USA.

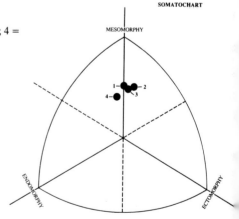

In the higher weight groups the Hungarian boxers too were increasingly endo-mesomorphic. Also, they were more endomorphic than other samples, especially those heavier than 75 kg. A similar trend is seen among the Cubans, who are perhaps the world's best boxers. The small group of young Cuban boxers are balanced mesomorphs.

As Fig. 6.13 shows, the overall trend is consistent. The lightest boxers are ecto-mesomorphs, the middleweights tend toward balanced mesomorphy, and the majority of the heavyweight classes are endo-mesomorphs.

Additional data on boxers are provided by Caldeira *et al.* (1986c), Carter 1970), Shin (1985), and Tanner (1964).

Canoeing (Table 6.4, Fig. 6.14)

The Olympic canoeists in Mexico City in 1968 competed in kayak and Canadian style events. About two-thirds of these were kayakers. There were no somatotypic differences between the two groups, although the kayakers were taller and heavier than those in the Canadian events.

Table 6.4. *Canoeing – male*

Sample, reference	N	Statistic	Age (yr)	Height (cm)	Weight (kg)	HWR	Somatotype
Mexico City Olympics, 1968 (De Garay *et al.*, 1974)	49	X̄	24.2	178.5	74.4	42.4	1.9-5.5-2.5
		SD	4.3	7.9	7.7	1.3	0.7 1.0 0.9
Kayak	34	X̄	25.0	180.3	75.9	42.6	1.9-5.4-2.6
		SD	4.3	7.5	8.0	1.4	0.7 1.0 1.0
'Canadian'	15	X̄	22.3	174.2	71.0	42.0	1.9-5.6-2.3
		SD	3.7	7.4	5.8	1.1	0.4 0.9 0.8
Montreal Olympics, 1976 (Carter *et al.*, 1982)	12	X̄	25.1	185.4	79.1	43.2	1.5-5.2-3.1
		SD	5.5	5.1	5.9	1.2	0.5 0.8 1.0
Mexico + Montreal (Carter, 1984b)	61	X̄	—	—	—	42.3	1.8-5.4-2.6
		SD				1.1	0.5 0.9 0.8
Czechoslovakia 'speed' (Štěpnička, 1974a)	26	X̄		178.7	75.8	42.2[a]	2.0-5.8-2.1
		SD		6.0	5.6	—	0.5 1.0 0.9
'slalom' (Štěpnička *et al.*, 1979a)	23	X̄		178.8	75.7	42.3[a]	2.1-5.7-2.3
		SD		5.7	7.4		0.6 0.9 0.8
Hungary ('paddlers') (Mészáros & Mohácsi, 1982a)	26	X̄	—	174.5	69.0	42.6[a]	2.2-5.5-2.5
USA (slalom) (Vaccaro *et al.*, 1984)	13	X̄	20.1	179.9	76.3	42.4[a]	2.9-5.2-2.4
		SD	2.1	6.8	6.1		0.6 1.0 0.9

[a] Calculated from mean height and weight.

X̄ = mean.

Inasmuch as 9 of the 12 competitors at the Montreal Olympics were kayakers, no analysis of differences was attempted. The combined mean of their somatotypes was 1.8-5.4-2.6, similar to that at Mexico City. The majority were ecto-mesomorphs and balanced mesomorphs. The somatotype distributions were elliptical, and their narrow range on endomorphy was among the lowest among the male Olympic athletes.

The Czechoslovak 'speed' and 'slalom' canoers had similar somatotypes. As a group they tended toward higher mesomorphy and lower ectomorphy than the Olympic contestants. The somatotypes of the 'speed' canoers were scattered in and around the balanced mesomorphy category. The somatotypes of Hungarian and USA samples were similar to the Czechoslavak samples, except that the USA athletes were more endomorphic.

Cycling (Table 6.5, Fig. 6.15)

The large sample of cyclists somatotyped at the Mexico City Olympics consisted of Black, White, Mestizo, and Oriental competitors in sprint, pursuit, and road events. Contrary to the expectation that track cyclists in the sprint and pursuit events would be more mesomorphic than the long distance road cyclists, there were no somatotypic differences by event. The Whites were less ectomorphic than the Orientals.

In the small sample of cyclists at the Montreal Olympics, the track cyclists were more mesomorphic and less ectomorphic than the road cyclists. Most of the road cyclists, including three from Hong Kong, were meso-ectomorphic. The mean for the track cyclists was 1.9-5.4-2.5, and for the road cyclists 1.6-3.9-3.8.

The somatotype mean for the master-class Czechoslovak track cyclists was 1.6-5.4-2.5, close to the Olympic track means and similar to the Cuban (track) and South Australians. The mid-season somatotype means for track and road cyclists in the British Olympic training squads showed higher endomorphy and mesomorphy and lower ectomorphy for the track cyclists than for the road cyclists. The studies also showed lowered endomorphy during the competitive season. Somatocharts (with no numerical data) for 14 French professional racers (Boennec *et al.*, 1980) show most subjects were ecto-mesomorphs and balanced mesomorphs with a mean close to the Olympic, Czechoslovak and British track samples. The subjects were not identified as either track or road cyclists. The authors reported slightly lowered endomorphy when they re-measured eight cyclists after six months.

The somatotypes of the national team from Indian, as well as another sample with an identical mean (Singh & Malhotra, 1986), were significantly less mesomorphic and more endomorphic than other samples. The subjects were not identified by road or track events. An earlier sample of Jat-Sikh cyclists had a more endomorphic mean of 4.4-4.7-2.1 (Singh & Sidhu, 1982).

(a)

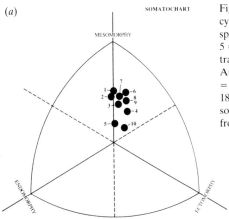

Fig. 6.15(*a*) Mean somatotypes for male cyclists. 1 = British, track; 2 = Cuba, sprints; 3 = Cuba, road; 4 = British, road; 5 = Cuba, youth; 6 = Czechoslovakia, track; 7 = Bolivar Games; 8 = South Australia; 9 = Olympics 1968 and 1976; 10 = India. (*b*) Track 'sprint' cyclist. Height = 181.3 cm; weight = 80.3 kg; HWR = 42.0; somatotype 2-6-2. (Photo with permission from J. Štěpnička.)

(b)

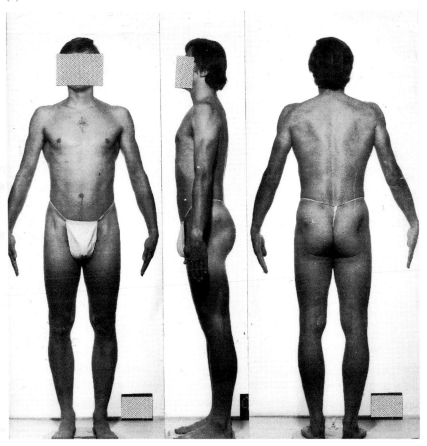

Table 6.5. *Cycling – male*

Sample, reference	N	Statistic	Age (yr)	Height (cm)	Weight (kg)	HWR	Somatotype
Mexico City Olympics,	100	$\overline{X}$	23.6	174.9	68.9	42.7	1.8-5.0-2.7
1968		SD	3.4	6.5	7.7	1.1	0.5 0.8 0.8
Sprints	14	$\overline{X}$	22.7	174.0	69.8	42.3	1.8-5.2-2.4
		SD	3.0	5.7	9.1	1.0	0.4 0.8 0.7
Pursuit	19	$\overline{X}$	22.9	174.6	68.8	42.6	1.8-5.1-2.6
		SD	3.5	6.7	9.5	1.4	0.5 1.1 1.0
Road	67	$\overline{X}$	24.0	175.1	68.7	42.8	1.8-4.9-2.7
(De Garay *et al.*, 1974)		SD	3.4	6.7	6.9	1.0	0.5 0.8 0.7
Montreal Olympics, 1976	18	$\overline{X}$	23.0	177.1	69.6	43.2	1.7-4.8-3.1
(Carter *et al.*, 1982)		SD	4.0	6.2	10.2	1.6	0.4 1.2 1.2
Mexico + Montreal	118	$\overline{X}$	—	—	—	42.8	1.8-5.0-2.8
(Carter, 1984*b*)		SD				1.2	0.5 0.9 0.9
Czechoslovakia 'track'	28	$\overline{X}$	22.2	178.3	74.2	42.4[a]	1.6-5.4-2.5
(Štěpnička *et al.*,		SD		5.6	7.7		0.5 0.5 0.6
1979*a*; Stejskal &							
Náprstková, 1975)							
India (Sodhi & Sidhu,	15	$\overline{X}$	25.1	172.6	63.1	43.4[a]	2.4-3.8-3.2
1984)		SD	4.8	6.8	7.6		0.8 0.7 0.9
British Olympic squad							
'road'	14	$\overline{X}$	22.4	176.2	68.4	43.1[a]	1.6-4.2-2.8
		SD		6.9	5.8		0.3 1.1 0.8
'track'	8	$\overline{X}$	21.1	175.4	74.0	41.7[a]	2.1-5.3-2.1
(White *et al.*, 1982*a,b*)		SD	0.3	5.3	6.6	—	0.3 1.1 0.8
Cuba, 1976–80							
Sprints	17	$\overline{X}$	21.5	171.0	68.1	41.9[a]	2.2-5.2-2.1
		SD	4.7	4.3	5.4		0.5 1.0 0.8
Road	16	$\overline{X}$	21.5	171.9	65.8	42.6[a]	2.0-4.8-2.5
(Rodríguez *et al.*, 1986)		SD	3.9	4.3	4.8		0.4 0.7 0.6
Bolivar Games, 1981	15	$\overline{X}$	26.0	167.7	63.8	42.0[a]	1.7-4.9-2.2
(Brief, 1986)		SD	5.6	5.1	6.2		0.4 1.0 0.8
South Australia (Withers	11	$\overline{X}$	22.2	176.4	68.5	43.1	2.0-5.2-2.9
et al., 1986)		SD	3.6	7.1	6.4	0.8	0.4 0.4 0.6
Cuba, youths (Alonso,	14	$\overline{X}$	12.5	155.0	45.7	43.4[a]	3.2-4.4-3.3
1986)							

[a]Calculated from mean height and weight.
$\overline{X}$ = mean.

Overall, there seem to be some ethnic differences among cyclists. It would be well to look within ethnic groups for possible differences between track and road cyclists. The data from within-country studies suggest that track cyclists are more mesomorphic than road cyclists.

Fencing (Table 6.6, Fig. 6.16)

The somatotype variation of the Olympic, Czechoslovak, Hungarian and Bolivar Games fencers is seen across the full width of the endo-mesomorph and ecto-mesomorph categories. The Cuban fencers are more ecto-mesomorphic and the Hungarians are more endo-mesomorphic than the Olympians. Except for slightly higher ectomorphy for sabre fencers in these two samples, there is little difference among events.

Table 6.6. *Fencing – male*

Sample, reference	N	Statistic	Age (yr)	Height (cm)	Weight (kg)	HWR	Somatotype
Montreal Olympics, 1976	9	$\bar{X}$	26.5	183.6	77.6	42.1	2.8-4.2-2.9
(Carter *et al.*, 1982)		SD	6.3	7.4	8.1	1.6	1.4 0.9 1.2
Czechoslovakia, 1971	17	$\bar{X}$	21.0	178.5	71.1	43.1[a]	2.5-4.5-3.0
Zrubák & Hrčka, 1976)							
Bolivar Games, 1981	18	$\bar{X}$	23.9	170.7	66.1	42.3	2.5-4.8-2.3
(Brief, 1986)		SD	4.7	5.2	7.1	1.4	1.1 0.9 1.0
Cuba, 1976–80	21	$\bar{X}$	23.3	174.9	69.6	42.5[a]	2.1-5.0-2.6
		SD	2.0	6.4	7.2		0.4 0.8 0.6
Foil	8	$\bar{X}$	22.8	172.3	66.0	42.6[a]	2.0-4.9-2.6
		SD	2.1	6.4	7.2		0.4 0.9 0.7
Epee	8	$\bar{X}$	23.0	175.4	71.8	42.2[a]	2.1-5.4-2.4
		SD	2.0	4.5	2.8		0.4 0.8 0.3
Sabre	5	$\bar{X}$	24.2	178.3	71.6	42.9[a]	2.2-4.6-2.8
(Rodríguez *et al.*, 1986)		SD	1.5	5.2	5.8		0.6 0.8 0.7
Hungary	91	$\bar{X}$	24.9	176.1	72.2	42.3	2.7-5.2-2.0
		SD	6.2	5.5	7.2	1.3	0.9 0.7 0.9
Foil	34	$\bar{X}$	24.9	173.4	70.4	42.1	2.8-5.2-1.8
		SD	6.1	4.6	7.0	1.2	1.0 0.8 0.8
Epee	33	$\bar{X}$	25.5	177.4	73.5	42.3	2.8-5.2-2.0
		SD	6.5	5.3	7.3	1.4	0.8 0.7 1.0
Sabre	24	$\bar{X}$	23.8	178.4	72.9	42.7	2.5-5.2-2.3
(Eiben, 1980, & unpubl.)		SD	6.0	5.7	7.2	1.2	0.7 0.7 0.8
Cuba, youths (Alonso, 1986)	21	$\bar{X}$	12.5	157.2	39.2	46.3[a]	3.0-3.2-3.5

[a]Calculated from mean height and weight.
$\bar{X}$ = mean.

Fig. 6.16. Mean somatotypes for male
fencers. 1 = Cuba, epee; 2 = Bolivar
Games; 3 = Cuba, foil; 4 = Cuba, sabre;
5 = Czechoslovakia, 1971; 6 = Montreal
Olympics; 7 = Cuba, youths.

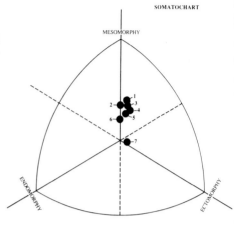

Football

The generally accepted usage of the word *football* includes the four forms of the sport known as: American and Canadian 'gridiron', rugby, soccer or 'true' football, and Australian Rules. Each has its own rules and national or international following.

(a) American and Canadian football (Table 6.7, Fig. 6.17(a))

(i) *Professional football.* (Wilmore *et al.* (1976a,b), in a study of 64 professional veterans and new players from several teams in the National Football League, reported that the players were endo-mesomorphic, with an approximate mean of $4\frac{1}{2}$-$6\frac{1}{2}$-$1\frac{1}{2}$. Comparison of somatotypes by players' positions showed that defensive and offensive linemen were the most endo-mesomorphic and defensive backs the least endo-mesomorphic. In a study of a Canadian (Edmonton) professional football team, Bagnall *et al.*

Fig. 6.17(a). Mean somatotypes for male
American football players. The line encloses
the distribution of players from three studies
of professional players – from Edmonton ▲,
San Diego ■, and other USA teams ●. The
means and keys by playing position are: OL
= offensive linemen; DL = defensive
linemen; LB = linebackers; DB = defensive
backs; RB = running backs; OB = offensive
backs; QBK = quarterbacks and kickers;
WR = wide receivers. The open circles and
numbers are for university and college
means. 1 = San Diego State University;
2 = University of Iowa; 3 = Oregon
colleges; 4 = University of Oklahoma.

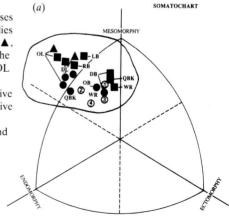

Table 6.7. *Football – male*

Sample, reference	N	Statistic	Age (yr)	Height (cm)	Weight (kg)	HWR	Somatotype
(a) *American and Canadian football*							
(1) Professional							
USA							
Defensive backs	26(9)[a]	$\bar{\mathrm{X}}$	24.5	182.5	84.8	41.5[b]	3.2-5.6-1.9
Offensive backs, Wide receivers	40(15)	$\bar{\mathrm{X}}$	24.7	183.8	90.7	40.9	3.5-6.2-1.6
Linebackers	28(9)	$\bar{\mathrm{X}}$	24.2	188.6	102.2	40.3	4.6-6.7-1.3
Offensive line & tight ends	38(11)	$\bar{\mathrm{X}}$	24.7	193.0	112.6	40.1	5.5-6.7-0.9
Defensive line	32(11)	$\bar{\mathrm{X}}$	25.7	192.4	117.1	39.3	5.4-6.9-1.2
Quarterbacks, Kickers	16(9)	$\bar{\mathrm{X}}$	24.1	185.0	90.1	41.3	5.0-5.9-1.3

Total 180(64)
(Wilmore *et al.*, 1976*a,b*)
([a]N in parenthesis is for those somatotyped.)

Sample, reference	N	Statistic	Age (yr)	Height (cm)	Weight (kg)	HWR	Somatotype
Canada							
Defensive backs	8	$\bar{\mathrm{X}}$	—	179.6	82.1	41.3	2.4-6.1-1.6
Wide receivers	8	$\bar{\mathrm{X}}$	—	182.4	82.8	41.9	2.5-5.6-2.1
Linebackers	4	$\bar{\mathrm{X}}$	—	184.4	99.4	39.8	3.1-7.3-0.8
Running backs	5	$\bar{\mathrm{X}}$	—	182.1	96.6	39.7	4.1-7.1-0.8
Offensive line	7	$\bar{\mathrm{X}}$	—	190.0	117.7	38.8	5.3-8.1-0.7
Defensive line	7	$\bar{\mathrm{X}}$	—	191.0	114.7	39.3	4.7-7.8-0.8
Quarterbacks, Kickers	5	$\bar{\mathrm{X}}$	—	182.1	86.0	41.3	2.6-6.0-1.7
Total	44	$\bar{\mathrm{X}}$		184.6	96.7	40.2	3.5-6.8-1.3
(Bagnall *et al.*, unpubl.)		SD					1.5 1.3 0.7
San Diego							
Offensive line	8	$\bar{\mathrm{X}}$	27.6	191.5	125.0	38.3	4.9-8.2-0.1
Defensive line	8	$\bar{\mathrm{X}}$	24.3	194.5	123.1	39.1	3.9-7.3-0.5
(Roberts & Carter, unpubl.)							
(2) *College*							
San Diego State University (Carter, 1970)	35	$\bar{\mathrm{X}}$	21.3	184.4	94.4	40.5	4.6-6.3-1.4
		SD	2.1	4.8	11.8	—	1.3 1.0 0.8
University of Iowa (Carter, 1970)	20	$\bar{\mathrm{X}}$	19.9	182.1	86.1	41.2	3.2-6.2-1.6
		SD	1.2	4.1	8.0	—	1.1 0.6 0.7
Oregon colleges (Allen, 1965)	66	$\bar{\mathrm{X}}$	20.3	181.6	84.9	41.3	3.6-5.5-2.1
		SD	2.8	5.3	9.5	—	0.4 0.6 0.8
University of Oklahoma (Votto, 1976)	23	$\bar{\mathrm{X}}$	20.5	188.2	93.5	41.5	4.2-5.4-1.9
		SD	1.0	6.1	13.7	—	1.5 1.2 1.1

(*continued*)

Table 6.7. (*Continued*)

Sample, reference	N	Statistic	Age (yr)	Height (cm)	Weight (kg)	HWR	Somatotype
(b) *Rugby football*							
South Africa, 1976							
National trials	47	X̄	25.9	183.2	86.2	41.5[a]	3.2-5.9-1.8
		SD	3.9	6.7	12.7		
Forwards	27	X̄	26.5	188.0	96.2	41.0[a]	3.8-6.1-1.6
		SD	4.5	5.3	11.5		
Backs	20	X̄	25.2	178.5	78.2	41.7[a]	2.6-5.7-2.0
		SD	3.1	6.8	8.9		
Retired players	37	X̄	44.3	177.8	85.5	40.4[a]	4.7-5.8-1.3
(Smit *et al.*, 1979*a*)		SD	3.8	4.4	7.2		
British Polytechnic club champions (Reilly & Hardiker, 1981)	28	X̄ SD	18–22	—	—		3.6-5.4-2.1 0.7 1.0 0.9
South Australia (Withers *et al.*, 1986)	16	X̄ SD	25.3 4.0	178.3 6.4	75.2 8.6	42.3 1.0	2.3-5.6-2.4 0.7 0.8 0.7
Brighton Polytechnic (Bale, 1986)	12	X̄ SD	—	177.1 6.0	76.6 8.0	42.0 1.0	2.3-5.4-2.4 0.7 0.7 0.7
(c) *Soccer football*							
Czechoslovakia, 1968 (Štěpnička, 1974*a*, Štěpnička *et al.*, 1979*a*)	72	X̄ SD	—	176.1 5.6	73.5 6.2	42.0[a]	2.3-5.9-2.0 0.9 0.8 0.7
Bratislava, 1970–71 (Chovanová & Zrubák, 1972; Zrubák & Hrčka, 1976)	20	X̄	23.5	178.4	74.9	42.3[a]	2.5-4.5-2.5
Brighton Polytechnic (Bale, 1986)	11	X̄ SD	—	181.7 5.2	76.9 7.0	43.0 1.3	2.7-4.7-3.2 0.6 0.6 0.8
South Australia (Withers *et al.*, 1976)	12	X̄ SD	25.3 4.0	178.3 6.4	75.2 8.6	42.3 1.0	2.3-5.6-2.4 0.7 0.8 0.7
Nigeria, 1984 (Toriola *et al.*, 1985)	15	X̄ SD	25.5 1.7	169.3 9.8	64.8 7.5	42.2[a]	2.5-4.7-2.9 0.6 0.5 0.7
India, State level							
Forwards	32	X̄ SD	—	167.1 4.6	56.5 4.5	43.6[a]	2.3-3.8-3.3 0.8 0.8 1.0
Halves	16	X̄ SD	—	169.0 4.7	57.0 3.3	43.9[a]	2.4-3.4-3.5 0.9 0.7 1.1
Backs	13	X̄ SD	—	167.7 4.9	56.9 3.7	43.6[a]	2.4-4.0-3.4 0.7 0.9 1.0
Stoppers	14	X̄ SD	—	172.6 3.8	61.1 4.4	42.8[a]	2.3-3.5-3.6 0.8 0.7 0.8
Goalkeepers (Sodhi & Sidhu, 1984)	8	X̄ SD	—	175.4 3.8	62.6 4.7	44.2[a]	3.1-3.3-3.8 1.2 0.5 0.8

Table 6.7. (*Continued*)

Sample, reference	N	Statistic	Age (yr)	Height (cm)	Weight (kg)	HWR	Somatotype
India, University	151	X̄	—	168.9	55.2	43.4[a]	1.9-3.6-4.2
(Kansal et al.,		SD		5.7	6.3		0.9 0.6 1.1
1986)							
Brazil, professional	29	X̄	25.1	173.0	69.3	42.1[a]	2.8-4.2-2.1
(Pinto, 1978)		SD	4.2	5.2	5.5		1.0 1.1 1.0
Brazil, professional	25	X̄	25.0	174.3	70.9	42.1[a]	2.2-4.8-2.3
(Matsudo, 1986)		SD		6.2	7.2		
Brazil, youths	30	X̄	13.0	155.1	44.2	43.9[a]	2.4-4.3-3.6
(Matsudo, 1986)		SD		9.0	7.6		
Cuba, 1976–80	19	X̄	23.0	174.5	70.2	42.2[a]	2.1-5.2-2.4
(Rodríguez et al.,		SD	2.0	4.8	5.5		0.4 0.5 0.4
1986)							
Bolivar Games, 1981	29	X̄	21.8	169.2	68.9	41.3	2.3-5.4-1.7
(Brief, 1986)		SD	2.8	5.6	5.4	1.1	0.7 0.9 0.7
Cuba, youths (Alonso,	33	X̄	12.5	149.7	41.8	43.1[a]	2.8-4.5-3.1
1986)							
(d) *Australian Rules Football*							
(Withers et al., 1986)	23	X̄	24.5	182.7	79.9	42.5	2.1-5.7-2.5
		SD	4.3	8.2	8.8	1.1	0.5 1.0 0.8

[a]Calculated from mean height and weight.
X̄ = mean.

(unpublished) found similar differences by position, but their players were less endomorphic in all positions than those in the Wilmore study. Their overall mean was $3\frac{1}{2}$-7-$1\frac{1}{2}$. In 1985 Roberts & Carter (unpublished) measured the somatotypes of eight offensive and six defensive linemen from the San Diego (California) professional football team. The means were similar, but the offensive linemen were more mesomorphic than the defensive specialists. The San Diego means were closer to the Edmonton players than to other US players. In the former two samples almost all linemen were extreme endo-mesomorphs. The lower mesomorphy and higher endomorphy in Wilmore's study may be due to differences in method or subject selection. Somatoplots of the means by playing position for the three studies show the similarities between the pairs of offensive and defensive linemen, running backs and linebackers, and defensive backs and wide receivers. Although the techniques of the paired positions differ, the physical attributes needed by position are similar. They are also similar in body size (height and weight).

Many of the players in the above studies are in the far northwest segment of the somatochart and also are very large. Most are beyond the 90th percentile in height and weight compared with reference norms (Wilmore *et al.*, 1976*a*). Most of the weight is attributable to high mesomorphy with a generous 'padding' of endomorphy overlaying it.

(*ii*) *US College football.* There are three levels of college football samples; small colleges, division II college level, and top level, university.

Some of the heavier linemen (over 105 kg) on the Iowa team were unavailable for testing. The Iowa and San Diego teams are similar in mesomorphy and ectomorphy, higher in mesomorphy than the Oregon teams, and lower in ectomorphy. The San Diego team's extreme endo-mesomorphs were linemen and linebackers, with a distribution similar to the professionals, but slightly lower in mesomorphy. Compared with United States norms, the majority of the San Diego team, like the professionals, are above the 90th percentile in height and weight (Carter, 1968). In all samples, the few ecto-mesomorphs that are found are wide receivers, defensive backs, or kickers. Some differences by position can be inferred from data on the Oklahoma team (Votto, 1976).

Carter (1968) summarized Sheldon's (1954) colourful descriptions of college and professional football players. In general these clinical statements have been supported by subsequent studies. Of course, when the Heath–Carter open scale is applied the resulting higher ratings show that the physiques are even more extreme than Sheldon suggested. In addition, the trend is toward greater size and higher mesomorphy among contemporary college and professional players. Sheldon (1940) observed that a coach who cannot distinguish between a $5\frac{1}{2}$ and a 6 in mesomorphy might not win many football games. Perhaps this is true today of the small college level teams, but at higher levels it appears that well trained, rugged 6's, 7's and 8's are mandatory for even a glimmer of success.

(*b*) *Rugby football* (Table 6.7, Fig. 6.17(*b*)).

Rugby players were mostly endo-mesomorphs and balanced meso-morphs. Among the South African trialists (some of the best players in the world) the forwards were more endo-mesomorphic than the backs. How-ever, Reilly & Hardiker (1981) found no difference in somatotype between the two positons among the Polytechnic players. Boennec *et al.* (1980) found that French rugby forwards were more endo-mesomorphic (3-6-1) than backs ($2\frac{1}{2}$-5-$2\frac{1}{2}$). The rugby players are somatotypically fairly similar to American football players. They are quite mesomorphic, with the lowest mean at 5.4. In a comparison of somatotypes of active players with those who had retired and were 18.4 years older, Smit *et al*, (1979*a*) found that retirees were the more endomorphic, and there was no difference by former playing position.

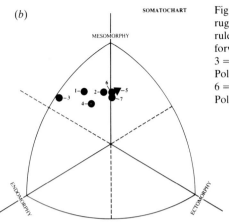

Fig. 6.17(*b*). Mean somatotypes for male rugby players (●), and for an Australian rules sample (▼). 1 = South Africa, forwards; 2 = South Africa, backs; 3 = South Africa, 'retired'; 4 = British Polytechnics; 5 = Australian Rules; 6 = South Australia; 7 = Brighton Polytechnic.

(c) *Soccer football* (Table 6.7, Fig. 6.17(*c*)).

In general, the somatotypes of European soccer players are distributed over much of the mesomorphic sector of the somatochart, with means close to $2\frac{1}{2}$-5-$2\frac{1}{2}$. The Czechoslovaks, Brazilians and the Bolivar Games players are slightly more endo-mesomorphic, and the Cubans, Nigerians, and English are more ecto-mesomorphic. French players (Boennec *et al.* 1980) are most like the Czechoslovaks.

Among the few studies that examined somatotype differences by position, Withers *et al.* (1986) cited the 1975 finding of Bell and Rhodes that among Welsh college players goalkeepers were less ectomorphic than strikers. Two studies from India showed that soccer players are much less mesomorphic and more ectomorphic than in other countries. Sodhi & Sidhu (1984) found small differences by playing position. The State level players tended toward greater endomorphy and mesomorphy and less ectomorphy than their

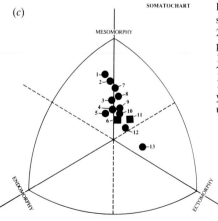

Fig. 6.17(*c*). Mean somatotypes for male soccer players. 1 = Czechoslovakia, 1968; 2 = Bolivar Games; 3 = Brazil, 1986, professionals; 4 = Bratislava; 5 = Brazil, 1978, professionals; 6 = Cuba, youths; 7 = South Australia; 8 = Cuba; 9 = Nigeria; 10 = Brighton Polytechnic; 11 = Brazil, youths; 12 = India, State; 13 = India, university. ● = adults, ■ = youths.

counterparts at the university level. Kansal *et al.* (1986) found goalkeepers were less mesomorphic and more ectomorphic than players in other positions.

Young Brazilians and Cubans have similar somatotypes. In an indirectly related study of 33 Brazilian referees, De Rose *et al.* (1979) found their mean somatotype of 4.1-5.7-1.4 to be much more endomorphic than the soccer players.

(d) Australian Rules football (Table 6.7, Fig. 6.17(*b*)).

In the only study of this form of football, Withers *et al.* (1986) reported a mean somatotype of 2.1-5.7-2.5 for South Australian players.

Gymnastics (Table 6.8, Fig. 6.18)

The means are near the 1½-6-2 somatotype for most of the adult samples. The majority of gymnasts are either ecto-mesomorphs or balanced mesomorphs. Only a few are higher in endomorphy than ectomorphy. The Danish gymnasts were more endo-mesomorphic than other samples, but they were a touring group who performed exhibitions of educational gymnastics, not Olympic gymnasts. The ratings were made from photographs in Cureton (1951).

There were no differences in height, weight, or somatotype among Whites, Mestizos or Orientals in the sample from the Mexico City Olympics (De Garay *et al.*, 1974). Furthermore, there were no differences between the 1968 and 1976 Olympic gymnasts (Carter, 1984*b*). Carter *et al.* (1971) found that better performers were more mesomorphic than other competitors in the 1968 AAU Championships. Overall, top class gymnasts have very low endomorphy, with little variation, and are as low as any group of male athletes. The Chinese national squad are the smallest gymnasts, are lowest in endomorphy, and are the most ecto-mesomorphic. Many gymnasts have an upper body mesomorphic dysplasia, i.e. the upper body is more mesomorphic than the lower, an undoubted favourable adaptation for several events in this sport. Some gymnasts have ratings of 7 or 8 in mesomorphy.

As expected, the young gymnasts are much less mesomorphic than their adult counterparts. In addition to the three samples in Table 6.8, Alonso (1986) reports a mean of 1.4-4.3-3.9 for three 12 year old Cubans. Thorland *et al.* (1981) found a combined mean of 2.3-5.0-3.2 for young gymnasts and divers at a Junior Olympic training camp. Based on these data, it could be anticipated that the young male gymnast may be about 4½ in mesomorphy at age 12 and reach 5½ or 6½ at optimal development.

Studies by Caldeira *et al.* (1986*c*), and Vivolo, *et al.* (1986*b*) provide limited data on Brazilian gymnasts. Zrubák *et al.* (1981) describe young Slovak gymnasts.

(a)

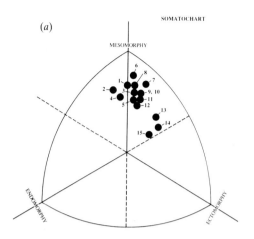

(b)

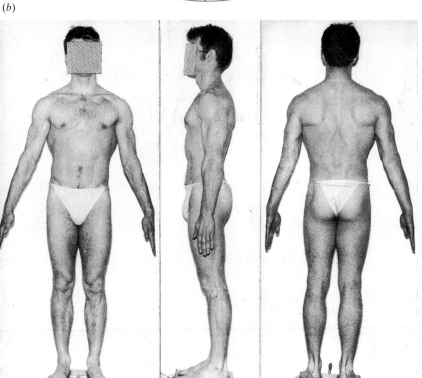

Fig. 6.18(*a*). Mean somatotypes for male gymnasts. 1 = AAU, USA, 1968; 2 = Denmark; 3 = Belgium; 4 = USSR; 5 = University of Iowa, 1958; 6 = Czechoslovakia, 1968; 7 = China, 1984; 8 = Cuba 1976–80; 9 = Olympics, 1968 and 1976; 10 = South Australia; 11 = Cuba, 1977; 12 = Venezuela; 13 = Oregon, Class II–IV; 14 = Brno, Czechoslovakia; 15 = Brazil, club. (*b*). Gymnast. Height = 169.0 cm; weight = 66.3 kg; HWR = 41.8; somatotype = 1-7½-2. (Photo with permission from J. Štěpnička.)

Table 6.8. *Gymnastics – male*

Sample, reference	N	Statistic	Age (yr)	Height (cm)	Weight (kg)	HWR	Somatotype
Mexico City Olympics, 1968 (De Garay *et al.*, 1974)	28	X̄	23.6	167.4	61.5	42.4	1.4-5.9-2.4
		SD	3.8	5.7	5.5	1.2	0.3 0.8 0.9
Montreal Olympics, 1976 (Carter *et al.*, 1982)	11	X̄	25.4	169.3	63.5	42.5	1.4-5.8-2.5
		SD	4.6	5.6	6.9	1.0	0.5 0.8 0.8
Denmark (Carter, 1970)	15	X̄	24.6	172.7	74.5	41.0	2.6-6.2-1.5
		SD	2.8	5.3	6.4	0.9	0.8 0.4 0.5
University of Iowa, 1958 (Carter, 1970)	10	X̄	22.3	176.5	71.8	42.5	2.0-5.8-2.6
		SD	2.1	5.8	6.1	1.2	0.5 0.7 0.9
USSR (Heath, in Carter, 1970)	5	X̄	—	172.7	72.2	41.5	2.6-6.0-2.1
		SD		2.8	1.2	0.5	0.7 0.6 0.4
AAU placers, 1968 (Carter *et al.*, 1971)	11	X̄	22.3	165.1	61.8	41.9	1.9-6.4-2.0
		SD	3.2	6.1	6.3	0.5	0.5 0.5 0.6
Czechoslovakia, 1968 (Štěpnička, 1974*a*, 1986)	58	X̄	—	169.7	66.5	41.9[a]	1.5-6.9-2.1
		SD		4.0	4.2		0.6 0.7 0.6
Venezuela (Pérez, 1981)	13	X̄	17.2	164.4	58.0	42.4	1.7-5.4-2.5
		SD	2.1	7.0	6.9	0.9	0.4 0.7 0.6
Belgium (Clarys & Borms, 1971)	14	X̄	—	170.2	65.8	42.2[a]	1.7-5.9-2.2
Cuba, international 1977 (Lopez *et al.*, 1979)	33	X̄	21.6	167.9	61.6	42.6	1.8-5.9-2.7
		SD	3.4	4.8	6.0	1.0	0.4 0.9 0.5
Cuba, 1976–80 (Rodríguez *et al.*, 1986)	15	X̄	22.4	168.0	64.2	42.0[a]	1.6-6.2-2.3
		SD	2.8	5.0	5.3		0.3 0.7 0.5
South Australia, 1981–83 (Withers *et al.*, 1986)	8	X̄	20.2	169.6	63.8	42.5	1.9-6.1-2.5
		SD	2.7	5.1	6.1	1.0	0.3 0.6 0.8
China, 1984 (Zeng, 1985)	19	X̄	19.8	157.4	52.5	42.1[a]	1.1-6.3-2.5
		SD	3.3	8.7	9.3		0.3 1.1 1.0
Oregon, Class II–IV (Broekhoff *et al.*, 1986)	17	X̄	12.5	149.6	40.2	43.7[a]	1.3-4.8-3.6
		SD	8–15	12.2	10.5		0.6 0.6 1.2
Czechoslovakia, Brno (Štěpnička, 1976*b*)	10	X̄	12.5	151.1	39.7	44.3[a]	1.5-4.4-3.9
		SD		7.1	5.6		0.6 0.6 0.6
Brazil, club (Araújo & Moutinho, 1978)	11	X̄	13.6	153.9	42.5	44.1[a]	2.2-4.2-4.0
		SD	2.3	12.8	9.6		0.4 0.8 1.0

[a]Calculated from mean height and weight.
X̄ = mean.

Table 6.9. *Handball – male*

Sample, reference	N	Statistic	Age (yr)	Height (cm)	Weight (kg)	HWR	Somatotype
Czechoslovakia, 1968	21	X̄	—	180.9	79.1	42.1[a]	2.4-5.6-2.6
(Štěpnička, 1974a)		SD		5.4	6.0		0.8 0.6 0.9
Czechoslovakia, 1977,	16	X̄	—	188.1	85.4	42.7[a]	2.0-5.1-2.8
national team (Štěp-		SD		4.2	4.3		0.5 0.9 0.7
nička, 1986)							
Hungary, 1970s							
1972–75	30	X̄	—	180.1	74.4	42.8[a]	1.8-5.5-2.8
1979–80	21	X̄	—	182.9	76.9	43.0[a]	2.9-5.1-3.0
(Mészáros &							
Mohácsi, 1982b)							
Brazil (N = 58)							
Pará	—	X̄	—	—	—	—	3.3-4.8-2.2
Rio de Janeiro	—	X̄	—	—	—	—	3.3-5.7-2.1
Rio Grande do Sul	—	X̄	—	—	—	—	2.4-4.9-3.1
Bahia	—	X̄	—	—	—	—	3.1-4.6-2.3
Sergipe	—	X̄	—	—	—	—	2.6-3.6-3.1
(Oliveira *et al.*, 1986)							
Cuba, youths (Alonso, 1986)	35	X̄	12.5	155.9	46.5	43.4[a]	3.0-4.4-3.3

[a] Calculated from mean height and weight.
X̄ = mean.

Handball (Table 6.9, Fig. 6.19)

Top European team handball players tend toward balanced meso-morphy and ecto-mesomorphy. A comparison of Class I Hungarian players from early to late 1970s showed recent increased size and endomorphy. In contrast, Czechoslovak players became slightly more ecto-mesomorphic

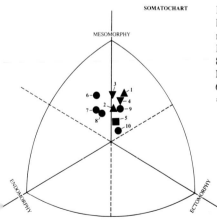

Fig. 6.19. Mean somatotypes for male team handball players. The names after Brazil refer to teams from the given region. Hungary (▲) 1 = 1972–75, 1977; 2 = 1979–80; Czechoslovakia (▼) 3 = 1968; 4 = National; Cuba (■) 5 = youths; Brazil (●) 6 = Rio de Janeiro; 7 = Para; 8 = Bahia; 9 = Rio Grande do Sul; 10 = Sergipe.

between 1968 and 1979. In Brazil, where team handball is a developing sport, mean mesomorphy varied from 3.6 to 5.7 for the top players from five states. Young Cubans were balanced mesomorphs.

Hockey
Hockey is presented as field hockey and ice hockey.

(a) Field hockey (Table 6.10, Fig. 6.20(*a*),(*b*)).
Olympic field hockey players were studied for the first time at the 1976 Montreal games. Thirty-three of the sample of 47 players were from Argentina, Kenya, Malaysia and Australia; 14 from New Zealand were measured before they left for Montreal (where they won the gold medal). The 14 New Zealand gold medallists were lower in endomorphy than the other 33 competitors. Possibly their lower endomorphy was due in part to their rigorous physical training program, designed to increase caloric expenditure and lessen excess fat weight.

The White players were more mesomorphic and less ectomorphic than those of Indian–Pakistani origins. Carter *et al.* (1981) suggested that physique differences among field hockey teams are due in part to ethnic differences as well as training differences. Most of the players for Kenya and Malaysia were of Indian–Pakistani origin. In field hockey India has been a leading country for several decades. Sidhu & Wadhan (1975) somatotyped 25 players at the National Institute of Sports in Patiala. Another group of 30 state level players at a National coaching camp were somatotyped according to playing position (Sodhi & Sidhu, 1984). The overall mean for the two studies was 3.5-4.0-3.0. Endomorphy gradually increased from forwards to halves, fullbacks and goalkeepers. Both samples were more mesomorphic and less endomorphic and ectomorphic than their respective control groups of non-athletes. At the Montreal Olympics the mean for the players of Indian–Pakistani origin was 2.6-4.0-3.2, which suggests that with similar ethnic backgrounds players at the highest level of competition are slightly less endomorphic and more ectomorphic than those at a lower level.

The Nigerian, New Zealand, Argentine and Australian players were more mesomorphic than the Indians and Pakistanis. The Nigerians were slightly more endo-mesomorphic than White players who competed in the 1976 Olympics.

(b) Ice Hockey (Table 6.10, Fig. 6.20(*c*))
There are several samples of ice hockey players, both youths and adults, mainly from Canada and Czechoslovakia. Among the three samples of CSSR players the national team was the most mesomorphic. Most top players are endo-mesomorphs or balanced mesomorphs with a mean near

(a)

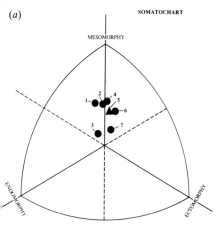

Fig. 6.20(a). Mean somatotypes for male field hockey players. 1 = Nigeria; 2 = Argentina; 3 = India; 4 = South Australia; 5 = Montreal Olympics; 6 = New Zealand; 7 = Kenya and Malaysia. (b). Field hockey player. Height = 178.0 cm; weight = 60.1 kg; HWR = 45.4; somatotype = $1\frac{1}{2}$-$3\frac{1}{2}$-$4\frac{1}{2}$. (c). Mean somatotypes for male adult ● and youth ■ ice hockey players. 1 = Czechoslovakia, national, 1976; 2 = Czechoslovakia, league, 1972; 3 = Czechoslovakia, elite; 4 = University of Western Ontario; 5 = University, Intramural; 6 = USA + Finland, Bantam; 7 = London, recreational; 8 = Quebec, Nordiques; 9 = Quebec, Ramparts; 10 = Canada, Bantam; 11 = London, All Stars; 12 = Kladno, youths.

(b)

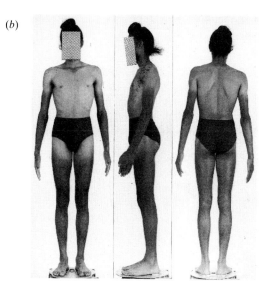

(c)

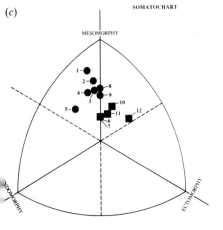

Table 6.10. *Hockey – male*

Sample, reference	N	Statistic	Age (yr)	Height (cm)	Weight (kg)	HWR	Somatotype
(a) *Field hockey*							
Montreal Olympics,	47	X̄	25.6	176.1	70.4	42.7	2.3-4.6-2.7
1976		SD	3.5	5.6	7.0	1.2	0.8 0.9 0.8
Argentina	16	X̄	24.6	177.0	73.6	42.3	2.4-5.1-2.3
		SD	4.6	6.5	5.2	0.9	0.8 0.8 0.7
New Zealand	14	X̄	26.3	176.0	70.3	42.7	1.9-4.5-2.7
		SD	2.5	6.4	7.9	1.0	0.4 0.8 0.8
Kenya–Malaysia	15	X̄	26.3	174.6	65.6	43.3	2.7-4.0-3.2
(Carter *et al.*, 1981,		SD	3.2	3.2	4.3	1.1	1.0 0.6 0.8
Carter, 1984*b*)							
Jat-Sikh, India (Sidhu	25	X̄	21.6	—	—	—	3.5-4.0-3.0
& Wadhan, 1975)							1.3 0.6 0.7
India, Patiala	30	X̄	23.7	—	—	—	3.2-3.8-2.7
		SD	2.7				
Forwards	13	X̄	—	169.8	63.6	42.5[a]	3.0-3.9-2.6
		SD		2.8	5.7		0.9 0.5 0.6
Halves	8	X̄	—	170.6	63.1	42.9	3.1-3.8-2.8
		SD		5.5	6.4		0.8 0.6 0.6
Fullbacks	4	X̄	—	175.5	68.5	42.9	3.5-4.0-2.9
		SD		7.2	11.7		0.7 1.1 1.4
Goalkeepers	5	X̄	—	168.2	59.9	43.0	3.7-3.5-2.9
(Sodhi & Sidhu, 1984)		SD		6.9	6.6		1.2 0.9 0.7
Nigeria (Toriola *et al.*,	14	X̄	25.7	167.2	65.2	41.5	2.8-5.1-2.0
1985)		SD	1.1		4.6	4.1	0.6 0.7 0.8
South Australia	14	X̄	23.7	178.6	73.9	42.6	2.4-5.4-2.6
(Withers *et al.*,		SD	3.6	5.9	5.7	0.9	0.6 0.8 0.8
1986)							
(b) *Ice Hockey*							
(i) *Adults*							
Czechoslovakia, elite	55	X̄	23.0	176.9	78.0	41.4[a]	2.6-5.7-1.9
		SD		4.5	7.6		
Forwards	33	X̄	22.7	176.2	76.4	41.5[a]	2.3-5.8-2.0
		SD		4.3	6.6		
Backs	16	X̄	22.6	179.8	82.6	41.1[a]	2.9-5.8-1.7
		SD		4.2	7.4		
Goalkeepers	6	X̄	25.1	174.7	74.8	41.5[a]	2.9-5.3-2.0
(Chovanová & Zrubák,		SD		5.5	8.9		
1972, Chovanová,							
1976*a*)							
Czechoslovakia,	91	X̄	—	176.8	78.6	41.3[a]	2.3-6.0-1.7
League, 1972		SD		5.2	6.9		0.8 0.8 0.7
National team, 1976	24	X̄	—	178.8	85.1	40.7[a]	2.4-6.4-1.3
(Štěpnička, 1974*a*, 1986)		SD		4.0	5.8		1.1 0.9 0.6
Quebec							
Nordiques,	12	X̄	25.3	175.2	75.9	41.4[a]	2.1-5.4-1.8
professional		SD	5.3	5.0	5.0		0.4 0.7 0.8

Table 6.10. (*Continued*)

Sample, reference	N	Statistic	Age (yr)	Height (cm)	Weight (kg)	HWR	Somatotype
Ramparts,	24	X̄	18.2	177.3	77.0	41.7[a]	2.1-5.2-2.0
major junior		SD	1.1	5.4	6.0		0.5 0.9 0.9
(Bouchard *et al.*, 1974)							
University of Western Ontario							
Varsity	21	X̄	—	176.0	75.5	41.6[a]	3.1-5.8-2.0
		SD		5.7	4.7		1.2 0.8 0.7
Intramural	21	X̄	—	175.4	73.7	41.8[a]	4.3-5.3-2.2
(Pirie, 1974)		SD		4.3	8.9		1.7 0.8 0.9
(ii) *Youths*							
USA + Finland,	59	X̄	14.6	169.1	59.5	43.4	3.2-4.8-3.2
Bantam, 4 teams		SD	0.5	7.9	8.5	1.4	1.0 1.0 1.1
(Newton, 1978)							
Canada, Bantam	13	X̄	14.5	—	—	—	2.0-4.7-3.0
(Larivière *et al.*, 1978)							
London, Ontario							
All Stars	15	X̄	10.0	141.4	34.5	43.4[a]	2.6-4.7-3.3
		SD		6.7	5.6		0.9 0.7 0.8
Recreational	15	X̄	10.0	140.8	34.7	43.2[a]	3.1-4.6-3.1
(Pirie, 1974)		SD		4.9	5.8		1.3 0.9 1.1
Czechoslovakia, Kladno	15	X̄	10.5	138.6	31.3	44.0[a]	1.6-4.3-3.9
(Štěpnička, 1976*b*)		SD		5.8	4.3		1.0 0.8 0.8

[a]Calculated from mean height and weight.
X̄ = mean.

$2\frac{1}{2}$-6-2. Compared by playing position, backs tend to be more endomorphic than forwards, and goalkeepers less mesomorphic than forwards or backs. It is of particular interest that goalkeepers have an exceedingly wide distribution, suggesting that somatotype is not an important factor in this highly specialized position. The Canadian professional, junior and university teams are somatotypically close to the Czechoslovak league and elite samples. Marcotte & Herminston (1978) also reported a mean of 3.1-5.8-2.0 for another university sample. The recreational university players are much more endomorphic and less mesomorphic than most of the high playing level adult samples.

Clearly, young players are less mesomorphic than adults. The ten- and fourteen-year-olds are fairly similar in mesomorphy, but the Kladno ten-year-olds are less endomorphic and more ectomorphic than the others.

Selection methods and growth differences may account for some of the differences.

Judo (Table 6.11, Fig. 6.21(*a*), (*b*))

The majority of the 13 judoists somatotyped at the Montreal Olympics were endo-mesomorphs (54%) and balanced mesomorphs (34%). When divided into two weight groups, the heavier group was slightly more endo-mesomorphic than the lighter group. Other studies have shown a similar trend toward higher endo-mesomorphy.

The trend toward increased endo-mesomorphy with increase in weight class seen in the Olympians is also observed in the Brazilian, Cuban, Hungarian, Pan American and World Championship competitors. Judoists

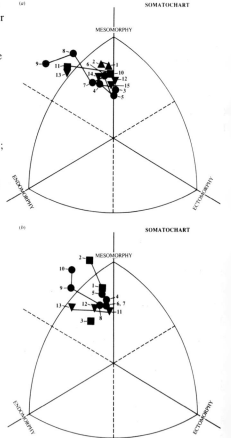

Fig. 6.21. Mean somatotypes for male judo competitors. Symbols connected by a line or lines show the trend between adjacent weight groups from the same study. In general, the lighter weight groups are to the right of the somatochart and the heavier groups are to the left. (*a*) Montreal Olympics (▲) 1 = 60–79.9 kg; 2 = 80–89.9 kg; Brazil (●) 3 = <59.9 kg; 4 = 65.0 kg; 5 = 70.5 kg; 6 = 78.0 kg; 7 = 85.6 kg; 8 = 92.3 kg; 9 = 136.4 kg; Hungary (■) 10 = 60–71 kg; 11 = >71 kg; 12 = Belgium; 13 = Korea; 14 = Bolivar Games; 15 = Czechoslovakia, 1977. (*b*) Cuba (■) 1 = 60–80 kg; 2 = >80 kg; 3 = youths; Pan Am Games (●) 4 = <60 kg; 5 = 60–65 kg; 6 = 65–71 kg; 7 = 71–78 kg; 8 = 78–86 kg; 9 = 86–95 kg; 10 = >95 kg; World Championships (▼) 11 = <71 kg; 12 = 71–86 kg; 13 = >86 kg.

Table 6.11. *Judo – male*

Sample, reference	N	Statistic	Age (yr)	Height (cm)	Weight (kg)	HWR	Somatotype
Montreal Olympics, 1976							
total	13	X̄	23.4	173.1	76.5	40.9	2.0-6.4-1.3
		SD	3.0	7.8	11.0	0.7	0.6 0.8 0.4
60–79.9 kg	9	X̄	—	—	—	41.0	1.9-6.4-1.5
		SD				0.8	0.3 0.9 0.5
80–89.9 kg	4	X̄				40.6	2.2-6.5-1.2
(Carter *et al.*, 1982;		SD				0.4	0.8 0.4 0.3
Carter, 1984*b*)							
Brazil							
<59.9 kg[a]	5	X̄	22.4	164.0	59.2	42.1[b]	2.1-5.5-2.2
		SD	2.7	3.1	0.5		0.2 0.4 0.6
65.0 kg	5	X̄	24.5	165.8	64.4	41.4	2.9-5.9-1.7
		SD	8.3	3.7	0.7		0.9 0.8 0.6
70.5 kg	5	X̄	27.6	172.2	70.0	41.8	2.9-5.4-2.0
		SD	7.0	4.8	0.6		1.4 1.0 0.9
78.0 kg	4	X̄	25.5	174.3	76.9	41.0	2.1-6.1-1.4
		SD	5.8	2.9	0.8		0.2 0.4 0.6
85.6 kg	5	X̄	25.5	177.8	81.9	40.9	2.0-5.9-1.4
		SD	6.8	6.0	3.3		0.9 0.8 0.8
92.3 kg	5	X̄	24.4	179.4	91.1	39.9	3.2-7.5-0.8
		SD	4.8	5.5	0.9		0.5 1.5 0.7
136.4 kg	4	X̄	21.5	186.7	110.4	38.9	4.9-7.1-0.2
(Araújo *et al.*, 1978*b*)		SD	2.4	10.5	17.2		2.6 0.8 0.5
Pan American Games, 1979							
<60 kg	8	X̄	21.3	161.6	59.7	41.4[b]	2.1-5.9-1.6
		SD	3.2	7.4	3.5		
60–65 kg	8	X̄	21.6	166.6	65.2	41.4[b]	2.2-6.1-1.6
		SD	3.1	2.9	0.9		
65–71 kg	9	X̄	22.2	172.7	72.0	41.5[b]	2.4-5.9-1.8
		SD	3.0	5.2	3.1		
71–78 kg	9	X̄	20.6	177.2	78.3	41.4[b]	2.3-5.8-1.6
		SD	1.7	4.6	1.3		
78–86 kg	8	X̄	24.1	180.5	83.5	41.3[b]	2.8-6.0-1.7
		SD	5.9	5.4	3.6		
86–95 kg	4	X̄	21.8	179.4	91.8	39.8[b]	3.6-7.0-0.9
		SD	1.3	2.9	1.3		
>95 kg	3	X̄	27.3	185.7	107.6	39.0[b]	3.7-7.7-0.7
(Chernilo *et al.*, 1979)							
Cuba, 1976–80							
60–80 kg	13	X̄	21.7	171.0	71.3	41.2[b]	2.0-6.2-1.1
		SD	2.1	5.1	5.2		0.5 0.8 0.7
>80 kg	8	X̄	23.6	180.8	95.2	39.6[b]	2.8-8.1-1.0
(Rodríguez *et al.*, 1986)		SD	2.4	3.8	12.6		1.3 1.2 0.4

(*continued*)

Table 6.11. (*Continued*)

Sample, reference	N	Statistic	Age (yr)	Height (cm)	Weight (kg)	HWR	Somatotype
Bolivar Games, 1981	15	X̄	24.2	163.5	73.3	40.8	2.6-6.0-1.4
(Brief, 1986)		SD	3.6	6.0	10.2	1.0	0.9 0.8 0.6
Czechoslovakia, 1977	10	X̄	—	179.4	83.0	41.1[b]	2.0-6.5-1.8
(Štěpnička, 1986)		SD		8.8	16.6		0.9 0.9 0.6
Hungary							
Total	18	X̄	21.5	174.6	81.3	40.3	3.6-7.0-1.6
		SD		7.7	18.6		1.9 1.5 0.9
60–71 kg	7	X̄	22.0	167.8	66.7	41.4	2.5-6.6-1.8
		SD	3.8	6.1	3.7		0.5 1.3 1.0
>71 kg	11	X̄	21.2	178.6	90.5	39.8	4.3-7.2-1.4
(Farmosi, 1980*a*)		SD	2.0	5.4	18.4		2.1 1.6 0.7
Belgium (Claessens	24	X̄	21.9	175.2	74.3	42.2	1.9-5.8-2.0
et al., 1986*a,b*)		SD		7.3	11.0	2.1	0.5 0.9 0.8
World Championships, 1981							
<71 kg	18	X̄	24.9	169.0	65.7	41.9[b]	2.3-5.6-1.9
		SD	4.0	4.4	4.3		0.4 0.5 0.4
71–86 kg	9	X̄	25.2	177.8	81.2	41.1[b]	3.0-6.0-1.7
		SD	4.7	4.3	3.7		0.5 0.7 0.7
>86 kg	11	X̄	25.8	186.8	108.3	39.2[b]	4.1-6.2-1.3
(Claessens *et al.*, unpubl.)		SD	3.6	7.8	15.1		0.9 0.6 0.4
Korea, Dong A University (Shin, 1985)	14	X̄	19.7	172.9	82.3	39.8[b]	4.3-7.0-1.3
		SD	1.1	5.4	19.4		2.0 1.1 0.8
Cuba, youths (Alonso, 1986)	28	X̄	12.5	144.7	46.4	40.3[b]	3.7-5.5-1.9

[a]Highest weight for each group.
[b]Calculated from mean height and weight.
X̄ = mean.

from Belgium and Czechoslovakia were balanced mesomorphs and those in the Bolivar Games and those from Korea were endo-mesomorphs.

Using the Leuven somatotype method, Claessens *et al.* (1986*b*) showed that typical Belgian judoists were endo-mesomorphs and that there was a shift to higher mesomorphy and lower ectomorphy in the higher of the two weight groups.

The somatotype means for the same weight classes were almost identical for Olympic judoists and wrestlers. This finding is in keeping with some similarities between the two sports.

Rowing (Table 6.12, Fig. 6.22)

Although most information is on heavyweight rowers, recently lightweight (average 70 kg, 72.5 kg maximum per boat) rowing has been recognized internationally. Two samples are presented here. The somatotypes of rowers from the 1968 and 1976 Olympics had almost identical distributions with a combined mean of 2.2-5.2-2.5. The most common somatotypes were balanced mesomorphs, with almost equal numbers of endo-mesomorphs and ecto-mesomorphs. In the Olympic samples there were no differences by ethnic origin, event, stroke, or geographic region, except for the greater endomorphy of the West German rowers compared with the Americans at the Montreal Olympics (Carter, 1984*b*). The mean somatotypes of other top level rowers were close to the Olympic means and averaged around 2-5½-2½. Even the San Diego State University crews have somatotypes within the Olympic distribution. In contrast, the New Zealand club, or recreational, rowers were more endomorphic than the national squad (Lewis, 1969) and other samples. At the USA Olympic trials, Sutorius (1969) found some differences among four eight-oar crews in mesomorphy and ectomorphy, but no differences between singles and eights. Rodríguez *et al.* (1986) reported means of 2.0-5.6-2.4 for singles and 2.5-5.2-2.5 for fours plus eights in lightweight rowers. Based on the somatoplots only, Class I Hungarian rowers were similar to Olympic rowers in distribution and mean somatotype (Mészáros & Mohácsi, 1982*a*). The two samples of lightweight rowers, from the World Championships and from South Australia have quite similar means and are less mesomorphic and more ectomorphic than heavyweight rowers.

In other studies of rowers Lewis (1969) and Williams (1977) used the M.4 method, and Jones *et al.* (1965) used the Sheldon method.

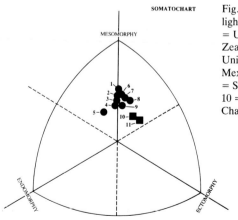

Fig. 6.22. Mean somatotypes for male lightweight ■ and heavyweight ● rowers. 1 = USA, singles; 2 = Cuba; 3 = New Zealand, national; 4 = San Diego State University; 5 = New Zealand, club; 6 = Mexico City Olympics; 7 = USA, eights; 8 = South Australia; 9 = Montreal Olympics; 10 = South Australia; 11 = World Championships, 1985.

Table 6.12. *Rowing – male*

Sample, reference	N	Statistic	Age (yr)	Height (cm)	Weight (kg)	HWR	Somatotype
Mexico City Olympics,	85	X̄	24.3	185.1	82.6	42.4	2.1-5.3-2.4
1968 (De Garay		SD	3.3	5.9	7.4	1.1	0.6 0.9 0.8
et al., 1974)							
Montreal Olympics, 1976	65	X̄	24.2	191.3	90.0	42.7	2.3-5.0-2.7
1976 (Carter *et al.*,		SD	3.3	5.7	5.6	1.0	0.6 0.9 0.8
1982)							
Cuba, 1976–80	24	X̄	22.2	187.4	86.3	42.4[a]	2.3-5.3-2.5
(Rodríguez *et al.*,		SD	2.3	3.6	4.0		0.6 0.8 0.6
1986)							
South Australia							
Heavyweight	7	X̄	24.7	192.3	88.7	43.1	2.0-5.2-3.0
		SD	1.9	5.3	4.7	0.7	0.5 0.7 0.5
Lightweight	5	X̄	21.3	181.9	72.0	43.8	2.1-4.3-3.4
(Withers *et al.*, 1986)		SD	2.4	3.1	4.1	1.0	0.2 0.4 0.7
World Championships,							
Hazenwinckel, 1985							
Lightweight	144	X̄	24.3	180.7	70.3	43.7	1.6-4.0-3.4
(Rodríguez, 1986)		SD	3.3	4.5	1.9	1.0	0.3 0.8 0.7
San Diego State	21	X̄	20.2	183.6	79.8	42.7[a]	2.7-5.1-2.6
University (Sutorius							
& Carter, 1967)							
USA Olympics trials,							
1968							
Singles	11	X̄	25.0	186.3	84.9	42.3	2.3-5.9-2.5
		SD	4.0	8.1	6.7	1.0	0.7 0.8 0.7
Eights (4 crews)	32	X̄	21.6	191.2	88.6	42.9	2.1-5.3-2.7
(Carter, 1971; Sutorius,		SD	1.4	3.4	4.7	0.9	0.4 0.5 0.6
1969)							
New Zealand, 1967–68							
National	16	X̄	24.8	187.0	85.6	42.4	2.5-5.2-2.5[b]
		SD	2.3	5.5	5.1	0.5	
Club	8	X̄	26.0	184.0	82.5	42.3	3.5-4.9-2.4[b]
(Lewis, 1969)		SD	4.3	3.3	4.8	1.5	

[a]Calculated from mean height and weight.
[b]Heath–Carter anthropometric somatotype calculated from means.
X̄ = mean.

Skiing and ski jumping (Table 6.13, Figs. 6.5, 6.23(*a*), (*b*))

The four events considered in this section are cross-country skiing, downhill skiing, Nordic combined (15 km cross-country and 70 m ski jump), and ski jumping. In an extensive study, Chovanová (1976*b*) somatotyped 14 athletes from twelve European countries that competed in championship events in the Tatras, Slovakia, in 1970–71 (Fig. 6.5). She divided the competitors into A and B groups based on performance, and included the CSSR national ski jumping team, which was best in the world at that time. In cross-country and downhill the A groups were slightly more mesomorphic, but there were no A–B differences in Nordic and jumping events. Among all the samples, cross-country skiers are more ectomorphic and less mesomorphic than those in other events, downhill skiers and ski jumpers are the most mesomorphic, and Nordic athletes are intermediate. The USA cross-country skiers are similar to the B group Europeans. The Czechoslovak and

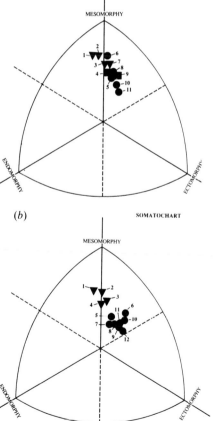

Fig. 6.23(*a*). Mean somatotypes for male skiers in cross-country (●) and Nordic combined (■) events, and ski jumpers (▼). 1 = Bulgaria, elite; 2 = Czechoslovakia, national; 3 = European, Group A; 4 = European, Group B; 5 = Bulgaria, elite; 6 = Czechoslovakia, 1968; 7 = European, Group B; 8 = European, Group B; 9 = European, Group A; 10 = USA, national; 11 = European, Group B. (*b*). Mean somatotypes for male downhill skiers. ▼ = adults; ● = youths. 1 = Bulgaria, elite; 2 = European, Group A; 3 = Czechoslovakia, 1969–71; 4 = European, Group B; 5 = Canada, Thunder Bay; 6 = Czech, club; 7 = Star, club. Bratislava; 8 = Slovak, club, 15.6 yr; 9 = Slovak, club, 17.6 yr; 10 = Canada, Rossland; 11 = Slovak, club, 13.3 yr; 12 = Slovak, club, 11.6 yr.

Table 6.13. *Skiing and ski jumping – male*

Sample, reference	N	Statistic	Age (yr)	Height (cm)	Weight (kg)	HWR	Somatotype
(a) *Cross-country*							
Czechoslovakia, 1968	46	X̄	25.2	174.6	70.6	42.3[a]	1.7-6.3-2.0
(Štěpnička, 1974a,		SD		4.5	5.1		0.6 0.7 0.7
1977, 1986)							
European							
Total	26	X̄	22.1	175.8	69.3	42.8[a]	1.9-5.3-2.8
		SD		5.2	6.3		
Group A	8	X̄	23.1	174.6	69.6	42.5[a]	1.8-5.5-2.6
		SD		4.7	6.0		
Group B	18	X̄	20.2	178.1	68.8	43.5[a]	2.1-4.8-3.3
(Chovanová, 1976b)		SD		5.6	7.1		
U.S. National, 1975	11	X̄	22.8	179.0	71.8	43.1[a]	2.0-4.5-3.0
(Sinning et al.,		SD	1.9	5.0	5.4		0.5 0.4 0.3
1977)							
Bulgaria, elite	30	X̄	21.7	176.8	70.0	42.9[a]	2.1-5.4-2.7
(Toteva & Sumanov,							
1984)							
(b) *Downhill (Alpine)*							
Adults							
European, 1971							
Total	43	X̄	23.0	174.0	70.3	42.2[a]	2.3-5.7-2.4
		SD		6.2	6.7		
Group A	23	X̄	23.3	173.6	71.1	41.9[a]	2.1-6.0-2.2
		SD		6.5	6.5		
Group B	20	X̄	22.7	174.5	70.1	42.3[a]	2.6-5.5-2.6
(Chovanová, 1976b)		SD		6.2	7.3		
Czechoslovakia, 1969–71	12	X̄	—	174.6	70.2	42.3[a]	2.3-5.6-2.6
(Štěpnička, 1974a,		SD		5.6	5.9		0.8 0.8 0.7
1977, 1986)							
Bulgaria, elite	35	X̄	22.4	173.4	71.8	41.7[a]	2.6-6.2-1.9
(Toteva & Sumanov,							
1984)							
Youths							
Czech, club (Štěp-	18	X̄	12–14	161.2	51.3	43.4[a]	1.5-5.0-3.6
nička & Broda, 1977)		SD		10.9	7.4		0.1 0.4 0.5
Slovak, club, 4 age	18	X̄	11.6	149.9	36.8	45.1[a]	2.3-4.4-4.1
groups (Chovanová,		SD	0.6	5.9	5.7		0.6 0.8 0.8
1981)	18	X̄	13.3	156.5	44.7	44.1	2.3-4.9-3.8
		SD	0.5	5.5	5.6		0.7 0.5 0.5
	17	X̄	15.6	169.6	57.9	43.8	2.4-4.7-3.6
		SD	0.5	5.8	5.7		0.3 0.8 0.9
	22	X̄	17.6	176.0	64.3	43.9	2.2-4.5-3.6
		SD	0.6	6.0	5.7		0.4 0.7 0.7
Star Club, Bratislava	11	X̄	13–14	156.0	46.9	43.3	2.5-4.5-3.2
(Znášik, 1979)		SD		8.9	10.0		0.9 1.0 1.2

Table 6.13. (*Continued*)

Sample, reference	N	Statistic	Age (yr)	Height (cm)	Weight (kg)	HWR	Somatotype
Thunder Bay, Ontario	9	X̄	16.5	173.1	65.5	42.9[a]	1.8-4.3-2.7
Juniors (Song, 1982)		SD	1.6	5.9	5.2		0.6 0.5 0.6
Rossland, British	26	X̄	10.9	143.0	34.6	43.9[a]	1.7-4.6-3.7
Columbia (Ross & Day, 1972)		SD	1.7	13.2	9.9		0.9 0.9 1.0
c) Nordic combined							
European, 1970–71							
Total	18	X̄	21.7	175.1	69.4	42.6[a]	2.0-5.5-2.7
		SD		6.6	5.3		
Group A	8	X̄	23.7	178.1	71.6	42.9	1.9-5.5-2.9
		SD		5.2	3.9		
Group B	10	X̄	20.0	172.7	67.6	42.4	2.1-5.6-2.5
Chovanová, 1976b)		SD		6.9	5.8		
d) Ski jumpers							
European, 1970–71							
Total	56	X̄	22.4	174.0	71.3	42.0[a]	2.1-6.0-2.3
		SD		6.4	7.8		
Group A	36	X̄	23.1	174.1	71.8	41.9	2.1-6.1-2.3
		SD		8.0	6.2		
Group B	20	X̄	21.1	172.8	68.2	42.3	1.9-5.9-2.5
Chovanová, 1976b)				4.6	10.2		
Czechoslovakia,	12	X̄	25.7	173.8	74.0	41.4	2.3-6.5-1.8[b]
national, 1970–71		SD		7.7	10.2		
Chovanová, 1976b)							
Bulgaria, elite	25	X̄	21.3	172.8	72.8	41.4	2.5-6.5-1.7
(Toteva & Sumanov, 1984)							

Calculated from mean height and weight.

This value for ectomorphy was calculated from data provided by the author. The value 2.4 in Chovanová (1976b) is in error.

= mean.

Bulgarian samples seem to be slightly more mesomorphic than other European samples.

Young skiers, from Czechoslovakia and Canada (Fig. 6.23(b)), are represented in downhill events only. They are markedly less mesomorphic and more ectomorphic than the championship level adults, as might be expected. Among the young Canadians, Ross & Day (1972) and Ross et al. (1976) found that the majority of the best skiers were ecto-mesomorphs. They noted that 'there were no "fat" skiers in the samples and no linear skiers who did not have a secondary muscular dominance'. The data on the Slovak club age groups show similar findings, with slightly higher ectomorphy than the Canadians.

Table 6.14. *Swimming – male*

Sample, reference	N	Statistic	Age (yr)	Height (cm)	Weight (kg)	HWR	Somatotype
Mexico City Olympics, 1968 (De Garay *et al.*, 1974; Hebbelinck *et al.*, 1975)	65	X̄ SD	19.2 2.4	179.3 6.2	72.1 6.8	43.0 1.0	2.1-5.0-2.9 0.5 0.8 0.7
Montreal Olympics, 1976 (Carter *et al.*, 1982)	33	X̄ SD	19.3 2.4	178.6 4.7	73.0 8.0	42.8 1.1	2.1-5.1-2.8 0.6 0.9 0.8
London Olympics, 1948 (Carter, 1970)	21	X̄ SD	21.4 2.7	183.4 5.6	79.6 8.7	42.8 1.3	2.9-5.4-2.7 0.9 0.7 1.0
Channel swimmers (Heath, unpubl.)	13	X̄ SD	32.0 9.8	173.1 7.0	85.0 11.0	39.5 1.8	5.0-6.2-1.1 1.2 0.8 0.5
San Diego State University, 1965 (Carter 1970, 1971)	24	X̄ SD	19.9 1.1	179.3 5.3	74.9 6.7	42.6 1.1	2.4-5.4-2.6 0.8 0.8 0.8
San Diego State University, 1978 (Atchley, unpubl.)	15	X̄	20.7	182.6	77.1	42.9[a]	2.1-5.3-2.9
San Diego State University, 1980 (Atchley-Carlson, 1981)	18	X̄ SD	21.1 1.5	179.7 4.5	74.5 4.6	42.5[a]	2.3-5.0-2.7 0.5 0.6 0.6
Munich Olympics, 1972 (Novak *et al.*, 1978)	14	X̄ SD	19.9 2.3	179.5 5.7	74.9 5.5	42.6[a]	2.8-5.0-2.6[b]
Venezuela (Pérez, 1981)	17	X̄ SD	17.2 2.5	175.6 3.8	68.0 7.3	42.2 1.7	2.2-4.9-3.0 0.4 1.0 1.2
Caracas, club (Pérez, 1977)	22	X̄	13.8	158.2	46.5	44.0[a]	1.8-4.3-3.7 0.8 0.9 1.0
Brazil, 1975 (Araújo, 1978)	25	X̄ SD	18.6 2.6	178.3 5.8	71.7 7.1	42.9[a]	2.2-4.5-2.9 0.9 0.7 0.7
Indiana University (Araújo 1978)	21	X̄ SD	19.5 1.3	183.6 6.2	77.3 6.5	43.1[a]	2.6-4.6-3.0 0.9 0.7 0.8
Bolivar Games, 1981 (Brief, 1986)	15	X̄ SD	17.8 2.8	174.4 5.3	68.6 5.7	42.6 1.0	2.2-5.3-2.6 0.6 0.8 0.8
California, University, freestyle (Murphy, 1975)	28	X̄ SD	20.5 1.5	183.2 5.6	77.5 5.9	43.0 1.4	2.5-5.1-2.9
Belgium, elite (Vervaeke & Persyn, 1981)	47	X̄ SD	10–22	—	—	—	2.0-4.3-4.3 0.6 1.2 0.8
China, national (Zeng, 1985)	19	X̄ SD	18.9 2.5	178.2 4.8	71.4 7.1	43.0[a]	2.5-5.0-3.1 0.8 0.9 0.8
Brighton Polytechnic (Bale, 1986)	7	X̄ SD	—	179.9 5.8	71.3 5.6	43.3 1.0	2.1-4.9-3.3 1.0 0.7 1.0[b]

Table 6.14. (*Continued*)

Sample, reference	N	Statistic	Age (yr)	Height (cm)	Weight (kg)	HWR	Somatotype
South Australia (Withers *et al.*, 1986)	6	X̄ SD	16.8 1.4	177.8 4.5	66.2 3.1	44.0 0.7	1.9-4.7-3.6 0.8 0.5 1.5
Cuba (Alonso, 1986)	14	X̄	12.5	149.1	40.7	43.4[a]	2.2-4.0-3.3
Manchester, club (Bagnall & Kellett, 1977)	9	X̄ SD	15.7 13–20	173.6	62.8	43.7[a]	1.6-4.6-3.4 0.6 1.0 1.0
USA, Junior Olympic, 1980 (Thorland *et al.*, 1983)	39	X̄ SD	17.3 0.9	180.7 7.6	72.7 7.4	43.3[a]	2.8-4.5-3.3 0.8 0.9 0.9

[a]Calculated from mean height and weight.
[b]Calculated from means of anthropometric variables.
X̄ = mean.

Swimming (Table 6.14, Fig. 6.24)

The means for the top class swimmers from the 1960s to 1980s centre around the 2½-5-3 somatotype. Comparisons showed that the 1948 Olympians were more endomorphic than those in the 1968 and 1976 Games. The combined mean for the 1968 and 1976 Olympic swimmers was 2.1-5.0-2.9. The backstroke swimmers in the 1968 Games were less mesomorphic and more ectomorphic than those in other strokes. There were no differences between White and Mestizo swimmers in the 1968 sample. The lower endomorphy of swimmers in the past two decades may be due to greater duration and intensity of training in recent times (Carter, 1984*b*).

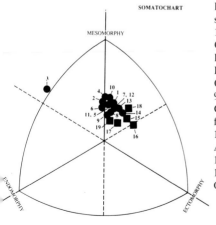

SOMATOCHART

Fig. 6.24. Mean somatotypes for male swimmers. ● = adults; ■ = youths. 1 = 1968 and 1976 Olympics; 2 = 1948 Olympics; 3 = Channel swimmers; 4 = San Diego State University, 1965; 5 = San Diego State University, 1978; 6 = 1972 Olympics; 7 = Venezuela; 8 = Brazil, 1975; 9 = University of Indiana; 10 = Bolivar Games; 11 = California universities freestyle; 12 = China, national; 13 = Brighton Polytechnic; 14 = South Australia; 15 = Caracas, club; 16 = Belgium, elite; 17 = Cuba, youths; 18 = Manchester, club; 19 = USA Junior Olympics.

Araújo (1978) found differences by stroke in Brazilian and University of Indiana swimmers. Breaststroke swimmers were more mesomorphic than freestyle distance swimmers, but in the Brazil and Indiana samples freestyle distance swimmers and backstrokers were more ectomorphic than other swimmers. Breaststroke swimmers had the lowest endomorphy values. Murphy (1975) found no somatotype differences between sprint and distance freestyle swimmers at California universities. In addition, Thorland *et al.* (1983) did not find any differences between sprint and distance events in Junior Olympians.

The Channel swimmers, rated by Heath from photographs provided by J. M. Tanner, are the most endomorphic swimmers. About half are extreme endo-mesomorphs; only one is a balanced mesomorph. Ratings by Sheldon's method also emphasized high endomorphy (Pugh *et al.*, 1960; Renson & Van Gerven, 1968–69).

As Araújo (1979) found, young swimmers tend to be less mesomorphic and more ectomorphic than adults. He showed that among Brazilian swimmers at ages 8, 11, 15 and 19, the youngest were balanced mesomorphs ($\overline{S}$ = 3.0-4.9-2.9) and the older groups were ecto-mesomorphic. When they followed 11 Brazilian 15 year olds over a season, Araújo *et al.* (1978c) found a slight decrease in endomorphy (final somatotype = 2.2-4-3.7). In a study of 12-year-old Hungarian swimmers, Mohácsi & Mészáros (1982) found that they became more mesomorphic and ectomorphic as a result of a training programme. Their mean at the end of training was 2.5-4.4-4.4.

In a study of 42 older swimmers (40 to 59 years) at the 1972 USA Masters national championships, Rahe & Carter (1976) found similar somatotypes for those in their forties and those in their fifties. The champions had a mean somatotype of 3.0-5.4-2.3 and the non-champions had a mean of 3.5-5.5-1.9, but there were no differences by components. The differences by stroke were small. Their somatotypes were remarkably similar to those of the 1948 Olympic swimmers.

Using the M.4 method, Leek (1969) found that New Zealand swimmers and San Diego State University swimmers (Carter, 1966) had similar somatotypes, but that there were some differences between strokes within the New Zealand sample.

Track and field (Table 6.15, Fig. 6.25(*a*),(*b*))
Track and field is a sport consisting of 22 Olympic events (not counting relays) that involve running, walking, jumping, vaulting, hurdling and throwing. Although analysis of data for so great a number of diverse events is difficult, somatotype findings are remarkably consistent in a large number of studies of different levels of competition. A review of findings for each event is too long for the available space, therefore the summary of track

(a)

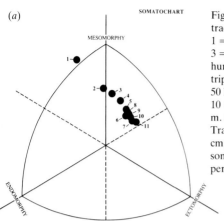

SOMATOCHART

Fig. 6.25(a). Mean somatotypes for male track and field athletes in 11 event groups. 1 = shot, discus, hammer; 2 = javelin; 3 = decathlon; 4 = 100 m, 200 m, 110 m hurdles; 5 = pole vault; 6 = high, long, triple jump; 7 = 800 m, 1500 m; 8 = 20 km, 50 km walk; 9 = 400 m, 400 m hurdles; 10 = marathon; 11 = 3000 m, 5000 m, 10000 m. (Redrawn from Carter, 1984b.) (b). Track and field decathlete. Height = 188.8 cm; weight = 87.3 kg; HWR = 42.6; somatotype = 2-$5\frac{1}{2}$-$2\frac{1}{2}$. (Photo with permission from J. Štěpnička.)

(b)

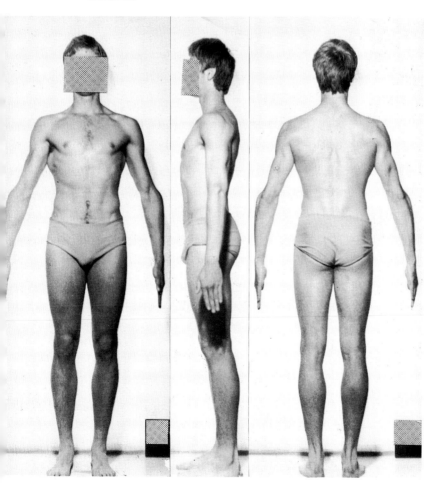

Table 6.15. *Track and field – male*

Event	N	Somatotype
100 m, 200 m, 110-metre hurdles	107	1.7-5.2-2.8
400 m, 400-metre hurdles	64	1.5-4.6-3.4
800 m, 1500 m	56	1.5-4.3-3.6
3000 m, 5000 m, 10 000 m	58	1.4-4.2-3.7
Marathon	32	1.4-4.4-3.4
20 km, 50-km walk	25	1.6-4.7-3.4
Shot, discus, hammer	28	3.2-7.1-1.1
Javelin	13	2.3-5.9-2.1
Decathlon	10	1.8-5.6-2.4
Pole vault	8	1.6-4.9-3.3
High, long, triple jump	59	1.7-4.6-3.4

[a]Olympic athletes, from Carter (1984*a*,*b*). For additional information by event see De Garay *et al.* (1974) and Carter *et al.* (1982).

and field is more global than for other sports. For in-depth analyses by event or groups of events, the reader is referred to the sources cited.

Table 6.15 and Fig. 6.25(*a*) give the somatotypes of 452 athletes in 11 event groupings from the 1960, 1968 and 1976 Olympics, summarized by Carter (1984*b*). Athletes in the same events from different Olympics were similar to one another. There appears to be a slight increase in mesomorphy for the shot, discus and hammer throwers (Carter, 1970, 1984*b*; Carter *et al.*, 1982; De Garay *et al.*, 1974). The mean somatoplots are on a relatively straight line running from extreme endo-mesomorphy of the throwers to ectomorph–mesomorphy of the distance runners. Walkers and marathon runners, a minor exception to this trend, are slightly more mesomorphic and less ectomorphic than the 3, 5 and 10 km runners. This line is NW to SE, and, except for the throwers, within each event the distribution is elliptical with little variation in endomorphy, and greater variation in mesomorphy and ectomorphy. The rough functional factors associated with this trend are power for the throwers, and endurance for the distance runners. There is a concomitant decrement in relative girths and breadths as physiques become more linear. The 19 athletes in a variety of events at the 1948 Olympics show essentially the same pattern (Carter, 1970), except for the slightly higher endomorphy in 1948.

Findings similar to those in the Olympic samples were obtained in studies of track and field athletes in Brazil (Guimarães & De Rose, 1980), China (Zeng, 1985), Cuba (Rodríguez *et al.*, 1986), Czechoslovakia (Štěpnička, 1974*a*, 1986), England (Bale, 1986), India (Sodhi & Sidhu, 1984), South America (Brief, 1986), South Australia (Withers *et al.*, 1986), USA (Carter,

1970, 1971), and Venezuela (Pérez, 1981). Other authors have confined themselves to examining fewer events and different levels of competition (Caldeira *et al.*, 1986*c*; Sharma & Dixit, 1985; Shin, 1985; Toriola *et al.*, 1985; Travill, 1984). Cureton (1951), Parnell (1958), Parsons (1973) and Tanner (1964) used other somatotype methods in their studies of track and field athletes.

A few studies of young athletes indicate that they have somatotype patterns similar to those of adults by the time they are adolescents, but are less mesomorphic than adults (Alonso, 1986; Chovanová & Pataki, 1982; Guimarães & De Rose, 1980; Hayley, 1974; Thorland *et al.*, 1981).

Volleyball (Table 6.16, Fig. 6.26)
There is rather a large range of means and distributions of somato-types of national level volleyball players. The majority of their somatotypes are ecto-mesomorphic, but they range from endo-mesomorphy to meso-ectomorphy. The Czechoslovaks, Hungarians and Cubans are the most mesomorphic; the Indians are the most ectomorphic and least mesomor-phic. The distribution of somatotypes of the Venezuelan team was clustered in two groups: 32% were meso-ectomorphs, and the remainder were in a circular pattern around 2-5-2 (Pérez, 1981). The USA Olympic team, subsequently the Olympic and World champions, centred around the $2\frac{1}{2}$-$4\frac{1}{2}$-$3\frac{1}{2}$ somatotype. The Venezuelan, Bolivarian, and South Australian teams were closest to the USA team. The young Cubans were mesomorph–ectomorphs. In most samples there was little variation in endomorphy, but greater variation in mesomorphy and ectomorphy.

Between the early and late 1970s the Hungarian Class I players became more endo-mesomorphic. In their study of teams from five countries at the South American Youth Championships. Caldeira *et al.* (1986*a*) found a

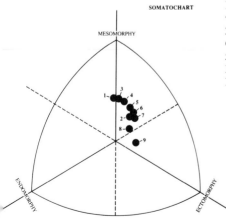

Fig. 6.26. Mean somatotypes for male volleyball players. 1 = Hungary, 1979–80; 2 = USA, national, 1983; 3 = Czechoslovakia, 1967; 4 = Cuba, 1976–80; 5 = South Australia; 6 = Venezuela; 7 = Bolivar Games; 8 = Cuba, youths; 9 = India, national.

Table 6.16. *Volleyball – male*

Sample, reference	N	Statistic	Age (yr)	Height (cm)	Weight (kg)	HWR	Somatotype
Czechoslovakia, Class I, 1967 (Štěpnička, 1974a, 1986)	108	X̄ SD	—	184.4 6.3	81.4 6.9	42.6[a]	2.5-5.5-2.6 0.9 0.8 0.9
Hungary, Class I 1972–75	32	X̄	—	182.7	76.4	43.1[a]	1.9-4.2-3.0
1979–80 (Mészáros & Mohácsi, 1982b)	18	X̄	—	183.4	80.4	42.5[a]	2.7-5.6-2.6
Venezuela (Pérez, 1981)	22	X̄ SD	22.0 3.4	182.3 5.2	73.6 7.2	43.5 1.8	2.0-4.6-3.3 0.7 1.3 1.3
Bolivar Games, 1982 (Brief, 1986)	8	X̄ SD	22.1 3.1	179.6 7.0	70.4 7.5	43.6 1.6	1.9-4.2-3.4 0.4 0.9 1.2
Cuba, 1976–80 (Rodríguez et al., 1986)	16	X̄ SD	23.7 2.4	186.9 6.6	83.4 7.2	42.8[a]	2.1-5.1-2.8 0.3 0.8 0.8
Cuba, youths (Alonso, 1986)	25	X̄	12.5	165.9	50.9	44.8[a]	2.5-3.9-3.7
South Australia (Withers et al., 1986)	11	X̄ SD	20.9 3.7	185.3 10.2	78.3 12.0	43.4 0.7	2.1-5.0-3.2 0.7 0.5 0.5
India National	14	X̄ SD		185.6 5.9	71.1 7.4	44.8[a]	2.5-3.1-4.2 0.7 0.7 0.9
Camp	13	X̄ SD	24.4 3.0	179.1 4.0	66.8 4.6	44.2[a]	2.5-3.3-3.8 0.6 0.5 0.6
State (Sodhi & Sidhu, 1984)	16	X̄ SD		178.7 5.3	65.8 5.4	44.3[a]	2.7-3.4-3.9 1.2 0.8 0.9
USA Olympic, 1983 (Carter, unpubl.)	13	X̄ SD	25.8 2.9	194.0 5.4	87.8 5.5	43.6 0.8	2.3-4.4-3.4 0.5 0.9 0.6

[a]Calculated from mean height and weight.
X̄ = mean.

mean of 2.0-4.3-3.2, which is close to the mean for the Bolivar Games players.

Water Polo (Table 6.17, Fig. 6.27)
   Olympic water polo players are primarily endo-mesomorphs, and their somatotype distribution extends over most of the mesomorphic sector. About half are somatotypically similar to swimmers; the remainder are more endo-mesomorphic, i.e. to the left of the swimmers. The Belgian and Olympic players had similar distributions; the Cubans were more ectomor-

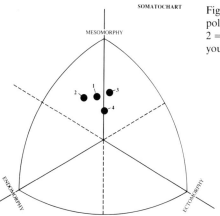

SOMATOCHART

Fig. 6.27. Mean somatotypes for male water polo players. 1 = Mexico City Olympics; 2 = Belgium; 3 = Cuba, 1976–80; 4 = Cuba, youths.

Table 6.17. *Waterpolo – male*

Sample, reference	N	Statistic	Age (yr)	Height (cm)	Weight (kg)	HWR	Somatotype
Mexico City Olympics, 1968 (De Garay *et al.*, 1974; Hebbelinck *et al.*, 1975)	71	X̄ SD	22.9 4.2	179.9 6.9	77.8 8.5	42.2 1.2	2.9-5.3-2.3 1.2 0.9 0.8
Belgium (Clarys & Borms, 1971)	44	X̄	—	178.2	80.2	41.3[a]	3.4-5.3-1.8
Cuba, 1976–80 (Rodríguez *et al.*, 1986)	14	X̄ SD	22.3 2.5	181.0 6.4	77.1 6.2	42.5[a]	2.1-5.4-2.6 0.6 1.0 1.1
Cuba, youths (Alonso, 1986)	31	X̄	12.5	153.2	45.1	43.0[a]	2.8-4.3-3.0

[a] Calculated from mean height and weight.
X̄ = mean.

phic and like Olympic swimmers. The young Cubans were lower in mesomorphy, centering around balanced mesomorphy. In his study of New Zealand water polo players (by the M.4 method), Leek (1968) found that senior players were more mesomorphic and less ectomorphic than juniors, and that goalkeepers did not differ from other players.

Weight lifting (Table 6.18, Fig. 6.28)

As expected, weight lifters, requiring great muscle size and strength, have among the highest known values in mesomorphy. There appear to be no ethnic somatotype differences among weight lifters, but considerable differences according to weight category (De Garay *et al.*,

Table 6.18. *Weight lifting – male*

Sample, reference	N	Statistic	Age (yr)	Height (cm)	Weight (kg)	HWR	Somatotype
Olympics, 1960–76							
<60 kg	16	X̄				40.5	1.4-6.9-1.0
		SD				1.0	0.3 0.7 0.5
60–79.9 kg	47	X̄				40.4	1.8-7.0-1.1
		SD				0.5	0.7 0.8 0.5
80–99.9 kg	27	X̄				39.0	2.7-7.8-0.7
		SD				0.8	0.9 0.7 0.3
100+ kg	8	X̄				37.0	5.1-9.1-0.4
(Carter, 1984*b*)		SD				0.9	0.8 0.8 0.1
Czechoslovakia, 1969	48	X̄	—	169.0	77.8	38.7[a]	3.4-7.2-1.3
(Štěpnička, 1974*a*, 1986)		SD		7.1	13.4		1.4 1.3 0.5
Czechoslovakia, 1981							
'Younger'	46	X̄	—	168.8	—	—	3.2-5.8-2.0
adolescents		SD		8.2			1.5 1.4 1.1
'Older'	28	X̄	—	171.5	—	—	3.1-6.1-1.8
adolescents		SD		6.0			1.3 1.1 0.9
(Chovanová *et al.*, 1983*a*)							
USSR (Heath, in	54	X̄	—	164.6	77.2	38.7[a]	4.2-6.6-1.0
Carter, 1970)		SD		12.4	18.9		1.3 0.6 0.3
Venezuela							
55–67 kg	9	X̄	22.0	163.7	61.6	41.1	1.8-6.2-1.7
		SD	4.0	5.7	4.4	0.8	0.4 0.7 0.6
69–122 kg	19	X̄	22.6	173.4	87.3	39.0	3.9-7.6-0.8
(Pérez, 1981)		SD	3.4	6.1	15.7	1.4	1.8 1.0 0.5
Cuba, 1976–80							
'Lightweight'	4	X̄	24.4	160.2	60.7	40.8[a]	1.9-6.0-1.5
		SD	4.6	7.6	4.4		0.3 1.0 0.9
'Middleweight'	7	X̄	23.9	169.4	76.8	39.9[a]	2.6-6.7-0.9
		SD	4.8	4.2	5.4		0.5 0.3 0.5
'Heavyweight'	4	X̄	27.4	178.2	106.5	37.7[a]	3.4-8.6-0.5
(Rodríguez *et al.*, 1986)		SD	5.3	5.1	14.5		1.9 1.3 0.0
Cuba, youths (Alonso,	9	X̄	12.5	149.4	46.9	41.4[a]	3.3-5.6-1.9[b]
1986)							
India (Sodhi &	33	X̄	—	166.7	66.3	41.2[a]	3.0-5.0-1.9
Sidhu, 1984)		SD		7.2	13.0		1.3 1.0 0.9
China, 1984 (Zeng,	24	X̄	21.1	166.5	72.4	40.0[a]	3.2-7.0-1.0
1985)		SD	3.2	8.5	14.6		1.3 1.2 0.5
Brighton Polytechnic	4	X̄	—	177.7	79.1	41.0	2.3-6.3-1.5
(Bale, 1986)		SD		5.2	9.3	1.3	0.5 1.3 0.6
South Australia	3	X̄	26.0	162.5	71.7	39.3	2.7-7.9-0.6
Power lifters		SD	1.5	10.7	17.7	0.8	0.6 0.9 0.4
(Withers *et al.*, 1986)							

[a]Calculated from mean height and weight.
[b]Calculated from anthropometry (error in reference).
X̄ = mean.

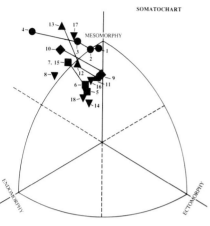

SOMATOCHART

Fig. 6.28. Mean somatotypes for male weight lifters. Symbols connected by a line or lines show the trend between adjacent weight groups from the same study. In general, the lighter weight groups are to the right and the heavier groups are to the left. Olympic Games (●) 1 = 1960–76, <60 kg; 2 = 1960–76, 60–79.9 kg; 3 = 1960–76, 80.0–99.9 kg; 4 = 1960–76; 100+ kg; Czechoslovakia (■) 5 = 1981, 'younger adolescents'; 6 = 1981, 'older adolescents'; 7 = 1969; 8 = USSR; Venezuela (◆) 9 = 55–67 kg; 10 = 69–122 kg; (▲) 11 = 'lightweight'; 12 = 'middleweight'; 13 = 'heavyweight'; 14 = India; 15 = China; 16 = Brighton Polytechnic; 17 = South Australia, power lifters; 18 = Cuba, youths.

1974). The large standard deviations for weight are noteworthy in both large and small samples in the studies of USSR, Indian, Chinese, English, and South Australian weight lifters, although they were not analysed by weight categories. When Carter (1984b) combined the data from the 1960, 1968 and 1976 Olympics into four weight categories, he found the higher the weight class the higher the endomorphy and mesomorphy and the lower the ectomorphy. This trend is seen in the data for each Olympics and for data from Venezuela, Cuba and CSSR. Mesomorphy averages about 6 to 7 in the lighter weight classes and 8 to 9 in the heaviest classes. The top class lifters are balanced mesomorphs at the lighter weights and extreme endo-mesomorphs in the heaviest weights.

Chovanová-Orvanová and her colleagues investigated the development and selection of weight lifters in a study of 174 Czechoslovak weight lifters from age 14 years. They found that during a training season somatotypes became less endomorphic (Chovanová et al., 1983b); that older adolescents were more mesomorphic than younger adolescents and that there were differences by weight class (Chovanová et al., 1983a); and that the lower the ectomorphy the better the performance in higher weight classes (Slamka et al., 1983). The somatocharts of Orvanová et al. (1984) for the ten weight classes showed a preponderance of ecto-mesomorphy and low endomorphy with little variability in the 52 kg class. The distributions for the subsequent classes shifted progressively to the left of the mesomorphic axis; and the somatoplots for the three upper weight classes (over 100 kg) are for extreme endo-mesomorphs. The authors noted that mesomorphy had the highest rating throughout the series, that with increase of weight class ectomorphy decreased as endomorphy and mesomorphy increased. The highest endomorphy was $8\frac{1}{2}$ and the highest mesomorphy was 13 (a value close to the 12 in mesomorphy recorded by Ross et al. (1974) for a super heavyweight lifter).

In general, they agreed with the findings on Olympic lifters, i.e. that major somatotypic differences were shown best among three weight groups: 52–75 kg, 82.5–100 kg and 100 plus kg. They also showed that somatotype components contributed to discriminant analyses of differences between the best and worst performers by weight classes (Orvanová *et al.*, 1984; Orvanová, 1986).

The somatochart in Boennec *et al.* (1980) shows that ten out of twelve French weight lifters were endo-mesomorphs or extreme endo-meso-morphs. The approximate mean was $2\frac{1}{2}$-$6\frac{1}{2}$-$1\frac{1}{2}$. The young Cubans were endo-mesomorphic and their 5.6 in mesomorphy is the highest of all the 12-year-old sports groups.

Tappen (1950) reported a mean of 3-$6\frac{1}{2}$-1 in his study of the 1947 AAU Championship weight lifters, somatotyped by the Sheldon method. The somatotypes clustered in the upper left of the somatochart, where they were of course restricted in mesomorphy by the Sheldon upper rating limit of 7. De Pauw & Vrijens (1972) and Tanner (1964) also reported on weight lifters rated by the Sheldon method.

Wrestling (Table 6.19, Fig. 6.29(*a*),(*b*),(*c*))

Although wrestling, like other weight-classified sports, is best examined by weight class, only the Olympic, India and California high school samples have been analysed in this manner. It is clear from these

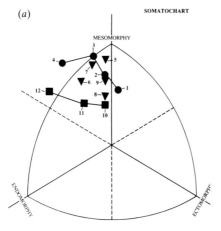

(*a*)

Fig. 6.29(*a*). Mean somatotypes for male adult wrestlers. Symbols connected by a line or lines show the trend between adjacent weight groups from the same study. In general, the lighter weight groups are to the right and the heavier groups are to the left. Olympic games (●) 1 = 1960–76, <60 kg; 2 = 1960–76, 60–79.9 kg; 3 = 1960–76, 80–99.9 kg; 4 = 1960–76, 100 + kg; 5 = Czechoslovakia, 1973; 6 = USSR; 7 = Cuba, 1976–80; 8 = Puerto Rico; 9 = Korea; India (■) 10 = 'light'; 11 = 'medium'; 12 = 'heavy'. (*b*) Greco-Roman wrestler. Height = 174.0 cm; weight = 82.2 kg; HWR = 40.0; somatotype = $2\frac{1}{2}$-8-1. (Photo with permission from J. Štěpnička.) (*c*). Mean somatotypes for young male wrestlers. (For an explanation of the lines see (*a*).) California (●) 1 = 52.2–58.9 kg; 2 = 59.0–63.9 kg; 3 = 64.0–71.1 kg; 4 = 71.2–80.6 kg; 5 = 80.7–88.0 kg; Puerto Rico (■) 6 = 6–8 yr; 7 = 9–11 yr; 8 = 12–13 yr; 9 = 14–15 yr; 10 = 16–18 yr; 11 = USA Juniors; 12 = Hartford, USA; 13 = Cuba.

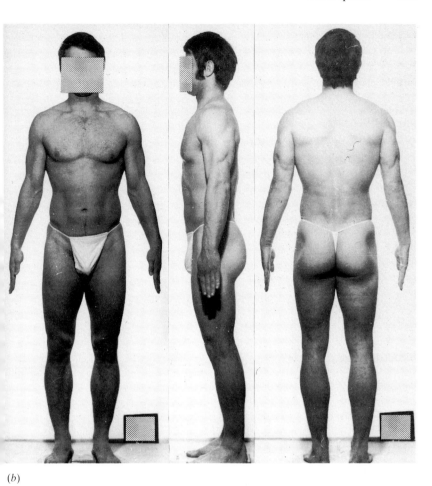

(b)

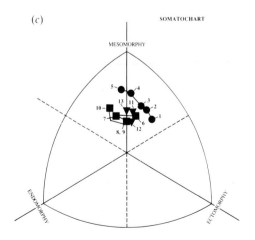

(c)

Table 6.19. *Wrestling – male*

Sample, reference	N	Statistic	Age (yr)	Height (cm)	Weight (kg)	HWR	Somatotype
(a) *Adults*							
Olympic, 1960, 1968, 1976							
<60 kg	26	X̄	—	—	—	42.4	1.6-5.6-2.3
		SD				1.2	0.4 0.8 0.9
60–79.9 kg	77	X̄	—	—	—	41.3	2.1-6.4-1.6
		SD				0.9	0.8 0.8 0.6
80–99.9 kg	26	X̄	—	—	—	40.5	2.6-7.0-1.2
		SD				1.0	0.8 1.0 0.5
100+ kg	8	X̄	—	—	—	39.4	4.2-7.3-0.8
(Carter, 1984b)		SD				1.1	1.3 0.8 0.5
Czechoslovakia, 1973							
Greco-Roman	22	X̄	—	171.9	78.4	40.2^a	2.6-6.8-1.6
		SD		11.1	26.5		1.9 1.0 0.8
Freestyle	24	X̄	—	170.8	74.4	40.6^a	1.8-7.1-1.4
(Štěpnička *et al.*, 1976; Štěpnička, 1986)		SD		11.0	14.6		0.8 0.7 0.5
USSR (Heath, in Carter, 1970)	34	X̄	—	167.1	77.1	39.3^a	3.5-6.4-1.3
		SD		9.4	24.5		1.0 0.7 0.5
Cuba, 1976–80							
Greco-Roman	14	X̄	21.0	170.2	76.8	40.0^a	2.9-6.6-1.3
		SD	2.1	11.2	20.9		1.8 1.4 0.5
Freestyle	17	X̄	23.8	171.3	79.8	39.8^a	2.8-6.9-1.2
(Rodríguez *et al.*, 1986)		SD	2.7	9.4	19.3		1.4 1.3 0.5
Puerto Rico	6	X̄	20.1	168.4	71.2	40.6^a	2.3-5.3-1.9
(Rivera *et al.*, 1986)		SD	3.1	8.2	12.2		1.0 0.5 1.8
India							
'Light'	12	X̄	22.4	159.5	55.8	41.7^a	2.5-4.9-2.0
		SD	3.8	6.8	4.7		0.6 0.8 0.8
'Medium'	16	X̄	22.9	170.7	72.8	40.9^a	3.6-5.3-1.5
		SD	3.1	6.7	7.2		1.1 0.9 0.8
'Heavy'	11	X̄	28.1	179.9	93.8	39.6^a	5.2-6.2-1.0
(Sodhi & Sidhu, 1984)		SD	5.9	6.2	9.2		1.7 0.7 0.6
Pusan, Korea (Shin, 1985)	19	X̄	20.3	172.2	71.8	41.4^a	2.2-6.2-1.7
		SD	1.2	5.3	8.6		1.0 0.9 0.6
(b) *Youths*							
California, high school, 1978							
52.2–58.9 kg	18	X̄	17.6	168.0	55.1	44.2	1.7-4.9-3.7
		SD	1.0	3.5	2.2	1.0	0.4 0.7 0.8

Table 6.19. (*Continued*)

Sample, reference	N	Statistic	Age (yr)	Height (cm)	Weight (kg)	HWR	Somatotype
59.0–63.9 kg	16	X̄	17.7	170.7	59.8	43.7	1.8-5.4-3.4
		SD	0.5	4.7	4.9	1.3	0.3 0.7 1.0
64.0–71.1 kg	15	X̄	17.8	174.5	66.3	43.1	1.8-5.4-3.0
		SD	0.8	4.4	2.1	1.3	0.4 0.9 0.9
71.2–80.6 kg	20	X̄	17.9	176.4	72.4	42.3	2.1-6.2-2.4
		SD	0.7	5.0	2.6	1.0	0.4 0.8 0.8
80.7–88.0 kg	14	X̄	18.0	181.7	81.0	42.0	2.7-6.5-2.2
(Carter & Lucio, 1986)		SD	0.5	4.7	4.6	0.9	0.8 0.9 0.7
USA, Juniors (Thorland *et al.*, 1981)	18	X̄	16.9	167.1	60.2	42.6[a]	2.6-5.4-2.9
		SD	1.2	7.8	11.2		1.0 1.0 1.2
Hartford, USA, high school, 1974 (Sinning *et al.*, 1976)	13	X̄	16.7	172.9	67.1	42.5[a]	2.7-5.0-2.8
		SD		8.3	12.9		1.0 1.2 1.1
Puerto Rico							
6–8 yr	4	X̄	7.7	131.0	30.3	42.0[a]	1.8-4.6-2.4
		SD	1.0	9.2	7.6		0.8 0.6 0.4
9–11 yr	25	X̄	10.6	136.0	34.9	41.3[a]	3.1-5.2-2.3
		SD	0.9	7.2	7.1		1.5 0.7 1.2
12–13 yr	26	X̄	13.0	149.2	44.5	42.1[a]	2.8-4.9-2.9
		SD	0.6	10.9	11.2		1.8 1.0 1.5
14–15 yr	36	X̄	15.0	163.5	58.0	42.2[a]	2.6-4.7-2.7
		SD	0.6	6.0	12.7		1.5 1.4 1.1
16–18 yr	24	X̄	17.3	167.7	68.2	41.1[a]	3.2-5.4-1.9
(Rivera *et al.*, 1986)		SD	0.8	5.2	13.1		1.6 1.8 1.3
Cuba, youths (Alonso, 1986)	58	X̄	12.5	142.5	39.2	42.0[a]	2.5-4.9-2.6

[a] Calculated from mean height and weight.
X̄ = mean.

studies and that of Štěpnička *et al.* (1976), that endomorphy and mesomorphy increase and ectomorphy decreases with increases in weight class. The lighter classes tend to balanced mesomorphy, and the heavier classes to endo-mesomorphy, just as in weight lifting and judo. In general, wrestlers are not quite as mesomorphic as weight lifters and body builders. There were no somatotypic differences between Greco-Roman and freestyle wrestlers at the 1968 Olympics (De Garay *et al.*, 1974); this also appears to be true of the CSSR and Cuban wrestlers.

Heath's ratings of Tanner's 1960 British Empire Games sample of 33 wrestlers resulted in a mean of 2.1-6.2-1.6 (Carter, 1970). As there appeared to be little difference among combined Olympic samples from 1960, 1968

Table 6.20. *Miscellaneous sports – males*

Sample, reference	N	Statistic	Age (yr)	Height (cm)	Weight (kg)	HWR	Somatotype
Badminton							
South Australia	7	$\overline{X}$	24.5	180.0	71.2	43.4	2.5-4.6-3.2
(Withers *et al.*,		SD	3.6	5.2	5.6	0.5	0.6 0.5 0.4
1986)							
Baseball							
University of Iowa	10	$\overline{X}$	20.3	180.3	80.7	41.7[a]	3.8-5.2-2.2
(Carter, 1970)		SD	1.2	5.1	7.6		1.0 0.3 0.9
Cuba, 1976–80	35	$\overline{X}$	26.5	177.9	78.5	41.6[a]	2.7-5.3-1.9
(Rodríguez *et al.*,		SD	3.6	6.8	8.3		0.9 1.0 0.8
1986)							
Dong-A University,	16	$\overline{X}$	19.7	177.5	73.4	42.4[a]	4.0-3.8-2.5
Korea (Shin, 1985)		SD	0.9	3.9	5.1		1.0 0.9 0.7
Cuba, youths (Alonso,	25	$\overline{X}$	12.5	153.9	46.6	42.8[a]	3.0-4.5-2.8
1986)							
Diving							
Olympics, 1968 (De	16	$\overline{X}$	21.3	172.1	65.5	42.7	1.9-5.4-2.7
Garay *et al.*,		SD	3.7	5.5	5.0	1.0	0.5 0.7 0.7
1974)							
Cuba, youths (Alonso,	4	$\overline{X}$	12.5	137.2	32.3	43.1[a]	1.9-4.9-3.0
1986)							
Figure Skating							
Canada							
Senior–Junior	12	$\overline{X}$	18.2	164.4	56.5	42.8[a]	1.7-5.0-2.9
		SD	3.6	7.3	8.2		0.3 0.9 0.6
Novice	7	$\overline{X}$	15.1	157.4	50.3	42.6[a]	1.7-5.0-2.7
(Ross *et al.*, 1977*b*;		SD	2.5	12.2	8.6		0.2 0.6 0.8
Faulkner, 1976)							
Golf							
San Diego State	9	$\overline{X}$	21.1	171.5	65.8	42.5[a]	4.1-5.0-2.3
University (Carter,		SD	1.5	6.9	6.9		0.2 0.7 0.8
1970, 1971)							
Brighton Polytechnic	5	$\overline{X}$	—	178.4	70.1	43.3	2.0-4.5-3.4
(Bale, 1986)		SD		5.0	7.0	1.3	1.0 0.0 0.9
Karate							
Belgium (Claessens *et*	24	$\overline{X}$	24.6	177.8	73.2	42.6	2.6-5.2-2.6
et al., 1986*a*)		SD		8.1	7.4	1.0	0.8 0.9 0.7
Dong-A University,	15	$\overline{X}$	19.5	172.5	69.0	42.1	3.1-5.4-2.3
Korea (Shin, 1985)		SD	1.4	6.0	9.6		0.9 0.7 0.7
Lacrosse							
South Australia	26	$\overline{X}$	26.7	177.6	74.0	42.4	2.9-5.4-2.5
(Withers *et al.*,		SD	4.2	5.5	8.6	1.5	0.9 1.1 1.0
1986)							

Modern Pentathlon							
Mexico City Olympics,	24	$\overline{X}$	24.9	174.7	69.6	42.4	2.0-5.3-2.4
1968 (De Garay		SD	3.1	5.0	5.1	1.0	0.8 0.8 0.7
et al., 1974)							
Orienteering							
Czechoslovakia, 1975	18	$\overline{X}$	—	179.4	70.4	43.5[a]	1.9-4.6-3.3
(Štěpnička, 1986)		SD		5.2	1.8		0.5 0.9 0.8
South Australia	7	$\overline{X}$	25.9	176.2	64.7	43.9	2.0-4.7-3.6
(Withers *et al.*, 1986)		SD	8.5	6.8	5.0	1.3	0.3 0.7 1.0
Parachuting							
Brazil (Pinto,	25	$\overline{X}$	32.2	170.2	68.7	41.6	3.4-4.9-1.8
1978)		SD	5.8	3.1	5.1	0.6	0.7 0.5 0.6
Shooting							
Cuba, youths (Alonso,	12	$\overline{X}$	12.5	150.9	43.7	42.8[a]	3.9-4.1-2.9
1986)							
Squash, Racquets							
Brighton Polytechnic	10	$\overline{X}$	—	178.5	71.1	43.6	2.5-4.7-3.6
(Bale, 1986)		SD		4.8	5.1	1.0	0.7 0.7 0.8
South Australia	9	$\overline{X}$	22.6	177.5	71.9	42.8	2.5-5.2-2.8
(Withers *et al.*,		SD	6.8	4.1	8.3	1.1	0.6 1.0 0.8
1986)							
Surfboard riders							
International, 1978	76	$\overline{X}$	22.2	173.6	68.0	42.5[a]	2.6-5.2-2.6
(Lowdon, 1980)		SD	3.2	5.9	7.2		0.7 0.9 0.8
Table Tennis							
Europe, 1973	48	$\overline{X}$	23.5	174.2	69.7	42.3[a]	3.5-3.9-2.5
(Eiben & Eiben,		SD	—	6.6	8.5		
1979)							
Cuba, youths (Alonso,	14	$\overline{X}$	12.5	145.4	39.8	42.6[a]	2.5-4.3-2.8
1986)							
Tennis							
South Africa Open, 1977							
Professionals	33	$\overline{X}$	27.1	182.8	76.5	43.1[a]	2.2-4.6-3.0
		SD	5.7	6.3	7.5		0.9 1.0 0.9
Amateurs	28	$\overline{X}$	23.8	178.5	69.9	43.3[a]	2.2-4.3-3.2
(Copley, 1980*a,b*)		SD	4.3	8.4	10.9		1.1 1.0 0.7
Cuba, youths (Alonso,	10	$\overline{X}$	12.5	147.4	39.2	43.4[a]	2.4-4.1-3.3
1986)							
Triathlon							
USA, national, 1984	7	$\overline{X}$	28.9	183.0	75.6	43.3	1.7-4.3-3.1
(Dolan, 1987)		SD	3.2	3.3	4.3	0.8	0.6 0.3 0.6
Yachting							
Czechoslovakia, 1977	15	$\overline{X}$	—	180.0	75.9	42.5[a]	2.3-5.2-2.6
(Štěpnička, 1986)		SD		5.4	6.4		0.7 0.7 0.6

Calculated from mean height and weight.
$\overline{X}$ = mean.

and 1976, by weight class, the overall mean of $2\frac{1}{2}$-$6\frac{1}{2}$-$1\frac{1}{2}$ for 137 wrestlers is representative of their means (Table 6.19). In a comparison of California high school wrestlers and Olympic wrestlers in similar weight classes Carter & Lucio (1986) found that Olympic wrestlers were older, shorter, more mesomorphic and less ectomorphic than high schoolers. The somatotypes of wrestlers from India differ by weight group and are much less mesomorphic than other national wrestlers. Nevertheless, they are the most mesomorphic of the Indian sportsmen and have some ratings of 5 and 6 in mesomorphy, which is high in a predominantly mesopenic population.

The California sample of young wrestlers, which, like other samples, is less mesomorphic and more ectomorphic than the adults; is the only one analysed by weight class. The majority of the lightest weight group are ecto-mesomorphic, but wrestlers are more endo-mesomorphic in the heaviest weight groups. The Puerto Ricans tended to become more endo-mesomorphic with age, but the data are inconsistent and no weight class analysis was available.

Kroll (1954) reported somatotypes of USA university wrestlers using Cureton's method, and Tanner (1964) gave Sheldon ratings to British Empire Games wrestlers.

Miscellaneous sports–male (Table 6.20, Fig. 6.30(*a*),(*b*))

In addition to the above studies, there is limited somatotype information on a number of other sports. The team sports, baseball and lacrosse, are included, but the others are individual sports. The means are plotted in the endomorph–mesomorph, endo-mesomorph, balanced meso-morph, and ecto-mesomorph categories. Most of the distributions (as shown in the original sources) are fairly restricted in each sport, with the exception of the markedly scattered distributions for baseball and table tennis. In baseball, pitchers are less mesomorphic and more ectomorphic than many of those in other positions (Rodríguez *et al.*, 1986). Imlay (1966) found differences by playing position using the M.4 method.

Two or more samples can be compared in several sports. The means in karate, orienteering, tennis, figure skating and two of the baseball samples are quite similar. The Korean baseball sample is less mesomorphic than the other two, and there are differences between samples in golf and squash racquets. Although Boennec *et al.* (1980) included no statistics in their study of sailing, they found about two-thirds of their helmsmen, crew, and solo sailors were ecto-mesomorphic and similar to the Czechoslovaks. Holzer *et al.* (1984) showed a wide distribution ($\overline{S} = 3\frac{1}{2}$-4-$2\frac{1}{2}$) of somatotypes among Austrian table tennis players, similar to that of other European players (Eiben & Eiben, 1979). Young Brazilian tennis players (Fernandes & Corazza, 1986) had more endo-mesomorphic somatotypes ($\overline{S} = 3.4$-4.7-2.1)

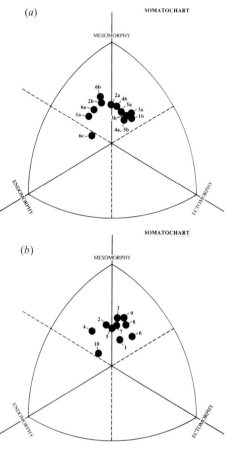

(*a*)

(*b*)

Fig. 6.30. Mean somatotypes for males in a variety of sports. Note the similarities or differences between means within and between sports. (*a*) 1a = golf – SDSU; 1b = golf – Brighton; 2a = karate – Belgium; 2b = karate – Korea; 3a = orienteering – Czechoslovakia; 3b = orienteering – South Australia; 4a = squash – Brighton; 4b = squash – South Australia; 5a = tennis – professional; 5b = tennis – amateur; 6a = baseball – Iowa; 6b = baseball – Cuba; 6c = baseball – Korea. (*b*) 1 = badminton; 2 = lacrosse; 3 = modern pentathlon; 4 = parachuting; 5 = surfboarding; 6 = triathlon; 7 = yachting; 8 = figure skating; 9 = diving; 10 = table tennis.

than players in the South African championships, who were more ecto-mesomorphic (Copley, 1980*a,b*)

The somatotypes of 12-year-old Cuban athletes invite speculative comparisons with older athletes in their respective sports, i.e. diving, baseball, tennis and table tennis, and shooting (Alonso, 1986).

Sports studied using other somatotype methods include: baseball (Imlay, 1966), climbing (Renson & Swalus, 1970), cricket and golf (Jones *et al.*, 1965), lumberjacks (Eränkö & Karvonen, 1955), surf life saving (Gayton, 1975), surfboard riding (Crooks, in Gayton, 1975), and table tennis (Murphy, 1972).

Female sports

Prior to the 1970s there were few somatotype data on female athletes. In the Philadelphia area Morris (1960) made a somatotype study of

151 female athletes from nine sports, rated by Heath according to Sheldon's (1954) criteria. Their mean somatotype was 4½-4-3, and the mean for a comparable college sample was 5-3-3. Carter's (1970) review reported only five sports groups that had been somatotyped by the Heath-Carter method, but in a later review (1981) he presented data for 16 sports, with a total of 87 samples and 1770 subjects in studies from the 1967–1980 period. The following material is taken from that review and from further studies.

Table 6.21. *Basketball – female*

Sample, reference	N	Statistic	Age (yr)	Height (cm)	Weight (kg)	HWR	Somatotype
Venezuela (Pérez, 1981)	19	$\bar{X}$	18.4	168.2	59.0	43.2	3.2-3.8-3.1
		SD	3.0	6.9	6.5	1.5	1.5 0.8 1.0
Canada (Alexander, 1976)	53	$\bar{X}$	20.0	170.6	63.9	42.7	4.0-3.5-2.7
		SD	1.6	5.9	5.5	1.5	0.7 0.9 1.0
Maryland (Vaccaro et al., 1979)	15	$\bar{X}$	19.4	173.0	68.3	42.3[a]	4.3-3.9-2.4
		SD	1.1	9.1	7.8	—	1.2 0.5 0.9
USSR (Heath, in Carter, 1970)	10	$\bar{X}$	NA	173.0	71.4	41.7[a]	4.3-4.5-3.0
		SD		2.0	3.8		0.6 0.4 0.7
San Diego State University, 1978 (Robinson & Carter, unpubl.)	9	$\bar{X}$	21.2	173.1	66.3	42.8	3.3-3.5-2.8
		SD	1.3	8.1	7.6	1.1	0.9 0.9 0.8
Brighton Polytechnic (Bale, 1986)	5	$\bar{X}$	NA	169.2	66.8	42.0	3.9-4.3-2.5
		SD		4.9	8.2	1.0	0.5 0.4 0.9
Hungary, 1972–75	33	$\bar{X}$	NA	172.6	64.2	43.1[a]	3.1-4.0-3.0
1979–80	20	$\bar{X}$	NA	175.7	64.0	43.9[a]	3.4 4.3 3.3
(Mészáros & Mohácsi, 1982b)							
Hungary, 1970s, national (Eiben, 1981)	30	$\bar{X}$	21.4	176.6	67.2	43.4[a]	3.8-3.8-3.5
		SD	17–31	6.7	8.5		
Bolivar Games, 1981 (Brief, 1986)	20	$\bar{X}$	20.7	167.6	65.0	41.7	4.3-3.8-2.1
		SD	3.7	7.7	6.4	1.4	1.1 0.8 0.9
South Australia (Withers et al., 1987)	18	$\bar{X}$	22.9	175.2	68.0	43.0	3.7-4.0-2.9
		SD	2.6	6.9	8.2	1.1	0.8 1.0 0.8
Austin (Shoup, 1978; Shoup & Malina, 1985)	18	$\bar{X}$	16.3	165.2	57.2	42.8[a]	3.9-4.0-2.5
		SD	1.1	6.7	4.9		0.9 0.9 1.8
Cuba, youths (Alonso, 1986)	30	$\bar{X}$	12.5	163.3	51.7	43.8[a]	3.3-2.8-3.6

[a]Calculated from mean height and weight.
$\bar{X}$ = mean.

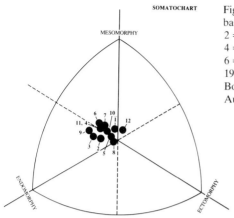

Fig. 6.31. Mean somatotypes for female basketball players. 1 = Venezuela; 2 = Canada; 3 = University of Maryland; 4 = USSR; 5 = San Diego State University; 6 = Brighton Polytechnic; 7 = Hungary, 1979–80; 8 = Hungary, national; 9 = Bolivar Games; 10 = South Australia; 11 = Austin high schools; 12 = Cuba, youths.

Basketball (Table 6.21, Fig. 6.31)

There was a large somatotype variation in basketball, partly due to the different functions of the playing positions. In adult samples the mean somatotypes are to the left of the somatochart centre, around the 4-4-3 somatotype. In a study of girls from 13 to 18 years of age Chovanová & Zapletalová (1980) found that increasingly with age guards were the most mesomorphic and centres the most ectomorphic. Shoup & Malina (1985) found differences between Whites, Blacks and Mexican Americans. Shoup (1978) reported a mean of 4-4-3 for University of Texas players; Mészáros & Mohácsi (1982*b*) found no differences between Hungarian players from early and late 1970s; and the mean of 3.5-2.0-3.7 for Brazilian players (age was not reported) found by Caldeira *et al.* (1986*c*) was similar to that of young Cubans. Bale (1981) found a mean of 3.7-3.4-3.3 for Brighton Polytechnic players using the M.4 method.

Body building (Table 6.22, Fig. 6.32(*a*),(*b*))

There are no published studies to confirm that women who participate in body building are high in mesomorphy and low in endomorphy, as might be expected. However, Table 6.22 and Fig. 6.32(*a*) and (*b*) present preliminary data from several body building competitions.

There are anthropometric somatotypes for eight competitors in the California State Body Building Championships, San Diego, 1985. Carter used height, weight, HWR and a video tape of the event to make photoscopic ratings of 28 competitors in the National Body Building Championships in Detroit, 1985. The women competed in four weight classes; 17 of the sample were placed first to fifth in their weight classes, this included all first and second placers. The same procedure was used to rate the women in the first five places in the International Federation of Body Building Champion-

Fig. 6.32(a). Mean somatotypes for female body builders. 1 = California, 1985; 2 = Detroit, 1985, lightweight; 3 = Detroit, 1985, middleweight; 4 = Detroit, 1985, light heavyweight; 5 = Detroit, 1985, heavyweight; 6 = Toronto, 1984. (b). Body builder. Height = 153.7 cm; weight = 47.8 kg; HWR = 42.4; somatotype = 3-$4\frac{1}{2}$-$2\frac{1}{2}$. (At competition time three months later, this athlete weighed 45.4 kg and with lowered skinfolds and a higher HWR of 43.1 her somatotype was $2\frac{1}{2}$-$4\frac{1}{2}$-3.)

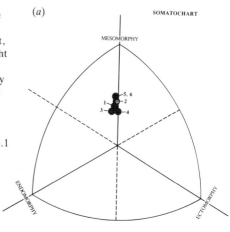

(a)

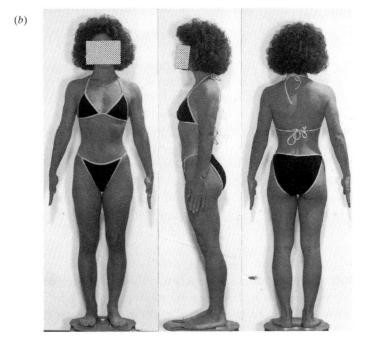

(b)

ships in Toronto, 1984. The means were in the balanced mesomorphy category, between 2-5-2, 3-5-3, and 3-5-2 somatotypes. The range of endomorphy across all samples was $1\frac{1}{2}$-3, of mesomorphy $3\frac{1}{2}$-$6\frac{1}{2}$ and of ectomorphy 1-4. Somatotype means were close together for light and heavy weights, and for middle and light heavy weights. The overall mean was $2\frac{1}{2}$-5-$2\frac{1}{2}$. Endomorphy was as low for body builders as for some of the gymnasts and distance runners, and mesomorphy was as high or higher than

Table 6.22. *Body building – female*

Sample, reference	N	Statistic	Age (yr)	Height (cm)	Weight (kg)	HWR	Somatotype
California, 1985 (Powell	8	X̄	27.5	160.7	56.8	41.9	2.4-4.8-2.1
& Carter, unpubl.)		SD	4.3	3.8	5.7	1.1	0.3 0.7 0.7
Detroit, 1985							
Lightweight	6	X̄	26.8	152.0	46.6	42.2	2.5-5.3-2.3
		SD	4.7	5.2	1.2	1.2	0.0 0.5 0.8
Middleweight	7	X̄	26.5	158.6	51.0	42.8	2.4-4.9-2.8
		SD	1.8	3.8	1.2	1.0	0.3 0.4 0.7
Light heavyweight	7	X̄	26.3	162.6	55.1	42.7	2.6-4.9-2.7
		SD	2.9	2.9	0.7	0.7	0.2 0.3 0.5
Heavyweight	8	X̄	30.3	165.7	61.6	42.0	2.4-5.3-2.2
		SD	6.0	3.0	3.8	1.0	0.2 0.6 0.8
Total	28	X̄	27.5	160.2	54.1	42.4	2.5-5.1-2.5
(Carter, unpubl.)		SD	4.2	6.2	6.0	1.0	0.2 0.5 0.7
Toronto, 1984 (Carter	5	X̄	—	—	—	—	2.3-5.2-2.1
unpubl.)		SD					0.3 0.3 0.2

X̄ = mean.

for some shot and discus throwers. Several former gymnasts have become top level body builders.

Gymnastics (Table 6.23, Fig. 6.33(*a*),(*b*))

Gymnastics is unique in the world of sports in that the best competitors in the world are likely to be pre- or mid-adolescent. As a result, many of the subjects are younger than those in other sports. In addition, the changes in style and demands of Olympic gymnastics have contributed to physique variation in the past several decades, as Carter (1981; 1988) noted. The mean somatotypes for the youngest samples (means ages 10 to 17 years) have been ecto-mesomorphic or central, near to a 2-4-3 somatotype, and the older samples are more endo-mesomorphic or central, near to 3-4-3. The remarkably consistent and restricted distributions of the young gymnasts suggests that in gymnastics the somatotype is a highly important factor from an early age. The distributional pattern seems to be well established even at the club level. The somatoplot distributions of most samples of gymnasts are elliptical in shape, on a NW to SE axis, with little variation in endomorphy. The mean somatotype of a combined sample of US Junior Olympic gymnasts and divers (mean age = 15.2 yr), who did not differ from each other, was 2.7-3.8-3.3 (Thorland *et al.*, 1981) – close to the gymnast-only samples reported herein. Studies of Caldeira *et al.* (1986*c*), Vívolo *et al.* (1986*b*), Bevans (1977) and Bale's (1981) M.4 study, present limited data on gymnasts.

Table 6.23. *Gymnastics – female*

Sample, reference	N	Statistic	Age (yr)	Height (cm)	Weight (kg)	HWR	Somatotype
Mexico City Olympics, 1968 (De Garay *et al.*, 1974)	21	X̄	17.8	156.9	49.8	42.9	2.9-4.2-2.8
		SD	3.7	5.1	4.5	0.7	0.7 0.5 0.5
Montreal Olympics, 1976 (Carter *et al.*, 1982)	15	X̄	17.0	161.5	50.9	43.7	2.2-4.0-3.4
		SD	2.0	5.7	6.0	1.0	0.4 0.6 0.7
Venezuela (Pérez, 1981)	10	X̄	13.0	151.7	43.4	43.0	2.5-4.4-3.4
		SD	1.6	6.9	7.2	1.2	0.7 0.6 0.7
Gama Filho (Araújo & Moutinho, 1978)	9	X̄	12.2	145.8	35.1	45.0	2.1-3.5-4.1
		SD	1.9	8.3	5.4	—	0.4 0.4 0.7
Springfield (Sinning, 1978)	44	X̄	19.4	160.6	53.7	42.6[a]	2.9-3.6-2.6
		SD	1.1	4.4	5.9	—	—
USA Colleges, best (Falls & Humphrey, 1978)	14	X̄	19.4	161.5	55.1	42.4[a]	2.7-4.4-2.6
		SD	1.5	4.7	5.8	—	0.8 0.8 0.7
USSR (Heath in Carter, 1970)	5	X̄	—	157.0	53.9	41.6[a]	3.8-5.2-1.6
San Diego, club (Strong, 1980)	20	X̄	14.8	156.0	46.8	43.3	2.7-3.5-3.4
		SD	1.1	8.1	6.7	1.3	0.5 0.7 1.0
California, best (Strong, 1980)	28	X̄	13.6	153.8	44.3	43.5	2.2-3.9-3.3
		SD	0.9	6.2	4.8	1.0	0.5 0.5 0.8
Munich Olympics, 1972 (Novak *et al.*, 1977)	5	X̄	19.0	163.5	52.5	43.7[a]	2.6-3.8-3.4[b]
		SD	3.5	2.3	1.2		
French-Canada (Salmela, 1979)	7	X̄	15.3	147.8	39.7	43.3[a]	1.7-4.3-3.1[a]
		SD	1.8	9.3	7.4		
Czechoslovakia, Brno (Štěpnička, 1976*b*)	17	X̄	12.5	147.8	37.0	44.4[a]	1.5-3.8-3.9
		SD	—	5.7	5.1	—	0.6 0.6 0.7
Ontario, club, Ontario, University (Yuhasz *et al.*, 1980)	13 8	X̄ SD	14.3 21.1	153.1 154.7	41.8 55.0	44.1[a] 40.7[a]	2.0-3.0-3.8 3.7 4.3 1.2
Belgium (Beunen *et al.*, 1981*b*)	23	X̄	16.6	158.4	49.6	43.1[a]	2.4-3.7-3.1
		SD	11–21	9.2	9.7		
Cuba, international, 1977 (Lopez *et al.*, 1979)	24	X̄	15.2	151.9	42.4	43.5[a]	2.0-3.9-3.5
		SD	2.5	5.6	6.1		
Cuba (Alonso, 1986)	6	X̄	12.5	140.8	35.7	42.8[a]	1.9-3.5-2.5
Bolivar Games, 1981 (Brief, 1986)	7	X̄	11.9	146.6	37.6	43.9	1.9-4.3-3.6
		SD	5.5	7.6	6.1	1.3	0.4 0.8 0.9
San Diego, club, 1984 (Carter & Brallier, 1988)	7	X̄	10.3	133.4	28.1	44.0	2.3-4.3-3.6
		SD	0.8	4.7	2.6	0.8	0.6 0.5 0.7
Oregon (Broekhoff *et al.*, 1986)	18	X̄	13.3	150.2	42.1	43.2[a]	1.7-3.9-3.3
		SD	9–17	11.1	10.0	—	0.6 0.5 0.7

[a]Calculated from mean height and weight.
[b]Calculated from mean anthropometry.
X̄ = mean.

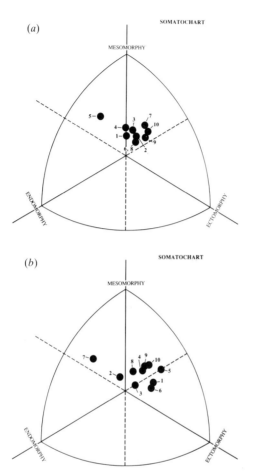

Fig. 6.33(*a*). Mean somatotypes for female gymnasts at the national or international level. 1 = 1968 Olympics; 2 = 1976 Olympics; 3 = Venezuela; 4 = USA colleges (best); 5 = USSR; 6 = 1972 Olympics; 7 = French-Canada; 8 = Belgium; 9 = Cuba, International; 10 = Bolivar Games. (*b*). Mean somatotypes for female gymnasts at club, state and university levels. 1 = Gama Filho Club; 2 = Springfield; 3 = San Diego Club; 4 = California (best); 5 = Brno; 6 = Ontario, club; 7 = Ontario, university; 8 = Cuba, youths; 9 = San Diego Club, 1984; 10 = Oregon.

Rowing (Table 6.24, Fig. 6.34(*a*),(*b*))

Before the 1976 Olympics there were no data available on women rowers. Analyses showed 60% of 51 rowers from the 1976 Olympics had central or balanced mesomorphy somatotypes. The subgroups were similar when compared by geographic origin, rowing style, and event. In part this may have been due to the wide range of somatotypes in rowing. Fig. 6.34(*a*) shows that somatotypes of most of the rowers in the eights event are

(a)

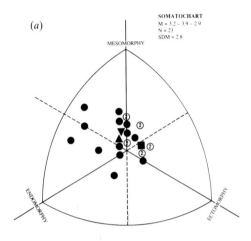

(b)

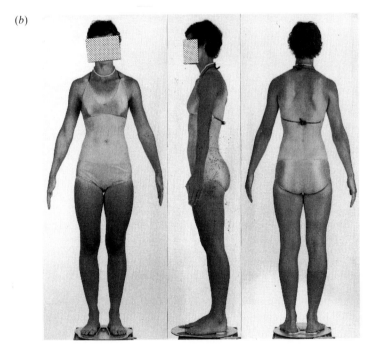

Fig. 6.34(a). Mean somatotypes for female Olympic eights ▲, all rowers at the 1976 Montreal Olympics ▼, and lightweight rowers at the 1985 World Championships ■. The distribution (dots and circled numbers) of the eights at Montreal is also shown. (b). Rower. Height = 179.2 cm; weight = 66.6 kg; HWR = 44.2; somatotype = 3-4-4.

Table 6.24. *Rowing – female*

Sample, reference	N	Statistic	Age (yr)	Height (cm)	Weight (kg)	HWR	Somatotype
Montreal Olympics, 1976							
Total	51	X̄	23.8	174.3	67.4	42.9	3.1-3.9-2.8
		SD	2.7	4.8	5.3	1.1	0.8 0.9 0.8
Eights only	23	X̄	23.1	175.8	68.3	43.1	3.2-3.9-2.9
(Carter *et al.*, 1982;		SD	2.3	4.4	5.8	1.0	1.0 0.5 0.6
Hebbelinck *et al.*, 1980)							
World Championships, Hazenwinckel, 1985							
Lightweights	50	X̄	24.1	167.1	57.1	43.9	2.4-3.0-3.5
(Rodríguez, 1986)		SD	3.7	—	2.0	1.3	0.8 1.1 1.0

X̄ = mean.

concentrated around the centre of the somatochart, with some 'outliers' to the left and below. The standard deviation for endomorphy is twice that for mesomorphy and ectomorphy. Rowers at the 1985 World Championships for lightweights were less mesomorphic and more ectomorphic than the Olympic heavyweights. Rodríguez (1986) found no difference between medallists and non-medallists, or among events. The estimated mean from the somatoplots of a sample of Hungarian rowers is $3\frac{1}{2}$-5-3 (Mészáros & Mohácsi, 1982*a*).

The overall distribution of Olympic rowers is similar to that of swimmers, but the rowers are much larger. Coaches recruiting athletes for rowing may be interested in knowing that many of the North American rowers said they had been competitive swimmers before they took up rowing.

Skating and skiing (Table 6.25, Fig. 6.35)

Most of the somatotype studies in both sports are on young athletes. The means for the skaters lie close together, between 2-4-4 and 3-4-3, and are similar to those of young gymnasts. Karen Magnussen, a world champion, had a $3\frac{1}{2}$-$4\frac{1}{2}$-$2\frac{1}{2}$ somatotype, reflecting higher endomorphy than that of the younger skaters. Ross *et al.* (1977*b*) suggested that increased endomorphy reflects normal female growth that accompanies physical maturation.

The best young Canadian skiers from Rossland and Nelson, British Columbia (Ross *et al.*, 1976) were ecto-mesomorphic. The Czech skiers were a little older, more endomorphic and mesomorphic and less ectomorphic than the Rossland skiers. The somatotypes of the five-woman USA Nordic ski team were endo-mesomorphic. Elite Canadian skiers were also endo-mesomorphic, with a mean of 3.0-4.9-1.9 (Ross *et al.*, 1977*a*).

Fig. 6.35. Mean somatotypes for female
skaters ● and skiers ■. Skaters: 1 = Canada
– senior–junior; 2 = Canada – novice;
3 = Canada – 12+ yr; 4 = Canada – <12 yr.
Skiers: 5 = USA Nordic, 1976;
6 = Rossland downhill; 7 = Czech downhill.

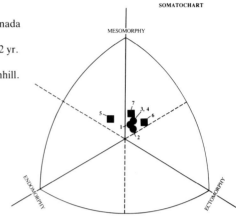

Table 6.25. *Skating and skiing – female*

Sample, reference	N	Statistic	Age (yr)	Height (cm)	Weight (kg)	HWR	Somatotype
Skating							
Canada, senior-junior	18	X̄	15.7	156.8	48.6	43.0[a]	2.6-3.8-3.0
(Ross *et al.*, 1977*b*)		SD	1.6	5.2	6.1	—	0.7 0.6 0.9
Canada, novice (Ross	9	X̄	13.2	153.5	42.1	44.1[a]	2.1-3.7-3.9
et al., 1977*b*)		SD	1.4	8.9	5.7	—	0.4 0.7 0.7
Canada, >12 yr	21	X̄	14.0	154.1	45.5	43.2[a]	2.5-4.1-3.2
(Faulkner, 1976)		SD	1.7	8.5	8.2	—	0.7 0.5 0.9
Canada, <12 yr	12	X̄	10.7	136.7	31.2	43.4[a]	2.5-4.2-3.3
(Faulkner, 1976)		SD	1.1	9.4	5.6		0.7 0.6 1.0
Skiing							
USA Nordic, 1976	5	X̄	23.5	164.5	53.9	42.8[a]	3.5-4.3-2.3
(Sinning *et al.*,		SD	4.7	3.3	1.1	—	0.7 0.5 0.8
1977)							
Rossland, downhill, 1970	15	X̄	10.7	139.7	32.7	43.7[a]	2.0-4.0-3.6
(Ross & Day, 1972)		SD	2.2	13.9	8.8	—	0.5 0.9 1.1
Czechoslovakia, downhill	11	X̄	13.0	156.4	48.2	3.0[a]	2.5-4.3-3.0
(Štěpnička & Broda,		SD	—	6.8	7.6	—	0.9 0.2 0.7
1977)							

[a]Calculated from mean height and weight.
X̄ = mean.

Swimming (Table 6.26, Fig. 6.36)

In general, like gymnasts, top level swimmers are younger than
athletes in the majority of other sports. In some samples, as Table 6.26
shows, there is a wide range of means for age (11.6 to 21.6 years). The mean

Table 6.26. *Swimming – female*

Sample, reference	N	Statistic	Age (yr)	Height (cm)	Weight (kg)	HWR	Somatotype
Mexico City Olympics, 1968 (Hebbelinck *et al.*, 1975)	27	X̄	16.3	164.4	56.9	43.1	3.1-4.0-3.0
		SD	2.9	7.1	9.1	1.2	0.9 0.7 0.9
Montreal Olympics, 1976 (Carter *et al.*, 1982)	32	X̄	16.6	166.9	57.8	43.2	3.2-3.8-3.0
		SD	2.6	5.7	6.8	1.2	0.8 0.7 0.8
Munich Olympics, 1972 (Novak *et al.*, 1977)	7	X̄	17.7	167.0	60.1	42.6[a]	3.2-4.6-2.6[b]
		SD	2.3	8.9	7.7		
Venezuela (Pérez, 1981)	14	X̄	14.8	163.7	55.2	42.7	3.2-4.1-2.8
		SD	1.2	4.5	4.4	1.4	1.1 0.9 0.8
Brazil, 1975 (Rocha *et al.*, 1977b)	15	X̄	16.3	166.0	59.6	45.5	3.3-3.5-2.9
		SD	1.7	4.5	4.6	1.1	0.7 0.5 0.9
Manchester, club (Bagnall & Kellett, 1977)	12	X̄	15.3	165.9	54.3	43.8[a]	2.1-3.8-3.5
		SD	13–17	—	—	—	0.5 0.9 0.8
Austin, club (Meleski *et al.*, 1982)	41	X̄	17.1	168.2	56.0	44.0	2.9-3.7-3.6
		SD	2.4	6.3	4.7	1.1	0.8 0.8 0.8
Caracas, club (Pérez, 1977)	12	X̄	13.8	158.2	46.5	44.0[a]	2.3-3.8-3.4
		SD	—	—	—	—	1.1 1.0 1.6
Chula Vista, club, 1976 (Krogh & Keska, unpubl.)	15	X̄	14.2	161.3	52.2	43.1	2.4-3.4-2.8
		SD	1.0	5.3	2.1	0.9	0.5 0.7 0.9
San Diego State University, 1976 (Krogh & Keska, unpubl.)	15	X̄	19.5	169.0	65.9	42.0	3.9-3.8-2.2
		SD	1.3	7.0	9.8	1.8	1.1 1.1 1.3
San Diego State University, 1978 (Atchley unpubl.)	15	X̄	19.5	169.6	61.8	42.9	3.9-4.1-2.8
		SD	1.1	6.8	6.8	1.7	0.9 1.0 1.2
San Diego State University, 1980 (Atchley-Carlson, 1981)	15	X̄	20.4	168.2	63.2	42.2[a]	3.8-3.9-2.3
		SD	1.6	5.7	5.7	—	1.0 0.7 0.6
Belgium (Vervaeke & Persyn, 1981)	47	X̄	10–22	—	—	—	3.3-3.4-3.8
		SD					0.9 0.8 0.6
Ontario, university, Ontario, club (Yuhasz *et al.*, 1980)	14	X̄	21.6	167.3	63.3	42.0[a]	4.5-2.6-2.2
	17	X̄	11.6	154.4	43.6	43.9[a]	3.1 3.1 3.7
Bolivar Games, 1981 (Brief, 1986)	12	X̄	14.5	160.0	54.9	42.2	3.4-4.5-2.4
		SD	0.9	7.3	7.0	0.9	0.8 0.7 0.8
Cuba (Alonso, 1986)	9	X̄	12.5	148.2	42.0	42.6[a]	2.8-3.3-2.6

(*continued*)

Table 6.26. (*Continued*)

Sample, reference	N	Statistic	Age (yr)	Height (cm)	Weight (kg)	HWR	Somatotype
Friendship Games, Havana, 1977							
Cuban swimmers	8	$\bar{X}$	13.2	154.3	47.6	42.6[a]	3.0-3.8-2.7
Foreign swimmers	57	$\bar{X}$	14.0	165.1	53.0	44.0[a]	2.5-3.3-3.6
(Pancorbo & Rodríguez, 1986)							
USA, Junior Olympic,	67	$\bar{X}$	15.8	168.2	58.5	43.3[a]	3.6-3.4-3.3
1980 (Thorland et al., 1983)		SD	1.4	6.6	5.9		0.8 1.1 1.1

[a]Calculated from mean height and weight.
[b]Calculated from means of anthropometric variables.
$\bar{X}$ = mean.

for Olympic swimmers is close to 17 years. The mean somatotypes fall near the centre of the somatochart, bounded by 4-4-2, 3-4-3, 2-4-4 and 4-4-4. The youngest are the club swimmers who are more ectomorphic and less endomorphic than the older university swimmers. The Olympic and Bolivar Games swimmers had means near 3-4-3 and 3½-4-3. The combined mean (N = 59) of the 1968 and 1976 Olympic samples, which did not differ, was 3.2-3.9-3.0 (Carter, 1984*b*). Most of the distributions about the mean were circular.

The Belgian and Ontario club swimmers, and foreign swimmers at the Friendship Games, were more ectomorphic than other samples; and the University of Western Ontario sample was much higher in endomorphy and lower in mesomorphy than other samples.

The three San Diego State University teams from 1976, 1978 and 1980

Fig. 6.36. Mean somatotypes for female swimmers. 1 = 1968 and 1976 Olympics; 2 = 1972 Olympics; 3 = Venezuela; 4 = Brazil, 1975; 5 = Manchester; 6 = Austin; 7 = Caracas; 8 = Chula Vista; 9 = San Diego State University; 10 = Belgium; 11 = Ontario, university; 12 = Ontario, club; 13 = Bolivar Games; 14 = Cuba, youths; 15 = Friendship Games, 1977, Cuban; 16 = Friendship Games, 1977, Foreign; 17 = USA Junior Olympics.

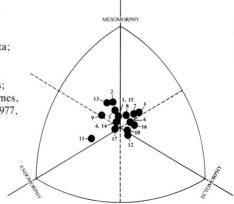

were somatotypically remarkably consistent, with an overall mean of 4-4-2½. For the 1980 team there was no somatotype change before and after five months of training (Atchley-Carlson, 1981). Araújo *et al.* (1978c) found no somatotype differences in 15-year-old Brazilian swimmers at four stages of training ($\bar{S}$ = 3.4-3.8-2.6). In another study Araújo (1979) found no differences by age (overall $\bar{S}$ = 3-3½-3½) in a study of four age groups (9.5, 12.0, 15.3, 16.9 years). However, Mohácsi & Mészáros (1982) found increased mesomorphy and ectomorphy in 12-year-old Hungarian swimmers, who had undergone training. At the end of training the mean somatotype was 4.6- 4.5-3.1, much higher in endomorphy and mesomorphy than other young swimmers. It is not clear whether there are real differences by swimming stroke or event. Small differences between strokes were found in the 1968 Olympic swimmers, but not in the 1976 Olympic swimmers (Carter, 1981) nor in the 1975 Brazilian selection (Rocha *et al.*, 1977b). Araújo (1978) found breaststrokers were the least endomorphic. However, the small samples in stroke and event groupings and multi-stroke and multi-distance training limit possible analyses and interpretation.

The somatotype of the three long distance Channel swimmers was 5-5½-1½. (Ratings were made by Heath from photos provided by J. M. Tanner.)

Tennis (Table 6.27, Fig. 6.37)

The somatoplots of tennis players cover a fairly wide circular area, with most means just to the left of the centre of the somatochart. There have been three studies, of different levels of play. Pallulat (1984), found no differences among high, middle and low ranked professional players. Copley (1980a,b) found that amateur players were less mesomorphic and more ectomorphic than professional players. In a study of three levels of USA players, Lebedeff (1980) noted that the two highest level players

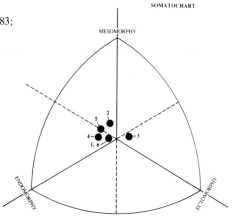

Fig. 6.37. Mean somatotypes for female tennis players. 1 = WTA professional, 1983; 2 = South Africa, 1977, professional; 3 = South Africa, 1977, amateur; 4 = USA, university; 5 = USA, juniors; 6 = Cuba, youths.

Table 6.27. *Tennis – female*

Sample, reference	N	Statistic	Age (yr)	Height (cm)	Weight (kg)	HWR	Somatotype
World Tennis Association, professional, 1983 (Pallulat, 1984)	50	X̄	25.4	170.5	61.6	43.2	3.8-3.5-3.1
		SD	5.4	7.7	5.4	1.4	0.9 1.0 1.0
South Africa, 1977							
professional	19	X̄	24.1	167.3	60.7	42.6[a]	3.1-3.9-2.6
		SD	4.5	4.7	6.0		1.3 1.0 0.9
amateur	11	X̄	24.9	167.9	55.4	44.0[a]	2.6-3.2-3.6
(Copley, 1980*a*,*b*)		SD	11.5	5.0	4.3		0.8 1.0 0.8
USA							
university, 1976	14	X̄	20.4	166.4	57.9	43.1	4.6-3.9-3.0
		SD	0.8	7.2	7.7	0.8	1.0 1.0 0.6
juniors, 1977	21	X̄	16.7	166.3	59.0	42.8	4.0-4.0-2.7
(Lebedeff, 1980)		SD	1.4	7.3	7.5	1.3	0.9 0.9 0.9
Cuba, youths (Alonso, 1986)	7	X̄	12.5	149.4	42.5	42.8[a]	3.6-3.3-2.9

[a]Calculated from mean height and weight.
X̄ = mean.

(National Collegiate and National Juniors) did not differ from one another, but were more mesomorphic and less endomorphic than other university players and a control group. Four of the samples had means around the 4-4-3 somatotype; South African amateurs were more meso-ectomorphic. It is interesting that the means of young Cubans (12.5 yr) and USA juniors (16.7 yr) are similar to that of Women's Tennis Association professionals.

Robson's (1974) and Bale's (1986) reports on 'racquet sports' presumably include tennis. Fernándes & Corazza (1986) report a mean of 5.2-4.5-2.5 for 17 young players in Brazil.

Track and field (Table 6.28, Fig. 6.38(*a*),(*b*),(*c*))

As in the case of males, the large number of track and field events precludes extensive individual reviews. Before 1968 there was far less information on females than males, and until recently little on distance runners.

Analysis of the data for the 1968 Olympic athletes (De Garay *et al.*, 1974) showed that shot and discus throwers were more endomorphic and meso-morphic and less ectomorphic than sprinters, middle-distance runners, jumpers and pentathletes. Javelin throwers were more endomorphic than middle distance runners and jumpers; and sprinters were more mesomor-phic and less ectomorphic than middle distance runners and jumpers. A smaller number of athletes measured at the Montreal Olympics showed

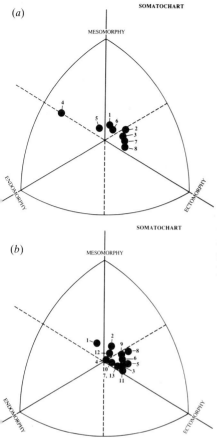

(a)

(b)

Fig. 6.38(a). Mean somatotypes for female Olympic track and field athletes in the 1968 and 1976 Olympics. 1 = sprinting and hurdling; 2 = 400 m; 3 = 800 and 1500 m; 4 = shot and discus; 5 = javelin; 6 = pentathlon; 7 = long jump; 8 = high jump. (b). Mean somatotypes for female track athletes in non-Olympic samples. 1 = Czechoslovakia, 100, 200 m; 2 = Czechoslovakia, 100 m hurdles; 3 = Venezuela, 100 m; 4 = Bolivar Games, runners; 5 = England, marathon 'elite'; 6 = Brighton Polytechnic; 7 = Midwest, high school; 8 = Europe, 800–3000 m; 9 = Belgium, 800–3000 m; 10 = San Diego, 1981, NMR; 11 = San Diego, 1981, OAR; 12 = San Diego, 1982, >40 yr; 13 = Cuba, youths. (c). High jumper. Height = 177.0 cm; weight = 59.0 kg; HWR = 45.5; somatotype = $2\frac{1}{2}$-$3\frac{1}{2}$-5. (Photo with permission from J. Štěpnička.)

(c)

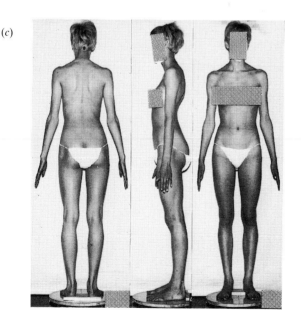

Table 6.28. *Track and field – female*

Sample, reference	N	Statistic	Age (yr)	Height (cm)	Weight (kg)	HWR	Somatotype
Olympics, 1968 + 1976							
Sprinting–hurdling	35	X̄	—	—	—	43.0	2.6-3.8-3.0
		SD				1.1	0.9 0.7 0.8
400 m	18	X̄	—	—	—	44.0	1.9-3.3-3.6
		SD				1.0	0.5 0.8 0.7
800 + 1500 m	13	X̄	—	—	—	44.1	2.1-3.3-3.7
		SD				0.8	0.7 0.7 0.7
Shot + discus	10	X̄	—	—	—	40.8	5.3-5.3-1.6
		SD				2.0	1.9 1.3 1.1
Javelin	6	X̄	—	—	—	43.0	3.4-4.0-2.9
		SD				1.8	1.3 1.0 1.2
Pentathlon	12	X̄	—	—	—	43.3	2.5-3.7-3.1
		SD				0.9	0.8 0.7 0.7
Long jump	8	X̄	—	—	—	44.5	2.3-3.2-4.0
		SD				1.0	0.1 0.7 0.8
High jump	14	X̄	—	—	—	44.5	2.3-2.9-4.0
(Carter, 1984*b*)		SD				0.8	0.5 0.6 0.6
Czechoslovakia							
100, 200 m, 1967	49	X̄	—	166.7	60.1	42.7[a]	3.4-4.3-2.7
		SD		4.6	6.2		0.8 0.4 1.0
100 m hurdles, 1977	26	X̄	—	170.8	61.8	43.2[a]	2.4-3.8-3.1
(Štěpnička *et al.*,		SD		5.2	5.3		0.5 0.8 0.8
1979*a*)							
Venezuela, 100 m	7	X̄	20.7	165.9	51.7	44.2	2.3-2.9-3.8
(Pérez, 1981)		SD	4.2	5.3	4.2	1.1	0.4 0.7 0.7
Bolivar Games, 1981,	8	X̄	20.3	159.9	51.2	43.1	2.7-2.9-2.9
track runners (Brief,		SD	4.2	3.3	2.4	1.0	0.8 1.0 0.7
1986)							
England, 1983							
marathon, 'elite'	11	X̄	29.4	166.4	54.7	44.0	2.8-3.6-4.6
		SD	7.6	3.9	5.6	1.3	0.5 0.8 0.6
'good'	12	X̄	29.8	163.9	51.2	44.3	3.2-3.4-4.6
		SD	8.4	6.0	2.2	1.7	0.8 0.9 1.0
'moderate'	13	X̄	29.7	161.3	53.0	43.0	3.5-4.0-3.4
(Bale *et al.*, 1985*b*)		SD	5.3	6.6	5.4	1.0	0.6 0.6 0.7
Brighton Polytechnic,	5	X̄	—	166.9	55.0	44.0	2.7-3.6-4.1
middle and long		SD		2.6	5.8	1.3	1.0 0.9 1.0
distance (Bale, 1986)							
Midwest USA, 1980,	72	X̄	15.6	164.2	51.1	44.3[a]	2.8-3.2-4.0
cross-country,		SD	1.1	6.0	5.7		
(Butts, 1982)							
Europe, 800–3000 m	33	X̄	24.1	164.9	52.3	44.1[a]	1.7-3.4-3.7
(Day *et al.*, 1977)		SD	4.2	4.7	3.9		0.5 0.7 0.8
Belgium, 800–3000 m	33	X̄	19.1	163.8	51.6	44.0[a]	2.3-3.5-3.7
(Day *et al.*, 1977)		SD	3.2	5.3	5.0		0.5 0.6 0.8

Table 6.28. (*Continued*)

Sample, reference	N	Statistic	Age (yr)	Height (cm)	Weight (kg)	HWR	Somatotype
San Diego, 1981							
Distance, NMR[b]	23	X̄	29.4	165.9	54.2	43.9	3.0-3.2-3.5
		SD	6.6	5.9	5.9	1.1	1.0 0.9 0.8
Distance, OAR[c]	16	X̄	25.1	164.9	50.4	44.7	2.6-2.8-4.1
(Mittleman, 1982)		SD	6.3	6.6	6.2	1.5	0.8 1.1 1.0
San Diego, 1982,	21	X̄	45.8	165.1	53.7	43.7	3.0-3.8-3.5
>40 yr, 5000 m to		SD	4.2	4.9	2.6	1.4	0.9 0.8 0.9
marathon (Wichary, 1984)							
Cuba, youths (Alonso, 1986)	39	X̄	12.5	156.0	44.7	44.0[a]	2.6-2.9-3.8

[a]Calculated from mean height and weight.
[b]Normally menstruating distance runners.
[c]Oligoamennorheic distance runners.
X̄ = mean.

similar trends (Carter *et al.*, 1982). Table 6.28 and Fig. 6.38(*a*) summarize the data for the two Olympic samples combined as eight event groups. There were no indications of change in somatotype between 1968 and 1976, except in the case of sprinters and hurdlers. In the 1976 samples there were no somatotypes to the left of the mesomorphy axis, but there were 10 (36%) to the left side in the 1968 sample (Carter, 1984*b*). It may be that today's female sprinter-hurdlers are less endomorphic than earlier ones. Arranged by events, the means are above and approximately parallel to the ectomorphic axis and to the male means.

There are several studies that report differences between events: Carter (1981; 1984*b*); Carter *et al.* (1982); De Garay *et al.* (1974); Guimarães & De Rose (1980); Sodhi & Sidhu (1984); Thorland *et al.* (1981) and Westlake (1967). Carter (1988) reviewed the studies of young athletes in different levels of competition. Overall, the findings are fairly consistent with those at the Olympic level.

Table 6.28 and Fig. 6.38(*b*) summarize recent reports of track events that include sprinting, hurdling and marathon running. All means are to the right of the mesomorphic axis, except for the 1967 CSSR sprinters. Some are in the central and some in the meso-ectomorphic categories. In an extensive study of English marathon runners, Bale *et al.* (1985*b*) found that 'moderate' runners were less ectomorphic than those in the two faster groups. The overall mean for the 36 runners was 3.2-3.7-4.2. The somatotypes of other distance runners are similar, including those of high school cross-country runners (Butts, 1982) and young Cubans (Alonso, 1986). Mittleman (1982) found that oligoamenorrheic distance runners were more ectomorphic than normally menstruating runners. Women over 40 years of age, who habitually run 64 km or more per week and are concerned about their figures, may be encouraged by the idea that they can maintain endomorphy at the level of younger distance runners. They are also less endomorphic and more ectomorphic than matched non-runners of similar age and exercise frequency, who play tennis, swim and participate in aerobics, and have a mean somatotype of 4.2-3.6-2.6 (Wichary, 1984).

Other studies of track and field athletes are: Caldeira *et al.* (1986*e*); Robson (1974); Shoup (1978); Shoup & Malina (1985) and Wilmore *et al.* (1977). Hay & Watson (1970) found differences between events for endomorphy and ectomorphy, but not for mesomorphy, using the M.4 method in a study of New Zealand athletes.

Volleyball (Table 6.29, Fig. 6.39(*a*),(*b*))

There are studies of National volleyball teams from Brazil, Hungary, Japan, USA and Venezuela. The USA 1985 team was less endomorphic and mesomorphic and more ectomorphic than the 1975 team. The Japanese were slightly more mesomorphic and less endomorphic than the Brazilians (Vívolo *et al.* 1986*a*). The team at the national training centre in Sao Paulo had a mean of 4.1-2.5-3.2 and was less mesomorphic than the national team (Caldeira *et al.*, 1986*e*). In other studies in Brazil, players from five countries at the South American Youth Championships had a mean of 3.6-3.6-2.5 (Caldeira *et al.*, 1986*a*); juvenile, university and national players differed from each other (Caldeira *et al.*, 1986*c*); and university teams from various regions of Brazil differed from one another (Caldeira *et al.*, 1986*d*).

Somatotypes for most of the top teams were distributed in and around the central category, with some means both above and below 4-4-4. The distributions of the Venezuelan and US teams were elongated in the NW–SE direction, due to the relatively small variation in endomorphy compared to variations in mesomorphy and ectomorphy. There were somatotype differences by playing level in the Hungarian and USA samples. In Hungary, Mészáros & Mohácsi (1982*b*) found the late 1970s teams were more endomorphic than those of the early 1970s, and the latter were similar to the national team measured by Eiben (1981). Texas high school and university players were meso-endomorphs and more endomorphic than the 1975 USA national players, whose somatotypes were more central. Between 1975 and 1978 the university teams became less endomorphic (Malina & Shoup, 1985). The junior varsity teams were more meso-endomorphic than the varsity teams (Kovaleski *et al.*, 1980). The young Cubans and the juvenile and university players from Brazil were the least mesomorphic.

Table 6.29. *Volleyball – female*

Sample, reference	N	Statistic	Age (yr)	Height (cm)	Weight (kg)	HWR	Somatotype
Hungary, Class I							
1972–75	31	X̄	—	170.6	61.5	43.2[a]	2.9-3.6-3.1
1979–80	18	X̄	—	170.9	61.2	43.4[a]	3.6-3.8-3.2
(Mészáros & Mohácsi, 1982b)							
Hungary, national,	25	X̄	20.3	174.8	68.3	42.8[a]	3.7-4.0-3.1
1970s (Eiben, 1981)		SD	16–33	4.8	5.3		
Venezuela (Pérez,	11	X̄	20.1	165.1	58.8	42.4	3.3-4.1-2.5
1981)		SD	2.9	3.6	3.7	1.4	0.6 1.0 1.0
Japan, 1977 (Vívolo	13	X̄	22.9	175.5	68.1	43.0[a]	3.4-3.7-2.9
et al., 1986a)		SD	2.0	3.3	4.7		
Brazil, 1977 (Vívolo	12	X̄	21.8	174.3	67.0	42.9[a]	4.1-3.4-2.9
et al., 1986a)		SD	4.0	3.9	5.2		
Brazil							
University	—	X̄	21.3	170.2	61.3	43.2[a]	3.2-2.7-3.0
Juvenile		X̄	17.4	171.9	61.4	43.6[a]	3.7-3.0-3.4
(Caldeira et al., 1986c)							
South Australia	11	X̄	22.8	173.0	61.8	43.8	3.0-3.5-3.5
(Withers et al., 1987)		SD	3.4	5.1	5.4	1.5	1.0 0.9 1.1
Austin, Texas							
High school, 1977	43	X̄	16.1	165.0	59.4	42.3[a]	4.7-4.0-2.6
(Shoup, 1978)		SD	1.1	8.5	9.8		1.3 1.0 1.4
University, 1973	15	X̄	19.9	166.0	60.6	42.3[a]	4.9-4.3-2.5
		SD	1.4	7.8	5.4		1.0 1.3 1.3
1978	14	X̄	20.0	171.9	63.4	43.1[a]	4.6-3.7-3.0
		SD	1.1	8.9	5.6		1.0 0.9 1.0
USA, national, 1975	18	X̄	22.5	174.4	67.4	42.9[a]	3.6-3.8-2.9
(Malina & Shoup, 1985)		SD	2.4	6.5	7.2		1.0 0.8 0.8
Central Michigan							
University, 1978							
Varsity	10	X̄	20.5	171.0	61.0	43.4[a]	3.7-3.5-3.2
		SD	1.2	8.6	6.2		0.8 0.8 0.9
Junior varsity	9	X̄	19.2	173.4	67.5	42.6[a]	4.5-4.3-2.6
(Kovaleski et al., 1980)		SD	0.8	6.8	6.6		0.8 1.0 0.6
San Diego State	12	X̄	20.5	178.4	68.5	43.6[a]	3.4-3.1-3.3
University, 1982		SD	1.9	6.2	5.9		1.0 0.8 0.7
(Carter, unpubl.)							
USA, national, 1985	19	X̄	22.7	179.6	71.0	43.4	3.1-3.4-3.2
(Carter, unpubl.)		SD	1.7	5.3	4.9	1.5	0.8 1.2 1.1

[a] Calculated from mean height and weight.
X̄ = mean.

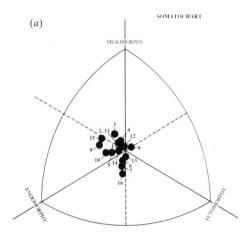

(a)

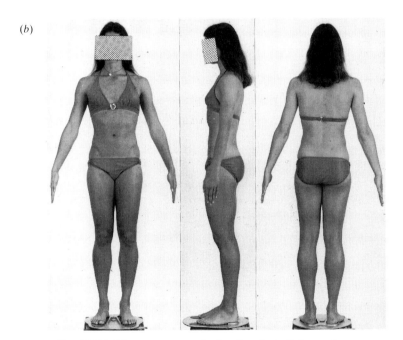

(b)

Fig. 6.39(a). Mean somatotypes for female volleyball players. 1 = Hungary, Class I; 2 = Hungary, national; 3 = Venezuela; 4 = Japan, 1977; 5 = Brazil, 1977; 6 = Brazil, university; 7 = Brazil, juvenile; 8 = South Australia; 9 = Austin, high school; 10 = Austin, university; 11 = USA, national, 1975; 12 = USA, national, 1985; 13 = San Diego State University, 1982; 14 = Central Michigan, varsity; 15 = Central Michigan, junior varsity; 16 = Cuba, youths. (b). Volleyball player. Height = 172.9 cm; weight = 62.6 kg; HWR = 43.6; somatotype = 3-$4\frac{1}{2}$-$3\frac{1}{2}$.

Miscellaneous Sports – Females (Table 6.30, Fig. 6.40(*a*),(*b*), (*c*),(*d*))

Ten team (Fig. 6.40(*a*)) and 12 individual (Fig. 6.40(*b*)) sports are included in this section. Although some events are for individuals, canoeing and synchronized swimming are included in team sports. Often in team sports there is considerable somatotype variation because of different requirements for the various playing positions. Comparisons within and between team sports are considered first.

There were large differences between the 1968 and 1976 Olympic samples in canoeing. Since both samples were small and sampled different events, they were combined to give a more representative sample (Carter, 1984*b*).

The Czechoslovak team handball samples, which were 10 years apart, differ in height and somatotype. The 1977 sample is more elite, taller, and less endo-mesomorphic than the 1967 sample. The Hungarian samples showed the opposite trend – the late 1970s sample was more endo-mesomorphic, but there were no differences in height and weight. Eiben (1981) reported a mean somatotype of 4.5-4.1-2.2 for another sample of Hungarian players. Perhaps differences in sample selection and methods account for the inconsistency in these findings. In a study of young Hungarian handball players, Temesi & Szmodis (1982) found no difference in somatotypes of 15–16-year-olds who had played the sport for different lengths of time. The authors suggested that different selection factors (including somatotype) have more influence than previous playing experience. In a study of young international handball players, Štěpnička, *et al.* (1979*b*) found an overall mean somatotype of 3.3-4.1-2.5. The somatotype differences were greater for playing positions than between teams.

In field hockey the means for the English Brighton and Liverpool teams and the South Australian team were close together, tending toward endo-mesomorphy. In the Liverpool study there were no differences between elite and county players and none by playing position. In a somatotype study of top New Zealand players, using the M.4 method, Johnston & Watson (1968) found a mean of $4\frac{1}{2}$-4- $2\frac{1}{2}$ and no differences by level of play; but at the highest level there was a reduced variation in components.

In the majority of team sports the means were slightly endo-mesomorphic and lay in the area bounded by 5-5-2, 4-5-2, 3-4-2, 3-4-3 and 4-4-3 and centered around the 4-$4\frac{1}{2}$-$2\frac{1}{2}$ somatotype. The exceptions were the slightly more meso-endomorphic cricketers, the more balanced mesomorph canoers, the synchronized swimmers almost in the centre, and the central netball players, who also have more ectomorphy than endomorphy.

In rugby, forwards were more endomorphic and mesomorphic, and less ectomorphic than backline players. There was wide somatotypic variation among the synchronized swimmers. The somatotypes of the solo and duet

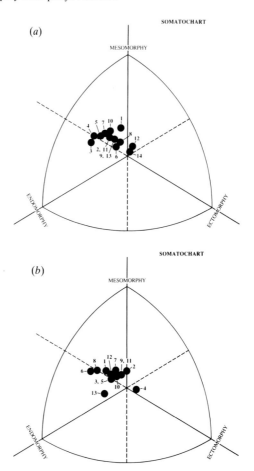

Fig. 6.40(*a*). Mean somatotypes for females in a variety of team sports.
1 = canoeing – Olympic; 2 = canoeing – Hungary; 3 = cricket – South
Australia; 4 = rugby – South California; 5 = soccer – South Australia;
6 = handball – Czechoslovakia; 7 = handball – Hungary; 8 = field
hockey – Liverpool; 9 = field hockey – Brighton; 10 = field hockey –
South Australia; 11 = lacrosse – South Australia; 12 = netball – South
Australia; 13 = softball – South Australia; 14 = synchronized swimming
– Canada. (*b*). Mean somatotypes for females in a variety of individual
sports. 1 = badminton – South Australia; 2 = diving – Olympic;
3 = fencing – Bolivar games; 4 = fencing – Cuba, youths; 5 = golf –
professional, USA; 6 = golf – amateur, San Diego; 7 = artistic
gymnastics – Czechoslovakia; 8 = judo – Bolivar Games; 9 = racquet
sports – Brighton; 10 = squash – South Australia; 11 = surfboarding –
International; 12 = table tennis – European; 13 = triathlon – South
California. (*c*). Artistic gymnast. Height = 169.5 cm; weight = 65.0 kg;
HWR = 42.2; somatotype = 3-5-2½. (Photo with permission from
J. Štěpnička.) (*d*). Handball player. Height = 166.0 cm; weight = 62.0
kg; HWR = 41.9; somatotype = 3½-4½-2. (Photo with permission from
J. Štěpnička.)

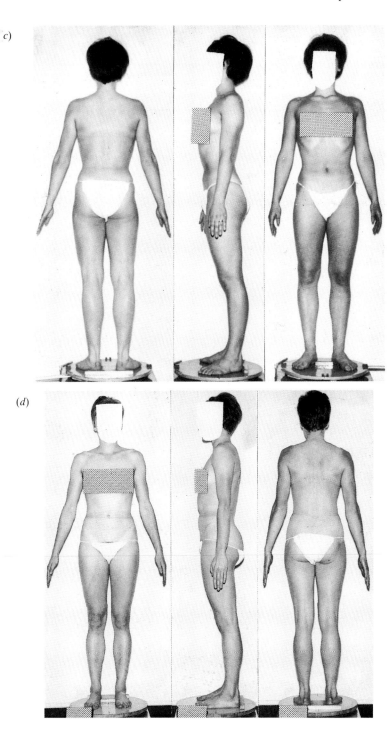

Table 6.30. *Miscellaneous sports – female*

Sample, reference	N	Statistic	Age (yr)	Height (cm)	Weight (kg)	HWR	Somatotype
Badminton							
South Australia, 1987	6	$\overline{X}$	23.0	167.7	61.5	43.5	4.1-4.4-2.5
(Withers *et al.*, 1987)		SD	5.3	2.5	2.6	0.7	0.7 0.8 0.5
Canoeing							
Olympics, 1968 + 1976	12	$\overline{X}$	21.1	168.2	62.3	42.5	3.0-4.5-2.5
(Carter, 1984*b*)		SD	6.6	5.8	6.6	0.7	0.5 0.7 0.5
Hungary, 'paddlers' (Mészáros & Mohácsi, 1982*a*)	30	$\overline{X}$	—	163.0	56.5	43.5[a]	4.0-4.5-2.5
Cricket							
South Australia	12	$\overline{X}$	25.4	165.8	63.4	41.7	4.9-4.4-2.0
(Withers *et al.*, 1987)		SD	5.1	6.7	7.8	1.7	1.3 1.1 1.1
Diving							
Olympics, 1968 + 1976	8	$\overline{X}$	21.1	160.9	52.8	42.9	2.9-4.1-2.9
(Carter, 1984*b*)		SD	7.0	3.0	3.9	0.7	0.7 0.7 0.5
Fencing							
Bolivar Games, 1981	7	$\overline{X}$	27.1	159.0	54.0	42.0	3.6-3.6-2.4
(Brief, 1986)		SD	5.0	6.0	3.7	1.3	0.6 1.0 1.0
Cuba, youths (Alonso, 1986)	5	$\overline{X}$	12.5	151.8	41.8	43.7[a]	3.0-3.2-3.5
Football							
(a) *Rugby*							
Southern California							
Total	56	$\overline{X}$	26.8	165.4	64.7	41.2[a]	4.5-4.5-1.8
		SD	3.9	5.4	7.2		1.3 0.9 1.0
Forwards	27	$\overline{X}$	27.4	166.0	68.9	40.5[a]	5.2-4.9-1.3
		SD	4.4	5.4	6.1		1.3 0.8 0.8
Backs	29	$\overline{X}$	26.2	164.9	60.7	42.0[a]	3.8-4.2-2.3
(Williams, 1984)		SD	3.3	5.4	5.9		0.8 0.9 0.9
(b) *Soccer*							
South Australia	11	$\overline{X}$	22.1	164.9	61.2	42.0	4.2-4.6-2.2
(Withers *et al.*, 1987)		SD	4.1	5.6	8.6	1.7	1.3 1.0 1.2
Golf							
USA, professional	26	$\overline{X}$	27.8	167.6	62.4	42.5	4.1-4.0-2.7
(Carter, 1971)		SD	7.9	7.2	8.2	0.5	1.1 1.1 1.1
San Diego, amateur	26	$\overline{X}$	40.5	164.8	62.9	41.3	4.9-4.6-2.1
(Carter, 1971)		SD	8.7	4.9	6.5	0.6	1.0 1.3 1.1
Gymnastics, artistic							
Czechoslovakia, 1967	71	$\overline{X}$	—	163.5	57.1	42.5[a]	3.6-4.3-2.6
(Štěpnička *et al.*, 1979*a*)		SD		4.2	4.4	—	0.9 0.5 0.8

Table 6.30. (*Continued*)

Sample, reference	N	Statistic	Age (yr)	Height (cm)	Weight (kg)	HWR	Somatotype
Handball							
Czechoslovakia, 1967	78	X̄	—	165.6	62.0	41.9[a]	4.1-4.3-2.3
		SD		4.9	6.0	—	1.0 0.5 0.9
Czechoslovakia, 1977	15	X̄	—	171.2	63.6	42.9[a]	3.7-3.9-2.7
(Štěpnička *et al.*, 1979*b*)		SD		5.4	6.4	—	0.9 0.7 1.0
Hungary, 1972–75	35	X̄	—	166.9	60.0	42.6[a]	3.1-4.1-2.6
Hungary, 1979–80	24	X̄	—	166.5	62.2	—	4.1-4.6-2.2
(Mészáros & Mohácsi, 1982*b*)							
Field Hockey							
Liverpool (Reilly &	24	X̄	23.4	162.8	60.0	41.6[a]	3.2-3.7-2.5
Bretherton, 1986)		SD	3.3	5.3	4.5		0.7 0.5 0.9
Brighton Polytechnic	11	X̄	—	162.7	59.4	42.0	3.8-4.4-2.6
(Bale, 1986)		SD		4.6	7.5	1.7	0.8 0.8 1.1
South Australia	17	X̄	22.6	166.5	62.3	42.1	3.7-4.5-2.2
(Withers *et al.*, 1987)		SD	2.3	7.5	7.3	1.2	1.0 0.8 0.9
Judo							
Bolivar Games, 1981	6	X̄	19.4	158.0	55.4	41.5	4.1-4.1-1.8
(Brief, 1986)		SD	2.7	5.9	6.7	0.8	1.1 0.7 0.5
Lacrosse							
South Australia	17	X̄	22.4	165.2	60.6	42.2	4.1-4.5-2.4
(Withers *et al.*, 1987)		SD	4.5	7.4	7.3	1.8	1.5 1.2 1.2
Netball							
South Australia	7	X̄	23.7	176.2	66.5	43.5	3.0-3.8-3.3
(Withers *et al.*, 1987)		SD	4.2	3.9	6.5	0.6	0.9 0.4 0.5
Orienteering							
Czechoslovakia, 1975	13	X̄	—	163.7	56.4	42.7[a]	3.2-4.0-2.8
(Štěpnička *et al.*, 1979*a*)		SD		5.7	6.2	—	0.8 0.6 0.8
Racquet Sports							
Brighton Polytechnic	9	X̄	—	166.9	61.9	42.6	3.6-3.9-2.6
(Bale, 1986)		SD		5.1	6.6	1.0	0.9 0.4 1.0
Softball							
South Australia	22	X̄	22.7	166.9	60.2	42.7	3.8-4.3-2.7
(Withers *et al.*, 1987)		SD	3.8	5.3	6.2	1.2	1.3 0.8 0.9

(*continued*)

Table 6.30. (*Continued*)

Sample, reference	N	Statistic	Age (yr)	Height (cm)	Weight (kg)	HWR	Somatotype
Squash							
South Australia	6	X̄	27.4	167.7	60.3	42.8	3.4-4.0-2.8
(Withers et al.,		SD	5.6	5.8	6.5	0.8	0.4 0.4 0.6
1987)							
Surfboarding							
International (Lowdon,	14	X̄	21.6	165.7	59.3	42.5	3.9-4.1-2.6
1980)		SD	3.4	4.9	6.7	—	1.1 0.7 0.8
Synchronized Swimming							
Canada, 1978 (Ross	136	X̄	16.8	163.8	53.2	43.7	3.3-3.6-3.4
et al., 1982)		SD	2.3	7.2	7.5	1.6	1.0 0.9 1.1
Table Tennis							
European, 1973 (Eiben	31	X̄	21.2	163.6	56.9	42.5[a]	4.5-3.3-2.7
& Eiben, 1979)		SD	15–38	7.0	7.4		
Triathlon							
Southern California	16	X̄	24.2	162.1	55.2	42.6	3.1-4.3-2.6
(Leake, 1987)		SD	4.3	6.3	4.6	1.2	1.0 0.8 0.9

[a]Calculated from mean height and weight.
X̄ = mean.

gold medallists at the 1976 Olympics were almost identical ($3\frac{1}{2}$-$4\frac{1}{2}$-2) (Carter, unpublished). Their somatoplots were at the upper edge of the distribution of the Canadian synchronized swimmers.

There were studies of team sports, using the M.4 method; by Johnston & Watson (1968) of netball players; by Bale (1981) of field hockey players; and of individual sports (golf) by McClure (1967).

The distribution of mean somatotypes for individual sports is similar to that of means for team sports. However, the main concentration of means is in a slightly smaller area bounded by 4-4-2, 3-4-2, 3-4-3 and 4-4-3, centered around $3\frac{1}{2}$-4-$2\frac{1}{2}$. The means for amateur golfers and judo competitors are more endomorph–mesomorph than other athletes; European table tennis players are meso-endomorphs, and young Cuban fencers are central somatotypes.

Amateur golfers are more endomorphic and mesomorphic than the professionals. However, the large numbers of endomorph–mesomorphs and the wide range of physiques in both groups suggests that somatotype is less important in this sport than in those with more limited distributions. In a study of world professional surfers, Lowdon (1980) found that the top three riders were more ectomorphic than less successful competitors. He also found some differences by country of origin.

The means for the racquet sports of badminton, squash and some tennis (see earlier) are fairly similar, but differ from those of table tennis and lower level tennis. In the triathlon, competition in swimming, cycling and running in sequence implies that the physiques of competitors should be compatible to all three sports. Leake (1987) found that somatotypically triathletes resembled Olympic swimmers more than Olympic runners. There were no data on Olympic cyclists for suitable comparisons. Robson (1974) reported that the somatotypes of 11 English cyclists centred around 3-5-2, but gave no statistics. Singh & Malhotra (1986) reported only the mean somatotype for 9 Indian cyclists – 5.2-3.2-2.6 – considerably different from the mean for English women cyclists.

At the Olympic level (De Garay *et al.*, 1974) and at the USA Juniors level (Thorland *et al.*, 1981) divers and gymnasts had similar somatotypes. The modern (artistic) gymnasts from the CSSR are slightly more endomorphic than the best Olympic style gymnasts. Bale (1986) found the mean for four 'sports acrobats' was 3.4-4.9-2.1 (Heath–Carter method), and was 3.5-3.9-2.9 (M.4 method) for seven members of a modern educational gymnatics display team.

In his study of 12-year-old Cuban athletes, Alonso (1986) reported a mean somatotype of 3.9-3.9-2.2 for seven shooters, the opposite side of the central category to the fencers. Barrell & Cooper (1982) gave a mean of 2.5-3.1-3.2 for five British orienteerers.

Champions versus others

Increasingly coaches and athletes recognize that the somatotype is an important variable in performance. Somatotype studies clearly show that distributions for a specific sport differ significantly from distributions for other sports. Because athletes, would-be athletes and coaches naturally hope to identify the optimal somatotypes for an event, often they are tempted to duplicate the physique, training and technique of a proved champion. While this approach has some merit, each champion has elusive qualities and characteristics that somatotype, training and technique cannot describe or account for. Experience teaches that sometimes champions succeed despite, as well as because of, unidentified differences.

Hypothetically, it is reasonable to look for the champion in the centre of the somatotype distribution for his event. Comparisons of medallists' somatotypes with those of other competitors in the 1960, 1968 and 1976 Olympic Games showed no consistent patterns by sports or events. (See Carter, 1984*b*; Carter *et al.*, 1982; De Garay *et al.*, 1974; Tanner, 1964). Some medallists were at the centres and some were near the boundaries of their event somatoplots. Although the distribution of somatotypes of successful competitors in a specific event is predictably limited, somatotype *per se* is but one element in performance–and a rather crude one at that. After all, competitions are won and lost by hundredths of points, seconds and metres; and vagaries of tactics, physical conditions of the arena, idiosyncrasies of judges' rulings, and just plain luck, all play a part on a particular day. But, all other factors being equal, a unique somatotype or unique aspect of a somatotype (such as a favourable dysplasia) of an athlete may give him or her a crucial advantage over other competitors.

It is disappointing to coaches and athletes to learn that two athletes with the same somatotype almost never have exactly the same overall capacities for high performance. Like two seemingly identical authomobiles that perform significantly differently when they are tuned differently, two apparently identical somatotypes perform differently under differing training methods. Even identical twins do not turn in identical competitive performances.

Somatotype changes over the years

Judging by photographs of athletes in the 1928 summer Olympics published by Kohlrausch (1930), and judging by somatotype photographs of Olympic athletes between 1948 and 1976, the somatotypes of successful Olympic athletes have changed. In this period, Carter (1984*b*) observed, somatotypes of female canoeists, gymnasts, track sprinters and hurdlers, and male swimmers have changed. Štěpnička (1986) observed differences between the somatotypes of male Czechoslovak athletes in the 1960–1970 period and 1978 onward. Male swimmers were less endomorphic; body builders, shot and discus throwers were more mesomorphic; high jumpers and 400 m runners were less mesomorphic and more ectomorphic, but track sprinters had not changed. He attributed the change in high jumpers to changes in technique. In 1970 the mean was 1.6-5.5-2.8, when most of the jumpers used the straddle technique. But in 1978 when the jumpers used the 'flop' technique the mean was 1.2-3.3-4.5, a dramatic shift from ecto-mesomorphy to meso-ectomorphy. Štěpnička also noted a shift toward higher mesomorphy in cross-country skiing, judo and basketball, sports for which increased strength relative to technique is advantageous.

Sport selection and youth sport

Arnot & Gaines (1984) say in their book, *Sportselection*, tha somatotyping is useful for grossly describing the structural requirements fo various sports. Somatotyping is also useful in helping to guide both childre and adults to sports appropriate for their present and potential somatotypes

Some raise questions like: does the athlete excel in a sport because he or she has the right somatotype, or do they have the right somatotype because they excel in the sport? and, does genetic predisposition to a somatotype determine the performance of the athlete? or, does the athletic training and participation in a sport alter the somatotype to conform to the demands of that sport? It appears that both factors contribute to athletic performance. Self-selection for a sport is common when athletes find they are better in some sports than others. And coaches often select for special training an athlete with modest skills because he 'looks like a good athlete or player'. Parnell (1958) found a relationship between somatotype and voluntary participation in sports. He found that predominantly mesomorphic Oxford students were much more likely to take part in regular 'strenuous outdoor exercise' than those lower in mesomorphy.

Somatoplots from studies of athletes have established the most common somatotype distributions for individual sports. The distributions of young athletes who are still growing are similar to those of older athletes. The elaborations of immature somatotypes of athletes are reasonably predictable. In addition, appropriate dietary and exercise regimens help toward developing optimal somatotype characteristics in young and growing athletes. Existing somatotype data are useful as guidelines for sport selection and choice of training appropriate to the enhancement of desired somatotype characteristics.

As mentioned above, studies of school age athletes show that their somatotype distributions resemble those of mature, successful competitive athletes. However, as we saw in Chapter 4, the longitudinal sequence of somatotypes of many children is inconsistent and unpredictable. The guidance of coaches, aware of the varying patterns of somatotypic development, is crucial in helping young athletes to discover their aptitudes for particular sports. With the somatotypes of successful athletes for models, the objective is to predict the most likely adult somatotypes, and to estimate the influence of appropriate nutrition and training in modifying the somatotype for optimal performance in the chosen sport.

These steps are suggested as a protocol for monitoring somatotype development and change in young and mature athletes:

1. Obtain a somatochart of National or Olympic level athletes in the appropriate sports. (These are available in the figures and references in this chapter.)
2. Obtain measurements, photograph and somatotype of the athlete.
3. Plot the somatotype of the athlete on the sport somatochart.
4. If the athlete's somatoplot is inside the sport somatochart area, decide whether further physique modification is indicated. If it is, plan the appropriate training and dietary regimen.

5. If the athlete's somatoplot is outside the sport somatochart area, it is necessary to decide whether appropriate training and dietary regimen is likely to bring his somatoplot within the desired area.
6. Repeat steps 3–5 periodically.

Figure 6.41 is an example of this procedure. The athlete is a 16-year-old male gymnast, with the somatotype 3-4-4. His somatotype is far to the south of the Olympic prototype area shown in Carter (1984*b*). The objective is to increase mesomorphy and decrease endomorphy. Added muscle mass can be expected with normal growth and concomitant increased testosterone secretion. Mesomorphy also can be expected to increase with appropriate weight and gymnastic training. Dietary regulation should create a negative energy balance and a decrease in endomorphy. It may be possible for this athlete to become a 2-5½-3 in the course of two years, and a 1½-6½-2 after two to four years. Of course both the dietary and training regimens are necessary to maintenance of the desired somatotype. Both coaches and athletes intuitively understand the concept of somatotype and the interactions of growth, nutrition and training with endomorphy (relative fatness), mesomorphy (musculoskeletal robustness) and ectomorphy (linearity).

If the athlete fails to progress toward the target somatotype under the prescribed regimen, it may be that influences like genetic limitations on stature, fat and muscle cells, preclude the proposed changes. The alternative is to direct the athlete toward another sport.

De Rose & Guimarães (1980) used this general approach with young Brazilian athletes. They found that sports scientists, with coaches and athletes, could apply much of the regimen described above as a guide in training and developing athletes.

Fig. 6.41. Hypothetical example of the possible somatotype pathway of a male gymnast from 16 years (A) to 18 years (B) to 22 years (C). The area enclosed by the line is that occupied by 95 percent of Olympic gymnasts.

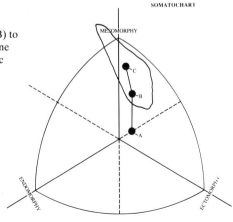

Sexual dimorphism

Male and female somatotypes differ consistently by sport and event just as males and females differ in the general population (Carter, 1981). Female Olympic athletes are more endomorphic and less mesomorphic than their male counterparts, and do not differ greatly in ectomorphy. The lines connecting male and female somatoplots generally lie at right angles to the ectomorphic axis of the somatochart, from northeast to southwest (Carter, 1984*b*). Bale's (1983), Pérez' (1981) and Štěpnička's (1972, 1974*a*) studies of male and female athletes show similar differences. It appears that in the same sports or events, higher levels of training or competition do not appear to significantly reduce sexual dimorphism. This may help to account for the continued differences in the performance of trained male and female athletes (Carter, 1985*b*).

Physical performance

Physical fitness and motor ability tests have been used as criterion measures in studying the relationships between somatotype and physical performance. The gross motor performance areas of strength, endurance, flexibility, speed, skill and balance are the most likely to be related to somatotype. Fine motor skills such as finger dexterity, typing, manual sorting and so on are unlikely to be related.

The early studies include those of Cureton (1947, 1951), Cureton & Hunsicker (1941), Everett & Sills (1952), Sills (1950), Sills & Everett (1953), Sills & Mitchem (1957), Thompson (1952) and several studies of men summarized by Willgoose (1961), and those on women by Carruth (1952), and Morris (1960). Later studies of men included those of Bale *et al.* (1984), Beunen *et al.* (1977, 1986), Caldeira *et al.* (1986*b*), Carter & Phillips (1969), Carter & Rahe (1975), Clarke (1971), Hebbelinck & Borms (1978), Hebbelinck & Postma (1963), Hebblelinck & Ross (1974*a,b*), Holopainen *et al.* (1984), Jensen (1981), Kozel (1978), Laubach (1969), Laubach *et al.* (1971), Laubach & McConville (1966*a,b* 1969), Olgun & Gürses (1986), Parnell (1958), Schreiber (1973), Slaughter *et al.* (1977*b*), Štěpnička (1974*b*, 1976*c*, 1986), Štěpnička *et al.* (1977, 1979*a*), Swalus (1967–68*a,b*, 1969), and Szmodis (1977).

A smaller number of authors have studied the somatotypes of females: Bale (1986), Caldeira *et al.* (1986*b*), Chytráčková (1979), De Woskin (1967), Farmosi (1980*b*), Hebbelinck & Borms (1978), Holopainen *et al.* (1984), Majors (1982), Slaughter *et al.* (1980), Sobral *et al.* (1986), Štěpnička (1972, 1976*c*), Štěpnička *et al.* (1977, 1979*a*), and Szmodis (1977).

Although general trends become apparent from collations of the above studies, the findings in some studies are unclear because sometimes two factors are not accounted for. First, somatotype is size dissociated while variables such as gross strength and jumping ability are related to absolute size, and maturity. Second, it is difficult to interpret results when the somatotype is disassembled into separate components for analysis. For example, in tests of speed, strength and distance separate component analysis does not account for the marked differences in performance for somatotypes 4-5-1 and 1-5-4, which have the same mesomorphy but different endomorphy and ectomorphy. Also, although physical performance might be expected to change with changes in the somatotype, the two are not necessarily related causally. And, performance may be significantly different for two persons with the same somatotype. Furthermore, large improvements in a performance test such as sit-ups, can be achieved without visible rating changes in the somatotype (Carter, 1980*b*).

Despite the above caveats, in general studies suggest that high mesomorphy is positively associated with most physical performance tests, and high endomorphy is negatively associated. Ectomorphy shows either slight positive association or none. Tests for strength, endurance, power, and speed are more highly related to somatotype than tests for flexibility, balance and fine motor skills, which seem unrelated. Štěpnička (1972, 1974*b*, 1986), who did some of the best studies of adult sportsmen, concluded that somatotype could account for 25% to 60% of the variance in physical fitness tests. He found that children's somatotype groupings were highly related to motor educability tests, competitive sports activity and ability, and organized voluntary physical education. Endomorphic children were the worst, and ecto-mesomorphic the best (Štěpnička, 1976*c*; Štěpnička *et al.*, 1977, 1981). Among adults, male ecto-mesomorphs and balanced mesomorphs and female ectomorph–mesomorphs and central somatotypes were the most versatile in performance tests. The more ectomorphic physiques do better in distance running tests and the more endo-mesomorphic physiques do better in strength or power tests.

Summary

The evidence in this chapter suggests that somatotypes are significantly related to success in physical fitness tests and in sports. It is apparent that persons in a finite (or limited) distribution of somatotypes are candidates to excel in sports and physical activities that require rapid movement of the body, movement of the body over long distances, or the need to move other persons or objects. There is substantial evidence that somatotype and success in sport and physical performance are positively related. An aspiring athlete needs to have (or train for attaining) a somatotype that is characteristic of those who have succeeded in the sport of choice. Often those who have somatotypes characteristic of their chosen sports are thought of as 'born athletes', and are believed to have been 'made' by their training for top performance.

Current knowledge can be generalized as follows:

1. Physical activities that place a premium on strength, power, speed or endurance, confine successful participation to the somatotypes best suited or best developed for the physical requirements of the activity.

2. In most sports top level athletes are more mesomorphic and less endomorphic than non-athlete reference groups.

3. Somatotype sexual dimorphism is consistent between athletes in the same sport, and is similar in direction and magnitude to non-athletes.

4. Some sports may have markedly different somatotype distributions, while other sports may have markedly similar somatotype distributions.

5. In track and field events there are significant differences among the somatotype distributions for individual events, although they belong to the same sport.

6. The range of somatotypes in team sports is usually wider than in individual sports, because of differences in playing position requirements.

7. As the level of competition increases, variation of somatotypes within a sport tends to decrease.

8. At the same level of competition there are some race-ethnicity somatotype differences within sports and events at the same levels of competition. However, these tend to be less than the differences between sports or events.

9. The somatotypes and somatotype distributions for successful young athletes are similar to those of their adult counterparts. They tend to be less mesomorphic (especially males), less endomorphic (especially females), and more ectomorphic than the adults.

10. Diet, growth and training can change somatotypes.

11. Changes in sport, including training, rules and techniques, may change the optimal somatotype for that sport.

12. Physical fitness test scores tend to correlate positively with mesomorphy, negatively with endomorphy, and variably with ectomorphy. Certain somatotypes appear to be superior in tests or sports that require strength, speed or stamina.

13. There is no relationship between somatotype and flexibility or tasks consisting primarily of fine neuromuscular skill.

14. Somatotypes are useful as a basis for counselling children and adults for participation in physical activities in which they are most likely to succeed.

7 Health, behavioural variables and occupational choice

Introduction

The inherently appealing idea that health, disease, behaviour and occupational choice are related to physique has long stimulated investigation of their possible relationships.

Health and Disease

Despite their limitations, early studies relating somatotype to health and disease provide useful guidance for further investigation. In the United States Draper (1924), following the Hippocratic typological tradition, identified morphological types e.g. "Gallbladder", "Ulcer" etc. associated with given diseases. Dupertuis initiated somatotype studies of disease entities at Draper's Constitution Clinic at Columbia Presbyterian Medical Centre, and these were also pursued by others elswhere e.g. Sheldon and Damon in the United States, and Parnell and Tanner in England. Morris & Jacobs (1950) provided interesting concepts and models for testing and interpreting endocrine pathways that might influence the basic constitution and development of the somatotype. Few have followed their lead.

As Damon (1970) pointed out, in constitutional medicine there are a number of reasons for studying physique. In the practice of preventive medicine known relationships or physical characteristics and disease can help to anticipate susceptibility, to recognize clues to underlying causes and to identify possible preventive and interventive measures. Although the terms constitutional medicine and constitutional anthropology are now rarely heard, there is continued research on relationships among morphological variation, general predisposition to certain diseases and the important influence of environment on suppression or enhancement of diseases. The phenotypic somatotype allows for noting somatotype changes which may be associated with onset or arrest of disease.

Although there are some promising clues to associations between somatotype, disease, syndromes and behaviour, many are negligible or tenuous at best. Bailey (1985) noted that many diseases (e.g. cardiovascular, cancer)

292

are multifactorial in their nature; their etiologies may not be separable into direct and indirect pathways, or causal and non-causal factors, so that a constitutional hypothesis is likely to be of limited help in diagnosing, treating or explaining them.

The following are brief references to a variety of studies, most of which used Sheldon and Parnell methods. Sheldon and colleagues (1949, 1954, 1969) presented empirical statements, some somatocharts, and virtually no statistics for associating individuals and groups with selected physical and mental diagnoses. There were somatocharts (Sheldon, 1949) for women with cancer of the breast and uterus, men and women with gallbladder disease and duodenal ulcer, and men with pulmonary tuberculosis. Little use can be made of these materials, because women and men were rated on different scales, and there were no age-matched controls.

Little or no association between the disease or condition and somatotype was found in studies of red blood cells and viscera (Ansley *et al.*, 1963); erythrocytes and schizophrenia (Ansley *et al.*, 1957); alcoholism (Bahamondes *et al.*, 1952; Damon, 1963; Hartl *et al.*, 1982; Monnelly *et al.*, 1983; Oyarzun, 1952; Parnell, 1956); acne vulgaris (Damon, 1957); and smoking and pulmonary function (Damon, 1961). There were slight associations found between somatotype and serum cholesterol (Garn *et al.*, 1950; Gertler *et al.*, 1950; Tanner, 1951*b*); serum pepsinogen and peptic ulcer (Damon & Polednak, 1967*a,b;* Niederman *et al.*, 1964); creatinine and 17-ketosteroids (Tanner *et al.*, 1959), otosclerosis and Meniere's disease (Damon *et al.*, 1955); blood volume and cardiac output (Gregerson & Nickerson, 1950); poliomyelitis, peptic ulcer and gallbladder disease (Draper *et al.*, 1944); hormones (Lapiccirella *et al.*, 1961); the 'chemotype' (Gertler, 1950); and uric acid (Gertler *et al.*, 1951*a*). Other studies include those on stuttering (Bullen, 1945); and nutritional status (Lasker, 1952).

Damon (1960) found no somatotypic difference between breast cancer patients and controls, but cervical cancer patients were less mesomorphic and patients with endometrial cancer were more endomorphic than controls. He also found that five-year survival rates for breast and cervical cancer slightly favoured more mesomorphic women.

Cardiovascular Disease

The relationship of somatotypes and cardiovascular disease has received considerable attention. There have been studies relating blood pressure to the Sheldon somatotype (Harlan *et al.*, 1962; Seltzer, 1966); axis deviation (Howard & Gertler, 1952); and rheumatic heart disease (Hellerstein *et al.*, 1969). Others have studied elevated risk and occurrence of coronary heart disease (Gertler *et al.*, 1950, 1951*a,b*; Gertler, 1967; Damon, 1965*a*; Damon 1970; Hellerstein *et al.*, 1969; Pomerantz, 1962; Spain *et al.*,

1953; 1955; 1963). Parnell (1959) noted that the dominantly mesomorphic most often succumb to coronary thrombosis in early middle life but the endo-mesomorphs carry the highest risk. The risk for lean muscular builds is less. Carter *et al.* (1965) found that middle-aged men enrolled in voluntary fitness programmes, primarily for coronary heart disease prevention, were primarily endo-mesomorphic (M.4 method). In general these studies have shown that individuals with somatotypes high in mesomorphy, high in endomorphy or both are at high risk. Increased body weight, also implicated as a high coronary heart disease risk factor, is probably reflected in increased endomorphy.

In a pioneering study of coronary disease in young adults, Gertler & White (1954) explored the causes of the leading fatal diseases in the United States. They made an anthropometric and morphological appraisal of physique, including the use of Sheldon somatotypes. The measurements used were stature, weight, 22 lengths and breadths and 12 anthropometric indices plus clinical evaluations, endocrine and biochemical tests, detailed family health, athletic and occupational histories. The subjects, 97 males and three females who had suffered myocardial infarctions at age 40 years or under, were followed from 1946 to 1953. The Sheldon somatotype ratings confirmed that high mesomorphy characterized the majority of their 97 subjects (over 85% were rated 4 and higher in mesomorphy, 42% were rated 5 and higher). In their control sample of 146 males less than half (20%) were dominant in mesomorphy and three times as many (21%) dominant in ectomorphy. The study took into account the role of excess weight and obesity in cardiovascular disease. Gertler & White noted that obesity could be a factor in coronary disease, but the acceptance of a theoretically stable somatotype obscured the importance of high endomorphy. However, because the Sheldon ratings in this study were based upon age-adjusted tables that assumed unchanging somatotypes, the somatotype data could not take account of significant gains in endomorphy. Heath re-rated, by the Heath–Carter method, the seven published somatotype photographs and showed that age-corrections for weight did not significantly alter ratings in mesomorphy, but greatly underrated endomorphy. SAD and comparative values for mean somatotypes and HWRs were as in Table 7.1.

Table 7.1. *Young males with coronary disease (N = 7)*

	Sheldon somatotype	Heath somatotype	HWR	$SAD_{(S-H)}$
Mean	3.4-4.8-2.1	6.1-4.6-1.4	11.81	2.80
SD	1.1 1.0 1.4	1.4 0.8 0.9	0.50	

The few somatotype studies using the Heath–Carter method, (which had not been developed at the time of the earlier investigations of cardiovascular disease), have emphasized the high incidence of cardiovascular disease among endo-mesomorphs. In a study of 146 cardiac infarction patients (mean age 52.7) in rehabilitation in West Germany, Smit *et al.* (1979*b*) found a mean somatotype of 4-5½-1. The majority were endo-mesomorphs and extreme endo-mesomorphs, with only 3 (2%) to the right of the mesomorphic axis. The authors concluded from the history of weight loss following infarction that the patients had lost the equivalent of 1 to 1½ units in endomorphy.

Each of 48 patients was asked to select from a series of somatotype photographs the somatotype most like himself. Forty- two saw themselves as more ectomorphic and much less endomorphic thirty years earlier. The mean choice was 2-5-4. A group of patients in a preventive treatment program had a mean somatotype plotted at 3-5-1 and 4-6-1. For comparison the investigators somatotyped a sample of 11 cardiac infarction patients from Sweden, mean age 54.5 years and mean somatotype 4.6-5.0-1.6, and a sample of 21 pulmonary disease patients from Hamburg, mean age 54.5 and mean somatotype 3.8-4.1-1.9. They also somatotyped 64 male offspring of the patients. The group under age 16 were close to 4-4-4; a group of 20 between ages 16 and 24.9 had a mean of 4.1-3.7-3.4, and 10 aged 25 or older had a mean of 4.5-4.8-2.5. The authors noted that although the offspring may catch up with their fathers' pre-infarction somatotypes, it is impossible to estimate their susceptibility to coronary disease. It is clear, however, that the sons, like their fathers, were markedly endo-mesomorphic.

Smit *et al.* (1979*b*) enrolled 12 patients in a post-clinical phase of myocardial infarction in a program of exercise and sports. They found no somatotype changes after eight months of training. In contrast, Carter & Phillips (1969) showed that normal middle-aged males in a cardiovascular training program decreased one-half unit in endomorphy in the course of two years. Control subjects increased endomorphy by the same amount. However, the intensity and duration of training was greater in the latter study.

In a study of risk factors for coronary heart disease Ryan (1980) compared a group of 17 boys aged 10–15, whose fathers or mothers had diagnoses of coronary heart disease with 15 boys whose parents had no evidence of coronary heart disease. There were no significant somatotype differences between the two groups, whose means were 2.7-4.8-3.6 for the experimental group and 2.6-4.4-4.2 for the control group.

In studies in Bratislava of young males with pathological Stage I juvenile hypertension, Palát *et al.* (1982) and Štukovský *et al.* (1983) found that the mean somatotype of 3.7-4.5-3.1 was similar to healthy controls. The mean was close to Štěpnička's (1974*b*) mean for Prague students (3.4-4.3-3.0), and, except for two endo-ectomorphs, were within the usual distribution for the students.

Allowing for differences in somatotype method, it is clear that middle-aged male endo-mesomorphs, endomorph-mesomorphs and meso-endomorphs are at increased risk for certain cardiovascular diseases. The risk may be higher for extremes in these categories. It is noteworthy that individuals with somatotypes high in ectomorphy and low in both endomorphy and mesomorphy, and those low in endomorphy but high in mesomorphy, are rarely found in distributions of coronary heart disease patients. Increasingly, high mesomorphy and high endomorphy in combination are believed to add to the risk of cardiovascular disease. Current medical practice emphasizes reduction of dietary lipids to prevent heart disease. Dietary restrictions and exercise help in weight reduction and lower endomorphy, but do not significantly alter mesomorphy.

Samples of middle-aged males of White northern European ancestry seem to show predominantly endo-mesomorphic physiques, with large proportions in categories indicated as 'cardiac disease prone'. For example, Bailey *et al.* (1982) found that male Canadians were more endo-mesomorphic with age (the mean somatotype of the 40–60-year-olds was $4\frac{1}{2}$-$5\frac{1}{2}$-$1\frac{1}{2}$). These findings encourage risk-reduction programmes emphasizing changes in dietary and exercise habits, and raise the question whether the incidence of cardiovascular disease would decline in a culture where physique changed from endo-mesomorphic to ecto-mesomorphic. The possibility invites appropriate studies.

Hyperbaric medicine

Dr John King (personal communication) included anthropometry and somatotype photographs in a study of 1313 divers and 1864 men working

Table 7.2. *Somatotypes of Dartford Tunnel workers*

Workers in compressed air Somatotype mean and SD N = 1864 3.0-4.9-2.1 1.2 1.2 1.2	Divers Somatotype mean and SD N = 1313 3.1-5.1-2.1 1.1 1.1 1.0
Type I divers (No Bends) 3.1-5.0-2.1 1.1 1.1 1.0	Type II divers (Bends) 3.3-5.4-1.8 1.1 1.0 1.1
Type I divers (Bends) 3.1-5.2-1.9 1.1 1.0 1.0	

in hyperbaric conditions (compressed air) during construction of the Dartford Tunnel near London. The men who were susceptible to the bends and degenerative processes in the hips were dominantly mesomorphic and had high ratings in endomorphy. King developed a policy that subjects with endomorphy higher than 5 or HWR lower than 12.00 (39.67) were unfit for employment involving hyperbaric pressures.

The divers were classified as Type I, subdivided as susceptible to bends and not susceptible to bends, and Type II, all susceptible to bends. There were significant differences in weight and almost significant differences in mesomorphy between those susceptible and not susceptible to bends. (Table 7.2).

Obesity

Obesity is commonly regarded as an important contributing factor to diabetes, hypertension and coronary heart disease. Although the definitions of obesity are equivocal, it is generally agreed that the obese have excess fat and high endomorphy ratings. Somatotype surveys of the general population are unlikely to reflect the incidence of obesity because the obese generally avoid such exposure.

Seltzer & Mayer (1964), included somatotypes in studies of 180 'healthy' white, obese adolescent girls, ages 12–17 years, with weights 48.0–150.0 kg, outpatients from the Childrens Hospital in Boston at a summer weight-control camp in Massachusetts. The mean somatotype for the sample was 8.5-4.3-1.2, compared with 4.3-3.5-3.3 for a sample of 67 fifteen-year-old public school girls. The obese were markedly more endomorphic, moderately more mesomorphic and less ectomorphic than the school girls (see Table 7.3 and Fig. 7.1). The two most endomorphic subjects were 17-5½-½ and 17-5-½. Endomorphy ratings ranged from 5 to 17, mesomorphy from 2½ to 6, and ectomorphy from ½ to 2½. Inter-observer reliability for

Fig. 7.1. Somatotype distribution of obese adolescent girls in Massachusetts, USA. ▲ = mean somatotype. Ratings by Heath from photographs provided by C. C. Seltzer.

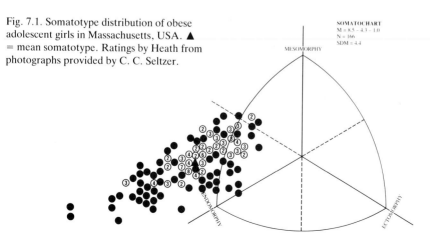

Table 7.3. Obese girls and women

Sample	N	Statistic	Age (yr)	Height (cm)	Weight (kg)	HWR	Somatotype	SDM	SAM
Adolescent girls, Massachusetts, USA (Seltzer & Mayer, 1964)	180	X̄	15.0	162.0	77.4	38.00	8.5-4.3-1.2	—	—
		SD	1.6	—	—	—	— 0.7 0.2	—	—
(Heath ratings, 1963)	166	X̄					8.5-4.3-1.0	4.4	2.1
		SD					2.3 0.7 0.5	2.9	1.4
Adult women, Massachusetts, USA (Seltzer & Mayer, 1969)	90	X̄	45.6[a]	159.5[a]	86.1[a]	36.13[a]	10.9-4.8-0.7	—	—
		SD	9.6	5.2	16.9	—	3.0 0.5 0.2	—	—
(Heath ratings, made in 1966)	102	X̄					8.4-4.7-0.8	3.2	1.5
		SD					1.6 0.6 0.3	1.8	0.7

[a] N = 94.
X̄ = mean.

Seltzer's and Heath's ratings, using Heath's (1963) open-ended scales, was approximately 0.9.

In a study (Seltzer & Mayer, 1969) of 90 adult, obese white women, ages 24–70 years and weights 61.7–155.5 kg, in a voluntary weight control organization in Boston, Massachusetts, Heath's and Seltzer's mean of $8\frac{1}{2}$-$4\frac{1}{2}$-1 was similar to the mean for the obese adolescent girls above (Table 7.3). Heath's later re-rating of the series applying Heath & Carter's (1967) modification of the scale for endomorphy showed an endomorphy range of $5\frac{1}{2}$ to 12, mesomorphy 3 to 6, and ectomorphy $\frac{1}{2}$ to 2 (Table 7.3, Fig 7.2).

The distributions in the Seltzer & Mayer (1964, 1969) studies are parallel to the somatochart endomorphic axis and outside its western border, with the majority of subjects markedly more endomorphic, less ectomorphic and more mesomorphic than comparable samples of 'normal' women. Although it is difficult to assess accurately bone diameters on the extremely obese, they generally have a high level of mesomorphy. The majority of the subjects in other samples (see Table 3.5. Chapter 3) are plotted within the borders of the somatochart. There are a few extreme meso-endomorphs in many female samples.

Studies of the obese that applied Sheldonian criteria found them to be dominant in endomorphy, but the investigators had some reservations about the ratings. In a study of 103 obese white women in an obesity clinic in Philadelphia, average age 39.2 years and average weight 101.4 kg. Angel (1949) found 90% of the women were dominant in endomorphy, with an average somatotype of 5.8-2.9-1.5. In a study of 27 obese white women in an obesity clinic in San Francisco, Craig & Bayer (1967) found no endomorphy rating less than 6, no ectomorphy greater than 3 and a mesomorphy range of 2–4.

Although there are no known studies of male obesity, there are examples

Fig. 7.2. Somatotype distribution of obese adult women in a voluntary weight reduction program in Boston, Massachusetts, USA. ▲ = mean somatotype. Ratings by Heath from photographs provided by C. C. Seltzer.

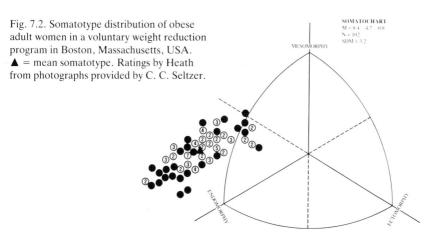

Fig. 7.3. Somatotype distributions of Type I and Type II diabetic women from London, England. ▲ = mean somatotype. Ratings by Heath from photographs provided by J. M. Tanner.

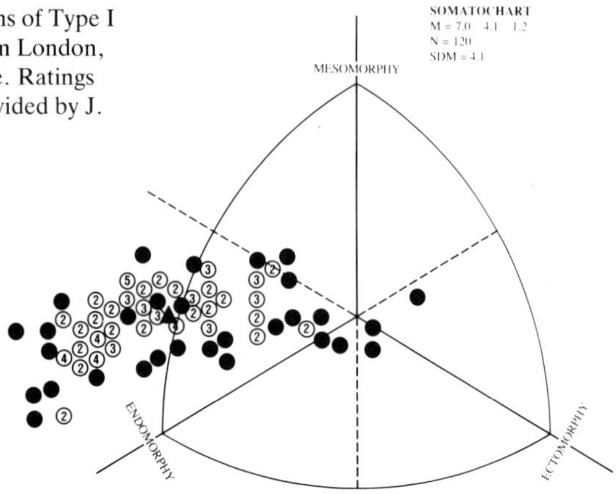

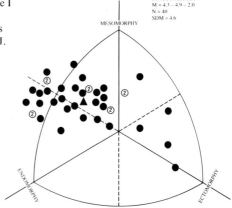

Fig. 7.4. Somatotype distributions of Type I and Type II diabetic men from London, England. ▲ = mean somatotype. Ratings by Heath from photographs provided by J. M. Tanner.

Fig. 7.5. Somatotype distribution of Type II diabetic women from San Diego, California, USA. ▲ = mean somatotype. (Redrawn from George, 1985.)

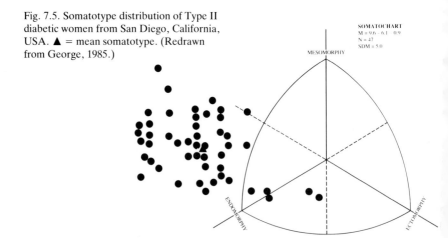

of extreme meso-endomorphs, mesomorph-endomorphs and endo-mesomorphs in samples described in Chapter Three (Tables 3.1, 3.2). The *Atlas of Men* (Sheldon, 1954) includes 24 photographs of men with HWRs lower than 11.00 (36.37), which Heath re-rated by Heath–Carter criteria (Carter 1985*a*). All but two of the somatotypes were extreme meso-endomorphs, with a mean of 9.4-5.3-0.5. (See Chapter 2, Fig. 2.5).

Diabetes

Lister & Tanner (1955) photographed and measured 40 men aged 28 to 71 and 115 women aged 16 to 76 in the diabetic clinic of the Royal Free Hospital in London. Sixteen men and 35 women were classified as 'acute onset' and 24 men and 80 women were 'non-acute onset' diabetics. The acute onset patients of both sexes were younger than the non-acute. Means for Sheldon somatotype ratings for women were 4.8-3.4-2.8 for early onset patients and 5.3-3.5-2.3 for late onset; for early onset men the mean was 2.7-4.0-3.9 and 3.7-4.5-3.7 for late onset.

Tanner (1956) suggested that physique reflected a gene complex that affected penetrance of genes predisposing the organism to particular disorders. He believed that some individual differences in physiology and physique of diabetics were at least partly genetically determined, and that although some people might carry the gene or set of genes that predisposed them to diabetes not all of them develop the disease.

In 1978 Heath re-rated the photographs from Tanner's study by Heath–Carter criteria, which appreciably altered the somatotypes. She did not know which subjects were acute onset and which were non-acute, but judging by the mean HWRs for Tanner's mean somatotypes, the mean Heath–Carter somatotypes would be about 6-4-1 (HWR = 40.33) for acute onset women, about 8-4½-1 (HWR = 38.48) for non-acute women, about 4-5-3 (HWR = 42.32) for acute onset men, and about 5-5-1 (HWR = 40.33) for non-acute men. (See Table 7.4, Figs. 7.3, 7.4).

The Heath–Carter ratings were considerably higher than the Sheldon ratings by Tanner, in mesomorphy as well as endomorphy, for both sexes. This is another example of the value of open-ended somatotype scales, since they make it possible to account for changes in somatotype over time and with changing lifestyle.

In San Diego, California, George (1985) somatotyped 47 Type II diabetic women aged 30 to 72 by the Heath–Carter anthropometric method. All subjects had fasting blood glucose concentrations greater than 140 mg/dl. The mean somatotype was 9½-6-1, with 91.5% dominant in endomorphy, and 85% meso-endomorphs (Table 7.4, Fig. 7.5). Because anthropometric mesomorphy ratings combined with high endomorphy tend to be higher than photoscopic ratings, a mean of 9-4½-1 is more likely with HWR 37.83.

Table 7.4. Diabetic women and men

Sample	N	Statistic	Age (yr)	Height (cm)	Weight (kg)	HWR	Somatotype	SDM	SAM
Type II women diabetics San Diego, USA (George, 1985)	47	$\overline{X}$	54.6	163.4	83.0	37.83	9.6-6.1-0.9	5.0	2.6
		SD	8.7	7.3	18.0	2.83	2.0 2.0 0.9	2.9	1.4
Type I and II women diabetics London, England (Heath ratings made in 1978)	120	$\overline{X}$	45.2[a]	156.5[a]	65.2[a]	38.88[a]	7.0-4.1-1.2	4.1	1.9
		SD	13.1	6.7	12.5	—	1.9 0.6 0.8	2.5	1.0
Type I and II men	40	$\overline{X}$	45.6	169.9	71.7	40.90	4.4-4.9-2.0	4.6	2.0
		SD	12.5	9.6	13.7	—	1.6 1.0 1.3	2.8	1.1

[a] N = 115.
$\overline{X}$ = mean.

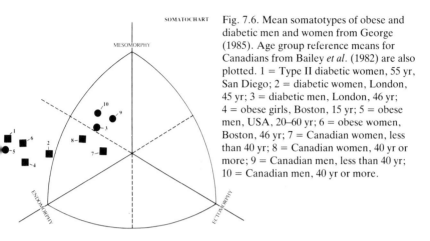

Fig. 7.6. Mean somatotypes of obese and diabetic men and women from George (1985). Age group reference means for Canadians from Bailey *et al.* (1982) are also plotted. 1 = Type II diabetic women, 55 yr, San Diego; 2 = diabetic women, London, 45 yr; 3 = diabetic men, London, 46 yr; 4 = obese girls, Boston, 15 yr; 5 = obese men, USA, 20–60 yr; 6 = obese women, Boston, 46 yr; 7 = Canadian women, less than 40 yr; 8 = Canadian women, 40 yr or more; 9 = Canadian men, less than 40 yr; 10 = Canadian men, 40 yr or more.

Fredman (1972, 1974) made studies of diabetic and non-diabetic female Tamil Indians in Cape Town, South Africa. In the first study, using Parnell's (1958) somatotype method, he found no difference between diabetics and non-diabetics. In the second study, using both Parnell's and Heath & Carter's (1967) methods, he found differences in age, blood glucose and mesomorphy. He suggested that lean body mass may be more important in the development of diabetes than obesity, but there has been no confirmation of this.

Mean somatotypes for obese males and females, diabetics, and reference Canadians (from Bailey *et al.*, 1982) are plotted in Fig. 7.6. At ages over and under 40, obese and diabetic females were significantly more endomorphic than women in the reference group, although there was some overlap in the distributions. The San Diego and London samples of females may not be strictly comparable in age, ethnicity and clinical diagnosis, but their somatoplots show a fair amount of overlap, and support the inference that extreme meso-endomorphy is characteristic of Type II diabetics. The obese men, like the obese women, are far to the southwest of Canadian reference group men over and under age 40. The diabetic men are closer to Canadian reference means; and include no more endomorphic somatotypes than expected (see Fig. 3.2, Chapter 3). The differences found between diabetic men and women compared with non-diabetic and obese subjects suggest further studies are needed.

Genetic Disorders
Down's Syndrome
Subjects with Down's syndrome, which is due to a chromosomal error, share well known, easily recognized physical characteristics and have

Table 7.5. *Down's Syndrome*

Sample	N	Statistic	Age (yr)	Height (cm)	Weight (kg)	HWR	Somatotype	SDM	SAM
Hungary, males (Buday & Eiben, 1982)	89	$\bar{X}$	17–55	154.3	60.6	39.39	5.9-5.9-1.1	4.0	
		SD	—	6.8	11.1				
	38	$\bar{X}$	17–24				5.1-5.6-1.4		
	47	$\bar{X}$	24–40				6.0-6.0-1.0		
	4	$\bar{X}$	40–55				6.5-6.0-0.8		
Hungary, females (Buday & Eiben, 1982)	48	$\bar{X}$	17–45	142.3	57.2	36.93	7.1-6.3-0.7	3.7	
		SD		5.0	12.6				
	20	$\bar{X}$	17–24				6.7-6.1-1.0		
	24	$\bar{X}$	24–40				7.5-6.8-0.5		
	4	$\bar{X}$	40–45				7.2-5.6-0.5		
Vista, California, males (Anderson, 1985)	14	$\bar{X}$	19.4	156.1	63.5	39.40	4.8[a]-5.7-1.1	5.2	2.6
		SD	2.1	5.1	13.5	2.3	2.0 2.1 0.6	1.5	1.1
Vista, California, females	18	$\bar{X}$	19.1	144.9	54.6	38.99	5.3-5.3-1.1	4.5	2.4
		SD	2.0	7.6	21.9	3.1	1.9 2.2 0.9	3.3	1.8

[a] Height-corrected endomorphy.
$\bar{X}$ = mean.

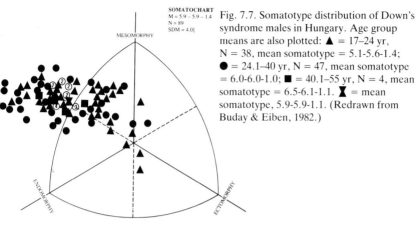

Fig. 7.7. Somatotype distribution of Down's syndrome males in Hungary. Age group means are also plotted: ▲ = 17–24 yr, N = 38, mean somatotype = 5.1-5.6-1.4; ● = 24.1–40 yr, N = 47, mean somatotype = 6.0-6.0-1.0; ■ = 40.1–55 yr, N = 4, mean somatotype = 6.5-6.1-1.1. ✗ = mean somatotype, 5.9-5.9-1.1. (Redrawn from Buday & Eiben, 1982.)

emarkably similar somatotypes. Both sexes are high in endomorphy and mesomorphy and low in ectomorphy.

In Hungary, Buday's & Eiben's (1982) study of 137 Down's syndrome patients included anthropometry and somatotypes. The 89 males and 48 females were grouped in ages 17–24, 24.1–40 and 40.1–55 years. The rounded mean anthropometric somatotype of the males was 6-6-1, and of the females 7-6½-1. The value of 6½ in mesomorphy seems a high estimate and is probably due to the difficulty of accurately measuring bone diameters and correcting girths from skinfolds on extremely obese subjects. Inasmuch as there is no record of female mesomorphy of 6½ and ratings of 6 are exceedingly rare, 5½ is probably more reasonable. The mean somatotype of the youngest male group was 5-5½-1½, compared with 4½-5-3 for a sample of young Hungarians (Gyenis, 1985); and the mean for the youngest female group was 6½-6-1 compared with 6-3-2 for a sample of young Hungarian women (Papai, 1980). (See Table 7.5, Figs. 7.7, 7.8, 7.9, 7.10)

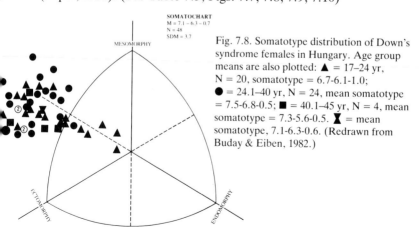

Fig. 7.8. Somatotype distribution of Down's syndrome females in Hungary. Age group means are also plotted: ▲ = 17–24 yr, N = 20, somatotype = 6.7-6.1-1.0; ● = 24.1–40 yr, N = 24, mean somatotype = 7.5-6.8-0.5; ■ = 40.1–45 yr, N = 4, mean somatotype = 7.3-5.6-0.5. ✗ = mean somatotype, 7.1-6.3-0.6. (Redrawn from Buday & Eiben, 1982.)

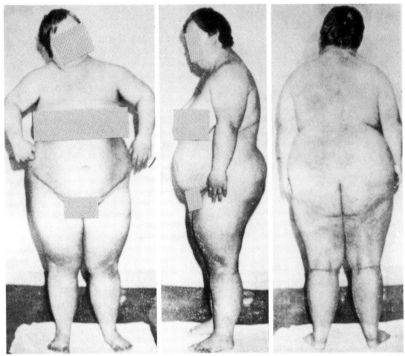

Fig. 7.9. A 29-year-old female Down's patient. Height = 142.0 cm; weight = 92.0 kg; HWR = 31.5; anthropometric somatotype = $10\frac{1}{2}$-11-$\frac{1}{2}$; anthropometric + photoscopic somatotype = $10\frac{1}{2}$-6-$\frac{1}{2}$, an extreme meso-endomorph. (From Buday & Eiben, 1982.)

Thirty-three males and 34 females were anthropometrically somatotyped in a study of 67 trainable mentally retarded (mean IQ 38) students, ages 16–23 years, in Sierra Vista High School, Vista, California, USA (Anderson, 1985). Thirty-two were Down's syndrome subjects; the other 35 included etiologies of phenylketonuria, fetal alcohol syndrome, birth trauma, cerebral palsy, rubella, hydrocephalus and idiopathies. The Down's males were more endo-mesomorphic and the females more endomorph–mesomorph than the other groups. Both Down's groups resembled the youngest Hungarian groups (Buday & Eiben, 1982). (See Table 7.5, Figs. 7.11, 7.12, 7.13.)

One female subject, weight 133.3 kg, height 154.3 cm and HWR 30.21, was rated 9-$12\frac{1}{2}$-$\frac{1}{2}$(!). Despite extremely wide humerus (7.75 cm) and femur (11.90 cm), re-measuring, and allowances for fat over muscular and bony parts, this somatotype is highly unlikely. This is an example of the necessity for a somatotype photograph to resolve discrepancies.

Somatotype studies indicate that Down's syndrome subjects are shorter

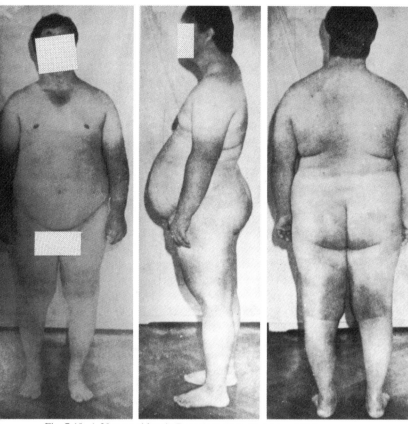

Fig. 7.10. A 28-year-old male Down's patient. Height = 154.5 cm; weight = 92.0 kg; HWR = 34.2; anthropometric somatotype = $9\frac{1}{2}$-$7\frac{1}{2}$-$\frac{1}{2}$; anthropometric + photoscopic somatotype = 9-7-$\frac{1}{2}$, an extreme meso-endomorph. (From Buday & Eiben, 1982.)

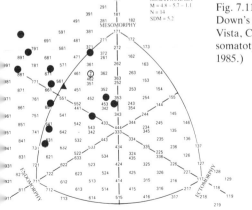

Fig. 7.11. Somatotype distribution of Down's syndrome males aged 16–23 years in Vista, California, USA. ▲ = mean somatotype. (Redrawn from Anderson, 1985.)

Fig. 7.12. Somatotype distribution
of Down's syndrome females aged
16–22 years in Vista, California,
USA. ▲ = mean somatotype.
(Redrawn from Anderson, 1985.)

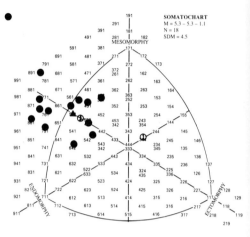

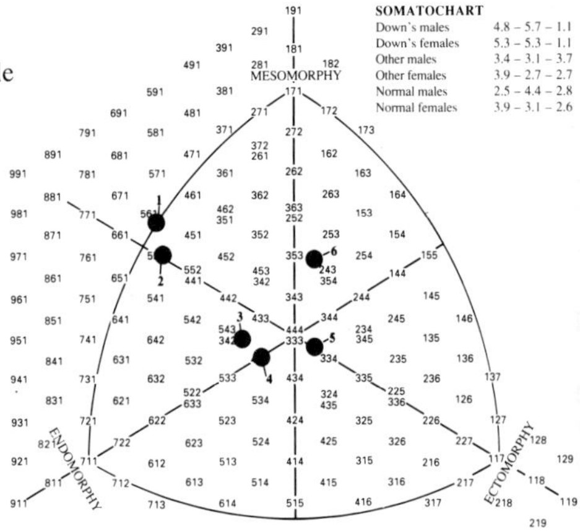

Fig. 7.13. Mean somatotypes of Down's
syndrome, other mentally retarded
etiologies, and normals. 1 = Down's male
(4.8-5.7-1.1); 2 = Down's female
(5.3-5.3-1.1); 3 = normal female
(3.9-3.1-2.6); 4 = other etiology female
(3.9-2.7-2.7); 5 = other etiology male
(3.4-3.1-3.7); 6 = normal male
(2.5-4.4-2.8).
(Redrawn from Anderson, 1985.)

and heavier, more endo-mesomorphic, endomorph–mesomorph and meso-
endomorphic, and less sexually dimorphic than reference groups. The
distributions of Down's somatotypes overlap with some obesity and diabetes
samples.

Streak gonad syndrome

Eiben *et al.* (1985) carried out a somatotype study of 70 patients
with streak gonad syndrome grouped according to their chromosomal
complement (i.e. 45,X; 46,XX; mosaics with a Y chromosome in some cells;
isochromosome Xq). They also compared them to fertile women of the same
age range (see Table 7.6 and Fig. 7.14). All groups were meso-
endomorphic. In endomorphy the 46,XX and mosaic Y groups were similar
to each other and the control group, but were higher in ectomorphy than the
controls. The 46,XX and controls were lower in mesomorphy than the other

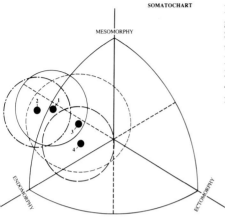

Fig. 7.14. Mean somatotypes of streak gonad samples with circles based on the SDM values which contain approximately two thirds of the subjects. 1 = 45, X subjects; 2 = 46, XX subjects; 3 = subjects with isochromosome; 4 = subjects with Y chromosome. (Redrawn from Eiben *et al.*, 1985.)

Table 7.6. *Reference Hungarian women and patients with streak gonad syndrome grouped according to their chromosomal complement*

Group	No.	Statistic	Somatotype		
			Endo	Meso	Ecto
Fertile women	164	$\overline{X}$	5.12	3.91	1.97
		SD	1.66	1.27	1.18
Patients with 45, X	30	$\overline{X}$	5.78	5.22	1.22
		SD	1.87	1.34	0.91
Patients with 46, XX	14	$\overline{X}$	5.32	3.64	2.89
		SD	1.29	1.25	1.30
Patients harbouring Y chromosome	10	$\overline{X}$	5.20	4.75	2.70
		SD	1.73	1.90	1.66
Patients with isochromosome Xq	16	$\overline{X}$	6.19	5.59	1.09
		SD	1.84	1.11	0.89

Adapted from Eiben *et al.* (1985).
$\overline{X}$ = mean.

three groups; and the 45,X and isochromosome groups were more meso-endomorphic than the other groups. The authors suggested that somatotypes shift toward meso-endomorphy when the short arm of the X chromosome is absent.

Growth disorders

In *Atlas of Children's Growth*, Tanner & Whitehouse (1982) presented somatotype photographs of 50 children with 20 growth disorders and syndromes. They included anthropometry for 19 of them but, unfortunately, somatotype ratings for none.

There are some extreme and apparently characteristic component values among the growth disorders: exceedingly *low* ectomorphy combined with exceedingly *high* mesomorphy in the achondroplasias; *low* mesomorphy and *high* ectomorphy in Marfan's and Klinefelter's syndromes; *high* endomorphy and *low* ectomorphy in Down's syndrome. Patients with the various syndromes could be expected to produce distinctive clusters of somatotypes on a somatochart, so that the somatotype photograph on its own identifies the proportional characteristics pathognomonic of many syndromes.

Other conditions
Blood lipids
Gordon (1984) explored the relationship between somatotype and various blood lipids in young male and female South African medical students. He found adverse lipid profiles in males with dominant endomorphy and mesomorphy. There were no associations with somatotype groups among females, but an adverse lipid profile for females with ratings of 4 or higher in mesomorphy (*per se*).

Injury
Khasigian *et al* (1978) examined rotational laxity and patellae alta in 40 young adult males divided into endo-mesomorphic (Mean = 5-6-1) and dominantly ectomorphic (Mean = 3-2-4) groups. They found increased external and total rotation in the ectomorphic group, but no association between patellae alta and somatotype.

In a study of 76 female university athletes (in basketball, softball, track and field, volleyball) followed for a season of training and competition, Greenlee (1986) found that somatotype and isokinetic strength deficits at knee and ankle joints helped to predict susceptibility to injury. Highly endomorphic and mesomorphic athletes suffered the majority of sprains and strains; and the more ectomorphic athletes suffered fewer sprains and strains, but a higher incidence of chronic injuries.

Reilly & Hardiker (1981) found no significant correlations between components and incidence of injury in rugby, but there were significant relationships between injuries and a soft tissue index based on the components.

Speech breathing
Hoit & Hixon (1986), in a study of the way body form might affect breathing function, explored the relationships between kinematic patterns of breathing among male subjects who were dominantly endomorphic, mesomorphic and ectomorphic. They found some differences between

somatotype groups and suggested some implications for evaluation and management of individuals with speech breathing disorders.

Legge–Calvé–Perthes syndrome
In a study of Legge–Calvé–Perthes syndrome (an osteochondrosis of the hip joint), Goff (1954) included somatotypes of 13 females and 54 males, ages 4 to 13 years. Heath used Sheldon criteria for photoscopic ratings of standard somatotype photographs. The data showed that these children were significantly more mesomorphic than children in 'normal' samples. At all ages the majority of LCPS children were also shorter than the norms for their age mates.

Posture
Sheldon (1940) described postural characteristics to be expected for certain somatotypes, but failed to quantify his observations. This is due, in part, to the subjective nature of many posture assessment methods. Brown (1960) and Kalenda (1964) found poor relationships between posture and somatotype components in women. On the other hand, Štěpnička (1976c, 1986) found that boys and girls who were more mesomorphic or ecto-mesomorphic had the best, and dominant endomorphs and ectomorphs the worst, posture. Young gymnasts (both boys and girls) had better posture than others. Adults with the highest mesomorphy tended to have the best posture. The author suggested that good posture may be significantly related to relatively harmonious development of muscles among athletic people.

There are interesting medical and physiological studies which use the Heath–Carter method to relate somatotype to 17-OH ketosteroids (Fleischmann *et al.*, 1977), thermoregulation and hypothermia (Docherty *et al.*, 1986; Hayward *et al.*, 1978, 1986), mental retardation in children (Drobný *et al.*, 1980; Chovanová *et al.*, (1982b)), and amenorrhoea (Bosze *et al.*, 1982).

Behaviour

To date, psychologists, psychiatrists and others have failed to find evidence of significant influence of the somatotype on personality, temperament or psychological well being.

Somatotype and temperament
In this and subsequent sections under behaviour the terminology used is borrowed from the original authors without discussion of appropriateness.

In Sheldon's (1940, 1942, 1949) concept, constitutional typology embraced morphological, temperamental, and psychiatric levels. Somatotype is the morphological level, temperament is the normal functional or behavioural level, and temperamental pathology is the psychiatric level.

From lists of alleged temperamental 'traits' he developed three temperamental components that he called viscerotonia, somatotonia and cerebrotonia, which roughly corresponded with the somatotype components endomorphy, mesomorphy and ectomorphy. Later the temperamental components were relabelled endotonia, mesotonia and ectotonia (Sheldon *et al.*, 1969; Hartl *et al.*, 1982). Twenty traits were listed for each temperamental component. For viscerotonia the emphasis was on relaxation, love of physical comfort, pleasure in eating and sociability; for somatotonia it was a tendency toward assertiveness, energetic activity, love of risk and power, and physical courage; for cerebrotonia it was a tendency toward restraint, introversion, love of privacy and solitude, and inhibition. Sheldon regarded these traits as underlying constitutional components and reported correlations of 0.79 between endomorphy and viscerotonia, 0.82 between mesomorphy and somatotonia, and 0.83 between ectomorphy and cerebrotonia. He saw these correlations as evidence of constitutional or innate relationships between physique (somatotype) and temperament or personality. Many critics have suggested that social stereotyping is at least as convincing an explanation.

Significant correlations were obtained in some other studies, but usually they were much lower than Sheldon's (Child, 1950; Fiske, 1944; Hanley, 1951; Kane, 1972; Sanford, 1953; Seltzer *et al.*, 1948; Smith, 1949). Some suggested that the correlations were spurious because of the halo effect, social stereotypes, self-concepts and similar factors (Anastasi, 1958). The original identification of components was based on a small sample of 33 college men. Later the traits were reworked and modified on a sample of 200 subjects aged 17 to 31 years. The unrepresentative nature of the original sample suggests that the relationships between traits and somatotype components were highly population specific and unsuitable as the basis of a universal system for rating temperament. Lubin (1950) found that some of Sheldon's reported intercorrelations were arithmetically impossible.

Several investigators examined the relationships between somatotypes and Sheldon's temperamental scales, or other tests of temperament (e.g. Alt, 1953; Arraj & Arraj, 1985; Bridges & Jones, 1973; Broekhoff, 1966 (Heath's ratings); Child, 1950; Cortes, 1961; C.W. Heath, 1954; Humphreys, 1957; Hunt, 1949; Lubin, 1950; Smithells, 1949; Smithells & Cameron, 1962; Zerssen, 1965); and between somatotypes and behaviour (e.g. Clarke, 1967 (Heath's ratings); Cortes & Gatti, 1965, 1966, 1970; Haronian, 1964; Janoff *et al.*, 1950; Lindzey, 1967; Parnell, 1958, 1984; Smith & Boyarsky, 1943; Sucec, 1979 (Heath–Carter ratings); Verdonk, 1972; Walker, 1962, 1963). (Unless otherwise noted, most of the studies used Sheldon's or

Parnell's methods, and not Heath's (1963) or Heath and Carter's (1967)). Others examined personality and somatotype relationships (e.g. Broekhoff, *et al.*, 1978 (Heath–Carter ratings); Cross, 1968 (Heath ratings); Deabler *et al.*, 1973; Ekman, 1951; Kane, 1961, 1964, 1972; Sanford, 1953; Sheldon, 1944*b*; Slaughter, 1970; Sleet, 1982; Stewart, 1980; Stewart *et al.*, 1973; Tanner, 1947; Tucker, 1983*c*). Little conclusive evidence linking somatotype and personality has been found.

There have been a number of studies that used self concept, self portrait and social stereotypes. Although there were no somatotype data for the subjects, they were asked to identify themselves with various visual materials, including actual somatotype photographs (rated by different methods), silhouettes, and drawing of 'somatotypes'. Self concept and portrait type studies include those of Borms *et al.* (1976) (Heath–Carter ratings), Broekhoff (1976) (Heath–Carter ratings), Dibiase & Hjelle (1968), Miller (1967) (Heath ratings), Sugarman & Haronian (1964), and Tucker (1982). Social stereotypes were assessed in other studies (for example: Broekhoff, 1976 (Heath–Carter ratings); Burian, 1969; Drowatzky, 1965 (Heath ratings); Felker, 1972; Greene, 1961, 1964 (Heath ratings); Hunt, 1951, 1952; Kiker & Miller, 1967; Miller *et al.* 1968; Miller & Stewart, 1968; Parnell, 1984; Powell *et al.*, 1974; Salokun & Toriola, 1985; Sleet, 1968, 1969; Staffieri, 1967; Strongman & Hart, 1968; Tucker, 1983*a,c*; Wells & Siegel, 1961; Winthrop, 1957; Yates & Taylor, 1978).

Other studies sought to establish relations between somatotype and psychological tests and characteristics (e.g. Child & Sheldon, 1941; Cortes & Gatti, 1966; Davidson *et al.*, 1955, 1957; Gyenis *et al.* 1980 (Heath–Carter ratings); Haronian, 1964; Jordan, 1964; Landers *et al.*, 1986 (Heath–Carter ratings); Lerner, 1969; Lerner & Korn, 1972; Lerner *et al.*, 1975; Reynolds, 1965 (Heath ratings); Schori & Thomas, 1973; Seltzer *et al.*, 1948; Sheldon, 1963; Smith, 1949; Somerset, 1953; Willgoose, 1952; Williams, 1977). But as Kane (1972) says, 'It may be that the relationship is general rather than specific, and may be observed experimentally only when reliable but rather global measures of psychological functioning are used.' (p. 100).

Tucker (1982) concluded that contemporary researchers have found little 'psychological predictive ultility' in somatotype variables. It might be added that some psychologists' inadequate understanding of somatotyping has contributed to disappointing findings. They have tended to regard as 'somatotyping' almost any method that recognizes body shape.

Inconclusive and often contradictory findings also may be due in part to using different somatotype methods. Sometimes characteristics have been correlated with components, and sometimes with whole somatotypes or somatotype categories. In some cases weak statistical approaches have been used. The wide variety of temperamental, behavioural, psychological, personality and social traits and tests that were used are not as well defined as those applied to sports and physical performance achievements and to medical diagnosis. However, despite obvious problems, there appears to be an element of constitutional or predictable behaviour associated with particular somatotypes.

But social stereotyping and behaviour modification also appear to influence the relationships under study. In any case, behaviourists believe that they are able to modify people's responses to their societies. Also different cultures admire different physiques and encourage varying social and behavioural characteristics. Sleet (1982) pointed out that merely labelling someone influences his personality; that in accepting a common social stereotype, one is led to behave so as to fulfil the expectations of the culture for that stereotype. In short, although there seem to be links between somatotype, temperament, personality and behaviour, it has not been possible to demonstrate clear cause and effect relationships.

Somatotype and psychiatry

By his own testimony, Sheldon worked out the hypotheses and completed a large part of the text for *The Varieties of Temperament* (1942) before he wrote *The Varieties of Human Physique* (1940). In the temperament book he proposed a three-component rating system for description and classification of temperament and psychiatric pathology. The subtitles of the first three books in the Human Constitution series *(An Introduction to Constitutional Psychology, A Psychology of Constitutional Differences,* and *An Introduction to Constitutional Psychiatry)* emphasize Sheldon's preoccupation with psychology and psychiatry. But for a variety of reasons, Sheldon's proposals led to little worthwhile research in these areas, and, except for the work of Hartl and his associates, there have been no innovative efforts to validate his hypotheses.

Sheldon's psychiatric hypothesis

It is plain that, as a corollary to somatotype method, description of temperament and psychiatric entities quantified in a three-component system might prove useful. In discussing 'The Psychiatric Variables' *Varieties of Delinquent Youth* (Chapter 2, pp. 41–96) Sheldon outlined a creative and potentially valuable hypothesis for psychiatry; he re-stated his concept of primary temperamental variables closely related to the three primary components of somatotype, and outlined a creative hypothesis for an 'operational psychiatry'. He suggested that the three psychiatric variables could be rated on a seven-point scale to yield a 'psychiatric index'. Accordingly, the first psychiatric component and the manic-depressive psychoses are closely related to the first and second somatotype components (endomorphy and mesomorphy); the second psychiatric component and the paranoid-schizophrenic psychoses are closely related to the second and third components (mesomorphy and ectomorphy); and the third psychiatric component and the hebephrenic-schizophrenic psychoses are closely related to the first and third somatotype components (endomorphy and ectomorphy).

He suggested that in fact the term 'cycloid psychosis' should *not* apply solely to the manic depressive cycle; that psychotics with balanced mesomorphy–ectomorphy alternated from paranoid somatic aggression to schizophrenic ideational hostility; that psychotics with balanced endomorphy–mesomorphy alternate between melancholia and euphoria; and that psychotics with balanced endomorphy–ectomorphy alternate between bizarre, irrelevant affect and bizarre, irrelevant ideation.

He noted that two ranges of somatotypes are likely to be given different diagnoses on successive hospital admissions *(Varieties of Delinquent Youth)*; 'the midrange somatotypes, and those falling near the morphological poles are most likely to have their diagnoses changed a half dozen times or more ... There is a clear tendency towards greater diagnostic agreement when the somatotype falls near any one of the three hypothetical psychiatric poles' (p. 57). He allowed for description of primary psychiatric components at both the psychotic and psychoneurotic levels, saying that 'psychiatric diagnosis and psychological description are aspects of one and the same process' (p. 85). Sheldon emphasized the importance of what he called 'the somatotype performance test' (i.e. strict conformity to standard procedures in handling and instructing subjects during somatotype photography). He observed that when the hebephrenic component is dominant it is almost impossible to induce the subject to hold his arms at full extension.

Sheldon's hypothetical distribution of various psychiatric diagnoses on a somatochart (see Figure 5 in *Varieties of Delinquent Youth*, p. 59) illustrates the most striking features of his proposal.

Sheldon (1949) applied this psychiatric schema to delinquents and to patients at Elgin State Hospital (Wittman *et al.*, 1948). He used categories that represented a pathological deficiency in temperamental components. He used the suffix '- penia' to characterize these as negative traits. Thus, *visceropenia* was the lack of compassion and relaxed, soft qualities; *somatopenia* was the lack of energy and drive for overt action; and *cerebropenia* was the lack of inhibition. Later these terms were re-labelled as *endopenia*, *mesopenia* and *ectopenia* (Sheldon *et al.*, 1969; Hartl *et al.*, 1982).

Other studies attempted to establish similar relationships between somatotype, psychiatric diagnoses and psychotic behaviour (for example: Adelson & Turner, 1963; Beulen, 1956; Brouwer, 1957; Eiben *et al.*, 1980

(Heath–Carter ratings); Kline & Tenney, 1950; Parnell, 1984; Sheldon 1952, 1965; Sheldon *et al.*, 1969; Skottowe & Parnell, 1962; Stewart, 1980; Watson, 1972; Zerssen, 1969). In general, these studies encountered the problems noted above for somatotype–'behaviour' relationships, and led to the conclusion that the findings were suggestive but inconclusive.

Hartl *et al.* (1982) presented a theoretical perspective on the constitutional method in their follow-up study of the 200 delinquent boys in Sheldon's (1949) study. They stated that they found relationships between somatotype, temperament and psychopathology that are useful and important, suggested some clinical applications of the method, and reinforced Glueck & Glueck's (1970) position that because human beings do not act in a vacuum, it is useful to consider the interaction of biological and sociocultural data. However, the question of a satisfactory method for solving the problems remains unanswered.

Delinquency and criminality

There have been several somatotype studies following Sheldon's (1949) landmark study of delinquent boys. These include the 30-year follow-up study (Hartl *et al.*, 1982) of Sheldon's 200 subjects, re-rated by the Trunk Index method; the controlled study (Glueck & Glueck, 1950, 1956) of 500 delinquent and 500 non-delinquent boys, rated by the Sheldon method; Seltzer's (1950, 1951) study using Sheldon's 1940 method; and Gibbens' (1963) study of 58 delinquent youths, somatotyped by Tanner according to Sheldon's 1954 method. Epps & Parnell (1952) used the M.4 method in a study of women delinquents; and Cortes & Gatti (1972) studied 100 delinquent youths and 100 controls using the M.4 method. Olson (1960) used Heath ratings in a study of Oregon boys. Although the studies included Parnell's M.4 ratings and Heath ratings, high ratings in mesomorphy were common to all of them.

Sheldon (1949) found mesomorphy and endo-mesomorphy associated with assertiveness and uninhibited action in a high proportion of the 200 youths. It is noteworthy that the above studies do not include 'white collar crime'. Sheldon, and later Parnell (1958) observed that similar somatotype and temperamental traits characterized men successful in business, politics and the military. Of course it is obvious that physique alone is not implicated in delinquency and criminality. When Sutherland (1951) re-analysed Sheldon's data he found no significant association of somatotype and degree of criminality. However, in a study of 283 men in Duell State Penitentiary in California the mean somatotype was 4.1-5.3-2.2 and the mean for the 14 'criminals' in the Hartl *et al.* (1982) study was 3.5-5.0-2.3. Heath (unpublished data) observed a high frequency of somatotypes close to 2-6-2 among the more violent criminals.

Heath estimated the somatotypes of Sheldon's (1949) 200 subjects, from the latest height and weight data (i.e. the HWR) as reported in Hartl *et al.* The subjects were not re-photographed in the follow-up study. As might be anticipated, the phenotypic Heath–Carter ratings yielded a mean somatotype of 3.7-4.5-2.6 for the 200 boys in the original study, and a mean of 5.7-4.4-1.6 for the follow-up study.

Dysplasias–physical and behavioural

Dysplasia is a disharmony of a somatotype component in different regions of the body, and can be expressed by making ratings region by region. Sheldon defined dysplasia as the degree of disagreement among the five regional somatotypes of a subject. In *The Varieties of Human Physique* he emphasized the potential significance of dysplasia in studies of morphological and psychological conflict. He repeated the definition of dysplasia and referred to its significance in each book, but reported no specific studies relating it to other variables. In *Varieties of Delinquent Youth* he merely noted dysplasias in the biographical sketches, but said nothing about their possible role in psychiatric disease in his chapter on psychiatric variables.

Parnell, in *Family Physique and Fortune* (1984), did not discuss dysplasia *per se*, but recognized that offspring often inherit disharmonious physical characteristics, and it appears that the offspring of parents of exceedingly different somatotypes are likely to have markedly dysplastic physiques. At the least it is safe to speculate that dysplasias or disharmonies of somatotype are related to physical performance, susceptibility to disease, and behaviour, and that possible relationships between dysplasia and these variables may deserve serious study. Certainly some disorders, for example Marfan's, Down's, Turner's, achondroplasia, Legge–Calvé–Perthes and others are easy to recognize by their characteristic dysplasias and somatotypes in photographs alone.

Occupational choice/preference

Damon & McFarland (1955) investigated the relationships between occupations and somatotype and the part played by genetic or constitutional factors in choices of employment in a free society. They noted that generalized descriptions of physique such as somatotype differentiate occupational groups more clearly than separate dimensions. They observed that the evidence favours selection and adaptation over occupational influences; that is, an occupation itself does not 'cause' mesomorphy or ectomorphy to be associated with success in it. They continued, ' ... jobs require specific patterns of physique and temperament for success, such patterns varying in strictness and in detail with the nature of the job. People choose jobs for various reasons, among them physical and temperamental affinity, so that there is self selection even among novices in an occupation. Those who succeed will on the whole tend to fit even more closely the pattern demanded by the job. This whole selection process will be reflected in the worker's physique to the extent (a) that the pattern is narrow; (b) that this system operates free from disturbing socio-economic influences; and (c) that physique and temperament are correlated' (p. 736).

The Armed Services

In the USA there have been several large surveys of somatotypes of army and air force military personnel, all applying Sheldon's method. (See Damon, 1955, Damon *et al.*, 1962; Damon & Polednak, 1967*a*; Dempster, 1955; Dupertuis, *et al.*, 1951*a*; Dupertuis & Emanuel, 1956; Hooton, 1948, 1951, 1959; McFarland, 1953; McFarland & Franzen, 1944; Newman, 1952; and Sheldon, 1943*a*, 1943*b*, 1949). Other studies include Sandhurst cadets in England (Tanner, 1954*b*), British Marines (M.4 method) (Donnan, 1959), New Zealand Air Force personnel (M.4 method) (Carter & Rendle, 1965), and Turkish, Greek and Italian military personnel (Dupertuis, 1963). Somatotype data for most of these are presented in Chapter 3, Table 3.7.

These studies showed that military personnel were more mesomorphic than college students, and mesomorphy increased with level of success. Likewise the variability of somatotypes decreased as the level of success increased. Also there were somatotype differences between specialties within the military population. The majority of somatotypes tended toward endo-mesomorphy. Donnan (1959) found that mesomorphy was higher among Royal Marines in the more hazardous occupations. Dupertuis (1963) found that Italian and Greek soldiers were more endomorphic than the Turks, and the Turks and Italians were more mesomorphic than the Greeks.

Using the Heath–Carter method, Carter & Rahe (1975) found that both successful and unsuccessful underwater demolition trainees had a restricted mesomorphic somatotype distribution. The successful trainees, with a mean of 2.5-5.9-1.9, became less endomorphic and more mesomorphic (by one-half unit each) during a four-month training program.

The selection processes and training in the armed services in most countries appears to favour the more mesomorphic in a given population. Candidates identified as having medical problems are rejected.

Other occupations

There are relatively few somatotype data for occupations other than military service personnel and students. Garn & Gertler (1950) (using the Sheldon method) observed that research workers were less endomorphic and mesomorphic, and more ectomorphic than general workers in the same plant. In Honshu, Japan, Kraus (1951) found that 93% of the male office workers were endo-mesomorphs with a mean of 3.5-5.5-1.4 (see Chapter 3, Table 3.7). He did not compare them with other Japanese. In a study of several groups of bus and truck drivers, Damon & McFarland (1955) found few lower than 4 in mesomorphy. On the whole they were more endo-mesmorphic than Air Force officers and as mesomorphic as combat pilots. Damon somatotyped the 40 subjects in Eränkö & Karvonen's (1955) study of Finnish men who competed in lumberjacking as a sport. They were ecto-mesomorphs with a mean somatotype of $2\frac{1}{2}$-$5\frac{1}{2}$- 3, which was more mesomorphic than the bus drivers. In a study of 79 scientific support personnel at Antarctic research bases, Leek (1970) used Parnell's M.4 method. They were generally high in mesomorphy (mean = $3\frac{1}{2}$-5-3), with no differences between the summer and winter groups.

In an extensive study of workers in various trades and occupations in the Dartford Tunnel project in England, King (personal communication) used the Heath–Carter anthropometric plus photoscopic method (ratings by Heath). At present there are no data available by work group. The overall somatotypes by age groups show that these workers are more mesomorphic than student groups, and are increasingly mesomorphic with age. They are somewhat more mesomorphic and less endomorphic than Canadians (Bailey *et al.*, 1982) of the same ages. (See Chapter 3, Tables 3.1, 3.2 and Fig. 3.4). Elsewhere in this chapter, more recent data on men working on the same project on compressed air (N = 1864) show the subjects to be primarily endo-mesomorphs, with a mean somatotype of 3.0-4.9-2.1. In a companion project, King somatotyped a large sample of professional divers employed in civil engineering and in overseas work, and in the North Sea. (See Figs. 7.15 and 7.16 for somatocharts of 18–19-year-old and 30–52-year-

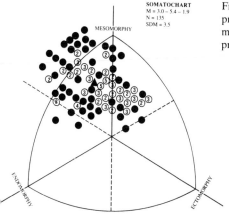

SOMATOCHART
M = 3.0 – 5.4 – 1.9
N = 135
SDM = 3.5

Fig. 7.15. Somatotype distribution of male professional divers aged 18–29 years. ▲ = mean somatotype. Plotted from data provided by J. D. King.

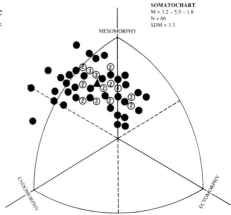

Fig. 7.16. Somatotype distribution of male
professional divers aged 30–51 years. ▲ =
mean somatotype. Plotted from data
provided by J. D. King.

old divers.) The majority of these are endo-mesomorphs. In the younger
group 15.5% are extreme endo-mesomorphs, and in the older group 9.1%.
There are few data for women; and the foregoing studies included none.
Gyenis *et al.* (1980) somatotyped (Heath–Carter method) 106 female
Hungarian transport drivers, ages 18 to 49 years. The drivers were extremely
endo-mesomorphic (mean = 6.8-2.9-1.6), compared with 165 fertile women
of similar age, with a mean somatotype of 4.7-3.5-1.5. The low mesomorphy
of the women drivers is especially surprising in its contrast to Damon &
McFarland's (1955) findings for men.

In Rio de Janeiro, Brazil, Pinto (1978) somatotyped 780 men and 459
women with an average age of 24.2 years. There were 18 groups of men and
8 groups of women in various occupations and professions. The results, in
Tables 7.7 and 7.8 and in Figs. 7.17 and 7.18 include students, military and
soccer groups which are not presented elsewhere in this book. The overall

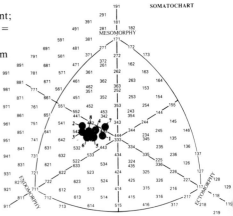

Fig. 7.17. Mean somatotypes of Brazilian
women in various occupations. 1 = student;
2 = computer technician; 3 = industry; 4 =
accountant; 5 = teacher; 6 = nurse; 7 =
office service; 8 = commerce. Plotted from
data in Pinto (1978).

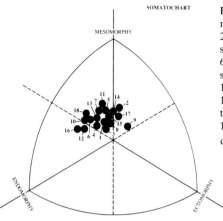

SOMATOCHART

Fig. 7.18. Mean somatotypes of Brazilian men in various occupations. 1 = student; 2 = military; 3 = office service; 4 = bank service; 5 = professional soccer; 6 = commerce; 7 = industry; 8 = public service; 9 = postman; 10 = accountant; 11 = salesman; 12 = teacher; 13 = telecommunications; 14 = computer technician; 15 = driver; 16 = executive; 17 = designer; 18 = nurse. (Plotted from data in Pinto, 1978.)

mean for women was $4\frac{1}{2}$-$3\frac{1}{2}$-$2\frac{1}{2}$. The student and commerce groups had a mean near 4-4-3, and the computer technician and industry groups had a mean near 5-4-2. The men, with a greater number of occupations, had a wider spread of means. Their overall mean was $3\frac{1}{2}$-$4\frac{1}{2}$-$2\frac{1}{2}$. The greatest differences were between the postmen and the accountant, teacher, executive and nurse groups. The salesmen were the most mesomorphic, and the

Table 7.7. *Various occupations. Young Brazilian women (N = 459)*

Occupation	N	Statistic	Age (yr)	Height (cm)	Weight (kg)	HWR	Somatotype
Student	270	X̄	20.3	159.8	53.0	42.5[a]	3.9-3.4-2.7
		SD	2.5	5.6	7.3		1.3 1.2 1.3
Computer technician	15	X̄	24.2	160.0	55.6	41.9	4.6-4.0-2.1
		SD	3.2	5.0	8.0		1.0 1.2 1.1
Industry	25	X̄	26.5	155.3	55.2	40.8	4.3-3.4-1.8
		SD	2.1	4.6	1.2		1.8 0.5 0.4
Accountant	8	X̄	22.7	158.0	54.1	41.8	4.0-3.5-2.3
		SD	1.6	0.7	1.7		0.5 1.6 1.5
Teacher	87	X̄	23.7	159.4	54.3	42.1	4.2-3.5-2.4
		SD	4.6	6.5	7.2		1.2 1.2 1.2
Nurse	8	X̄	23.3	154.0	52.5	41.1	4.8-3.7-2.7
		SD	4.9	3.0	1.0		0.3 0.7 1.0
Office service	18	X̄	22.9	159.8	53.7	42.4	3.6-3.8-2.7
		SD	4.1	6.0	5.3		1.8 1.1 1.0
Commerce	28	X̄	27.7	159.3	55.2	41.8	4.3-3.4-2.2
		SD	6.7	5.9	3.6		1.2 1.6 1.3

Adapted from Pinto (1978).
[a]Calculated from means.
X̄ = mean.

Table 7.8. *Various occupations. Young Brazilian men.* (N = 780)

Occupation	N	Statistic	Age (yr)	Height (cm)	Weight (kg)	HWR	Somatotype
Student	368	$\overline{X}$	21.0	172.5	66.4	42.6[a]	3.1-3.9-2.7
		SD	2.2	6.1	9.0		1.3 1.2 1.3
Military	113	$\overline{X}$	24.8	171.4	68.9	41.8	2.0-4.1-2.4
		SD	5.0	6.0	8.1		1.3 1.1 1.1
Office service	59	$\overline{X}$	23.4	171.7	65.2	42.7	3.2-3.7-2.6
		SD	2.3	5.6	8.7		1.3 1.1 1.2
Bank service	36	$\overline{X}$	24.6	170.8	65.9	42.3	3.5-4.0-2.4
		SD	5.3	2.9	8.2		1.2 1.2 1.3
Professional	29	$\overline{X}$	25.1	173.0	69.3	42.1	2.8-4.2-2.1
soccer		SD	4.2	5.2	5.5		1.0 1.1 1.0
Commerce	25	$\overline{X}$	24.4	170.0	67.4	41.8	3.8-4.2-2.1
		SD	3.7	6.2	9.0		1.6 1.3 1.0
Industry	22	$\overline{X}$	26.8	172.2	70.4	41.7	3.3-4.5-2.3
		SD	5.4	4.9	4.0		1.0 1.4 1.6
Public service	22	$\overline{X}$	27.3	171.5	66.5	42.3	3.2-3.8-2.6
		SD	5.1	6.2	10.1		1.1 1.0 1.0
Postman	20	$\overline{X}$	22.5	171.3	60.9	43.5	2.6-3.8-3.6
		SD	2.0	7.2	7.5		1.2 1.2 1.9
Accountant	14	$\overline{X}$	24.1	172.3	67.4	42.3	4.2-4.3-2.3
		SD	5.8	5.6	6.2		1.4 0.8 1.3
Salesman	13	$\overline{X}$	24.3	170.4	67.6	41.8	3.0-4.8-2.2
		SD	7.1	6.8	10.9		1.2 1.3 1.1
Teacher	12	$\overline{X}$	28.6	170.6	70.9	41.2	4.2-3.8-2.0
		SD	7.4	7.1	12.9		1.3 1.8 1.3
Telecommunication	10	$\overline{X}$	26.0	172.7	69.4	42.0	3.6-4.2-2.3
		SD	6.3	4.3	9.1		1.7 0.8 0.9
Computer technician	8	$\overline{X}$	22.9	167.9	60.2	42.8	2.6-4.2-2.6
		SD	2.4	6.1	8.1		0.7 1.3 0.9
Driver	8	$\overline{X}$	22.8	170.5	65.5	42.3	3.1-4.2-2.6
		SD	2.1	2.4	9.8		1.7 0.9 1.1
Executive	8	$\overline{X}$	25.0	174.3	70.7	42.2	4.7-4.1-2.2
		SD	4.8	6.6	9.1		1.6 1.5 1.4
Designer	7	$\overline{X}$	23.7	172.7	63.7	43.2	2.5-4.1-2.8
		SD	3.5	5.5	9.0		1.2 0.4 1.2
Nurse	6	$\overline{X}$	25.0	172.5	73.0	41.3	4.2-4.6-2.2
		SD	5.9	8.1	13.5		1.8 1.6 1.7

Adapted from Pinto (1978).
[a]Calculated from means.
$\overline{X}$ = mean.

office and public service workers were the least. The majority of means were endo-mesomorphic; only the postmen, the designers and the military personnel were more ectomorphic than endomorphic.

These studies indicate at least a moderate association between somato-type and occupation. While some jobs attract more mesomorphic physiques, there is considerable overlap of distributions. Somatotype distinctions between sports is greater, probably because physical demands for top level performance in sports are more restrictive than in the workplace. Likewise, although construction jobs exclude low mesomorphy, some who are domi-nantly mesomorphic work at office jobs. Indeed Ross *et al.* (1974) found one of the most mesomorphic men on record (somatotype $4\frac{1}{2}$-12-1) was employed in a library, spent most of his non-working hours in intensive strength training, and was a world record holder in weight lifting.

Naturally somatotype is only a partial index among the many factors that influence job success. Selection procedures and self-selection also play important roles. Damon & McFarland (1955) suggested that a young man lower than 4 in mesomorphy was probably ill-advised to choose heavy truck driving as an occupation. It may be that prospective employers and selection panels, despite lack of valid stereotypes, are subconsciously biased toward candidates with 'appropriate' physiques. They may select in their own image, and applicants may subconsciously aspire to certain occupations because they identify with the physiques and temperaments of their role models. It may be possible for employers to use physique and temperamen-tal characteristics as criteria that help to avoid failures due to physical limitations that lead to accidents and injuries.

Academic area of study

There are few somatotype data for college and university students that reflect distributions associated with given academic disciplines. Most of the existing studies reported in Chapter 3 are not differentiated by the field of the students. In the existing somatotype studies successful university students represent a small fraction of their age groups at large.

Tanner (1952, 1954*b*) applied the Sheldon method to compare somato-types of Oxford University students, medical students in London and Sandhurst military cadets. The medical students were lower in endomorphy and higher in mesomorphy than the Oxford students. Tanner suggested that the relatively low mesomorphy (mean = 3.6) of Oxford students might mean that mesopenes might do well in academia in comparison with the more mesomorphic medical students and Sandhurst cadets. He speculated that '... boys high in both mesomorphy and intelligence spend a good part of their time at school in athletic and other extracurricular activities, while those low in mesomorphy but equally intelligent are less easily drawn away from their books. Consequently at the highest reaches of examination lists, the mesopenes have it.' (Tanner, 1952).

In a study of 2866 students in nine faculties at Birmingham University, Parnell (1953, 1958) showed that there were considerable somatotype differences among them. Students of mining, medicine, dentistry and mechanical engineering had high frequencies in endo- and ecto-mesomorphy categories. Frequently physicists were meso-ectomorphic; the majority of mathematicians, lawyers, chemists and honours degree students in Arts were endo-ectomorphs or ecto-mesomorphs (except for the mathematicians).

Cockeram (1975), also using Parnell's method, found that male Arts students were less mesomorphic than physical education students. However, his Arts students were higher in mesomorphy than students in the same major at Oxford and Birmingham.

Jones *et al.* (1965), following Tanner's procedures in a study of 169 students at Loughborough College of Technology, found that 63% were dominant in mesomorphy, with a mean of $2\frac{1}{2}$- $4\frac{1}{2}$-$3\frac{1}{2}$. This finding was similar to Parnell's for mechanical engineers.

A group of South African medical students somatotyped by Gordon (1984) were more endomorphic and mesomorphic, and less ectomorphic than two samples of English medical students in London (see Chapter 3, Tables 3.1 and 3.2).

Heath's ratings of somatotypes of two female samples showed that a group of home economics students were less endomorphic and more ectomorphic than a group of music students in London (see Chapter 3, Tables 3.4 and 3.5).

Studies based on the Medford Boys' Growth Study show almost no relationship between physique and schoolboys' interests and vocations. Lynde (1968, 1969) found low correlations and no useful predictive value with somatotype components in samples of 10–12-year-old boys, and the occupational interests of 15–17-year-olds. In a study of somatotypes and vocational interests, Olson (1960) classified a group of 15-year-old boys as outstanding athletes, scientists, students of fine arts, leaders, scholars, poor students and delinquents, for comparison with other boys in the same age group. Apart from higher mesomorphy in athletes, there were no component differences among the groups. He concluded that somatotype by itself was not a determining factor at age 15.

Porter's (1958) study of 50 boys, aged 16–19 years, who volunteered to spend six weeks in central Iceland, found that their mean somatotype (rated by Tanner) was 2.8-3.9-4.1. They were less endomorphic and more ectomorphic than a comparable sample of English public school boys, whose mean somatotype was 3.1-3.8-3.6. Porter's vague inference was that the differences might have been related to '... the urge to join the expedition to Iceland.'

Overall, there seems to be some indication that among males choice of academic field and somatotype is related. However, there is considerable overlap between vocational groups. Clearly, somatotype is only one of a number of factors influencing such choices. There is insufficient evidence for women and schoolboys to indicate any positive relationships.

Physical education

Physical educators and students who aspire to be physical educators have been closely associated with somatotype research and the evolution of somatotype method. Sheldon's public and private utterances repeatedly expressed special appreciation of the physical educators, as in this unpublished fragment, which is a quotation from a proposed lecture, which Sheldon sent to Philip Smithells, professor of physical education at the University of Otago in New Zealand.

> 'The profession of physical education has always seemed to me the most religious profession. Their faith is secure, and their worship serene, for they have found the Immortal Soul. The soul is the body. Nor do they need to argue or preach about this. Rather they love to exercise it, to make it stalwart and straight, to render it redolent of sweet sweat, in short to save it by mesomorphic exhortation. Physical educators derive an ecstasy from seeing the soul stand up straight'.

Smithells & Cameron (1962), calling attention to the importance of physique in physical education, said 'Successful adults have all manner of build, but those drawn to physical education as a career tend to have some common characteristics. Certain builds are almost never attracted to this profession.' (p. 27). When Parnell (1958) somatotyped the 1955 entrants to Loughborough Teachers' College he found they were 'almost completely confined to dominant muscular builds'. He was impressed by the limited physique distribution and the associated vigour, enthusiasm and bonhomie of the changing room. He also questioned compulsory game systems, was concerned about communications among the varieties of temperaments and physiques, and asked: 'How are teachers of physical education to acquire a comprehensive outlook if their bias in physique and culture deprives them of "inborn" understanding of two-thirds of their pupils who have different somatotypes?' (p. 58). And: 'How is any one person, confined within the narrow limits, both physical and temperamental, of his own somatotype, to grasp the full range of experience necessary to manage personal contacts successfully?' (p. 59).

Not only might we expect physical educators to have good physiques themselves, but they should be able to help others to have insights into the potentialities and limitations of their physiques. Cureton (1947) aptly counselled, 'Anyone who has worked with boys knows the great appeal of a good physique. Boys idolize it. The characteristics of good physique should be exceedingly well known to physical educators. There should be no keener body build judges than the professional men and women in physical education because human flesh is more valuable than horse flesh and a remarkable amount of useful guidance is possible based upon a careful appraisal by an expert.' (p. 69).

Male physical education students

The studies of Cureton (1947), Thompson (1952), Finlay (1956), Brouwer (1957), Parnell (1958), Hebbelinck & Postma (1963), Carter (1964), Tanner (1964), Swalus (1967–68b), Cockeram (1975), Aitken *et al.* (1980), and Beunen & Van Hellemont (1980) used a variety of methods to investigate the somatotypes of male physical education students in a number of institutions and countries. The somatotype distributions showed fairly consistent patterns of dominant and moderately high mesomorphy and approximately equal proportions of endo-mesomorphs and ecto-mesomorphs. The height and weight data show that the samples also are similar in body size. Comparisons showed that physical education students were more mesomorphic and less endomorphic than other students.

Studies using the Heath–Carter somatotype method are summarized in Table 7.9 (N = 2049) and the means are plotted in Fig. 7.19. Mean somatotypes were largely balanced mesomorphs, with two ecto-mesomorphic means and one endo-mesomorphic. For all samples the relatively narrow ranges of means by component were 1.9 to 3.1 for endomorphy, 4.7 to 5.5 for mesomorphy, and 2.2 to 3.3 for ectomorphy.

Fig. 7.19. Mean somatotypes of male physical education students from institutions in different cities. 1 = Dunedin; 2 = Brussels; 3 = Prague; 4 = San Diego; 5 = Bratislava; 6 = Olomouc; 7 = Eugene; 8 = Sao Caetano; 9 = Rio de Janeiro; 10 = Amsterdam; 11 = Eastbourne; 12 = Budapest. For references see Table 7.9.

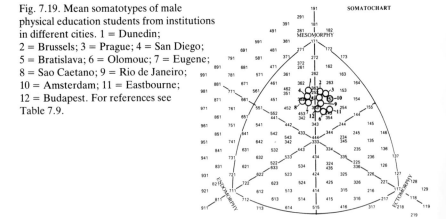

Table 7.9. *Male physical education students*

Sample, reference	N	Statistic	Age (yr)	Height (cm)	Weight (kg)	HWR	Somatotype
Dunedin	90	X̄	20.4	175.5	71.9	42.3	2.7-5.5-2.3
		SD	2.0	6.4	7.4	1.2	0.7 0.7 0.4
Brussels[a]	70	X̄	21.2	176.0	70.7	42.7	2.4-5.4-2.7
		SD	2.2	6.1	8.3	1.6	0.9 1.1 1.2
Prague[a]	95	X̄	20.4	176.5	71.4	42.6	2.1-5.5-2.6
		SD	1.6	5.1	5.9	1.1	0.6 0.7 0.8
San Diego (Carter *et al.*, 1973)	62	X̄	22.8	180.3	79.1	42.2	2.9-5.4-2.4
		SD	2.8	6.9	12.0	1.7	1.1 1.1 1.2
Bratislava (Hrčka & Zrubák, 1975)	175	X̄	21.3	177.7	73.8	42.4[b]	2.0-5.0-2.5
		SD	1.9	5.8	7.3	—	—
Prague (Štěpnička *et al.*, 1979a)							
1969/70	95	X̄	20.4	176.5	71.4	42.6[b]	2.1-5.5-2.6
1971/72	72	X̄	21.9	178.2	74.4	42.4	2.1-5.4-2.5
1972/73	73	X̄	21.3	178.4	73.4	42.6	2.0-5.2-2.7
1973/74	105	X̄	20.4	177.7	72.9	42.5	1.7-5.4-2.5
1974/75	80	X̄	21.2	178.0	74.0	42.4	2.0-5.2-2.5
1975/76	118	X̄	20.9	179.1	75.8	42.3	2.1-5.2-2.6
1969/76	543	X̄	21.0	178.0	73.5	42.5	2.0-5.3-2.6
Olomouc (Riegrová, 1978)	44	X̄	—	178.6	71.3	43.1[b]	2.5-4.7-3.0
		SD		6.3	6.6	—	0.7 0.8 0.7
Eugene (Broekhoff *et al.*, 1978)	74	X̄	22.0	179.0	76.3	42.2	3.0-5.1-2.2
		SD	2.5	6.6	10.3	1.3	1.2 1.1 0.9
Sao Caetano	15	X̄	—	—	—	—	3.1-5.0-2.2
Rio de Janeiro	31	X̄	—	—	—	—	2.2-4.7-2.6
Amsterdam (Araújo *et al.*, 1986)	52	X̄	—	—	—	—	1.9-5.0-3.0
Eastbourne (Bale, 1986)	74	X̄	20.4	179.8	73.3	43.3	2.3-4.7-3.3
		SD	1.5	5.9	7.4	1.3	0.7 0.9 1.1
Budapest[c] (Mészáros & Mohácsi, 1982a)	819	X̄	18.5	176	72	42.3[b]	3.0-5.0-3.0

[a]Brussels group also used in study by Araújo *et al.* (1986). Prague group same as 1969/70 sample in Štěpnička *et al.* (1979a). [b]Calculated from mean height and weight. [c]Means estimated from graphs.
X̄ = mean.

In a study of age, size and somatotypes of physical education students from Belgium, Czechoslovakia, New Zealand and the United States, Carter *et al.* (1973) found similar patterns of dominant mesomorphy and moderate ectomorphy. They suggested that differences in age, body size and endomorphy were due to selection factors at the different institutions.

The samples of Araujo *et al.* (1986) of physical education students from Sao Caetano do Sul and Rio de Janeiro in Brazil were similar to two samples from Amsterdam (Netherlands) and Brussels (Belgium). In a six-month follow-up the Brazilian samples had not changed. Rocha *et al.* (1977*a*) reported a mean somatotype of 3.1-4.5-3.3 for 627 male candidates for entry to the Escola de Educacao Fisica e Deportas da UFRJ, Rio de Janeiro. This is close to the two Brazilian samples in Table 7.9. Only about 10% of the candidates are accepted as students.

In a somatotype study of physical education students at the University of Oregon, Broekhoff *et al.* (1978) found no differences among the first to fourth year groups, and distributions similar to those in a San Diego State University sample (Carter *et al.*, 1973). They suggested that the greater variability in USA samples versus European and New Zealand samples might be due to less stringent entrance requirements for physical education students in universities in the USA.

Hrčka & Zrubák (1975) found the means for physical education students at Comenius University in Bratislava differed between years by no more than a half-unit in any component.

In Prague, Štěpnička *et al.* (1979*a*) studied 543 physical education students from 1969 to 1976 at Charles University. The somatotypes of the large number who were followed longitudinally were stable during the study. The more successful students were better in physical performance tests, more mesomorphic, less endomorphic and less ectomorphic than the less successful students (Fig. 7.20(*a*),(*b*)). The overall mean of 2.0-5.3-2.6 was close to the Bratislava sample. However, Riegrová's (1978) sample from Palacký University in Olomouc (Czechoslovakia) was slightly less mesomorphic and more endomorphic and ectomorphic than the Prague sample.

In comparisons of Heath–Carter somatotypes of physical education students and other students, the physical education students were consistently more mesomorphic and less endomorphic, but about the same in ectomorphy.

It is not surprising that the somatotype distributions of physical education students overlap those of outstanding athletes in many sports, as many of them are excellent all-round athletes and some are accomplished athletes. Athleticism is an important advantage in the selection process in some institutions. Also, many students are drawn to physical education because of their success in physical activities. High mesomorphy, especially ecto-mesomorphy, is an advantage in many sports (see Chapter 6).

Somatotype studies of physical educators show that high mesomorphy and low endomorphy characterize the students who join the profession. Six male faculty members at the University of Otago (Carter, 1964), ages 26 to 45, had a mean somatotype of 2.5-5.4-3.0. Štěpnička (personal communication) somatotyped 30 physical education teachers (ages 29 to 65) at Charles University. Their mean height was 176.3 cm, weight 76.4 kg, and mean somatotype 2.6-5.8-1.8. The somatotypes of the two 65-year-old

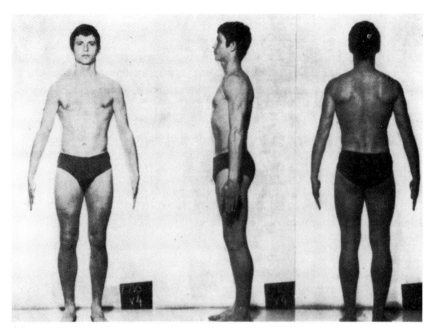

(a)

(b)

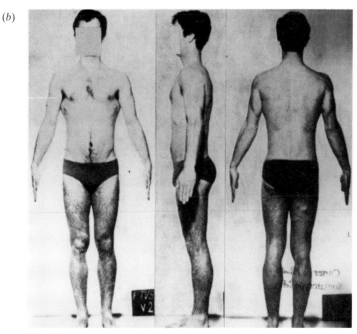

Fig. 7.20. Somatotypes of male physical education students from
Charles University, Prague, Czechoslovakia. (a). S = $2\frac{1}{2}$-6-$1\frac{1}{2}$; (b). S =
2-5-3. (From Štěpnička et al., 1979a.)

teachers were 2-7-1 and 1½-6½-2. At San Diego State University, Carter (unpublished study) found a mean somatotype of 3.3-5.8-2.0 in 14 of the physical education faculty, ages 29 to 54. The Prague teachers were slightly less endomorphic than their San Diego counterparts, and the Otago teachers were more ecto-mesomorphic than the other two.

In a study of 30 general subject teachers and 30 physical education teachers, who had graduated from ten to twenty years earlier, Borms *et al.* (1978) found no differences in somatotype distributions by category, but the physical educators (mean = 2.4-5.0-2.0) were lower in endomorphy than the general subject teachers (mean = 3.0-4.8-1.8).

When the physical education teachers are compared with 30–49-year-old Canadians the physical education teachers were less endomorphic and slightly more ectomorphic, but are about the middle of the samples in mesomorphy.

Female physical education teachers

Somatotype studies of women physical education students show that they tend to be more mesomorphic and less endomorphic than other students. Perbix (1954) and Brouwer (1957) used the Sheldon method; Carter (1965) and Bale (1969, 1979, 1980) used the Parnell M.4 method. The majority of somatotypes were meso-endomorphs, endo-mesomorphs, and ecto-mesomorphs, with distributions clustered around the ectomorphic axis northwest of the somatochart centre.

Heath *et al.* (1961) applied the Heath (1963) method to three samples of English students photographed by Tanner at various colleges of physical education. These and subsequent studies using the Heath–Carter method are summarized in Table 7.10 (N = 2016) and Fig. 7.21. The mean values for the large sample of Hungarian students were estimated from graphs in

Fig. 7.21. Mean somatotypes of female physical education majors from institutions in different cities. 1 = England; 2 = Brussels; 3 = Dunedin; 4 = Prague; 5 = San Diego; 6 = Olomouc; 7 = Eugene; 8 = Budapest; 9 = Lisbon; 10 = East-bourne. For references see Table 7.10.

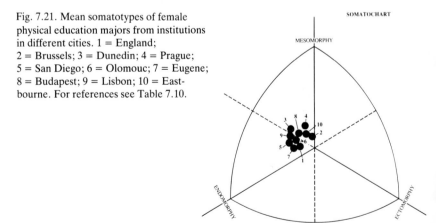

Table 7.10. *Female physical education students*

Sample, reference	N	Statistic	Age (yr)	Height (cm)	Weight (kg)	HWR	Somatotype
English							
I	112	X̄	18–22	164.3	59.9	42.1	4.5-4.3-2.7
		SD		5.4	5.9	1.2	0.5 0.6 0.9
II	57	X̄	—	166.6	60.0	42.5	4.2-3.8-3.4
		SD	—	5.1	5.0	1.5	0.7 0.8 1.3
III	136	X̄	—	164.3	56.2	43.0	4.2-3.6-3.4
(Heath *et al.*, 1961)		SD	—	—	—	—	—
Brussels (Day *et al.*, 1977)	33	X̄	—	—	—	—	3.4-4.1-3.0
Dunedin	103	X̄	19.2	164.5	59.6	42.1	4.2-4.3-2.3
		SD	1.1	5.9	6.3	1.3	0.8 0.6 0.9
Prague[a]	44	X̄	19.6	166.1	61.3	42.2	3.3-4.3-2.4
		SD	0.7	4.7	7.3	1.3	0.9 0.9 0.9
San Diego (Carter *et al.*, 1978)	75	X̄	22.2	165.2	59.5	42.4	4.4-4.0-2.6
		SD	2.0	6.1	7.5	1.7	1.1 0.8 1.1
Prague (Štěpnička *et al.*, 1979a)							
1969/70	44	X̄	19.6	166.1	61.3	42.1[b]	3.3-4.3-2.4
1970/71	50	X̄	19.7	167.1	61.8	42.3	3.1-4.3-2.6
1972/73	49	X̄	20.4	166.0	61.6	42.0	3.3-4.7-2.3
1973/74	55	X̄	20.5	165.4	61.4	41.9	3.2-4.5-2.2
1974/75	60	X̄	20.5	167.6	62.9	42.1	3.1-4.2-2.4
1975/76	63	X̄	20.4	165.7	59.6	42.4	2.9-4.1-2.6
1969/76	321	X̄	20.2	166.3	61.4	42.1	3.2-4.4-2.4
Olomouc (Riegrová, 1978)	43	X̄	—	165.5	59.7	42.4[b]	4.2-3.9-2.5
		SD		5.4	6.5	—	1.1 0.8 0.8
Eugene (Broekhoff *et al.*, 1978)	94	X̄	21.2	164.8	59.7	42.2	3.9-3.7-2.4
		SD	2.2	5.8	7.8	1.6	1.0 1.1 1.0
Budapest[c] (Mészáros & Mohácsi, 1982a)	831	X̄	18–19	164	57	42.6[b]	3.5-4.0-2.5
Lisbon (Sobral *et al.*, 1986)	174	X̄	18	159.8	55.5	41.9[b]	3.9-3.8-2.2
		SD	—	10.0	9.4	—	1.1 0.9 1.0
Eastbourne (Bale, 1980)	37	X̄	20.0	165.2	60.5	42.1[b]	3.6-4.2-2.7
		SD	1.3	5.2	7.4	—	0.8 0.8 1.1

[a]Prague group same as 1969/70 sample in Štěpnička *et al.* (1979a). [b]Calculated from mean height and weight. [c]Means estimated from graphs.
X̄ = mean.

Mészáros & Mohácsi (1982a). The means for endomorphy ranged from 2.9 to 4.5, for mesomorphy from 3.6 to 4.7, and for ectomorphy 2.2 to 3.4. Náprstková (1973) compared the Charles University students with sportswomen, but gave no means. Presumably the subjects were part of the larger study by Štěpnička *et al.*, (1979a).

Carter *et al.* (1978) found that there were somatotype differences, but no

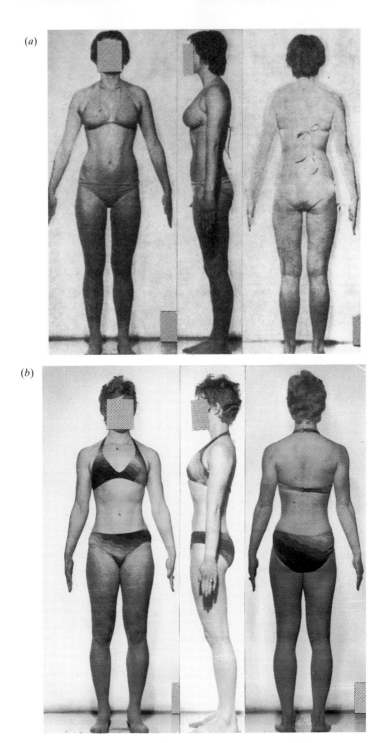

Fig. 7.22. Somatotypes of female physical education majors from
Charles University, Prague, Czechoslovakia. (a). S = 3-4-3;
(b). S = $2\frac{1}{2}$-$4\frac{1}{2}$-3. (From Štěpnička et al., 1979a.)

differences in body size among students from Prague, Dunedin and San Diego. The Prague sample was lowest in endomorphy, and the San Diego sample was lowest in mesomorphy. Broekhoff *et al.* (1978) found that the University of Oregon (Eugene) and San Diego physical education students were similar, but differed from physical education students in other countries. They concluded that the differences could be due in part to differences in entrance and curriculum requirements.

In addition to their male study, Štěpnička *et al.* (1979*a*) conducted a comprehensive study of female physical education students from 1969 through 1976 in their first year of study at Charles University in Prague (Czechoslovakia). See Fig. 7.22(*a*),(*b*)). The consistency of the means in both the male and female samples is striking. Although there were few significant differences between years, there was a trend toward decreasing endomorphy and mesomorphy with increasing ectomorphy from the 1972–73 group onward. When the 1972–73 group was re-measured in their fourth year their somatotype means differed from their first year, and endomorphy had decreased one-half unit (Tomášová, 1977). In tests of motor efficiency the high-scoring students were more ecto-mesomorphic and less endo-mesomorphic then the lower scoring students. The students rated as having good 'morphological precondition' for physical efficiency had endomorphy ratings of $1\frac{1}{2}$–3, mesomorphy ratings of 4–5, and ectomorphy of 2–$3\frac{1}{2}$. The students from Palacký University, Olomouc were more endomorphic than those from Charles University.

The mean somatoplots (see Figure 7.21) for the ten studies of physical education students are fairly similar, clustering west and northwest of the centre of the somatochart in the area bounded by somatotypes 4-4-2, 3-4-2, 3-4-3, 5-4-3. In all comparisons with students in other fields, physical education students were less endomorphic and more mesomorphic. It can be inferred that the differences between the two groups are related to the special physical skills and abilities required by the chosen field, and not to academic and teaching abilities required of both groups.

There is some indication that there was a trend toward lower endomorphy and higher ectomorphy in female physical education students in Czechoslovakia and England between the 1950's and the 1970's. The changes may be due to increasing participation in exercise and sports programmes. As was found with the male physical education students, there are similarities between the somatotypes of the female students and successful athletes.

The mean somatotype for 656 female candidates for entry to the Escola de Educacao Fisica e Deportes da UFRJ, Rio de Janeiro, was 4.8-4.1-3.2. Endomorphy is higher than in any of the samples in Table 7.9. About 10% of the candidates are subsequently accepted as students.

Except for Carter's (unpublished) data, there are no studies of successful

female physical education teachers. Carter found that Heath–Carter somatotypes ($\overline{S}$ = 3.7-4.0-2.7) of five women faculty members at the University of
Otago in 1962 were slightly less endomorphic and more ectomorphic than
their students, whose mean was 4.2-4.3-2.3.

Dance

Among the performing arts, physique is most likely to play a major
role in dance. However, serious dancers regard themselves as artists first and
athletes second, although sports scientists may see them in the reverse
order. Beyond their artistic gifts, ballet, jazz and modern dancers have
considerable athletic skills and train (or have trained) in several dance
forms. Some earn their living (often meagre) performing or teaching dance.
Many enjoy dance as recreation, avocation, and for fitness.

There are a few Heath–Carter method somatotype studies of both male
and female dancers that shed some light on what somatotypes are successful
at different forms and different levels of dance. (See Table 7.11 and Figs.
7.23, 7.24, 7.25). Johnson (1969) somatotyped San Diego area female
dancers, who had seven or more years of dance training and were in concert
performing groups. Samples of modern dancers and professional members
of the 'corps de ballet' of the San Diego Ballet Company were less
endomorphic and mesomorphic, and more ectomorphic than a control
group of college students whose mean was 4.6-3.6-2.9.

Farmosi (1978) somatotyped female members of the 'Ballet Ensemble of
Pecs' and the 'Hungarian State Folk Dance Ensemble', with ages ranging
from 19.3 to 37.6 years. Lavoie & Lèbe-Néron (1982) studied the physiques
of male and female professional and recreational jazz dancers in Montreal.
McClure (unpublished study, 1985) somatotyped samples of male and

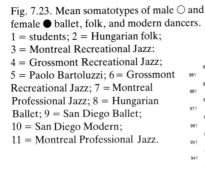

Fig. 7.23. Mean somatotypes of male ○ and
female ● ballet, folk, and modern dancers.
1 = students; 2 = Hungarian folk;
3 = Montreal Recreational Jazz;
4 = Grossmont Recreational Jazz;
5 = Paolo Bartoluzzi; 6 = Grossmont
Recreational Jazz; 7 = Montreal
Professional Jazz; 8 = Hungarian
Ballet; 9 = San Diego Ballet;
10 = San Diego Modern;
11 = Montreal Professional Jazz.

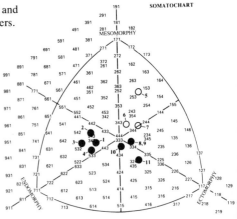

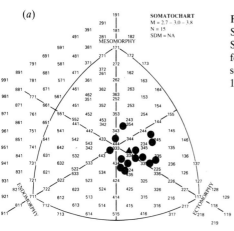

(a)

Fig. 7.24(a). Somatotype distributions of San Diego female ballet dancers. (b). Somatotype distributions of San Diego female modern dancers. ▲ = mean somatotype. (Redrawn from Johnson, 1969.)

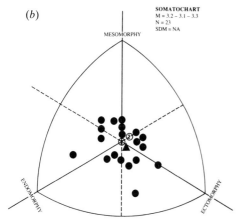

(b)

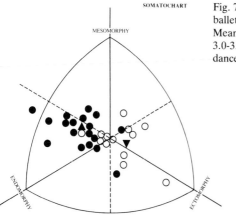

Fig. 7.25. Somatotypes of Hungarian female ballet (○) and folk dancers (●) dancers. Mean somatotype for ballet dancers (▼) = 3.0-3.2-4.0; mean somatotype for folk dancers (▲) = 4.7-4.1-2.7.

Table 7.11. *Ballet, folk, jazz and modern dancers*

Sample, reference	N	Statistic	Age (yr)	Height (cm)	Weight (kg)	HWR	Somatotype
Females							
San Diego ballet	15	X̄	16.5	162.8	50.1	44.2	2.7-3.0-3.8
		SD	3.1	6.4	7.3	0.9	0.6 0.5 0.6
San Diego modern	23	X̄	27.8	165.3	55.2	43.6	3.2-3.1-3.3
(Johnson, 1969)		SD	7.9	4.3	4.9	0.9	0.8 0.7 0.6
Hungarian ballet	13	X̄	19–37	161.1	47.5	44.5[a]	3.0-3.2-4.0
		SD		6.1	4.3		0.8 0.6 0.9
Hungarian folk (Far-	18	X̄	19–37	159.5	52.8	42.5	4.7-4.1-2.7
mosi, 1978)		SD		4.0	3.9		0.9 0.7 0.9
California ballet	7	X̄	21–32	162.6	50.1	44.1	2.9-3.4-3.7
(Werner & Carter,		SD		4.7	4.8	0.8	0.5 0.8 0.6
unpublished)							
Montreal pro jazz	6	X̄	19.0	162.0	47.6	44.7	3.1-2.4-4.4
		SD	1.7	2.4	5.1		1.2 0.5 1.0
Montreal rec jazz	17	X̄	24.7	162.0	59.1	41.6	5.0-3.5-2.4
(Lavoie & Lèbe-		SD	6.6	4.1	7.8		1.2 0.8 1.2
Néron, 1982)							
Grossmont rec jazz	11	X̄	27.0	162.3	54.5	42.8	4.7-3.1-2.8
(McClure, unpubl.)		SD	6.6	4.8	2.8		0.6 0.6 0.6
Males							
Montreal pro jazz	4	X̄	21.0	174.0	64.6	43.4	2.0-3.6-3.6
(Lavoie & Lèbe-		SD	4.0	6.0	9.0		0.6 1.0 1.4
Néron, 1982)							
Grossmont rec jazz	4	X̄	23.1	181.9	73.3	43.5	3.0-4.1-3.3
(McClure, unpubl.)		SD	2.5	8.4	8.5		0.7 1.0 0.7

[a]Calculated from means.
X̄ = mean.

female recreational jazz dancers at Grossmont College, El Cajon, California.

The San Diego and Hungarian ballet companies have almost identical mean somatotypes and somatochart distributions (Carter, 1971; Farmosi, 1978). It seems that the physical and training requirements limit the physiques, and the ballet mistress applies the criterion of physical compatibility, i.e. similar size and somatotype, with the 3-3-4 somatotype prototype. Female ballet dancers tend to a mesomorphic dysplasia, i.e. greater muscular development of the lower extremities. Seven trained ballet dancers at the California Ballet School in San Diego in 1980 had a mean somatotype of 2.9-3.4-3.7 (Werner & Carter, unpublished). Thus, the three samples of ballet dancers are similar.

The San Diego modern dancers and Montreal professional jazz dancers are close to the 3-3-4 prototype, and are more ectomorphic than the other three dance groups which have meso-endomorphic means. Both recreational jazz groups, are similar, have a combined mean of 4.9-3.4-2.6, and are similar to Johnson's control group. The Hungarian folk dancers are the most mesomorphic (by approximately a half-unit), with two members more ectomorphic than endomorphic. All performing groups, except the folk dancers, are more ectomorphic than endomorphic.

The mean somatoplots of four professional and four recreational male jazz dancers from Montreal are close to the Grossmont group. The only other known somatotype data are for Paolo Bortoluzzi, one of the foremost male dancers in the world. He is a 1.0-4.5- 2.5, an ecto-mesomorph (Ross *et al.*, 1974). A professional male ballet dancer from San Diego, California, had a somatotype of 2-5-4 (Werner & Carter, unpublished). It appears from general observation that most professional male dancers in ballet are ecto-mesomorphs, with occasional balanced mesomorphs (as Mikail Baryshnikov appears to be) and mesomorph–ectomorphs.

Other applications of somatotyping

Art

Somatotypers often observe that stereotypical somatotypes tend to characterize certain artists and artistic periods. Sculptors tended to present Zeus and other Greek gods as extreme mesomorphs. When Bok (1974, 1976, 1983) estimated the somatotypes of a variety of works of art by different artists in different periods, he found both consistency and variation when the same subjects were portrayed. Some sculptures tended toward high mesomorphy, as in the 1-8-1 for the Doryphoros Polycleitus and the 1-7½-2 for Michelangelo's David. Rubens' women were about 7½-3-½. Christ varied from 1-3-7 in the Pieta from Avignon to 1-5-3 in Rubens' Crucifixion. Venus was often portrayed as about 5-3-2, while portrayals of Adam and Eve varied a good deal.

Aesthetics

Štěpnička (1976*a*, 1983) attributed beauty to physiques that were harmonious and non-dysplastic. He found that the most aesthetically pleasing physiques were about 2-5-3 for males, and about 3-4-3 for females. In social status ratings ecto-mesomorphic males are often identified as the most admired. In Western societies fashion models are less mesomorphic than 3-4-3, and are often about 4-2½-4 or 3-2½-5. Sheldon (1940) introduced the idea of a t-index to rate the aesthetic characteristics of physique, which he later called the *t* component (Sheldon,

1954). Hartl *et al.* (1982) referred to this component as *SI* or structural integration. Many readers have viewed Sheldon's *t* component concept as pejorative.

Religion
In *Personality Types and Holiness,* Alexander Roldan, S. J. (1967) attempted to adapt the concepts of somatotype and related temperamental and behavioural characteristics to interpretations of holiness. Father Roldan's notion of three hagiotypes related to the three components of somatotype may be of interest to those with a religious bent.

Body Composition
Some of the discussion of studies of body composition/somatotype component relationships referred to earlier in this book bear repetition. There are a number of studies that have included somatotypes as well as body composition variables, or have referred to possible relationships (Behnke *et al.*, 1957; Brožek, 1965; Brožek & Keys, 1952; Bulbulian, 1984; Carter & Phillips, 1969; Clauser *et al.*, 1969; Dupertuis *et al.*, 1951a; Slaughter & Lohman, 1976, 1977; Slaughter *et al.*, 1977a,b, 1980; Somerset, 1953; Wilmore, 1970). Most studies have sought to validate the components against a criterion measure of body composition such as whole body density and its derivative, relative fat. The correlations between endomorphy and relative fatness are fairly high, from about 0.70 to 0.90, which is the range of intercorrelations between various methods of estimating relative fat. Correlations between mesomorphy and lean body mass, or fat free mass, are generally low. This is because most body composition methods cannot derive a valid measure of musculo-skeletal robustness relative to height, and also because absolute instead of relative values are sometimes used in the calculations. Correlations with ectomorphy are meaningless because ectomorphy is not a tissue but a ratio of mass to stature, and as such gives no direct body composition information. In view of recent criticisms of the validity of inconstancy of the fat free mass and its effect on estimates of any body tissue (Martin *et al.*, 1986), it is reasonable to suggest that it would be better to consider the criterion body composition variables to be endomorphy and mesomorphy (A plus P rating) and to test how good other variables are at estimating them. Even non-somatotypers are suspicious of percent fat estimations that do not agree with their subjective observations.

Summary

1. There are characteristic somatotypes associated with some diseases or conditions.

2. Some genetic syndromes are limited to a small range of somatotypes. In some cases the somatotype provides a reliable description of the syndrome.
3. The relationships between somatotype and temperament, behaviour, personality, self concept, social stereotypes, and psychological types are poor to moderate.
4. Sheldon's psychiatric hypothesis has not been tested extensively. Its application to delinquency has shown moderate evidence of clinical success.
5. Although the somatotypes of young male delinquents are markedly mesomorphic, this by no means distinguishes them from the many of similar somatotype whose behaviour and performance is highly acceptable.
6. Dysplasia is an unexplored aspect of somatotype that deserves attention in studies of disease, family structure, and growth or genetic disorders.
7. In some armed service categories, occupations, and areas of academic study there are some characteristic somatotype distributions. In general, somatotypic variation is greater in given occupations than in sports.
8. The somatotype distribution for female ballet dancers is restricted to the area around 3-4-4.

8 *Recapitulation and new directions*

Recapitulation

As we have seen, scientists motivated by persistent wishes to quantify and compare the infinite patterns of morphology and characteristics of *Homo sapiens,* have invented many systems for classifying physical variation. The early systems in general were verbal descriptions that classified persons in two, three or four extreme categories, which could not account for the divers midrange combinations. The need to describe heretofore unclassified persons led to more objective and comprehensive systems.

Sheldon's (1940) system, called somatotyping, was a radical departure from antipodal, extreme categories. Sheldon and his associates postulated three components of physique, which they rated on continuous scales and expressed as a three-digit number called a *somatotype.* As subsequently defined by Heath and Carter, a somatotype quantified relative fatness or *endomorphy,* relative musculoskeletal robustness or *mesomorphy,* and relative linearity or *ectomorphy.*

A somatotype is a quantitative overall appraisal of body shape and composition, an 'anthropological identification tag', a useful description of human physique. Sheldon derived the idea of somatotype from concepts of human morphology and constitutional medicine, and regarded it as an estimate of a fixed genotype that required but one rating for life. Although critics of the somatotype concept conceded that it offered considerable promise as a research tool, they doubted his claim that somatotype components derived from the three embryological layers, deemed his ratings subjective and questioned the permanence of the somatotype. Foreseeably, a good many investigators have tried to make somatotyping more objective and biologically sounder.

This book has presented an account of Heath's and Carter's modifications of somatotype method that began with Heath's (1963) and Heath & Carter's (1967) proposed modifications, designed to avoid some of the limitations of Sheldon's system. Heath and Carter modified the somatotype method to include different anthropometric as well as photoscopic assessments. Their phenotypic definition recognized that body form and composition are

340

subject to change, that a rating describes the present physique, not a past nor future physique. Their method applies universally to both sexes at all ages. During the past two decades the wide use of this method has greatly increased our knowledge of somatotype distributions of both sexes at all ages in many countries; somatotype changes during growth, aging, and special training and nutritional regimens; and relationships of somatotypes to many kinds of physical performance.

Somatotype is the generic word for the Sheldon method, for the Heath–Carter method, and for seven other methods (and their minor variations) which have had limited use. Some of the methods are based on the Sheldon genotype model; other methods are based on the phenotype model. Chapter 2 examines the differences among the various methods and shows how some are better suited than others for examining specific questions.

As this book emphasizes, ideally a somatotype is based on a standard somatotype photograph and specific anthropometry. Experienced somato-typers can assign satisfactory photoscopic somatotypes from photographs plus height and weight, and anthropometric somatotypes from 10 dimensions. The three numerals of the somatotype provide a unique summary of distinctive characteristics of individual morphology. Somatotypes plotted on a somatochart provide a graphic presentation of individual and group variation. Rated photographs help to identify different somatotypes and to identify them with their places on the somatochart.

The unusual nature of the somatotype rating calls for special kinds of analysis, although both parametric and non-parametric statistics can be applied to somatotype data. There are special techniques (see Appendix II) for analysing the three-numeral somatotype. It is better to calculate the three-dimensional distances between somatotypes, as a primary analysis, and separate component analysis should be used to supplement analysis of the somatotype as a whole, because the conceptual and logical problems arising from such analyses destroys the integrity of the somatotype.

Somatotype literature

A large variety of somatotype studies have yielded much useful information about human physique and characteristics associated with somatotypes. In the past twenty years increasing numbers of these have used the Heath–Carter method. They have dealt with somatotype distributions of different populations, with heritability of somatotypes, with changes of somatotype in growth and aging, with somatotypes in sports and physical performance and with medical, behavioural, occupational and other variables associated with somatotype.

The literature available in an area of study is a good indication of activity

Table 8.1. *Somatotype bibliographic entries for 1940–1979*

Decade	Group A	% of A	Group B	% of B	Total	% of total
1940s	41	7.6	—	—	41	6.5
1950s	137	25.6	18	18.6	155	24.5
1960s	159	29.7	53	54.6	212	33.5
1970s	199	37.1	26	26.8	225	35.5
Totals	536		97		633	

and interest in the area or subject. An analysis of the bibliography of somatotyping assembled in the course of writing this book showed the number of entries by decade, the type of entry, and the somatotype method used. Table 8.1 shows the literature for 1940 to 1979 tabulated as Group A – books (including individual chapters), journal articles and published reports; and Group B – theses and dissertations, papers presented and/or published abstracts, unpublished reports and studies. Deutsch & Ross (1978) combined a Canadian computerized retrieval system and manual searching when they compiled (1942–1972) their bibliography. Their analysis used stringent criteria for retaining 'relevant publications' and showed that there is an increasing trend toward publication in a greater number of journals and periodicals.

During the four decades there were 633 references to somatotyping, of which 536 (85%) were published articles, books and reports, and 97 (15%) were dissertations and theses, papers, abstracts and unpublished studies. In the 1950s there was a dramatic increase in references, which continued steadily in the next two decades. In the 1940s and 1950s almost all studies used Sheldon's system, and a few used Hooton's and Cureton's modifications. Use of Parnell's M.4 method also increased in the late 1950s and early 1960s. During the 1970s there was a sharp drop in studies using Sheldon's (1940, 1954) methods. A few studies used Sheldon's Trunk Index method (Sheldon *et al.*, 1969); and Parnell's method continued to be used to a limited extent.

Table 8.2 shows that the Heath–Carter method was used in 28% of all studies in the 1960s, and 70% in the 1970s. Two-thirds of these were published in the second half of the decade. During the 1960s 62% of the publications that used the Heath–Carter method were from Group B, of which 32 were theses and dissertations based on the Medford Boys' Growth Study, directed by H.H. Clarke. The findings of these studies are in Clarke's (1971) book; some were published separately.

Of the 58 theses and dissertations using the Heath–Carter method 36 were

Table 8.2. *Somatotype bibliographic entries using the Heath–Carter method for 1950–1979*

Decade	Group A	% of A	Group B	% of B	Total	% of total
1950s	1	0.7	4	6.2	5	2.1
1960s	25	14.5	41	63.1	66	27.7
1970s	147	85.0	20	30.8	167	70.2
Totals	173		65		238	

from the University of Oregon, 9 from San Diego State University and 13 from other universities. In the 1970s 85% of all studies using the Heath–Carter method were in Group A. There are 234 bibliography citations for 1980–1986, more than in each of the previous four decades. The use of somatotyping is increasing, and the fact that 89% of the studies used the Heath–Carter method shows clearly that the phenotypic method is chosen most often.

The following statements summarize some of the salient findings reported in the foregoing chapters.

Populations

Samples from populations around the world reflect extensive variations suggesting differences due to genetics, sex, nutrition, physical activity and aging. Varying degrees of sexual dimorphism are seen in paired samples of males and females, with females more endomorphic and less mesomorphic than males. There is distributional overlap between some somatotype samples and clear differences between others. For example, there is almost no overlap among the somatotype distributions of the Eskimos, Indians, Manus and Nilotes, and each population is confined to a relatively small number of somatotypes. The greater variability in European and North American somatotype samples is probably due to the combined effects of genetic heterogeneity, environmental differences, age and occupational differences, and variation in physical activity.

Genetics

It is clear that both genetics and environment influence the present or phenotypic somatotype. The extent of their respective contributions to the somatotype over a lifetime is uncertain, but the environment plays an important role. Understanding the role of genetics depends upon advances in assessing the heritable contribution to the development of the body tissues that make up the somatotype.

Growth

The range of childhood somatotypes is more limited than that of adults. Longitudinal studies show that children's somatotypes change in a generally consistent pattern that is different for each sex. Up to about age twelve there is less sexual dimorphism than in adolescents and adults. During growth some somatotypes change dramatically in an unexpected manner, while some are relatively stable and change predictably. Adult group and individual somatotypes can change with age. Change and stability of somatotype appear to be functions of nutrition, exercise, and health status.

Sport and physical performance

At high competition levels certain sports require well defined somatotypic prototypes. Usually the variation in somatotypes for specific sports narrows as the level of competition rises. The degree of sexual dimorphism in the same event or sport is parallel to that in the reference population. Some somatotypes are especially well suited to high performance in physical fitness tests of speed, strength or stamina. Training for some sports reduces endomorphy to lowest possible levels and increases mesomorphy to exceedingly high levels. Knowledge of somatotypes has proved useful in selection of the appropriate sport for serious training or for recreation guidance.

Health and behaviour

At best, studies thus far have shown only moderate relationships suggestive of group tendencies between somatotype and behaviour, temperament or mental disease. In general, the findings have not encouraged individual prediction. Delinquents as a group are more mesomorphic than their peers, many of whom are dominantly mesomorphic but non-delinquent. Likewise, it appears that some diseases are more common in certain somatotypes or groups of somatotypes than in others. The somatotype distributions for some of the genetic syndromes are markedly restricted and different from one another. However, there is no known cause and effect relationships between somatotype and medical or behavioural categories.

Preventive medicine and behavioural psychology have adopted intervention techniques designed to prevent some of the life-style induced diseases, and are likely to alter somatotypes. Although there often is wide somatotypic variation among workers in specific occupations, physical and psychological adaptations to the job are also important. Our somatotypes do to some degree influence what we can achieve for ourselves physically and behaviourally.

Purpose and uses of somatotyping

Somatotyping was designed to quantify in a three figure number components of human physique, which are known as endomorphy, mesomorphy and ectomorphy. A somatotype can be thought of as a *gestalt* rating of body shape and composition and a summary of photoscopic characteristics and a set of anthropometric dimensions. An endomorphy rating 'averages' overall relative fatness; a mesomorphy rating 'averages' overall relative musculoskeletal robustness; and an ectomorphy rating 'averages' overall relative linearity.

Somatotypes are useful for description and comparison of populations, and for monitoring growth and aging changes and changes due to exercise and physical training. Somatotype data can lead to hypotheses related to differences between samples, relationships between somatotype and performance in its broadest sense, incidence of disease, and behavioural characteristics of specific categories of subjects. Somatotypes are superior to individual anthropometric dimensions for general differentiation of physiques.

Limitations of somatotyping

Sheldon and his colleagues failed to fulfil their promise to reestablish the broken continuity between biological and social science. The concept of the genotypic, unchangeable somatotype and other unreasonable strictures fatally interfered with the optimistic expectations of early somatotype researchers.

Although the Heath–Carter phenotypic concept and other modifications freed somatotype method from former constraints, the somatotype loses some of the details of the physique, just as the calculated mean loses individual values. Somatotyping is but one technique and can yield only partial answers. The somatotype cannot, and should not, be expected to answer all questions related to physique.

The severe physical requirements of some sports and occupations impose a premium on certain somatotypes. Somatotypes play a smaller role in those activities that exact fewer requirements. In most contexts somatotypes contribute in varying degrees to the outcome, while other variables assume more prominent roles. Often it is important, limitations notwithstanding, to know the somatotype composition of a sample, as well as sex, age, height and weight of the subjects.

Despite some limitations and questions related to its theoretical and practical aspects, somatotyping as a research tool is useful and fulfils its purpose–and it works. This is true of the phenotypic methods in general and the Heath–Carter method in particular. The latter is currently the most used method of rating physique. It is the method of choice of investigators from different countries in a large number of disciplines. No other method has supplanted Sheldon somatotyping as a useful method for obtaining an overall, concise, quantification of physique.

New directions

In a letter to Thomas Cooper, Thomas Jefferson wrote on the importance of a knowledge of the body: 'No knowledge can be more satisfactory to a man than that of his own frame, its parts, their functions and actions'. The somatotype contributes to this kind of knowledge about the group as well as the individual. In the areas of theory, application and methods further contributions and progress can be anticipated.

Theory

The relationships between the components and whole body composition need to be explored. The simple and logical anatomical bases of the components need refinement. It is hoped that useful information will come from techniques like computerized axial tomography (CAT scans) and magnetic resonance imaging (MRI) of the whole body; and that techniques like these can take the place of whole body chemical analyses, now possible only in post mortem studies. For some kinds of research it may be that the quantification and integration of tissue masses available through these methods will replace standard somatotyping. Perhaps CAT scans and MRI will help to quantify internal adipose and other tissues that should be accounted for in somatotype component ratings. However, it seems likely that somatotyping and its relatively simple scaling system will continue to be useful.

Somatotype assessment makes assumptions regarding relative masses of internal tissues and organs and the ratings of endomorphy and mesomorphy. The HWR is related to total mass and is closely related to ratings of ectomorphy. Note that the masses referred to in somatotyping are relative, not absolute.

Girths and biostereometrics help to describe characteristics of shape and body volume distribution, and to model change in body shape. They do not 'see' the underlying tissues. And, although they are helpful in body description, they lack the three-dimensional perspective of the somatotype.

Increased knowledge of *in vivo* tissues of the body should help to evaluate the roles of environment and heredity in the evolution of the somatotype. At the same time, there should be studies of family resemblances with controls for socio-economic status, nutrition, physical activity, and disease. New knowledge of the effects of internal and external factors on tissues will help to elucidate the influence of growth and aging processes on somatotype changes. For example, successive measurements, somatotype photographs and mesomorphy ratings demonstrate and permit quantification of increases in muscle hypertrophy. But only anatomy, physiology and the biochemistry of neuromuscular adaptation can explain the processes of change in muscle tissue.

Dysplasia

In the study of somatotype dysplasia is an important and neglected aspect. Persons who have had no training in observing subjects and examining somatotype photographs can easily recognize dysplasia or disharmony in a physique. Somatotypes can be rated to reflect regional differences in physiques; and selected anthropometric dimensions can demonstrate dysplasias. It is possible to give a rating of 2-4½-3½ to a whole physique and to give ratings of 2-5-3 to the lower half of the body and 2-4-4 to the upper half of the body. This aspect of the somatotype is called a mesomorphic dysplasia of the lower body (i.e. the lower half of the body is significantly more mesomorphic than the upper half.) Addition of anthropometry can reveal the extent of the proportional contribution of bone breadths and muscle girths to the upper and lower body.

Some dysplasias are an advantage in certain sports. Some moderate muscular dysplasias can be produced by specific training, and may be temporary. Other dysplasias are regarded as negative, and may be associated with medical, psychiatric and social problems. There is some speculation that dysplasia is related to predisposition to certain diseases, illnesses and injury, and to psychopathy and behavioural disorders.

It is reasonable to hypothesize that dysplasias are expressions of human genetic diversity, reflections of the diversity of parental genetic bequests and of the many genes that contribute to patterns and variations in physique. Somatotype photographs teach that wholly non-dysplastic physiques are in the minority. It is also true that small to moderate dysplasias do not seem to hamper individuals seriously. The heritability and long-range, lifelong role of dysplasias are important areas for future research.

The range of possible change of individual somatotypes is interesting and important. It appears that the basic pattern is genetic and within predictable limits can be modified, or stabilized, by nutrition and exercise. One may conceive therefore of a *cluster* of somatotypes possible for a given individual, and this could be important in medical and guidance profiles. From a current somatotype photograph and appropriate anthropometric dimensions, a cluster of reasonably predictable somatotypes can be estimated from simple manipulation of weight relative to the stature of the subject.

As we know, for a given mesomorphy, endomorphy and ectomorphy are inversely related so that, for each postulated increase in fat weight, endomorphy increases and ectomorphy decreases. Also we can assume for the moment that unless there are drastic changes in physical activity mesomorphy is fairly stable. Similarly, for each postulated decrease in weight ectomorphy increases and endomorphy decreases. The plotting of the possible somatotypes resulting from the postulated weight changes produces a somatoplot of the cluster of possible somatotypes for the subject. This material could be an effective introduction to establishing an optimal weight as a goal, with a regimen of suitable nutrition and appropriate physical activity. It is an approach compatible with the growing awareness of the risks of weight gain and lack of exercise, the importance of appropriate diet, weight control, and suitable exercise.

In his work with students and with families Parnell (1984) found that few persons have a clear image of their own bodies and somatotypes. He counselled that realistic appreciation of one's physical potentials and limitations is desirable, and that somatotype photographs are an ideal medium for achieving it.

The cluster concept of possible individual somatotypes recognizes that subjects with extremely high mesomorphy, or with extremely high or low ectomorphy, have the smallest clusters of potential somatotypes. The midrange somatotypes have the largest clusters of potential somatotypes, largely because their range of possible ectomorphy is greatest. Manipulation of weight against height for a sample of midrange and extreme somatotypes demonstrates this point.

Applications

Somatotyping has proved to be a good descriptive and classification system for learning about relative shape and body composition, and their variation in samples or populations. Continuing records of somatotypes in the same and new populations are valuable to check against previous records, and special samples need to be examined for answering particular questions. Experimental studies are important for observing the influence of specified conditions in studies of growth, nutrition, exercise, and medication. The majority of studies reported in this volume fall in these categories. As a result we know a good deal about a number of somatotype categories.

Further exploration is needed not only in areas already examined, but also in many others in which we have little or no good somatotype information. Little is known about the role of somatotype in the behavioural sciences, particularly in psychiatry, for to date the results are equivocal and hampered by design and conceptual problems.

The relationship between disease and somatotype needs serious investigation. It is difficult to interpret existing work, partly because of problems of research design, and partly because the contribution of the somatotype to a multivariate problem has not been calculated. No one believes somatotype is causal. It is generally accepted that somatotypes change, but no one can predict magnitude of changes, or their influence on the somatotype-disease relationship. The *somatotype reflects the effect rather than the cause*. Longitudinal studies are needed to shed light on these questions.

Methods

The Heath–Carter method employs relatively simple methods and technology, with wide application in field work, yet it is sufficiently sophisticated for laboratory research and analysis. Sometimes field and laboratory research require two levels of techniques. There are, however, some improvements in methodology that might be made. The following observations, which could apply to other somatotype systems, are here confined to the Heath–Carter method.

In laboratory research and longitudinal growth studies, for example, we urge the combination of anthropometry and photoscopy. In retrospect, it is a pity that measurements (other than height and weight) were not routinely taken along with the early somatotype photographs. However, many investigators find the anthropometric estimate of the somatotype valid and useful in cross-sectional studies and surveys of large samples. Its extensive use has initiated rebirth of interest in somatotype as a research tool. It is not difficult to teach the necessary anthropometry and the hand calculations called for on the Heath–Carter rating form. Several improvements of the rating form have been suggested since the original presentation of the Heath–Carter method in 1967. Students quickly learn the basics of anthropometric somatotyping. Reliable photoscopic somatotype rating requires a reasonable amount of training and experience.

Studies of skinfold measurements, show that sums of skinfolds are more reliable than single skinfolds, and that there are different skinfold patterns for sex, age, race-ethnicity, some diseases, and perhaps levels of fatness. The sum of three upper body sites has served well in estimates of endomorphy. Measures of lower body skinfolds enhance estimates of total fatness and endomorphic dysplasias, but five to eight skinfold measurements at sites representative of the body regions are adequate. It appears that the endomorphy table (and equations) should incorporate the umbilical and front thigh skinfolds as well as the medial calf skinfold. When photoscopy is combined with anthropometry the discrepancies are accounted for in the criterion rating.

Experience shows that skinfolds corrected for height give a more valid estimate of endomorphy. The correction should be routine and should be noted. For simple comparisons of height-corrected studies with earlier ones it is feasible to height-correct the means of the latter.

The present anthropometric scales for mesomorphy use two limb breadths and two limb girths relative to height. This minimal sampling may be inadequate for some subjects. Four additional measurements are suggested: biacromial and biiliac breadths and girths of forearm and thigh (corrected for skinfolds). The two breadths require a large sliding anthropometer. Trunk girths are unsuitable for mesomorphy estimates because they encompass areas that contain mostly non-mesomorphy related tissues and organs. However, the photoscopic rating easily accounts for trunk musculature in low-fat subjects. Incorporating these dimensions in the anthropometric estimate of mesomorphy has not been developed.

The scale for ectomorphy for both anthropometric and photoscopic ratings, is unsatisfactory for discriminating among subjects with exceedingly low HWRs, such as the obese. In view of large volume of available somatotype data with ratings based on the HWR table, it seems preferable to keep the existing scale and to accept this limitation rather than face the interpretative chaos that would follow a change.

The representative selection of photographs and ratings in this book partially solves the need for a set of rated photographs accompanying an instructional methodology. The pending revision of the laboratory workbook *The Heath–Carter Somatotype Method* (Carter, 1980a) will provide an adequate teaching and training sequence.

(The reader is reminded that one is not a somatotyper because he or she can calculate somatotypes from anthropometric data alone. The somatotyper must also have a concept of what the somatotype looks like and where it is located on the somatochart.)

The Heath–Carter method is attractive for its very simplicity, and all concepts of the somatotype and its derivation should be kept simple. Although some of the above suggestions might improve the anthropometric somatotype estimate, the improvement may be only marginal. There should be no changes without appropriate statistical analysis and careful judgement. There are a limited number of answers to questions found in correlations and F-ratios. At all costs avoid the pitfall of trying to derive sample specific anthropometric equations for somatotypes. The generalized approach to anthropometric somatotyping proposed in 1967 has served us well and has avoided the numerous equations that have produced chaos in the body composition area. Recent reversion to generalized equations in body composition studies has improved estimates of body fat.

Overall it is an advantage that the Heath–Carter somatotype method applies to both sexes at all ages. Its application to young children is somewhat uncertain. The method needs some modification and validation in this area. Neither the photoscopic nor the anthropometric rating satisfactorily accounts for the developing tissues, particularly in the skeleton.

Computer programs for calculating somatotypes and resolving group data are well established. The programs include those written for both mini and mainframe computers. Existing graphic capabilities permit viewing of the somatotype in three-dimensional space and plotting in any plane. These graphs help us to understand the movement of somatotypes in space. Nevertheless, the two- dimensional somatochart remains the standard and most appropriate visual aid. We welcome development of new methods of analysis that will add new meaning to somatotype data.

Conclusion

The somatotype represents the physique of each subject as a unified whole. It reminds us of the important 'wholeness' and unique individuality of those persons studied for whatever reason, and that each part of a body, whether an organ or a cell, is related in probably unknown ways to the 'wholeness' we call a somatotype.

The technique of somatotyping evolved from attempts to classify human physique. The somatotype is an anthropological identification tag and provides a summary of the whole physique. The Heath–Carter method is the most widely used. It applies universally to both sexes at all ages. It helps to elucidate data from a broad spectrum of studies that include selected samples, population studies and longitudinal studies. Increasingly, anthropologists, human biologists, physical educators, sports scientists and behavioural and social scientists use somatotyping in their search for further understanding of the variation in human physique.

Appendix I[1]
The Heath–Carter somatotype method

Introduction

The Heath–Carter somatotype method is a modification of the system developed by Sheldon and his colleagues. It uses much of the original vocabulary and employs those criteria of his basic approach which are objective and straightforward. But it is distinguished by important fundamental modifications: (1) The somatotype rating is a phenotypic rating, which allows for changes over time. (2) The rating scales for the three components are open and have been redefined so as to apply to the physiques of both sexes at all ages. (3) Selected anthropometric dimensions help to objectify somatotype ratings.

The Heath–Carter somatotype is a semi-quantitative description of the existing relative shape and composition of a human body. It is expressed as a three-numeral rating with one numeral for each of three components of physique, always recorded sequentially. The three numerals written as 2-5-3, for example are a somatotype rating, in which 2 represents the amount of endomorphy or the first component, 5 represents mesomorphy or the second component, and 3 represents ectomorphy or the third component. Thus a somatotype gives a quantified expression of individual variations in morphology and of what a body looks like.

The three components of a somatotype rating are defined briefly as follows:

Endomorphy (or first component) is a rating on a continuum of relative fatness of a physique.
Mesomorphy (or second component) is a rating on a continuum of musculoskeletal robustness relative to stature.
Ectomorphy (or third component) is a rating on a continuum of relative linearity or 'stretched-outness' of a physique.

A low rating in endomorphy describes a markedly lean physique with a minimum of subcutaneous fat, or little relative fatness. High ratings in endomorphy characterize conspicuously obese physiques with large de-

[1] Much of material in this chapter is adapted from *The Heath–Carter Somatotype Method* (Carter, 1980a). Additional details are available in this source.

352

posits of subcutaneous fat, or noticeable relative fatness. An extreme would be the carnival 'fat lady'.

Low ratings in mesomorphy describe physiques with narrow bone diameters and small muscle mass relative to stature. High ratings in mesomorphy signify large muscle mass with wide bone diameters relative to stature. Mesomorphy may be thought of as a rating of lean body mass relative to stature on a continuum of markedly low to exceptionally high ratings. An extreme would be a 'Mr Universe' body builder.

Low ratings in ectomorphy denote physiques with great mass relative to stature. High ratings in ectomorphy denote physiques with little mass relative to stature and relatively elongated limb segments. Ectomorphy ratings are based largely, but not entirely, on a height–weight ratio (height divided by cube root of weight). Ratings in ectomorphy are inversely, to some degree independently, related to both endomorphy and mesomorphy. For example, with low ratings in ectomorphy either or both endomorphy and mesomorphy may be high or relatively high. Three subjects with the same heights and weights, the same height–weight ratios and ratings of one in ectomorphy may have the somatotypes 5-5-1, 3-7-1 or 7-3-1. However, with high ratings in ectomorphy, ratings in either endomorphy or mesomorphy are relatively low. That is, three subjects with the same heights and weights, height–weight ratios, and ratings of six in ectomorphy may have the somatotypes 1-3-6, 3-1-6 or 2-2-6. Extremes are seen in the Nilotes of Africa.

The criteria for rating the components and for allocating a somatotype rating are extensions of their definitions, and are presented in detail in the section 'Procedures for photoscopic rating'.

Rating scales

Ratings of each component begin theoretically at zero and have no fixed upper end points. No ratings less than one-half are given, and by definition no physique can have a zero rating of a component. Reliable ratings to the nearest half-unit can be made, which means that the scales are divided into half-unit intervals, e.g. $2\frac{1}{2}$-$4\frac{1}{2}$-3.

The following ranges of component ratings have been verified by experience, but do not preclude the possibility that higher ratings may occur in the human species:

Endomorphy	one-half to 16
Mesomorphy	one-half to 12
Ectomorphy	one-half to 9

In general, component ratings of $\frac{1}{2}$ to $2\frac{1}{2}$ are regarded as low values. Ratings of 3 to 5 are regarded as midrange. Ratings of $5\frac{1}{2}$ to 7 are regarded as high, and ratings higher than 7 as extremely high.

The word somatotype always implies the three-numeral rating. A component rating is always recorded together with the other two. Unless the three components are recorded together the unique concept and the meaning of the somatotype rating is lost. The words *endomorphy, mesomorphy* and *ectomorphy* refer to the separate components. To refer to a person as a mesomorph indicates his or her dominant component but says nothing about the other two. For example, the word *mesomorph* alone does not accurately describe and differentiate such somatotypes as 4-5-1, 1-5-4 and 2-5-2, although all of them are dominantly mesomorphic. It is important to think of the somatotype components as continuous variables recorded in quantitative shorthand to describe the variations in human physique. Experience has shown that the possible variations are limited; that a subject who is low in endomorphy and mesomorphy must be at least high midrange in ectomorphy (e.g. 1-2-5); that a subject who is high in both endomorphy and mesomorphy must be low in ectomorphy (e.g. 8-6-1); that there are no 1-1-1's and no 8-8-8's.

Obtaining a Heath–Carter somatotype
There are three methods for obtaining a Heath–Carter somatotype:
1. The photoscopic somatotype. Experienced somatotypers can make reliable photoscopic ratings by visual inspection using standard somatotype photographs, accompanied by the subject's height and weight, and with reference to the table of distribution of somatotypes according to ratios of height divided by cube root of weight (HWR) (Table. I.1). This method therefore requires a photograph taken according to standardized instructions, measurements of height and weight, and a table of somatotypes according to HWR. It also requires familiarity with the rating criteria, and experience in judging the relative amounts of the three components as observed in the photographs.
2. The anthropometric somatotype. Anthropometric somatotypes can be calculated from 10 anthropometric dimensions (height, weight, 4 skinfolds (triceps, subscapular, supraspinale, medial calf), 2 girths (flexed upper arm, and calf), and 2 breadths (biepicondylar humerus and femur).
3. The anthropometric plus photoscopic somatotype. This is the method of choice. As its description implies, it is based upon reference to a standard somatotype photograph and rating criteria, to the anthropometric somatotype, and to the table of distribution of somatotypes according to HWR.

I. *The Heath–Carter photoscopic somatotype*

It is possible to attain a reasonable degree of skill in somatotype rating after careful study of the criteria and after rating a series of photo-

Table I.1. *Distribution of somatotypes according to their height/cube root of weight (HWR)*

HWR (Imperial units)	Ectomorphy ratings									HWR (metric units)
	1	2	3	4	5	6	7	8	9	
15.40									119	50.91
15.20								118	129 219	50.25
15.00							117	128 218		49.59
14.80							127 217	138 318 228		48.93
14.60						126, 216	137 317 227			48.27
14.40						136, 316 226	237 327			47.61
14.20					135, 315 225	146, 416 236, 326				46.95
14.00				134, 314 224	145, 415 235, 325	246, 426 336				46.28

(continued)

Table I.1. (*Continued*)

HWR (Imperial units)	Ectomorphy ratings									HWR (metric units)
	1	2	3	4	5	6	7	8	9	
13.80				144, 414 234, 324	245, 425 335					45.62
13.60			233	154, 514 244, 424 334	255, 525 345, 435					44.96
13.40			153, 513 333	254, 524 344, 434						44.30
13.20		242, 422	163, 613 253, 523 343, 433	354, 534 444						43.64
13.00		162, 612 252, 522	263, 623 353, 533 443							42.98
12.80	341, 431	172, 712 262, 622 352, 532 442	363, 633 453, 543							42.32
12.60	171, 711 261, 621 351, 531 441	182, 812 272, 722 362, 632 452, 542								41.66

12.40	181 271, 721 361, 631 451, 541	282, 822 372, 732 462, 642 552	40.99
12.20	191 281, 821 371, 731 461, 641 551		40.33
12.00	291, 381 471, 741 561, 651		39.67
11.70	661, 391 481, 841 571, 751		38.68
11.40	491, 941 581, 851 671, 761 10-3-1 3-10-1		37.69
11.00	771 951, 591 861, 681 4-10-1 10-4-1		36.37

(continued)

Table I.1. (*Continued*)

HWR (Imperial units)	Ectomorphy ratings									HWR (metric units)
	1	2	3	4	5	6	7	8	9	
10.50	781, 871 691, 961 4-11-1 11-4-1 5-10-1 10-5-1									34.71
10.00	791, 971 5-11-1 11-5-1 6-10-1 10-6-1									33.06
9.50	891, 981 12-5-1 7-10-1 10-7-1 11-6-1									31.41
9.00	991 12-6-1 13-5-1 8-10-1									29.75

graphs. Consistently valid and reliable ratings require practice and experience, especially when ratings are based on visual cues. Such a skill is comparable to the skills required for rating atherosclerotic and arteriosclerotic changes in the retinal vessels of the eye, for judging wines by taste, or the scoring of such sports as gymnastics. It is easy to judge wide differences, but more training and knowledge of criteria are needed for judging fine differences. The somatotyper should establish the reliability and validity of his or her ratings against those of an experienced rater.

Photography and equipment

The standard somatotype photograph is a print showing front, side and back views of a subject, in standard poses.

Equipment

To obtain satisfactory somatotype photographs, in a permanent somatotype laboratory, the preferred equipment consists of a 12.7 x 17 cm portrait camera with 21.6 to 24.1 cm focal length lens, and a custom-made sliding back which allows exposure of one-third of the film for each view of the subject. Contact prints, which lessen errors inherent in enlarging, are a convenient size for rating. The lens to subject distance is 4.5 metres for a portrait camera with a 21.6 cm focal length lens.

In some circumstances, such as field work, a 35 mm camera is preferable to the rather unwieldy portrait camera. The greater the lens to subject distance and the longer the focal length of the lens (consistent with maximum image size), the less distortion of the subject along the optical axis. Though satisfactory photographs can be obtained with a 50 mm lens (the minimum) at 4.5 metres, lenses of longer focal length (80 mm at 5.8 metres, 110 mm lens at 8 metres, or 135 mm at 10 metres) are preferable.

Photographic prints

Fine grain film should be used. Lighting and exposure time should be adjusted to obtain good definition and to eliminate misleading shadows. Prints should be made on relatively shrink-proof paper to avoid distortion. For comparison of different series it is useful to include a scale. A metre marker at the same lens distance as the subject is useful, while a background grid of horizontal and vertical lines helps in posing.

Turntable

A freely moving turntable, built with a conventional 360 degree ball bearing assembly and mounted on a base 10–15 cm high, greatly helps the subject to maintain the same pose for the three photographs. Stops at 90 degree intervals help to standardize the three views.

Adaptations

A wide range of photographic assemblies, which preserve the important principle of a distortion-free image, have been reported by Sheldon (1954), Tanner & Weiner (1949), Dupertuis & Tanner (1950), Sills (1950), Brown & Mott (1951), Parnell (1954), Hertzberg *et al.* (1963), Tanner (1964), Jones & Stone (1964), Weiner & Lourie (1969) and Carter (1980*a*). Hooton criticized the general quality of somatotype photographs: 'With all the learned articles that have appeared on the subject of photography in the nude for somatotyping purposes and with all the thousands of subjects that have been photographed, the fact remains that somatotypers are constantly being asked to make morphological ratings on prints so inferior that accurate appraisal of the details required is impossible.' (Dupertuis & Emanuel, 1956, p.25). The comment is valid today.

Clothing

It is most informative if subjects can be photographed without clothing. If this is not feasible, minimal clothing is acceptable, such as nylon swimming trunks for males, and bikinis for females. However, estimates of sexual maturation, such as are required in growth studies, necessitate photographs of nude subjects.

The somatotype pose (adapted from Sheldon, 1954)

1. Front view. The poser stands in front of the subject and demonstrates to him the position of attention as follows: (a) fingers together; (b) thumbs parallel to and touching the forefingers; (c) fingers extended and pointed perpendicularly to the floor; (d) hands face the thighs, 10–15 cm away from them, with the middle fingers in the mid-frontal plane; (e) elbows are forcibly extended; (f) knees and hips extended, body upright; (g) face is straight toward the camera with the head held in the horizontal eye–ear plane. Just before taking each picture, the subject is told to hold a medium breath, and is warned before the turntable is rotated.

2. Side view. The arms should be in the mid-frontal plane, the right arm out of sight, and the left arm in forced extension to emphasize the triceps muscle. The arms should not obscure the back or front line of the body, the body and face should be in perfect profile, and the legs in perfect alignment with no flexion or hyperextension of the knees. Otherwise the pose is the same as the front view.

3. Back view. The position is the same as for the front and side views, but the position of the shoulders and the extension of the elbow should be checked.

Height–Weight ratio and somatotype distribution

To obtain a photoscopic somatotype rating, consider the graphic information revealed in the somatotype photograph, and the height and weight data for the subject; check Table I.1 (the table of the distribution of somatotypes by HWR); and compare the anthropometric rating with the photoscopic estimate. Note that exact somatotype ratings cannot be determined from the HWR table, which indicates the possible ratings associated with each ratio.

In Table I.1, adapted from the distributions of Heath (1963) and Heath & Carter (1967), HWRs ascend from the bottom to the top of the table, and ectomorphy increases from left to right. Metric HWR values are shown in the right hand column, Imperial HWR values in the left-hand column. The somatotypes are expressed in whole numbers; half-unit ratings may be interpolated.

Many of the relationships in Sheldon's (1954) tables of distributions are nonlinear and illogical (Heath, 1963). Heath reported that in the cases of fifteen somatotypes differing by one rating unit in one of the three somatotype components, there were corresponding increments or decrements of 0.20 in HWRs (Imperial units). Her photoscopic impressions matched the HWRs assigned to these fifteen somatotypes. She therefore constructed a HWR table on the basis of a difference of 0.20 in the HWR for one unit of change in the rating of a given component. (See Table I.2.) For example, if endomorphy is increased by one unit from 3-4-4 to 4-4-4, the HWR decreases by 0.2 units to 13.20 (43.64).[2] If mesomorphy is lowered by one unit from 3-4-4 to 3-3-4, the HWR increases to 13.60 (44.96).

Table I.1 allows for the influence on somatotype ratings of the apparent nonlinearity of increasing endomorphy and the influence of that skewness on somatotype ratings. The intervals between HWRs lower than 12.00 (39.67) are increased to 0.3 between 12.00 and 11.70 and between 11.70 and 11.40; to 0.4 between 11.40 and 11.00; and to 0.5 between 11.00 and 10.50, between 10.50 and 10.00, between 10.00 and 9.50 and between 9.50 and 9.00.

An experienced somatotype rater is as familiar with the relationships between HWRs and ratings in Table I.1 as the skilled chemist is with the periodic table of the elements. It is a bridge between photoscopic impressions and assignment of somatotype ratings; and it is a bridge between anthropometric measurements and the assignment of somatotype ratings.

The Heath–Carter photoscopic somatotype rating

Heath–Carter photoscopic somatotype ratings are based upon standard somatotype photographs together with a record of age, present

[2] Metric equivalents of the HWR are shown in parenthesis when appropriate.

Table I.2. *The relationships between the height–weight ratio and increasing or decreasing component values*

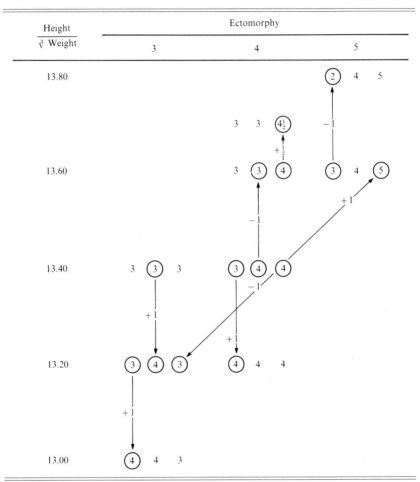

height and weight. Accurate rating depends upon skill in recognizing the probable rating for each component, and in reconciling photoscopic impressions with the appropriate somatotype–HWR relationships in Table I.1.

In order to acquire understanding and skills for making reliable photoscopic somatotype ratings it is important to:

1. Review the definitions of somatotype components.
2. Understand the general concepts and guidelines for the Heath–Carter photoscopic somatotype rating.

3. Be familiar with the rating criteria and procedures for examining somatotype photographs.
4. Study the reference photographs.
5. Rate the practice series for which accurate assessments are available, and compare results.
6. Rate and re-rate new series of photographs to establish consistency.

General concepts and guidelines

Ratings of each of the components and the final somatotype rating are based on these general concepts and guidelines:

1. A somatotype rating describes the physique as it is recorded in the present photograph, i.e. what is seen. History, previous body weight, shape, size, performance, occupation and ethnic group are disregarded.

2. A somatotype is *not* a measure of size but a shorthand description of relative body shape and apparent composition. The best background for rating somatotype components is a knowledge of gross human anatomy and body composition.

3. The same rating criteria apply to both sexes at all ages, within the limitations noted later. The same ratings are given to degrees of a given component that appear to be the same, regardless of difference in sex and age.

4. Endomorphy is rated from the relative amount of fatness suggested in the photograph, regardless of where it is deposited. The leanest subjects are assumed to have a small amount of essential fat stored in bone marrow, tissues of the brain and nervous system, and cells and internally around vital organs. If the skin were removed from the 1's in endomorphy there would be no visible subcutaneous fat laid over the musculoskeletal frame. The 1's in endomorphy are rare, as Table I.1 shows.

The majority of somatotypes have endomorphy ratings of 2, 3, 4, or 5, indicated by increasing accumulations of subcutaneous fat over muscles, especially over the abdomen and on upper arms and thighs. As endomorphy increases, muscle relief decreases and the bony promontories become less prominent. When endomorphy is high, the volume of the abdominal trunk equals or exceeds the volume of the thoracic trunk. The mass of the body appears to be concentrated in the trunk and in the proximal segments of the limbs. The distal segments of the limbs often appear to taper in comparison with the rest of the body.

5. Mesomorphy, or musculoskeletal robustness per unit of height, is easily assessed in subjects with little subcutaneous fat (the 1's in endomorphy). Its rating becomes more difficult as subcutaneous fat increases. If one imagines the body segments in cross section, low ratings in mesomorphy are

given if bones and muscles are small relative to their length. Dysplasias (regional variations) complicate ratings in mesomorphy. Ratings of 5 in mesomorphy *may* be given to subjects with relatively small muscles but relatively large bones, or with relatively small bones and well developed muscles.

Low ratings in mesomorphy (the 1s and 2s), are rare and hence difficult to differentiate. But because they may be high in either endomorphy or ectomorphy, the HWR is a valuable clue. The common, midrange somatotypes with almost equal degrees of endomorphy and mesomorphy are the most difficult to differentiate. High mesomorphy with low endomorphy and ectomorphy is the easiest to rate.

In evaluating the muscular contribution to mesomorphy it is particularly useful to note the relative bulk or development in the following muscles: trapezius, deltoid, pectoralis major, triceps, brachialis, biceps brachii, forearm flexors and extensors, erector spinae, latissimus dorsi, abdominal muscles, quadriceps, hamstrings and gastrocnemius-soleus.

6. Ectomorphy or linearity. Since ectomorphy indicates a body that as a whole appears to be stretched out, with no concentration of bulk, high ratings are given to subjects with low body bulk, those that resemble pencils. Dominant ectomorphy occurs with relatively low endomorphy and mesomorphy and high HWRs, greater than 13.60 (44.96). Dominant ectomorphs may be either short or tall. Their limbs appear long and somewhat cylindrical, without pronounced muscle definition or fat overlay.

7. Regional variation in somatotypes, when the distribution of components is different in various parts of the body (dysplasia) is a complicating element. It will be discussed later on.

8. In most studies it is assumed that somatotype subjects have normal health.

9. Two physiques with the same somatotype rating are rarely identical. But the physiques of two subjects rated as 2-5-3 look more alike than either looks like subjects rated 2-5-2 or 2-5-4.

10. Female breasts, which consist of alveolar glands, connective tissue and fat, influence overall ratings in endomorphy, in that the greater their mass, the lower the HWR, and hence the higher the rating in endomorphy. Allowance has to be made for this.

Procedures for photoscopic rating

The following procedures assume that the rater has photographs of 20 or more subjects. The procedures for smaller series can be simplified as follows:

1. Arrange the photographs from high to low according to the calculated HWR for each subject.

PHOTOSCOPIC SOMATOTYPE RATING FORM

1. ENDOMORPHY RATING SCALE AND CHARACTERISTICS (Relative fatness)

$1 \quad 1\frac{1}{2} \quad 2 \quad 2\frac{1}{2} \qquad 3 \quad 3\frac{1}{2} \quad 4 \quad 4\frac{1}{2} \quad 5 \qquad 5\frac{1}{2} \quad 6 \quad 6\frac{1}{2} \quad 7 \qquad 7\frac{1}{2} \quad 8 \quad 8\frac{1}{2} \quad 9 \ldots\ldots$

Low relative fatness; little subcutaneous fat; muscle and bone outlines visible.	Moderate relative fatness; subcutaneous fat covers muscle and bone outlines; softer appearance.	High relative fatness; thick subcutaneous fat; roundness of trunk and limbs; increased storage of fat in abdomen.	Extremely high relative fatness; very thick subcutaneous fat and high amounts of abdominal trunk fat; proximal concentration of fat in limbs.

2. MESOMORPHY RATING SCALE AND CHARACTERISTICS (Musculo-skeletal robustness relative to height)

$1 \quad 1\frac{1}{2} \quad 2 \quad 2\frac{1}{2} \qquad 3 \quad 3\frac{1}{2} \quad 4 \quad 4\frac{1}{2} \quad 5 \qquad 5\frac{1}{2} \quad 6 \quad 6\frac{1}{2} \quad 7 \qquad 7\frac{1}{2} \quad 8 \quad 8\frac{1}{2} \quad 9 \ldots\ldots$

Low relative musculo-skeletal development; narrow skeletal diameters; narrow muscle diameters; small joints in limbs.	Moderate relative musculo-skeletal development; increased muscle bulk and thicker bones and joints.	High relative musculo-skeletal development; wide skeletal diameters; bulky muscles; large joints.	Extremely high relative musculo-skeletal development; very bulky muscles; very wide skeleton and joints.

3. ECTOMORPHY RATING SCALE AND CHARACTERISTICS (Relative linearity)

$1 \quad 1\frac{1}{2} \quad 2 \quad 2\frac{1}{2} \qquad 3 \quad 3\frac{1}{2} \quad 4 \quad 4\frac{1}{2} \quad 5 \qquad 5\frac{1}{2} \quad 6 \quad 6\frac{1}{2} \quad 7 \qquad 7\frac{1}{2} \quad 8 \quad 8\frac{1}{2} \quad 9 \ldots\ldots$

Low relative linearity; great bulk per unit of height; round like a ball; relatively bulky limbs.	Moderate relative linearity; less bulk per unit of height; more stretched-out.	High relative linearity; little bulk per unit of height.	Extremely high relative linearity; very stretched-out; narrow like a pencil; minimal bulk per unit of height.

PHOTOSCOPIC RATING INFORMATION

I.D. No. .... Name .... Age .... Male .... Female .... Date ....

First estimate .... Height .... Weight .... HWR ....

Final rating .... NOTES:

Fig. I.1. Photoscopic somatotype rating form. This form is used along with the photoscopic criteria and reference photographs to help establish the rating for each component. It is not used to describe the somatotype.

2. Group the photographs by their proximity to given HWR rows in Table I.1. The HWRs of the group may range from highs of 14.40 (47.61), 14.20 (46.95) and so on to lows of 12.00 (39.67) and lower.

3. Beginning with the photograph(s) with the highest HWR's, arrange the photographs in each HWR category according to inspectional judgment of relative values or dominance of the three components. If, for example, there are several photographs with HWRs of 13.80 or higher, the rater soon learns that ectomorphy is almost always the highest component. Next arrange the photographs in order of endomorphy and/or mesomorphy (high to low) within each HWR group. This procedure gives the rater a *gestalt* impression of the relative dominance of the three components and of the general character of the distribution of the series.

4. Record on the photoscopic rating form (see Fig. I.1) the first impression of the relative values of the three components. This form shows the scales for the three components. Note that the scales read, in half-intervals, from the smallest rating on the left to the rating of 9 on the right. Write in values higher than 9 when they occur. The scales are continuous and are open at the upper end. At the lower end the scale effectively begins with 0.5 for ectomorphy only, as values lower than 1 are rarely, if ever, found in

endomorphy and mesomorphy. Values of 9 and higher are occasionally found in endomorphy and mesomorphy and are possible in ectomorphy.

Note the information on the photoscopic rating form. Recheck the general guidelines for ratings. Check a key file of typical ratings, Circle the appropriate unit on each of the three scales to obtain a first estimate of the somatotype. That is, a circle around 3 in endomorphy, around 4 in mesomorphy, and around 5 in ectomorphy denotes an estimated somatotype of 3-4-5.

5. On a separate sheet of paper write down the somatotypes with the same or similar relative dominance as the estimated rating and lying on the same HWR row. From the HWR rows above and below the estimated rating, write down the ratings with the same relative dominance. For example, from the rows adjacent to 3-4-5, there are 4-3-4 on the row below, 2-3-4, 3-3-5 and 2-4-5 on the row above, and 4-3-5 on the same row.

6. Reconcile the possible somatotypes with photoscopic impressions suggested by the photograph. Half-unit interpolations may be more appropriate than whole unit ratings. A first estimate is not always correct. Check the photograph with criterion examples.

7. After making ratings of all photographs at the same HWR, compare them to check their similarities and differences on each component for compatibility with the HWR.

8. Repeat the above process for each successive HWR group. Compare ratings of each group with ratings of preceding group.

9. When all HWR groups have estimated ratings and have been compared, arrange the photographs on a large surface, first, in order of increments of endomorphy. Repeat for mesomorphy and ectomorphy. Following these comparisons the rater will find it easier to make ratings that seemed difficult at first.

10. Calculate means for each of the three components and for the HWR. As a check for rating bias, consult Table I.1 to compare the mean somatotype with the mean HWR. (For details, see later in this chapter).

Additional rating hints

1. Take particular note of regional dysplasias in the first and second components. Watch for variations in individual fat deposition and distribution, and for variations between the sexes. Often females have much thicker deposits of subcutaneous fat over hips and thighs than elsewhere on the body. For an unbiased impression of the body, block off four of the five regions and estimate the somatotype region by region.

2. When endomorphy and mesomorphy, or ectomorphy and mesomorphy are almost equal, e.g. 4 or $4\frac{1}{2}$, it is often difficult to decide on a rating. Only experienced raters expect to make the distinction relatively easily. When in doubt, give the same rating to both components.

3. For ratings of 1 and 1½ in endomorphy, muscle outlines should be apparent at the borders of the large superficial muscles.

4. Because skin colour markedly affects perception of apparent size, light-skinned subjects appear larger than dark-skinned subjects.

5. High ratings and dominance in endomorphy (7 or higher) usually are accompanied by moderate to high ratings in mesomorphy (i.e. 3 to 6). 1's in mesomorphy are rare, and are rarer, if not nonexistent, with high endomorphy.

6. When the HWR is lower than 12.00 (39.67) and there is some evidence of linearity in the distal limb segments and/or the neck, a rating of 1 in ectomorphy is given, but when there is no such evidence, a rating of one-half in ectomorphy is given.

II. *The Heath–Carter anthropometric somatotype*

Several investigators (e.g. Cureton, 1947, 1951; Parnell, 1954, 1958; Damon *et al.*, 1962) demonstrated the feasibility of applying anthropometric measurements to estimating somatotype ratings based on Sheldon's criteria. Heath & Carter (1967) developed an anthropometric estimate based on the Heath–Carter method to lessen the subjective element of somatotype ratings based solely on photographs, age, height and weight.

Anthropometry greatly helps inexperienced somatotypers who lack confidence in their accuracy and ability to agree with criterion raters.

The inherent advantages of anthropometry are:

1. It provides an objective method of somatotyping.

2. It provides the best estimate of a criterion somatotype in the absence of a photograph.

3. It provides an objective starting point for an anthropometric plus photoscopic rating when a photograph is available.

4. It makes possible quick estimates of somatotypes in the field, where photographs are not yet available.

5. It provides the data for a reliable somatotype rating when minimal clothing is desirable. For example, subjects who refuse to be photographed with or without clothing can be included in a sample.

6. Measurements can be used for other analyses and evaluations of body structure as well as for somatotype rating.

7. Anthropometric measurements allow assessment of regional differences in somatotype components in a given subject, which only the most experienced somatotypers can recognize by inspection. Sometimes sums of skinfolds and HWRs can be used for discriminations finer than half-unit component ratings. The pattern of musculoskeletal diameters and girths can provide a measure of dysplasia, especially in mesomorphy.

Equipment for anthropometry

The anthropometric equipment includes a height scale and Broca plane, weight scale, small sliding caliper, a flexible steel or fibreglass tape measure and a skinfold caliper. The small sliding caliper is a modification of a standard anthropometric caliper or engineers' vernier type caliper. For accurate measuring of biepicondylar breadths the caliper branches must extend to 10 cm and the tips to 1 cm in diameter (Carter, 1980*a*). Skinfold calipers should have interjaw pressures of 10 g/mm^2 over the full range of openings. The Harpenden and Holtain calipers are highly recommended. The Slim Guide caliper produces almost identical results. Lange and Lafayette calipers also may be used.

Measurement techniques

Ten anthropometric dimensions are needed to calculate the anthropometric somatotype: height, weight, four skinfolds, two bone breadths and two limb girths.

Height. Take height with the subject standing straight, against an upright wall, touching the wall with heels, buttocks and back. Orient the head in the Frankfort plane (the upper border of the ear opening and the lower border of the eye socket on a horizontal line), and the heels together. Instruct the subject to stretch upward and to take and hold a full breath. Lower the Broca plane until it firmly touches the vertex.

Weight. The subject, wearing minimal clothing, stands in the centre of the scale platform. Record weight to the nearest tenth of a kilogram. A correction is made for clothing so that nude weight is used in subsequent calculations.

Skinfolds. Raise a fold of skin and subcutaneous tissue firmly between thumb and forefinger of the left hand and pulled away from the underlying muscle. Apply the edge of the plates on the caliper branches 1 cm below the fingers of the left hand and allow them to exert their full pressure before reading the thickness of the fold. Take all skinfolds on the right side of the body. The subject stands relaxed, except for the calf skinfold which is taken with the subject seated.

Triceps skinfold. With the subject's arm hanging loosely, raise a fold at the back of the arm at a level halfway along a line connecting the acromion and the olecranon processes.

Subscapular skinfold. Raise the subscapular skinfold adjacent to the inferior angle of the scapula in a direction which is obliquely downwards and outwards at 45°.

Supraspinale skinfold. Raise the fold five to seven centimetres above the anterior superior iliac spine on a line to the anterior axillary border and on a diagonal line going downwards and inwards at 45°. (This skinfold was formerly called suprailiac and anterior suprailiac. The name has been changed to distinguish it from other skinfolds called 'suprailiac', but taken at different locations.)

Medial calf skinfold. Raise a vertical skinfold on the medial side of the leg, at the level of the maximum girth of the calf.

Biepicondylar breadth of the humerus, right. The width between the medial and lateral epicondyles of the humerus, with the shoulder and elbow flexed to 90°. Apply the caliper at an angle approximately bisecting the angle of the elbow. Place firm pressure on the crossbars in order to compress the subcutaneous tissue.

Biepicondylar breadth of the femur, right. Seat the subject with knee bent at a right angle. Measure the greatest distance between the lateral and medial epicondyles of the femur with firm pressure on the crossbars.

Upper arm girth, flexed and tensed, right. The subject flexes the shoulder to 90° and the elbow 45°, clenches the hand and maximally contracts elbow flexors and extensors. Take the measurement at the greatest girth of the arm.

Calf girth, right. The subject stands with feet slightly apart. Place the tape around the calf and measure the maximum circumference.

Read height and girths to the nearest mm, biepicondylar diameters to the nearest 0.5 mm, and skinfolds to the nearest 0.1 mm (Harpenden caliper), or 0.5 mm on other calipers.

Traditionally, for the anthropometric somatotype, the larger of the right and left breadths and girths have been used. When possible this should be done. However, in large surveys it is recommended that all measures (including skinfolds) be taken on the right side. The measurer should mark the sites and repeat the complete sequence. For further calculations, the duplicated measurements should be averaged, including estimation of test–retest reliability of the measurer. For more reliable values relatively inexperienced measurers should take triplicate skinfold measurements and average them.

Reliability of measurements

The advantages of anthropometry are lost unless the measurements are accurate and reliable. It is essential to learn precise measurement techniques and accurate calculations. Although at first sight anthropometric

HEATH–CARTER SOMATOTYPE RATING FORM

NAME A.W. AGE 20yr 5mo SEX: (M) F NO: 573
OCCUPATION Student ETHNIC GROUP Black DATE 10 April, 1980
PROJECT: Track sprinters MEASURED BY: L.C.

Skinfolds mm

		SUM 3 SKINFOLDS (mm)
Triceps	= 6·4	Upper Limit 10.9 14.9 18.9 22.9 26.9 31.2 35.8 40.7 46.2 52.2 58.7 65.7 73.2 81.2 89.7 98.9 108.9 119.7 131.2 143.7 157.2 171.9 187.9 204.0
Subscapular	= 7·1	Mid-point 9.0 13.0 17.0 21.0 25.0 29.0 33.5 38.0 43.5 49.0 55.5 62.0 69.5 77.0 85.5 94.0 104.0 114.0 125.5 137.0 150.5 164.0 180.0 196.0
Supraspinale	= 4·6	Lower Limit 7.0 11.0 15.0 19.0 23.0 27.0 31.3 35.9 40.8 46.3 52.3 58.8 65.8 73.3 81.3 89.8 99.0 109.0 119.8 131.3 143.8 157.3 172.0 188.0
SUM 3 SKINFOLDS = 18·1		× (170.18)/(ht=175) mm (height corrected skinfolds)
Calf = 5·2		

Endomorphy	1	1½	2	2½	3	3½	4	4½	5	5½	6	6½	7	7½	8	8½	9	9½	10	10½	11	11½	12

Height cm	178·3	139.7 143.5 147.3 151.1 154.9 158.8 162.6 166.4 170.2 174.0 177.8 181.6 185.4 189.2 193.0 196.9 200.7 204.5 208.3 212.1 215.9 219.7 223.5 227.3
Humerus width cm	7·20	5.19 5.34 5.49 5.64 5.78 5.93 6.07 6.22 6.37 6.51 6.65 6.80 6.95 7.09 7.24 7.38 7.53 7.67 7.82 7.97 8.11 8.25 8.40 8.55
Femur width cm	9·75	7.41 7.62 7.83 8.04 8.24 8.45 8.66 8.87 9.08 9.28 9.49 9.70 9.91 10.12 10.33 10.53 10.74 10.95 11.16 11.36 11.57 11.78 11.99 12.21
Biceps girth 33·9 -7 33·3		23.7 24.4 25.0 25.7 26.3 27.0 27.7 28.3 29.0 29.7 30.3 31.0 31.6 32.2 32.9 33.6 34.3 35.0 35.6 36.3 37.0 37.6 38.3 39.0
Calf girth 37·6 -C 37·1 -·5		27.7 28.5 29.3 30.1 30.8 31.6 32.4 33.2 33.9 34.7 35.5 36.3 37.1 37.8 38.6 39.4 40.2 41.0 41.7 42.5 43.3 44.1 44.9 45.6

Mesomorphy	½	1	1½	2	2½	3	3½	4	4½	5	5½	6	6½	7	7½	8	8½	9

Weight kg = 69·2	Upper limit 39.65 40.74 41.43 42.13 42.82 43.48 44.18 44.84 45.53 46.23 46.92 47.58 48.25 48.94 49.63 50.33 50.99 51.68
Ht. / ∛Wt. = 43·4	Mid-point and 40.20 41.09 41.79 42.48 43.14 43.84 44.50 45.19 45.89 46.32 47.24 47.94 48.60 49.29 49.99 50.68 51.34
	Lower limit below 39.66 40.75 41.44 42.14 42.83 43.49 44.19 44.85 45.54 46.24 46.93 47.59 48.26 48.95 49.64 50.34 51.00

Ectomorphy	½	1	1½	2	2½	3	3½	4	4½	5	5½	6	6½	7	7½	8	8½	9

	ENDOMORPHY	MESOMORPHY	ECTOMORPHY
Anthropometric Somatotype	1½	5½	3
Anthropometric plus Photoscopic Somatotype			

BY: L.C.

RATER: ....

• Biceps girth in cm corrected for fat by subtracting triceps skinfold value expressed in cm.
▲ Calf girth in cm corrected for fat by subtracting medial calf skinfold value expressed in cm.

11/8 = 1·4   +4·0/5·4

Fig. I.2. The Heath–Carter somatotype rating form showing calculations for determining an anthropometric rating of 1½-5-3.

measurements appear to the beginning investigator to be easy, a high level of skill and reliability requires considerable practice.

Although calculation of the Heath–Carter anthropometric somatotype is an objective procedure, the validity of the rating depends on the reliability of the measurements used. Investigators should report test–retest reliability for measurements. In comparisons of distributions of two independent measures on the same subjects, the means should not differ significantly, and the Pearson product-moment r should be above 0.90. Specifically, height and weight should have test–retest values of $r = 0.98$. Girths and diameters should have r's between 0.92 and 0.98. For skinfolds r's between 0.90 and 0.98 are acceptable.

Calculating the Heath–Carter anthropometric somatotype

There are two ways to calculate the anthropometric somatotype. One is to enter the variables onto a specially constructed rating form, and the other is to enter them into equations derived from the rating form. The use of the rating form will be described first. Fig. I.2 shows an example of calculations using the rating form.

1. Record pertinent identification data in top section of rating form.

Endomorphy rating (steps 2–5)

2. Record the measurements from each of the four skinfolds.

3. Sum the triceps, subscapular, and supraspinale skinfolds; record the sum in the box opposite Sum 3 Skinfolds. Correct for height by multiplying this sum by 170.18/height in cm.

4. Circle the closest value in the Sum 3 Skinfolds scale to the right. The scale reads vertically from low to high in columns and horizontally from left to right in rows. 'Lower limit' and 'upper limit' on the rows provide exact boundaries for each column. These values are circled only when *Sum 3 Skinfolds* are within a few millimeters of the limit. In most cases circle the value in the row 'midpoint'.

5. In the row Endomorphy circle the value directly under the column circled in No. 4 above.

Mesomorphy Rating (steps 6–10)

6. Record height and diameters of humerus and femur in the appropriate boxes. Make the corrections for skinfolds before recording girths of biceps and calf. (Skinfold correction: Convert triceps skinfold to cm by dividing by 10. Subtract converted triceps skinfold from the biceps girth. Convert calf skinfold to cm, subtract from calf girth.)

7. On the height scale directly to the right, circle the height nearest to the measured height of the subject. (Note: Regard the height row as a continuous scale.)

8. For each bone diameter and girth circle the figure nearest the measured value in the appropriate row. (Note: Circle the lower value if the measurement falls midway between two values. This conservative procedure is used because the largest girths and diameters are recorded.)

9. Deal only with columns, not numerical values for the procedures below. Find the average deviation of the circled values for diameters and girths from the circled values in the height column, as follows:

(a) Column deviations to the right of the height column are positive deviations. Deviations to the left are negative deviations.

(b) Calculate the algebraic sum of the deviations (D). Use this formula: Mesomorphy = (D/8) + 4

(c) Round the obtained value of mesomorphy to the nearest one-half rating unit.

10. Circle the closest value for the mesomorphy obtained in No. 9 above. (If the point is exactly midway between two rating points, circle the value closest to 4 on the scale. This conservative regression toward 4 guards against spuriously extreme ratings.)

Ectomorphy Rating (steps 11–14)

11. Record weight (kg).

12. Obtain height divided by cube root of weight (HWR) from a nomograph or by calculation. Record HWR in the appropriate box.

13. Circle the closest value in the HWR scale. (See note in No. 4 above.)

14. Circle the ectomorphy value below the circled HWR.

The anthropometric somatotype

15. Record the circled values for each component in the row 'Anthropometric somatotype'.

16. Investigator signs name to the right of the recorded rating.

The identification data in the upper section of the rating form are somewhat arbitrary. Investigators may change these to suit their purposes. To establish test–retest reliability investigators may repeat measurements and record them at the side of the form. Record averages in the appropriate boxes.

Principles of the calculations

The two principles important in calculating mesomorphy from the rating form are: (1) When the measurements of bone diameters and muscle girths lie to the right of the circled height column, the subject has greater musculoskeletal robustness per unit of height (i.e. higher mesomorphy) than a subject whose values lie to the left of the height column. The average deviation of the circled values for diameters and girths is the best index of average musculoskeletal development relative to height. (2) The scale is

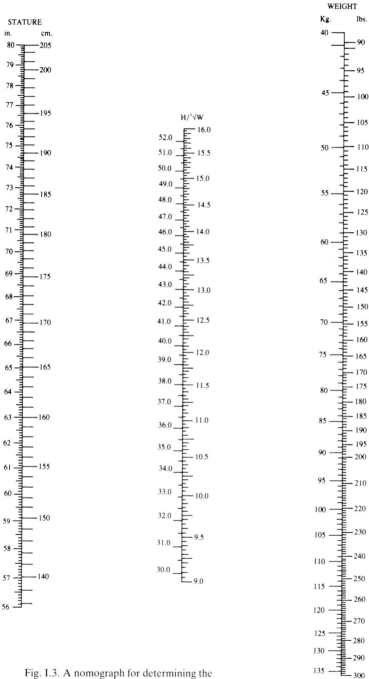

Fig. I.3. A nomograph for determining the height/cube root of weight (HWR), in Imperial and metric units.

constructed so that the subject is rated 4 in mesomorphy when the average deviation falls in the column under the subject's height, or when the four circled values fall in the subject's height column. That is, the average deviation to the right of the height column is added to 4.0 in mesomorphy, and the average deviation to the left is subtracted from 4.0 in mesomorphy.

An alternative way of calculating anthropometric somatotypes uses tables instead of the rating form (Hebbelinck & Ross, 1974*a*).

Height–Weight ratio calculation

The height–weight ratio may be obtained from a nomograph (Sheldon, 1954; Hebbelinck & Ross, 1974*a*; Carter, 1980*a*; Clarys *et al.*, 1970), from tables of height–weight ratios (Parnell, 1958), by slide rule, cube root tables, hand calculator with cube root functions, or calculator using height–weight conversion factors (Ross *et al.*, 1974; Carter, 1980*a*). See Figure I.3 for a nomograph for calculation of height divided by cube root of weight in imperial and metric units.

Limitations of the rating form

Although the rating form provides a simple method of calculating the anthropometric somatotype, especially in the field, it has some limitations. First the mesomorphy scales do not include some of the values found at the low and high ends with unusually small and large subjects. Second, some rounding error may occur in calculating the mesomorphy rating, because the subject's height often is not the same as the column height. If the anthropometric somatotype is regarded as an estimate these are not serious problems. Nevertheless the following procedures described by Carter (1980*a*) can correct these problems.

Equations for a decimalized anthropometric somatotype

1. *Endomorphy.* Endomorphy $= -0.7182 + 0.1451(X) - 0.00068(X^2) + 0.0000014(X^3)$.
   $X =$ sum of triceps, subscapular and supraspinale skinfolds. For height-corrected endomorphy, multiply X by 170.18/height in cm.
2. *Mesomorphy.* Mesomorphy $= [(0.858 \times \text{humerus breadth}) + (0.601 \times \text{femur breadth}) + (0.188 \times \text{corrected arm girth}) + (0.161 \times \text{corrected calf girth})] - (\text{height} \times 0.131) + 4.50$.
3. *Ectomorphy.* Ectomorphy $= \text{HWR} \times 0.732 - 28.58$.

If HWR is less than 40.75 but more than 38.25, Ectomorphy $= \text{HWR} \times 0.463 - 17.63$. If HWR is equal to or less than 38.25 give a rating of 0.1. HWR = height/cube root of weight.

These equations, derived from Heath & Carter (1967), use metric units. The equation for endomorphy is a third degree polynomial. The equations for mesomorphy and ectomorphy (when the HWR is greater than 40.74) are linear. When the HWR is below 40.75 a different equation is used. If the calculation for any component is zero or negative, a value of 0.1 is assigned as the component rating, because by definition ratings cannot be zero or negative. The photoscopic rating would be one-half. If such low values occur the raw data should be checked. Values less than 1.0 are highly unlikely to occur for endomorphy and mesomorphy, but are not unusual for ectomorphy. The component ratings should be rounded to the nearest tenth of a unit, or nearest one-half unit depending on their subsequent use.

III. *The Heath–Carter anthropometric plus photoscopic somatotype*

The anthropometric rating gives the best objective estimate of the somatotype. While the HWR table provides a basis, the final anthropometric plus photoscopic rating includes some subjective judgment and experience. Two somatotypers should achieve closer agreement using anthropometric plus photoscopic ratings than using the photoscopic ratings alone.

General procedures

The following are needed for the anthropometric plus photoscopic somatotype rating:

1. Standard somatotype photographs.
2. The table of distribution of somatotypes according to HWR (Table I.1).
3. The photoscopic rating form (Fig. I.1).
4. The Heath–Carter somatotype rating form (Fig. I.2).

The following general procedures are followed in assigning a somatotype rating:

1. A somatotype photograph is a record of morphological characteristics, including those sampled by anthropometric measurements.

2. The objective is to reconcile the criteria of photoscopic inspection, the anthropometric somatotype and the alternative possible somatotypes for a given HWR.

3. The number of steps and the length of time required for deciding on an appropriate somatotype rating vary from subject to subject and from rater to rater.

The HWRs and inspection readily identify extreme dominance of endomorphy and ectomorphy. Total skinfolds identify the most likely rating of endomorphy. Inspection, relatively low total skinfolds, and HWR indentify dominant mesomorphy. Midrange somatotypes, with their indeterminate HWRs present the greatest difficulties. In general, as a first step, total skinfolds are used to identify the endomorphy ratings. The next step is to compare inspectional impressions of endomorphy and mesomorphy with the anthropometric estimates. Ordinarily, no more than one-half unit up or down in either component is at issue. An example: if the HWR is 13.00 (42.98) we ask, 'is this $3\frac{1}{2}$-4-$2\frac{1}{2}$, 4-$3\frac{1}{2}$-$2\frac{1}{2}$, or 4-4-3?' Total skinfolds are 33.8 mm, almost dead centre for $3\frac{1}{2}$ in endomorphy. By inspection, 3 seems a little high for ectomorphy. The anthropometric values for mesomorphy fall between $3\frac{1}{2}$ and 4. By inspection, mesomorphy seems closer to 4 than $3\frac{1}{2}$. Therefore $3\frac{1}{2}$-4-$2\frac{1}{2}$ is a satisfactory rating.

Often there are no differences between the anthropometric somatotype and the combined rating. But for subjects high in endomorphy the difference may be one unit or larger.

Step-by-step

An experienced somatotyper should have no difficulties with the general procedures outlined above, particularly one who is accustomed to using phenotypic (present) somatotype ratings, regardless of the rating system used. The following steps, incorporating several built-in checks, are suggested:

1. (a) Arrange a series of somatotype photographs and their anthropometric somatotype rating forms in groups ranked by HWR, that is, by anthropometric ratings of ectomorphy.

(b) When any one-half unit category is large, subdivide the group in three parts–upper limit, midpoint and lower limit.

2. Beginning with the group highest in ectomorphy, arrange them from low to high according to the anthropometric ratings in endomorphy. At this point work only with three or four subjects. If all subjects are equal in endomorphy, or if there is a greater range in mesomorphy, rank them from low to high in anthropometric mesomorphy ratings.

3. Record on a worksheet the identification number and the anthropometric rating of the first subject–usually the one with the highest HWR. In a column headed combinations, if there are half-unit ratings in the anthropometric somatotype, record the possible combinations derivable from that rating. That is, for a rating of 2-$4\frac{1}{2}$-4, the two possible combinations are 2-4-4 and 2-5-4. If there are three half-unit values in the rating, as in $1\frac{1}{2}$-$4\frac{1}{2}$-$3\frac{1}{2}$, there are six possible combinations: 1-5-3, 2-5-3, 2-4-4, 2-4-3, 1-4-4 and 1-5-4. (It is not necessary to raise or lower all three components.)

4. For example, assume that the anthropometric somatotype of the first subject is $1\frac{1}{2}$-3-$6\frac{1}{2}$ and the HWR is 14.42 (47.67). See Table I.1 for somatotypes with HWR 14.40 (47.67). Note that these ratings have the same relative component dominance as the anthropometric somatotype. First, record the anthropometric somatotype. Second, record those somatotypes (HWR = 14.40) which differ from $1\frac{1}{2}$-3-$6\frac{1}{2}$ by no more than one-half unit in any component (1-3-6 and 2-3-7) Third, record those somatotypes which

differ from $1\frac{1}{2}$-3-$6\frac{1}{2}$ by one rating unit (3-1-6, 2-2-6 and 3- 2-7). Fourth, record those somatotypes in the two adjacent rows (HWR = 14.20 (46.95)) and HWR = 14.60 (48.27) which differ from $1\frac{1}{2}$-3-$6\frac{1}{2}$ by no more than one-half unit (1-3-7 at 14.60, and 2-3-6 at 14.20).

5. Reconcile photoscopic impression (previously recorded on Fig. I.1) with ratings possible according to HWR (14.40 (47.67)) and with the anthropometric data. If there appear to be no discrepancies among these ratings, record in the column Final Rating the somatotype $1\frac{1}{2}$-3-$6\frac{1}{2}$. However, if the total skinfold value is between $1\frac{1}{2}$ and 2 (e.g. 18.9 mm), consider a possible rating of 2 in endomorphy. Apply the same scrutiny to mesomorphy and ectomorphy. There should be little difficulty if the HWR and the anthropometric somatotype are in close agreement. In practice, the 'final somatotype' rarely differs from the anthropometric somatotype by more than one-half unit in any one component. Note that the half-unit ratings are interpolations of whole rating units in the same or adjacent rows in Table I.1.

6. Note that reconciling differences among anthropometric, HWR and photoscopic somatotypes requires careful checking of all data. First, check HWR, which may not have been accurately calculated and/or recorded. Consider the possibility that there are errors in recording height or weight. Note the possibility that skinfold values do not always accurately reflect distribution of body fat. Some women have greater fat deposition around the buttocks and thighs (an area not included in the skinfold measurements) than on the rest of the body (a dysplasia in endomorphy). Sometimes a relatively large calf skinfold value indicates that the other three skinfold values underestimate the rating in endomorphy for a given subject. Similarly, mesomorphy can be overestimated in subjects with hypertrophied upper limb and shoulder girdle musculature. Mesomorphy may be underestimated in subjects with highly developed musculature in the trunk and thigh, areas that are not measured. Photoscopic inspection is important in recognizing discrepancies in regional somatotypes. When in doubt about a 'final rating', it is best to move on to other photographs and to return later to reconsider the troublesome reconciliations.

7. Refer to the three or four photographs in the group being rated while reconciling the data for a photograph under consideration. Comparisons of somatotypes with similar HWR's and ratings in ectomorphy help to keep in perspective minor differences among somatotypes.

8. Continue the rating process for each subject. Occasionally, cross check with photographs already rated and glance at those to be rated. Note that as ectomorphy decreases, endomorphy and mesomorphy increase. These systematic changes help the rater to maintain an objective pattern. The need for cross checking for possible somatotypes lessens with the downward progression through the range of ectomorphy. When several subjects have the same somatotype, one can make ratings much more quickly. Note that

the higher the rating in endomorphy, the more difficult is the accurate rating of mesomorphy and the more important is Table I.1 and the anthropometric data. Remember that subcutaneous fat tends to increase bone breadths and limb girths, thus spuriously increasing mesomorphy.

One unit differences between anthropometric and photoscopic component ratings are rare. However, high ratings in endomorphy ($5\frac{1}{2}$, 6 and higher) sometimes differ by one unit or more.

9. When the series is completed, some somatotype ratings will remain questionable. Lay out the whole series of photographs in groups of 25 to 30, in ranked order of mesomorphy according to your tentative rating. Check the series for discrepancies in ranking. Recheck the data. Make a 'final rating'.

10. Repeat the steps in No. 9 for endomorphy. Endomorphy may be checked first, according to personal preference.

11. If the ratings given in Steps 1 through 10 are unsatisfactory, repeat the steps starting with a different component, i.e. endomorphy or mesomorphy. In most series the greatest variability is in ectomorphy. If this is not so, it may be preferable to start with either of the other two components. The HWR is probably the most objective criterion in the rating procedure. It seems logical to proceed from the most objective to the least objective criterion to arrive at a satisfactory somatotype rating.

Rating bias

There are several ways to check for bias in the anthropometric measurements and in the choice of ratings in Table I.1. The rater should look for (a) a trend toward low endomorphy ratings in the anthropometric somatotype, or (b) for a tendency to choose ratings low in endomorphy, or (c) to choose ratings in the lower row of the HWR table. To check for possible biases, calculate the mean HWR, the mean anthropometric somatotype, and the mean 'final rating'. Compare the possible somatotypes for the mean HWR with the two mean somatotype ratings. If the anthropometric and the 'final somatotype' have similar means, it may indicate that rating changes from the anthropometric ratings averaged out, or that the rater made few photoscopic changes.

If the anthropometric mean and the 'final' mean somatotype differ from the mean HWR, there may be some bias in one of the processes. For example, if the mean HWR is 13.20 (43.64), the mean anthropometric somatotype is 2-5-3, and the mean 'final rating' is 3-4-3, it would appear that the rater was biased toward increasing endomorphy and decreasing mesomorphy. The source of the bias probably can be established only if an equally experienced rater makes independent measurements and ratings. It would be easier to identify a discrepancy if the HWR were 13.20 (43.64), the anthropometric somatotype were 5-3-3, and the 'final rating' were 4-3-3.

Such a 'final rating' would indicate that the rater felt that the skinfold values were overestimated in a significant number of cases. When the HWR is 13.00 (42.98), the anthropometric rating is 5-3-3 and the 'final rating' is 4-3-3, the bias appears to lie with the rater rather than with the anthropometry.

Frequency polygons for the distribution of component ratings provide a quick check on bias. They show if the bias is distributed throughout the range of component ratings or if it is different in different samples. They also show if there is a tendency to choose whole number ratings over half-unit ratings. These checking procedures are recommended for both experienced and inexperienced raters.

In females with high endomorphy and moderate mesomorphy, raters tend to overestimate mesomorphy because of undercorrection for fat in the limb girth. Fat over the humerus and femur epicondyles spuriously increases measurements. An anthropometric somatotype of 7-6-$\frac{1}{2}$ with a HWR of 11.70 (38.68) is suspect, because female 6s in mesomorphy are exceedingly rare. Table I.1 and inspection of the photograph may suggest that a rating of 7-4-$\frac{1}{2}$ is more appropriate. Sometimes the anthropometric rating of endomorphy may be incorrect because of measurement errors in high skinfolds.

Subjects with low skinfold values (2 or less in endomorphy) and moderately high ectomorphy sometimes are given spuriously low mesomorphy ratings. A subject may have a HWR of 13.40 (44.30) together with an anthropometric rating of 2-4-4. Usually the photograph shows the subject to be more mesomorphic than a 4. Inspection shows greater mesomorphy in the trunk and thighs than is reflected in the anthropometric somatotype. A rating of 2-5-4 or 2-4$\frac{1}{2}$-4 is probably more appropriate.

Occasionally the data suggest an anthropometric somatotype that is incompatible with both the HWR and the photoscopic impression. For example, suppose the photoscopic impression is that the somatotype is 3-4-4; the HWR is 13.40 (44.30); and the anthropometric somatotype is 2-3-3 (for which the HWR should be 13.60). In such a case it is usually best to give the rating 3-4-4, which is compatible with the photoscopic impression and is confirmed by the HWR.

Other considerations

Although skinfold values are an excellent estimate of body fatness, factors such as fat patterns, compressibility, skin thickness, internal versus external fat stores, ethnicity, sex and age affect the relationship (Brožek, 1965; Martin *et al.*, 1985.)

Similarly, differences in skeletal density may affect second component ratings. According to Brožek (1965); 'Skeletal density shows a significant decrease during late maturity and old age, as well as significant race (higher in Blacks, both male and female, than in Whites) and sex differences (higher values in males). Consequently, the whole-body density of older individuals appears to be lower, in part as a result of skeletal demineralization, and does not simply reflect increases in fat content,' (p. 9).

Raters should consider the possible influence of these factors on both the anthropometric and photoscopic ratings, despite the fact that quantification of the variations in relation to somatotypes mentioned above has not been worked out.

Sometimes, the available data include the photograph, height, weight and anthropometric measures for calculating endomorphy alone, or for calculating mesomorphy alone. All available information should be used to help determine the final rating.

Somatotyping children

With minor modifications, the Heath–Carter somatotype method can be applied to children. The needed modifications are: (a) an additional HWR table that includes somatotypes more commonly found in children than in adults (see Table 4.1, Chapter 4); (b) extrapolations downward of the scales used on the rating form for the calculation of mesomorphy from anthropometry; and (c) a height correction applied to total skinfolds in calculating endomorphy from anthropometry.

When the methods of obtaining the somatotype are modified in these respects, the anthropometric method (most useful in surveys) can be applied to children from age six years. A criterion rating from a photograph adds validity to the somatotype rating. It is substantially more difficult to rate somatotypes of subjects under age six years. Such ratings may be of less value than those of subjects at later ages. At ages of six and younger, when the validity of the anthropometric somatotype is uncertain, ratings from photographs are essential.

Photoscopic ratings of children

The somatotype–HWR table modified for children is an essential adjunct to Table I.1. Although Heath has had wide experience in somatotyping children (e.g. Berkeley Growth Study, Gesell Institute Studies, Medford Growth Study) absence of published photographs presented a major problem in photoscopic ratings of children. There are two books containing somatotype photographs of children (Petersen, 1967; Tanner & Whitehouse, 1982), both of which allegedly used Sheldon's rating methods.

Petersen's *Atlas for Somatotyping Children* contains 560 photographs of children. Although Sheldon published neither rating criteria nor somatotype photographs of children, Petersen states that the somatotype ratings are based on Sheldon's criteria. Because he published no data for each child (not even age, height and weight) his atlas is of no practical use as a rating reference.

In their *Atlas of Children's Growth,* Tanner & Whitehouse (1982) supplied useful data with each photograph of 89 subjects followed from early childhood to early adulthood. For each subject Tanner used the Sheldon criteria and made one somatotype rating only, at approximately age 18. Heath has re-rated these photographs at all ages, using the Heath–Carter photoscopic criteria. A list of these ratings is given in Appendix III. These, together with the photographs, provide an excellent reference source.

Anthropometric somatotypes for children

The extended scales are required for calculating mesomorphy in children and other small subjects. These scales may be added to those on the rating form, or the rating may be calculated using the equation for mesomorphy (Carter, 1980*a*). In the anthropometric somatotype, an adjustment for size is made in mesomorphy and ectomorphy. The original description of the method did not contain a similar adjustment for endomorphy. Such an adjustment is logical and gives a weighting to the total skinfolds in relation to the height of the subject. This correction should be applied routinely to children.

Hebbelinck *et al.* (1973) recommended an adjustment based on a reference height standard of 170.18 cm, a combined universal mean for adult men and women. Before determining the rating for endomorphy the sum of the three skinfolds is multiplied by the reference height (170.18/height in cm), to adjust for body size.

For example:

Given: Sum of 3 skinfolds = 33.5 mm (i.e. endomorphy without correction = $3\frac{1}{2}$ on the rating form)

Height = 151.0 cm

Adjusted skinfolds = 33.5 × (170.18/151.0) = 33.5 × 1.13 = 38.8 mm

Adjusted endomorphy = 4 (an increase of a half-unit).

Using the regression equation for calculating endomorphy, the ratings are 3.4 versus 3.9 (an increase of 0.5 units) for the sums of three skinfolds of 33.5 mm and 37.8 mm, respectively.

The adjustment for endomorphy should be used when anthropometric ratings only are used. The somatotyper should report both values and should indicate which is used in the analysis. When anthropometric plus photoscopic ratings are made, the adjustment usually is logical from inspection of the photograph and comparison with alternative somatotypes in the HWR table.

The height correction for endomorphy, which has been used in several

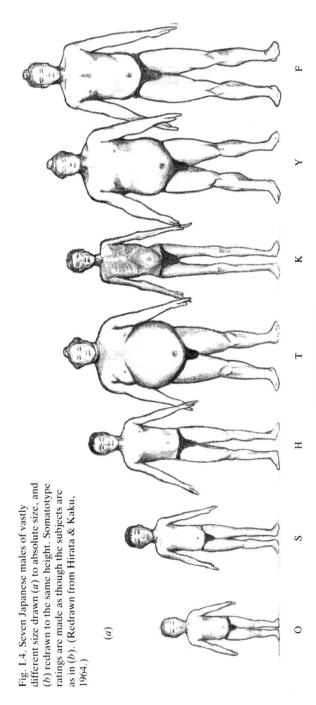

Fig. 1.4. Seven Japanese males of vastly different size drawn (a) to absolute size, and (b) redrawn to the same height. Somatotype ratings are made as though the subjects are as in (b). (Redrawn from Hirata & Kaku, 1964.)

(a)

O   S   H   T   K   Y   F

ABSOLUTE HEIGHT

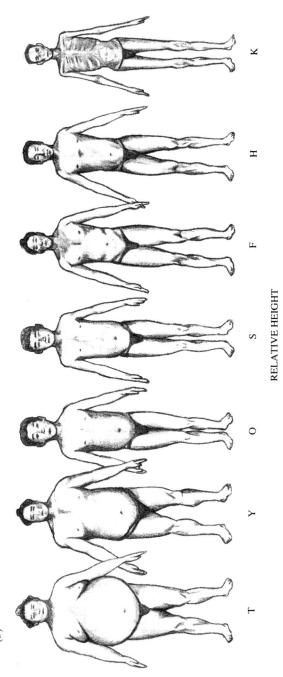

(b)

RELATIVE HEIGHT

T   Y   O   S   F   H   K

studies of children, seems to be sound. These studies examined the validity of the correction (see Chapter 4 for their results). Possibly other corrections of skinfolds for body size are better than that described above. A promising possibility is body surface area as the size adjustment with skinfolds, the method Katch *et al.* (1979) used.

Examples of somatotypes

As noted earlier, the somatotype describes the shape and composition of the body, *not* its size. It is size-dissociated. Ratings should be made as though everyone were adjusted to the same stature. This point is illustrated with drawings from photographs and data for seven Japanese adult males of greatly different sizes in Hirata & Kaku (1964). (See Fig. I.4 *a,b.*) In the upper figure the subjects are ordered from left to right by increasing absolute height. In the lower figure they are drawn to the same height and ordered from left to right from the most ponderous (lowest HWR) to the most linear (highest HWR). See Table I.3 for their height, weight, HWR and photoscopic somatotypes (these were rated from the photographs, not from the drawings.) The somatotypes show a wide range in all three components. The three heaviest subjects (T, Y, F), sumo wrestlers, are not only high in mesomorphy, but range from 3 to 10 in endomorphy. Subject K is clearly the most ectomorphic, with almost no subcutaneous fat, attenuated muscles and small bones for his height. Subjects O, S and H have similar mesomorphy ratings with progressively decreasing endomorphy.

Examples of standard somatotype photographs, with front, side and back views, and ratings are shown in Fig. I.5(*a*)-(*y*). The photographs illustrate some of the variation in adult somatotypes from different samples.

The somatotypes of these examples are plotted in Fig. I.6. It should be clear from the above examples that continuous variation in each component produces many combinations, both between and beyond those illustrated.

Table I.3. *Height, weight, height/weight $\sqrt[3]{}$ ratio, and somatotype of seven Japanese males of different absolute sizes*

Subjects	T	Y	O	S	F	H	K
Height (cm)	174.0	182.0	101.3	119.8	207.0	156.0	174.0
Weight (kg)	166.0	139.0	18.3	25.3	127.0	52.0	42.0
HWR	31.67	35.14	38.44	40.81	41.19	41.80	50.06
Somatotypes	10-7-$\frac{1}{2}$	6$\frac{1}{2}$-7$\frac{1}{2}$-$\frac{1}{2}$	5-5-1	4$\frac{1}{2}$-4$\frac{1}{2}$-1$\frac{1}{2}$	3-6$\frac{1}{2}$-2	3$\frac{1}{2}$-4$\frac{1}{2}$-2	1-1-7

Data from Hirata & Kaku (1964), for the subjects shown in Fig. I.4(*a*), (*b*).

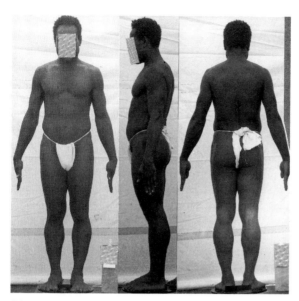

(*a*). 35 yr, 162.1 cm, 72.7 kg, 11.75 (38.83), 1½-9-0.5.

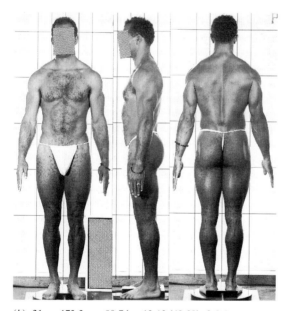

(*b*). 31 yr, 173.2 cm, 80.7 kg, 12.13 (40.09), 2-8-1.

Fig. I.5. Examples of somatotypes. The captions for each subject include age, height, weight, HWR (in Imperial and metric units), and the somatotype. Because the photographs are from different studies the sizes of the subjects are not directly comparable.

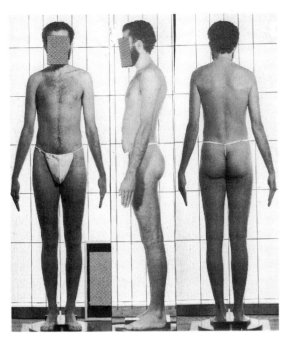

(*c*). 20 yr, 180.4 cm, 64.5 kg, 13.61 (44.99), 3-3½-5½.

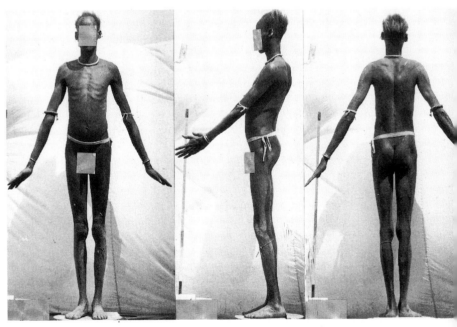

(*d*). No age, 182.5 cm, 51.4 kg, 14.85 (49.09), 1½-2-8.

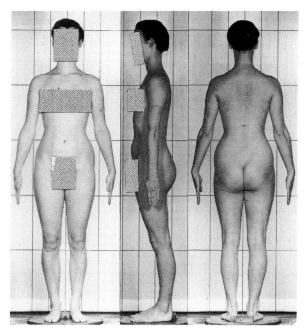

(*e*). 18 yr, 165.0 cm, 57.3 kg, 12.94 (42.80), $4\frac{1}{2}$-$3\frac{1}{2}$-3.

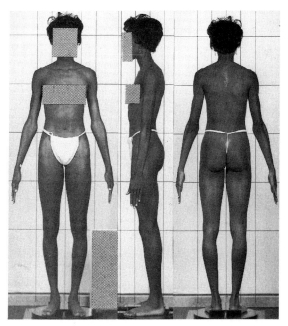

(*f*). 20 yr, 176.1 cm, 55.4 kg, 13.98 (46.20), $1\frac{1}{2}$-$3\frac{1}{2}$-$5\frac{1}{2}$.

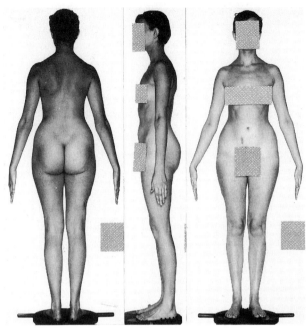

(*g*). 20 yr, 168.2 cm, 56.5 kg, 13.26 (43.84), 5-2½-4.

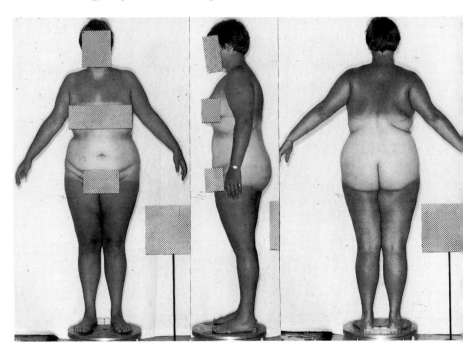

(*h*). 17 yr, 161.7 cm, 86.2 kg, 11.08 (36.61), 10-5½-½.

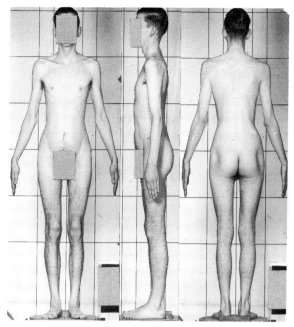

(*i*). 23 yr, 189.1 cm, 63.0 kg, 14.38 (47.53), 2-2-6.

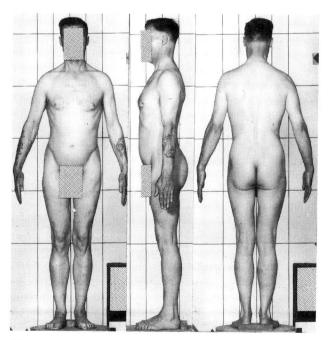

(*j*). 56 yr, 167.5 cm, 60.8 kg, 12.89 (42.60), $3\frac{1}{2}$-5-$2\frac{1}{2}$.

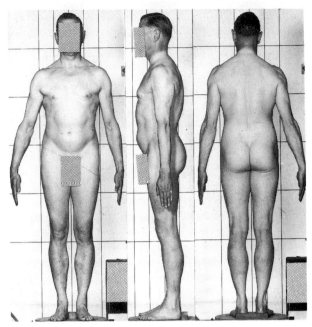

(*k*). 53 yr, 180.7 cm, 84.0 kg, 12.48 (41.27), 4-5½-2.

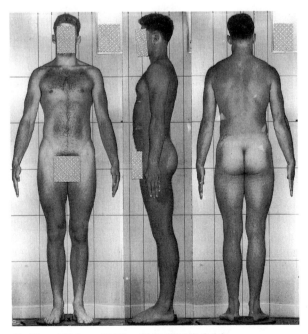

(*l*). 21 yr, 194.7 cm, 98.7 kg, 12.74 (42.14), 3-6-2.

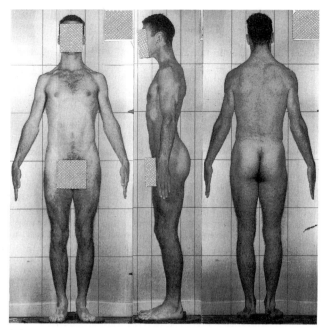

(*m*). 23 yr, 186.7 cm, 78.9 kg, 13.17 (43.54), 1-6-3.

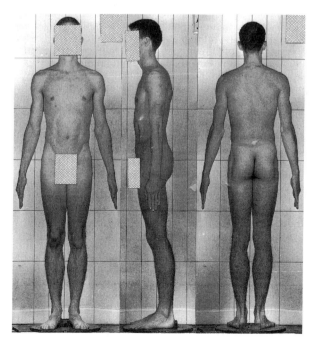

(*n*). 19 yr, 181.4 cm, 63.7 kg, 13.75 (45.43), $1\frac{1}{2}$-$4\frac{1}{2}$-$4\frac{1}{2}$.

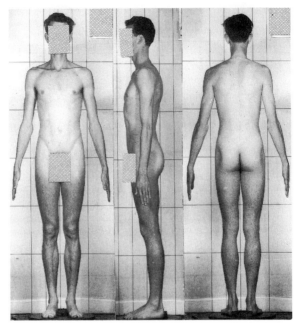

(*o*). 22 yr, 185.7 cm, 63.7 kg, 14.07 (46.50), 1-4-5.

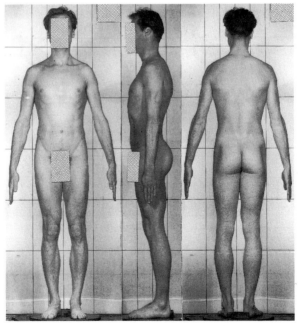

(*p*). 27 yr, 177.7 cm, 68.3 kg, 13.15 (43.47), 2-5-3.

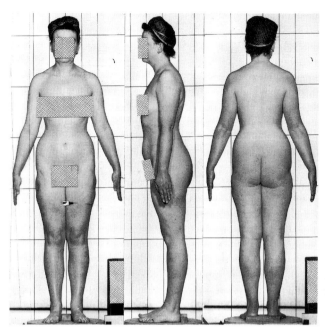

(*q*). 32 yr, 166.5 cm, 69.5 kg, 12.25 (40.50), 5½-4-1.

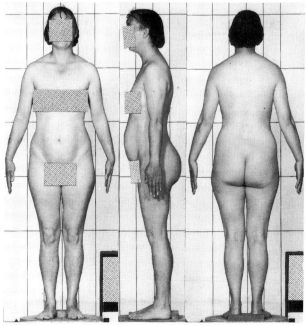

(*r*). 47 yr, 161.0 cm, 63.7 kg, 12.20 (40.32), 6-4-1.

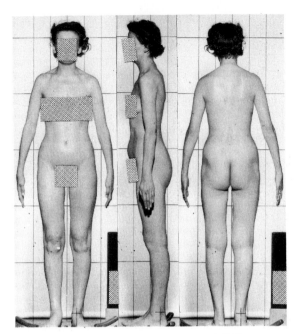

(s). 27 yr, 163.7 cm, 51.3 kg, 13.33 (44.06), 4½-2½-3½.

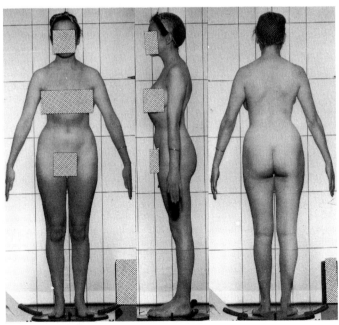

(t). 28 yr, 162.3 cm, 53.8 kg, 13.01 (43.00), 5-3-3.

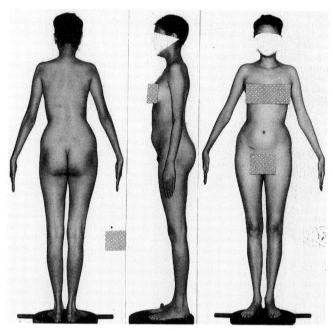

(u). 20 yr, 163.0 cm, 47.6 kg, 13.60 (44.98), 4½-1½-5.

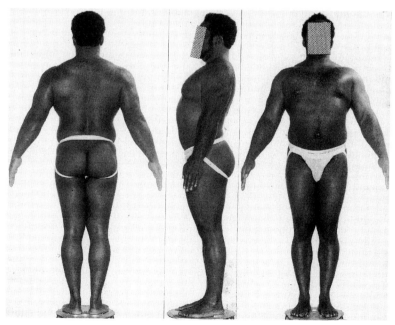

(v). 31 yr, 171.5 cm, 100.5 kg, 11.16 (36.89), 4-9-½.

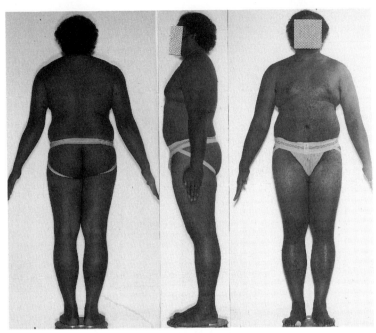

(*w*). 22 yr, 196.3 cm, 148.9 kg, 11.21 (37.06), 6-8-1.

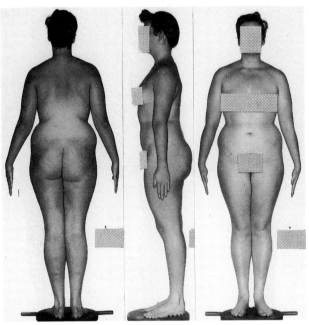

(*x*). 19 yr, 172.8 cm, 86.0 kg, 11.84 (39.15), $7\frac{1}{2}$-$4\frac{1}{2}$-$1\frac{1}{2}$.

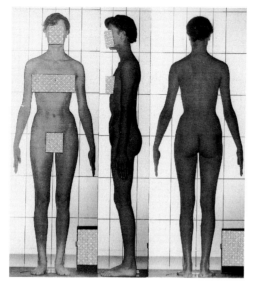

(*y*). 18 yr, 178.2 cm, 56.0 kg, 14.09 (46.58), 3½-2-6.

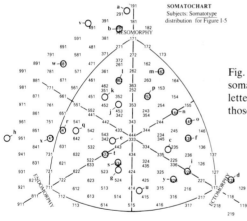

Fig. I.6. Somatotype distribution for the somatotypes of subjects from Fig. I.5. The letters next to each circle correspond with those on the photographs in I.5.

Summary

The Heath–Carter somatotype method is an objective method of evaluating the overall physiques of adults and children. It rates the present somatotype by anthropometric and/or photoscopic means. It can be amended for interpolation and extrapolation to record the whole range of variation in human physiques.

Appendix II
Analysis

Carter *et al.* (1983) devised new approaches to the analysis of the three-number somatotypes, which apply two- and three-dimensional concepts as well as traditional methods.

In keeping with definitions and meanings in Chapter 1 and Appendix I, the word *somatotype* refers to the three-number rating, and the words *component* and *components* refer to endomorphy, mesomorphy and ectomorphy. The three numerals of the somatotype describe a physique as a whole and the contribution of the three components. The relative magnitude of component ratings in a somatotype convey unique meaning in comparing somatotypes with one another, but separate treatment of components blurs the concept of the somatotype and often produces meaningless interpretation. For example, although the 5 in endomorphy in the somatotypes 5-7-1 and 5-2-3 is quantitatively the same, by itself it tells us nothing about the striking differences between the two somatotypes. Three-number somatotype ratings also have the advantage of amenability to statistical analysis and two-dimensional plotting on standard somatocharts.

The somatochart

History
Understanding the somatochart and two- and three-dimensional plotting of somatotypes is basic to statistical analysis of somatotype data. The word *somatochart*, an abbreviation for somatotype chart, presents diagrammatically the relationships among somatotypes. Parnell (1958) was first to use the word somatochart. Although Sheldon (1940) was the first to use a triangle-shaped figure to display somatotypes, he did not use the word somatochart. He regarded the three components as variables represented by orthogonal coordinates. He thought of the somatotypes as distributed in three-dimensional space, and said he drew the triangular figure (Fig. II.1) as a representation of the somatotype distribution 'from a point along a line almost perpendicular to the surface approximated by the somatotypes.' In this figure he allotted to each somatotype an area roughly proportional to its

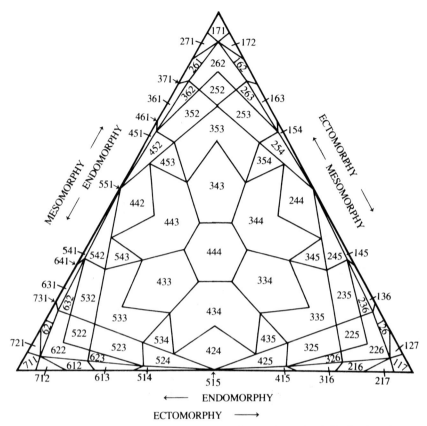

Fig. II.1. The first 'triangle' used to show the two-dimensional distribution of somatotypes. The area for each somatotype is roughly proportional to the incidence of the somatotype in a population of 4000 cases. (From Sheldon, 1940, p. 118.)

incidence in a sample of 4000 subjects, and also 'arbitrarily equalized' for 19 groups, or 'natural families' of somatotypes (p. 119).

Sheldon & Tucker (1938), in an unpublished manuscript, drew a similar, but inverted triangular figure, with the 1-7-1 placed at the bottom. They used the words *pyknosomia*, *somatosomia*, and *leptosomia* instead of endomorphy, mesomorphy and ectomorphy.

Sheldon (1949) continued to emphasize the three-dimensional distribution of somatotypes and submitted a revised 'schematic two-dimensional projection of the theoretical spatial relationships among the known somatotypes' (p. 16), which he called a 'cluster chart' (Fig. II.2). He neither gave a reason for its use nor mentioned its origin. This arc-sided triangle, with some modifications, is the basic model for subsequent somatocharts.

The similarity of the arc-sided somatotype triangle to the Reuleaux

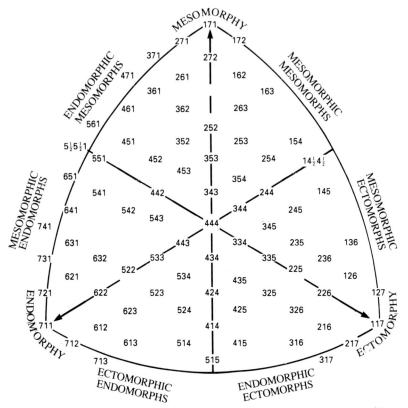

Fig. II.2. Sheldon's 'cluster chart' which is a revision of the first 'triangle'. (From Sheldon, 1949, p. 16.)

triangle suggests its possible, though unverified, origin. Franz Reuleaux (1829–1905), an engineer and mathematician who taught at the Royal Technical High School in Berlin, discovered that the arc-sided triangle is the simplest non-circular curve of constant width. Although earlier mathematicians were familiar with the curve itself, Reuleaux was the first to demonstrate its constant width properties. The 'curved triangle', as Reuleaux called it, has a constant width equal to the side of the interior triangle. The Reuleaux triangle is defined as the curve of constant width that has the smallest area for a given width. The area is (pi − square root of 3) times width squared. The corners are angles of 120 degrees, the sharpest possible on such a curve. There are many such curves (of constant width), and one of these makes it possible to drill square holes. A recently developed rotary automobile engine used the Reuleaux triangle in the design of its piston.

The lack of instructions for plotting intermediate or possible new somatotypes on the 'new' Sheldon (1949) triangle created a problem. Dupertuis &

Emanuel (1956, p. 24) reported that according to Hooton and his colleagues, 'Neither the contractors nor their expert draftsmen were able to figure out the system on which this "schematic two-dimensional projection" was drawn up.'

The Sheldon laboratory remedied this deficiency in some degree by superimposing a coordinate grid over the triangle. Although many copies of this revised figure were circulated, Sheldon did not publish it. There was in this figure a North (or Y) axis scale of 1 to 25 units, and an E (or X) axis scale of 1 to 13 units. Thus the somatotype 4-4-4 had coordinates of 13 for North and 7 for East. The following equations were used to find the coordinates of a given somatotype:

$$\text{North coordinate} = \text{N13} + 2\,(\text{mesomorphy}) - \text{endomorphy} - \text{ectomorphy}$$
$$\text{East coordinate} = \text{E7} - \text{endomorphy} + \text{ectomorphy}$$

This figure was published by Villanueva (1979). However, it was used extensively (though not published) to hand plot the somatotypes of the athletes measured at the 1968 Olympics in Mexico City, under the direction of Alfonso L. De Garay and Johanna Faulhaber.

Modifications

In the 1972 edition of *The Heath–Carter Somatotype Method* (Carter, 1980a) the author reoriented the somatochart by superimposing and scaling the X,Y coordinates so that the somatotype 4-4-4 has coordinates of 0,0 (Fig. II.3). This allowed simplification of the previous equations as follows:

$$X \text{ coordinate} = \text{ectomorphy} - \text{endomorphy}$$
$$Y \text{ coordinate} = 2\,(\text{mesomorphy}) - (\text{endomorphy} + \text{ectomorphy})$$

There have been several modifications of the original Sheldon triangle. Duquet (1980) cited Furst's and Stohr's 1941 publications of alternative methods of plotting three scales on a triangle. However, these were not used to plot somatotypes. In their studies Cureton & Hunsicker (1941) adapted Sheldon's triangle by reversing the endomorphy and ectomorphy corners. No one copied their idiosyncrasy.

Stephens & Taylor (1962) used a simple graticule for two-dimensional plotting of somatotypes. They observed that there was some distortion in somatotype distributions plotted on a flat diagram, and suggested that a satisfactory plot was possible with a system analogous to a map of a portion of the earth's surface. According to this system the projection is regarded as a plane parallel to the plane containing the points of the somatotypes 1-7-1,

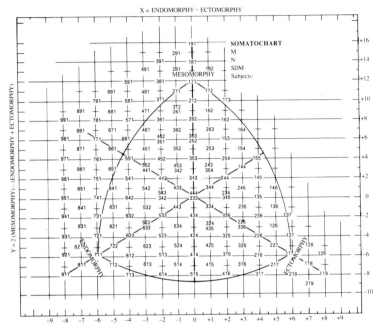

Fig. II.3. Somatochart and grid with equations for plotting individual somatotypes. By using the X, Y coordinates, somatoplots can be interpolated and extrapolated between and beyond those on the somatochart. (From Carter, 1980*a*.)

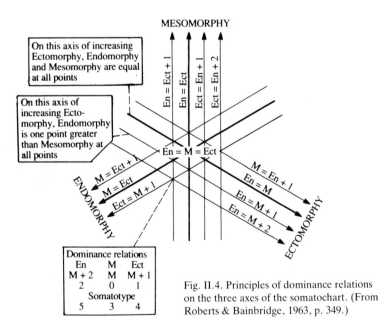

Fig. II.4. Principles of dominance relations on the three axes of the somatochart. (From Roberts & Bainbridge, 1963, p. 349.)

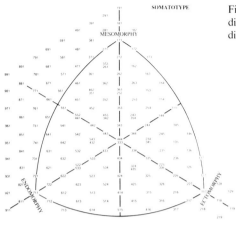

Fig. II.5. A somatochart for displaying the distribution of somatotypes in two dimensions.

7-1-1, and 1-1-7. The points of all somatotypes are plotted as perpendiculars dropped from their point in space onto this plane. Thus the somatochart shows the points as they would be seen from an infinite distance on the line through points 0-0-0 and 4-4-4.

In their study of Nilotic physiques, when Roberts & Bainbridge (1963) needed to plot new points outside the traditional somatochart, they developed principles based on dominance relationships (Fig. II.4) for construction of the somatocharts. They showed how the dominance relationships could explain mathematically the occurrence of more than one somatotype at the same point on the somatochart – in other words, the three-dimensionality of the somatotype distribution.

When Carter constructed a revised somatochart (Fig. II.5), he used the principles of Stephens & Taylor and of Roberts & Bainbridge, with the X, Y coordinate system shown in Figure II.3. (See Carter & Phillips, 1969, and Carter, 1970.)

The somatochart revised by Carter provides for extrapolation of somatotype plots outside the confines of Sheldon's seven-point scales and the area of the 'arc-sided triangle'. Although the printed somatochart shows somatotypes with ratings as high as 9 in a component, plotting is not limited to the somatotypes printed on the chart. Using the X, Y coordinates any somatotype can be plotted. In samples including extreme somatotypes extrapolations are necessary. This system also provides for analysis of somatotypes plotted in two dimensions and for plotting by computer graphics.

Plots in two and three dimensions

Carter *et al.* (1983) discussed the relationships between two- and three-dimensional plots of somatotypes. Ross & Wilson (1973) used whole

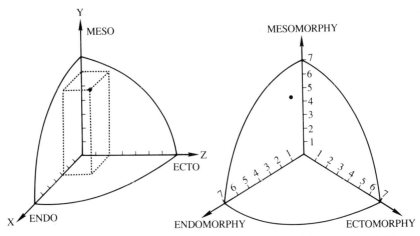

Fig. II.6. A three- and two-dimensional representation of the somatotype 3-6-2.
(Redrawn after Duquet & Hebbelinck, 1977.)

ratings and plotted the projected positions of the somatotypes on a two-dimensional (planar) grid system. Duquet & Hebbelinck (1977) plotted the positions in a three-dimensional system.

A point in three-dimensional space called a *somatopoint* is the best conceptual representation of the three-component somatotype. Thus a triad of X, Y, and Z coordinates can represent a somatotype. The X coordinate is the first component (endomorphy), Y is the second (mesomorphy) and Z is the third (ectomorphy). The scales on the coordinate axes are component units, with the hypothetical somatotype of 0-0-0 at the origin of the three axes. Although this three-dimensional representation is precise, it does not lend itself to simple graphic display of sample distributions. The three-dimensional model, a theoretical representation with the component scales at right angles to one another, facilitates comparisons of different sample plots. However, there are moderate intercorrelations between components, which differ from sample to sample. At present comparisons between samples require plotting relative to a model.

In general, investigators have used a two-dimensional projection for displaying somatotypes, despite the completeness of the three-dimensional representation. A *somatoplot* is the projection of a somatotype location in three-dimensional space on a two-dimensional grid or somatochart. Figure II.3 shows a somatochart for plotting individual or mean somatotypes from the component ratings, using two-dimensional X,Y coordinates. Figure II.6 illustrates the three- and two-dimensional representations of the somato-type 3-6-2. When the somatotypes of a sample are projected onto a somatochart (that is, from somatopoints to somatoplots) the original information about the distribution is somewhat diminished. The 'real' distance between somatoplots of two somatotypes often is less than the 'real' distance between somatopoints of the same two somatotypes. (Duquet & Hebbelinck (1977) discuss the limitations of the two-dimensional approach more fully.)

Despite moderate distortion, the two-dimensional somatochart is convenient and useful for visual display of a sample and for further analysis. The somatochart is an analogue of a two-dimensional Mercator projection of a three-dimensional hemisphere of the earth.

Somatochart areas and categories

Routine plotting of individual somatotypes on a somatochart produces an immediate visual picture of group concentrations and dispersions of a sample. The division of the somatochart into sectors and further division into 13 partitions open the way for subsequent analyses. The three axes, labelled endomorphy, mesomorphy, and ectomorphy, intersect at the centre and divide the somatochart into sectors. Somatotypes become more polar as values for a component rating increase from the centre of the chart along a given axis. The extremes of any of the three components are said to lie at the poles of endomorphic, mesomorphic or ectomorphic axes. In broad terms, we can say the somatochart is oriented along the mesomorphy axis in a north–south direction, with dominant mesomorphs in the northern section (including the northwest and northeast sectors), with dominant endomorphs in the southwest corner, and dominant ectomorphs in the southeast corner. Those falling in the centre of the somatochart are central somatotypes.

For finer distinctions, the thirteen partitions in Fig. II.7 show the distribution of the thirteen categories of somatotypes described in Table II.1.

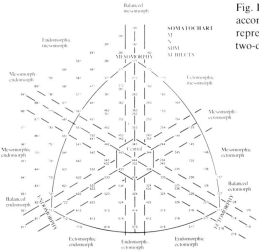

Fig. II.7. Somatotype categories labelled according to component dominance and represented as areas on the two-dimensional somatochart.

Table II.1. *Definitions of 13 somatotype categories based on areas of the somatochart*

Balanced endomorph:	endomorphy is dominant and mesomorphy and ectomorphy are equal (or do not differ by more than one-half unit).
Mesomorphic endomorph:	endomorphy is dominant and mesomorphy is greater than ectomorphy.
Mesomorph–endomorph:	endomorphy and mesomorphy are equal (or do not differ by more than one-half unit), and ectomorphy is smaller.
Endomorphic mesomorph:	mesomorphy is dominant and endomorphy is greater than ectomorphy.
Balanced mesomorph:	mesomorphy is dominant, endomorphy and ectomorphy are less and equal (or do not differ by more than one-half unit.)
Ectomorphic mesomorph:	mesomorphy is dominant, and ectomorphy is greater than endomorphy.
Mesomorph–ectomorph:	mesomorphy and ectomorphy are equal (or do not differ by more than one-half unit); and endomorphy is lower.
Mesomorphic ectomorph:	ectomorphy is dominant; and mesomorphy is greater than endomorphy.
Balanced ectomorph:	ectomorphy is dominant; endomorphy and mesomorphy are equal and lower (or do not differ by more than one-half unit).
Endomorphic ectomorph:	ectomorphy is dominant; and endomorphy is greater than mesomorphy.
Endomorph–ectomorph:	endomorphy and ectomorphy are equal (or do not differ by more than one-half unit); and mesomorphy is lower.
Ectomorphic endomorph:	endomorphy is dominant; and ectomorphy is greater than mesomorphy.
Central:	no component differs by more than one unit from the other two, and consists of ratings of 2, 3, or 4.

Modified from Carter, 1980*a*.

Somatotypes lying outside the arc-sided triangle may be classified according to the definitions in the table, or may be identified as extremes for their particular category (e.g. a 3-8-1 is an extreme endomorphic mesomorph.)

Figure II.7 shows that the somatotype 3-5-2 is in the endomorphic mesomorph category. It is called an endo-mesomorph, because mesomorphy is the dominant component, and endomorphy is second. A 1-6-3 is called an ecto-mesomorph; a 2-3-5 a meso-ectomorph, a 2-4-4 a mesomorph–ectomorph or an ectomorph-mesomorph; and a 2-5-2 a balanced mesomorph. A 4-4-3 and a 4-3-4 are central somatotypes. From these examples it can be seen that the noun describes the major area in which the somatotype lies, and the adjective describes the kind of somatotype within that area.

Approaches to somatotype analysis

The nature of the data, the numbers of subjects and the kinds of groups to be compared determine exact procedures of an analysis. Figures

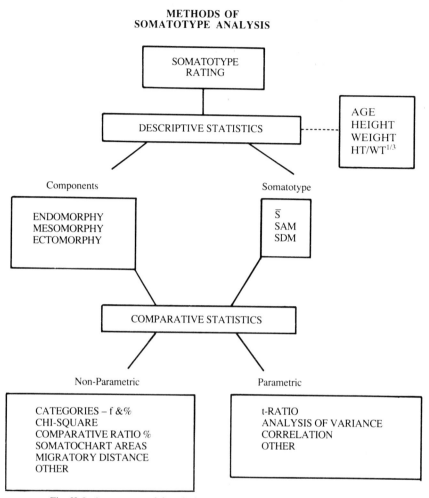

**METHODS OF
SOMATOTYPE ANALYSIS**

SOMATOTYPE
RATING

DESCRIPTIVE STATISTICS

AGE
HEIGHT
WEIGHT
HT/WT$^{1/3}$

Components

ENDOMORPHY
MESOMORPHY
ECTOMORPHY

Somatotype

$\bar{S}$
SAM
SDM

COMPARATIVE STATISTICS

Non-Parametric

CATEGORIES – f &%
CHI-SQUARE
COMPARATIVE RATIO %
SOMATOCHART AREAS
MIGRATORY DISTANCE
OTHER

Parametric

t-RATIO
ANALYSIS OF VARIANCE
CORRELATION
OTHER

Fig. II.8. A summary of descriptive and comparative statistics for use in somatotype analysis. $\bar{S}$ = mean somatotype; SAM = somatotype attitudinal mean; SDM = somatotype dispersion mean (Adapted from Carter *et al.*, 1983.)

II.8 and II.9 summarize and elaborate on some typical methods of analysis of somatotype data.

Descriptive statistics

As seen in Fig. II.8, for analysis of somatotype ratings for one or more samples, the mean, standard deviation, standard error of the mean, and low and high ratings are calculated for each component. To provide further information on each sample, the same statistics should be calculated routinely for age, height, weight and the HWR. Also a somatochart should

SOMATOTYPE COMPONENTS

	Endomorphy	Mesomorphy	Ectomorphy
Mean	$\overline{X}_{EN}$	$\overline{X}_{ME}$	$\overline{X}_{EC}$
Variance	s^2_{EN}	s^2_{ME}	s^2_{EC}
Standard Deviation	s_{EN}	s_{ME}	s_{EC}

BIPLANAR COORDINATES FOR SOMATOPLOTS

	x coordinate	y coordinate
Mean	$\overline{X}$	$\overline{Y}$
Variance	s^2_x	s^2_y
Standard Deviation	s_x	s_y

SOMATOTYPE AS A WHOLE

SOMATOTYPE

	Two dimensions	Three dimensions
Mean	$\overline{S}$	$\overline{S}$
Variance	s^2_D	s^2_A
Standard Deviation	s_D	s_A

SCATTER ABOUT $\overline{S}$

	Two dimensions	Three dimensions
Mean	SDM	SAM
Variance	s^2_D	s^2_A
Standard Deviation	s_D	s_A

Fig. II.9. Abbreviations for statistics used in somatotype analysis. Subscripts in the upper half of the figure are abbreviations for the column headings; in the lower half subscripts D and A refer to dispersion and attitudinal distances respectively. (From Carter *et al.*, 1983.)

be plotted. Examples of results of these calculations are shown in the upper part of Fig. II.10. (See Carter *et al.*, 1983, and Table II.2 for equations used in calculating these statistics.)

The mean somatotype, $\overline{S}$, is the mean for each of the components and is the measure of the central tendency for the sample. The dispersion or scatter of somatotypes about their mean is the other statistic of interest. The distances between the somatotypes and their means are used to calculate these statistics. They provide the average scatter about $\overline{S}$ and can be calculated from either two- or three-dimensional distances. Equivalent terms and methods that are more appropriate to the characteristics of the

Table II.2. *Equations for calculating and analysing somatotype data*

Equations for calculating the anthropometric somatotype

Endomorphy

Endomorphy* $= -0.7182 + 0.1451(X) - 0.00068(X^2) + 0.0000014(X^3)$
where X = sum of triceps, subscapular, and supraspinale skinfolds.
*(For height-corrected endomorphy, X is multiplied by 170.18/height in cm)

Mesomorphy

Mesomorphy $= (0.858$ humerus breadth $+ 0.601$ femur breadth $+ 0.188$ corrected arm girth $+ 0.161$ corrected calf girth) $-$ (height $\times 0.131) + 4.50$

Ectomorphy

Ectomorphy $=$ HWR $\times 0.732 - 28.58$

where HWR (height–weight ratio) $= \dfrac{\text{height}}{\sqrt[3]{\text{weight}}}$

Note: If HWR < 40.75, but >38.25, then Ectomorphy $=$ HWR $\times 0.463 - 17.63$.
If HWR $\leqslant 38.25$, then a rating of 0.1 is assigned for Ectomorphy.

Calculation of somatoplot coordinates
(i) Calculation of X-coordinate

$X =$ ectomorphy $-$ endomorphy

(ii) Calculation of Y-coordinate

$Y = 2(\text{mesomorphy}) - (\text{endomorphy} + \text{ectomorphy})$

Somatotype dispersion distance (SDD)
The somatotype dispersion distance is the difference between two somatoplots which have the coordinates (X_1, Y_1) and (X_2, Y_2) and is calculated as follows:

$$\text{SDD}_{1,2} = \sqrt{[3(X_1 - X_2)^2 + (Y_1 - Y_2)^2]}$$

The SDD is used for calculating distances on a two-dimensional somatochart, and is expressed in Y-units.

Somatotype dispersion mean (SDM)

$$\text{SDM} = \sum_{i=1}^{n} \text{SDD}_i/n$$

The somatotype dispersion mean is the mean SDD of all the somatoplots in a sample from its mean somatoplot $(\bar{S})$.

Somatotype attitudinal distance (SAD)
The somatotype attitudinal distance is the difference between *somatopoints* (points in three dimensional space which represents somatotypes), and is calculated in component units:

$$\text{SAD}_{A,B} = \sqrt{[(I_A - I_B)^2 + (II_A - II_B)^2 + (III_A - III_B)^2]}$$

where I, II, and III represent the endomorphic, mesomorphic, and ectomorphic components of a somatotype, and A and B are two somatotypes.
The SAD is the three-dimensional analogue of the somatotype dispersion distance (SDD).

Somatotype attitudinal mean (SAM)
The somatotype attitudinal mean is the average of the somatotype attitudinal distances (SADs) of each somatotype.

$$\text{SAM} = \sum_{i=1}^{n} \text{SAD}_i/n$$

The SAM is the three-dimensional analogue of the somatotype dispersion mean (SDM)

(*continued*)

Table II.2. (*Continued*)

Somatotype migratory distance (MD)
$$MD_{A,D} = SAD_{A,B} + SAD_{B,C} + SAD_{C,D}$$
where A to D are four somatotypes, and SAD is the somatotype attitudinal distance between pairs of somatotypes.

Somatotype intensity (INT)
$$INT_A = SAD_{O,A}$$
where the intensity of somatotype A is equal to the magnitude of the SAD from the origin O to A.

Somatotype t-ratio (independent samples)
$$t = \frac{SAD_{\bar{S}_1 - \bar{S}_2}}{\sqrt{\left[\frac{\Sigma(SAD_1^2) + \Sigma(SAD_2^2)}{n_1 + n_2 - 2} \times \left(\frac{1}{n_1} + \frac{1}{n_2}\right)\right]}}$$
where SAD = somatotype attitudinal distance and $\bar{S}$ = mean somatotype.

Somatotype analysis of variance (independent samples)
Sum of squares within sample:
$$SS_w = \sum_{j=1}^{k} \sum_{i=1}^{k} (SAD_1^2)$$
where SAD_1 is the SAD of each somatotype from its sample mean somatotype.
Sum of squares between samples:
$$SS_b = \sum_{j=1}^{k} n_j (\bar{S}_j - \bar{M})^2$$
where n_j = number in each group, $\bar{S}_j$ = mean somatotype of each group and $\bar{M}$ = mean somatotype of combined groups.
$$F\text{-ratio} = \frac{SS_b/df_b}{SS_w/df_w}$$
Comparative ratio of a proportion
$$CR_P = \frac{\text{Difference betweeen proportions}}{\text{Standard error of difference}}$$
$$= \frac{P_1 - P_2}{\sqrt{\left[\frac{P_1 \times Q_1}{N_1} + \frac{P_2 \times Q_2}{N_2}\right]}}$$
where 1 and 2 are the two samples; P and Q are the percentage with and without the characteristic in each sample; and N is the number in each sample.

---

somatotype are required, because of some difficulties in using conventional statistics. These statistics are designated as SAM and SDM in Figs. II.8 and II.9.

Somatotype dispersion distance and mean

The somatotype dispersion distance (SDD) is the distance in two dimensions between any two somatoplots (the plot of a somatotype in 2D). The SDD is calculated in the *Y*-units of the two-dimensional coordinate

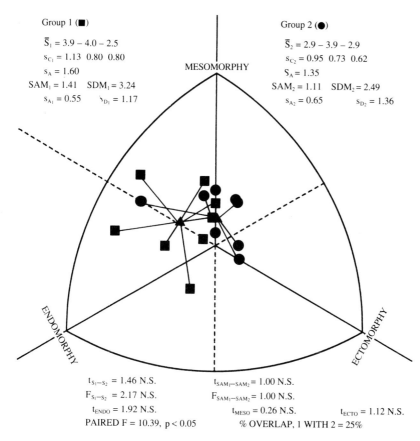

Group 1 (■)

$\bar{S}_1 = 3.9 - 4.0 - 2.5$
$s_{C_1} = 1.13 \ 0.80 \ 0.80$
$s_A = 1.60$
$SAM_1 = 1.41 \quad SDM_1 = 3.24$
$s_{A_1} = 0.55 \qquad s_{D_1} = 1.17$

MESOMORPHY

Group 2 (●)

$\bar{S}_2 = 2.9 - 3.9 - 2.9$
$s_{C_2} = 0.95 \ 0.73 \ 0.62$
$S_A = 1.35$
$SAM_2 = 1.11 \quad SDM_2 = 2.49$
$s_{A_2} = 0.65 \qquad s_{D_2} = 1.36$

ENDOMORPHY

ECTOMORPHY

$t_{S_1-S_2} = 1.46$ N.S.
$F_{S_1-S_2} = 2.17$ N.S.
$t_{ENDO} = 1.92$ N.S.
PAIRED F = 10.39, p < 0.05

$t_{SAM_1-SAM_2} = 1.00$ N.S.
$F_{SAM_1-SAM_2} = 1.00$ N.S.
$t_{MESO} = 0.26$ N.S. $\qquad t_{ECTO} = 1.12$ N.S.
% OVERLAP, 1 WITH 2 = 25%

Fig. II.10. Somatochart showing the somatoplots of two samples (Groups 1 and 2), with a summary of descriptive and comparative statistics. The subscripts c1 and c2 indicate that the standard deviations are for each of the components of their respective groups. In each sample the lines connecting each somatoplot to its respective mean (triangles) represent the SDDs. For other abbreviations see Fig. II.9. (From Carter *et al.*, 1983).

system (Ross & Wilson, 1973). The somatotype dispersion mean (SDM) is the average of these distances from $\bar{S}$ for a given sample.

Somatotype attitudinal distance and mean

The somatotype attitudinal distance (SAD), in three dimensions, is the distance between any two somatopoints (the plot of a somatotype in 3D). It is calculated in component units. The somatotype attitudinal mean (SAM) is the average of the SADs of each somatopoint from $\bar{S}$ of a sample (Duquet & Hebbelinck, 1977).

The following are examples of the SAD and the SDD as calculated for each of several pairs of somatotypes:

$$
\begin{array}{llll}
\text{1-6-2} - \text{1-6-3:} & \text{SAD} = 1.00; & \text{SDD} = 2.00 \\
\text{1-6-2} - \text{2-7-2:} & \text{SAD} = 1.41; & \text{SDD} = 2.00 \\
\text{2-4-2} - \text{4-5-2:} & \text{SAD} = 2.24; & \text{SDD} = 3.46 \\
\text{3-5-3} - \text{4-5-2:} & \text{SAD} = 1.41; & \text{SDD} = 3.46 \\
\text{3-3-3} - \text{4-4-4:} & \text{SAD} = 1.73; & \text{SDD} = 0.00 \\
\end{array}
$$

In a comparison of the 1-6-3 and the 2-7-2 with the somatotype 1-6-2 the SDD is 2.00, and they are equidistant on the two-dimensional somatochart. However, the SADs differ, which reflects the additional distance in the depth of the three-dimensional plot (i.e. away from the page surface). The somatotypes 2-4-2 and 3-5-3, having the same XY coordinates, are plotted at the same point on the somatochart, but they differ in depth. Thus, although the SDDs to somatotype 4-5-2 are the same (3.46) the SADs differ, showing a greater distance from the 2-4-2 than from the 3-5-3. The 3-3-3 and the 4-4-4 are plotted at the same point and therefore are zero SDD units apart, but they are at different levels, as shown by the SAD of 1.73.

Frequency and relative frequency of somatotypes by somatochart areas or categories, in addition to descriptive statistics, also describe a sample.

Comparative statistics

As shown in Fig. II.8, there is a variety of parametric and non-parametric statistics for comparing the somatotypes of two or more groups. Some of these procedures can be applied to the components individually, but are not recommended as the primary method of analysis. These procedures should be used only to support or clarify analyses of somatotype data.

Non-parametric statistics

The relative frequency of somatotypes by category or somatochart area is probably the most commonly used non-parametric statistic. Relative frequencies can be determined from somatochart areas (see Fig. II.11). It is possible to define arbitrarily the categories as other areas of the somatochart, or as combinations of logically adjacent categories.

Chi-square. A Chi-square test can be applied to test for significance of the differences between the distribution of somatotypes by category or area (Carter, 1980a; Parnell, 1958). The frequencies of somatotypes by categories for a sample (e.g. a reference population) are designated as theoretical frequencies, and those from the sample under examination are designated as the expected frequencies.

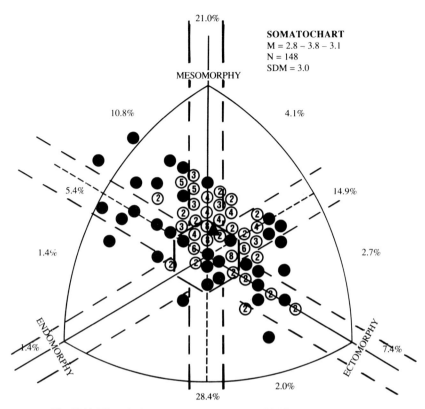

Fig. II.11. The relative frequencies of somatotypes in categories as defined in Table II.1. ▲ = mean somatotype. Based on data in Carter *et al.* (1982).

Comparative ratio. The comparative ratio of two proportions (Parnell, 1958) is appropriate for determining whether two samples differ with respect to a given characteristic. For example, Parnell used this method to see if in a given group the percentage of young men, who had a certain physique and suffered nervous breakdowns, differed from the percentage expected from the same physiques in a separately selected sample. If the ratio exceeds 100 the breakdowns are more common in the first group. If the ratio is lower than 100, the breakdowns are less common. The comparative ratio for a particular somatotype or category is equal to the percentage with the characteristic in sample one, divided by the percentage with the characteristics in sample two, times 100. Such ratios are especially unstable if the percentages are small, so they should be used with caution. A critical ratio for proportions is used to test the significance of their difference (see Table II.2).

Migratory distance. When somatotypes are followed over time the distance and direction of their paths can be quantified. The migratory distance between sequential somatotypes is the sum of the SADs or SDDS between them. The sum, which may be between the somatotypes of one subject followed over time or between consecutive means, has been used to quantify change in growth studies (Pařízková & Carter, 1976; Weese *et al.*, 1975).

Intensity distance. The intensity of a somatotype is the magnitude of the vector from the origin of the three component scales (the hypothetical 0-0-0 somatotype) to the somatopoint in three dimensions. It is an expression of the distance of the somatotype along an axis from the origin of the X,Y,Z coordinates. Every change in somatotype has an associated intensity change, as long as the change is not a switch in component values (Duquet, 1980; Carter *et al.*, 1983).

Figure II.12 is a schematic illustration of the relationships between measures of somatotype intensity and dominance. Thus, the somatotype may change in intensity (e.g. 2-4-1 and 3-5-2), in dominance (e.g. 3-5-2 and 2-5-3), in both intensity and dominance (e.g. 2-5-3 to 2-4-1) or in stability (e.g. 2-4-1 to 3-5-2 to 2-5-3; p1, p2, p3, where p1 to p3 are three somatopoints). Changes in direction also may occur.

The following are calculations for the above examples:

2-4-1 to 3-5-2:   SAD = 1.73;   SDD = 0.0;   INT = 1.58
3-5-2 to 2-5-3:   SAD = 1.41;   SDD = 3.46;   INT = 0.0
2-5-3 to 2-4-1:   SAD = 2.24;   SDD = 3.46;   INT = 1.58
$MD_{p1,p2,p3}$ = 1.73 + 1.41 = 3.14

Cutting line principle. Walker (1962) developed the 'cutting line principle' for comparing two groups of plotted somatotypes. He divided the two plots with a straight line, chosen by inspection, in order to yield the highest possible number of plots of each group on either side of the line. He computed a chi-square to test the homogeneity of the two plots, and a tetrachoric correlation as an index of relationship. He imposed some conditions on the procedures for locating the line, because of the possible arbitrary nature of line location.

Percent overlap. Two samples can also be compared graphically by circumscribing the limits of each of a pair of somatotype distributions on a somatochart and counting the frequency and relative frequency of somatoplots that overlap in each distribution (Carter *et al.*, 1983; Hebbelinck *et al.*, 1980). In Fig. II.10 this procedure was applied to groups 1 and 2, which showed a 25% overlap of group 1 with group 2.

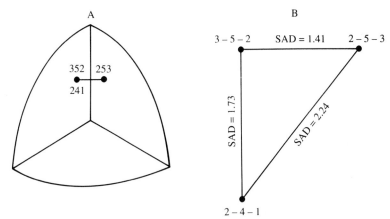

Fig. II.12. A schematic representation of the relationships among three somatotypes. A is a two-dimensional representation showing both the 3-5-2 and 2-4-1 somatotypes plotted at the same point; and B is a three-dimensional representation showing the somatotype attitudinal distances (SADs) among somatotypes as viewed from below (enlarged but drawn to scale). (From Carter *et al.*, 1983.)

Ross (1976) described the use of the basic concepts of the SDD and the SDI to derive an I-Index to show the amount of commonality between samples.

The I-Index is a non-parametric test based on a geometric model representing any pair of somatotype distributions (somatoplots). It is a ratio derived from the intersection of circles drawn with centres on the mean somatoplots and radii equal to the somatotype dispersion index (now called somatotype dispersion mean) of each sample. It is described as:

I = (area in common/area not in common) × 100
When I = 100 the circles are concentric and have equal radii.
When I = 0 there is no overlapping between the circles.

Although the concept is simple, the geometry is complicated. The I-Index is simply a model for estimating the commonality of samples. It is good only insofar as the distributions are approximately symmetrical about their mean. Analysis by the I-Index of elliptical and other forms of somatoplots will provide misleading conclusions. Further, the choice of one multiplied by the SDI for the radii of the circles is arbitrary; it could just as easily be 1.5 or 2.0 multiplied by the SDI. Although Ross (1976) made clear the principles of the I-Index and its application, there are several errors and geometrical inconsistencies in the article. Aubry (unpublished manuscript, San Diego State University, 1976) revised Ross's I-Index derivation and incorporated it into computer programs. Although the I-Index is not often calculated, simple graphic comparisons can be made by drawing the circles from their respective means on the somatochart to illustrate overlap between samples (Carter *et al.*, 1983).

Parametric statistics

Parametric statistics are appropriate for comparisons among groups of somatotypes, provided that the correct computational elements are used in the calculation of the variances and mean differences (Carter *et al.*, 1983). Analysis of variance and *t*-ratios apply to testing differences between samples. However, there are two groups of tests for analysis of somatotypes as a whole (Fig. II.9). One group of tests are for the differences between somatotype means and the other one for the scatter of somatotypes about their means. The appropriate equations are listed in Table II.2.

F- and t-ratios. *F-* and *t*-ratios may be used for comparisons between two or more mean somatotypes and their variances. In these comparisons the distances between the means (e.g. $\bar{S}_i - \bar{S}_{ii}$) in terms of SADs (for 3D) or SDDs (for 2D) are tested for significance. These tests determine whether there is a difference between the somatoplots or somatopoints seen on the somatochart.

The second test is for comparison between two or more somatotype dispersions as shown by the scatter of somatoplots or somatopoints about their respective means. Note that this comparison is independent of the position of the distributions on the somatochart, that is, whether or not they overlap or are completely separate distributions.

A two-way analysis of variance is appropriate in a repeated measures experiment, such as somatotypes before and after a physical training program. In these calculations only SAD values can be used, for which the two-way analysis of variance allows. There is no analogous somatotype *t*-test for correlated data.

The *F-* and *t*-ratios appear to be valid when the scatter of somatotypes about their mean is approximately uniform in all directions (i.e. circular). They are less likely to be valid when one or more samples of somatoplots is asymmetrical (i.e. elliptical) (Carter *et al.*, 1983).

See Figure II.10 for examples of applying some parametric statistics to two somatotype groups. The descriptive statistics for groups 1 and 2 respectively are in the upper left and right corners of the figure. (The symbols are the same as in Fig. II.9.) The comparative statistics are listed below the somatochart. Neither the *t*-ratios and *F*-ratios for comparisons between somatotype means ($\bar{S}$s), nor the *t*- or *F*-ratios between the somatotype attitudinal means (SAMs) are significant. The *t*-ratios for each com-

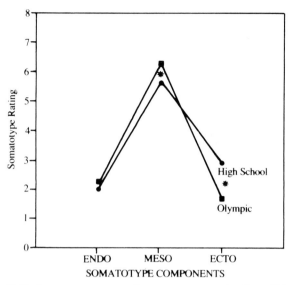

Fig. II.13. A compogram comparing mean component ratings of high school and Olympic wrestlers. The asterisk indicates a significant difference ($p < 0.05$) between means for mesomorphy and ectomorphy. Based on data from Carter & Lucio (1986).

ponent (t_{ENDO}, t_{MESO}, t_{ECTO}) are not different for the two groups. The PAIRED F reveals a significant difference when the two groups are assumed to be from two measurement occasions on the same subjects, not from two different samples. The per cent overlap of group 1 with group 2 is calculated as 25%. Conclusion: There are no statistical differences between the two groups.

Alternative parametric analyses have been proposed by Cressie *et al.* (1986). They say that the procedures of Carter *et al.* (1983) prematurely collapse the three component somatotype vectors into a scalar SAD value, leading to inappropriate degrees of freedom for the *F*-ratio. They propose a rationale for analysis using a one-way multivariate analysis of variance; and subsequent application of Hotelling's T^2, univariate *F*-ratios, discriminant function analysis and forward stepwise discriminant analysis.

Compogram. Araújo *et al.* (1978) described a simple method, which they called a compogram, for presenting the results of comparisons between the mean somatotypes of two samples. They plotted the means for the three components from two samples and connected the means for each. An asterisk on the graph indicates significant differences between means (as determined by an appropriate test) in any component (see Fig. II.13).

Correlation. Product–moment, partial and multiple correlations can be used to examine the relationships between components and other variables. However, interpretation of these statistics may be difficult or misleading, because each component is treated independent of its context in the somatotype. Also, although correlations among the three components provide descriptive information about the sample, they are sample specific and greatly influenced by the range of ratings and the precision of the scale (i.e. tenth, half- or one-unit ratings).

In correlation analyses use of the whole somatotype is preferred to separate components. Parnell (1958) and Walker (1962) successfully used tetrachoric correlation and subsequent chi-square analysis to examine the relationships between somatotypes and other variables such as success and failure, and having or not having a disease or other physical condition.

In an innovative approach to assessing the correlation between somatotype and motor performance variables. Araújo (1986) proposed correlating the distances (SADs) between the criterion somatotype and other subjects' somatotypes with other variables such as the time, place or distance scores. For example, it is possible to determine the correlation between the SADs between each runner and the winner of a 1500 m race and the time of finish. In essence this relationship can be drawn as a scattergram with the SADs on the Y-axis and time on the X-axis, with the winner's somatotype (SAD = 0) and time at the origin of the scattergram.

Orvanová *et al.* (1984) used regression lines to show the differences in direction of the somatoplots in ten different groups of subjects. They calculated the slopes of the regression lines from the X, Y coordinates of the somatochart. Although these authors did not do so, the r value and the standard error of the regression line could be used as measures of the scatter of the somatoplots. A high r value and small standard error would indicate a linear (or elliptical) distribution and the slope of the line would indicate the direction.

Calculations and computer programs

A small electronic calculator facilitates all the computations in this chapter. Before seeking computer solutions to the tedious process of analysing group data, it is advisable to do all procedures by hand and to test them against the examples in Carter (1980*a*) and Carter *et al.* (1983).

Personal computers and simple programming languages like BASIC provide easily available access to calculating somatotype variables and simple statistics suitable for data for individuals and small groups. Main frame computers and more powerful programming languages such as FORTRAN are desirable for larger samples and more complex statistics.

Table II.3. *Computer programs for somatotype analysis*
Prosoman computer programs by S. P. Aubry and J. E. L. Carter, San Diego State
University.

STYPE – calculates anthropometric somatotype;
CATE – frequencies, histogram, % by categories;
SPLOT – somatochart and plots;
SANOV – one-way anova, whole somatotypes;
TPAIR – paired *t*-ratio, whole somatotypes;
MIGDIS – distances between somatotypes, intensity;
INFREQ – N of somatotypes within selected circles or spheres.

The following list is a summary of the programs:
1. STYPE calculates the anthropometric Heath–Carter somatotype, and provides a listing
   by subject of the original and derived variables, as well as descriptive statistics for all
   variables.
2. CATE calculates and displays the frequencies in a histogram, and a table, and lists the
   subjects by category.
3. SPLOT draws a bi-dimensional somatochart, plots the frequencies of the somatoplots
   and provides descriptive statistics for the somatotypes and the somatoplot. A listing is
   given of the input and plotting information by subject.
4. SANOV calculates a one-way analysis of variance (ANOVA) for independent samples,
   using distances between the whole somatotype ratings. Descriptive statistics for the
   somatotype attitudinal distances (SAD) or the somatotype dispersion distance (SDD)
   are printed, as well as an ANOVA table and a distance matrix between means.
5. TPAIR calculates a *t*-ratio for paired somatotypes using distances between the whole
   somatotype ratings. Input data by subject and calculations for the differences are
   listed.
6. MIGDIS calculates the sum of the distances between sequential somatoplots, in two- or
   three-dimensions, and plots their positions. The output lists input somatotypes by
   subject and occasion, migratory distances (MD), and intensities (INT).
7. INFREQ calculates and lists the number of subjects from each somatotype sample (A
   and B) which lie inside, outside, or in the intersections of circles or spheres, based on
   the radii selected.

For further information on these programs, including purchase, write to: J. E. L. Carter,
Department of Physical Education, San Diego State University, San Diego, CA. 92182.
USA.

Whenever possible, standard programs should be used. To encourage
consistent analysis among somatotypers, S. P. Aubry and J. E. L. Carter
have developed a software package to use with large computers, such as
CYBER, IBM, and VAX. The package, called PROSOMAN (Computer
Programs for Somatotype Analysis), consists of programs for calculating
somatotypes, descriptive and comparative statistics, and plotting somatocharts. See Table II.3 for a summary of PROSOMAN.

Summary

The type and extent of analysis indicated for somatotype data
depends on the purpose of the study and the nature of the hypotheses to be

tested. Descriptive statistics and a somatochart should be the minimum for any study. The following list (Carter *et al.*, 1983) should be consulted before proceeding with analysis:

1. For sample descriptions:
   (a) means and standard deviations for each component, SAM, and SDM;
   (b) somatocharts showing individual values and mean somato-plots, as these are the best way to display somatotype data;
   (c) somatoplots by category or area on the somatochart.
2. For average differences between samples:
   (a) differences between somatotype means ($\bar{S}_1 - \bar{S}_2$);
   (b) differences between component means (endomorphy, meso-morphy, ectomorphy);
   (c) differences between the scatter of somatotypes about $\bar{S}$ ($SAM_1 - SAM_2$; variances about $\bar{S}_1 - \bar{S}_2$);
   (d) differences between categories or areas;
   (e) differences in intensity (INT_{p1p2}).
3. For differences between two somatotypes:
   (a) $\bar{S}_1 - \bar{S}_2$;
   (b) (INT_{p1p2});
   (c) differences between a particular somatotype (S_1) and a group somatotype ($\bar{S}_j$).
4. For change in somatotype:
   (a) SAD (total change);
   (b) INT (change in magnitude);
   (c) MD (consecutive change).

Some of the calculations may be made by using SDDs, but there may be some distortion of true values in three dimensions. SDDs should be used only when describing or comparing the somatoplots on the somatochart.

Some assumptions must be made when the SAD is used (and for MD and INT). The basic model uses perpendicular coordinate axes and equal units on all axes. This means that the model has 'equal appearing intervals', not only within each component but also between the components. For other recommended analyses, see Cressie *et al.* (1986).

Appendix III
Ratings of published somatotypes

Appendix III.1. *Ratings of somatotype photographs of children, from* Atlas of Children's Growth, *Tanner & Whitehouse (1982)*

No.	Age	Heath–Carter somatotype[a]	Tanner somatotype[b]
Male somatotypes			
1.	3.0	4.0-4.0-1.0	
	4.0	4.0-4.0-1.0	
	5.0	3.0-4.0-1.0	
	6.0	3.5-4.0-2.5	
	7.0	3.0-4.0-3.0	
	8.0	3.0-3.5-3.0	
	9.0	3.0-3.5-3.0	
	10.0	2.5-3.5-4.0	
	11.0	2.5-3.5-4.0	
	12.0	2.5-3.5-4.0	
	13.0	2.5-3.5-4.0	
	14.0	2.0-4.0-4.0	
	15.0	2.0-4.0-4.0	
	16.0	2.0-4.0-4.0	
	17.0	2.0-4.0-4.0	
	18.0	2.5-4.0-4.0	
	19.0	2.5-4.0-4.0	2.5-4.0-4.0
2.	2.5	6.0-3.0-0.5	
	3.0	6.0-3.0-0.5	
	4.0	6.0-3.0-0.5	
	5.0	6.0-3.0-0.5	
	6.0	5.0-3.0-1.0	
	7.0	5.0-3.0-1.5	
	8.0	5.0-3.0-2.0	
	9.0	4.5-3.5-3.0	
	10.0	4.0-3.5-3.0	
	11.0	4.0-3.5-3.0	
	12.0	4.0-3.5-3.0	
	13.0	4.0-3.5-3.0	
	14.0	4.0-3.5-3.0	
	15.0	6.0-3.5-2.0	
	16.0	6.0-4.0-1.0	
	20.0	5.0-4.0-2.0	6.0-3.0-1.0

(*continued*)

Appendix III.1. (*Continued*)

No.	Age	Heath–Carter somatotype[a]	Tanner somatotype[b]
3.	5.0	5.0-4.0-1.0	
	6.0	4.5-4.0-2.5	
	7.0	3.5-4.0-3.0	
	8.0	3.0-4.0-3.0	
	9.0	3.0-4.0-3.0	
	10.0	3.0-4.0-3.5	
	11.0	3.0-4.0-4.0	
	12.0	3.5-4.0-4.0	
	13.0	3.0-4.0-3.5	
	14.0	3.0-4.0-4.0	
	15.0	2.5-4.0-3.5	
	16.0	2.5-4.5-4.0	
	17.0	2.5-4.5-4.0	
	18.0	2.5-4.5-3.5	
	19.0	3.0-4.5-3.0	2.5-5.5-2.5
4.	3.5	8.0-4.0-0.5	
	4.0	8.0-4.0-0.5	
	5.0	6.5-4.0-0.5	
	6.0	4.5-4.5-1.0	
	7.0	4.0-4.5-1.5	
	8.0	3.0-4.5-2.0	
	9.0	3.0-4.5-2.0	
	10.0	3.0-4.5-2.0	
	11.0	2.5-4.5-2.5	
	12.0	2.5-4.5-2.5	
	13.0	2.5-4.5-3.0	
	14.0	2.5-4.5-3.0	
	15.0	2.5-4.5-3.0	
	16.0	2.5-5.0-3.0	
	17.0	2.5-5.0-3.0	
	18.0	3.0-5.5-2.5	
	22.0	2.0-5.5-2.5	2.5-5.5-2.0
5.	5.5	6.0-4.0-0.5	
	6.0	6.0-4.0-0.5	
	7.0	4.0-5.0-0.5	
	8.0	3.5-5.0-1.0	
	9.0	4.0-4.5-1.0	
	10.0	4.0-4.5-1.0	
	11.0	3.5-4.5-1.5	
	12.0	3.5-4.5-1.5	
	13.0	3.5-5.0-1.5	
	14.0	3.0-5.0-1.5	
	15.0	3.0-5.0-1.5	
	16.0	3.0-5.0-2.0	
	17.0	3.0-5.0-2.0	
	19.0	3.5-5.0-2.0	3.5-5.5-1.5
6.	6.0	2.0-4.0-2.0	
	7.0	2.0-4.0-2.0	

Appendix III.1. (*Continued*)

No.	Age	Heath–Carter somatotype[a]	Tanner somatotype[b]
	8.0	2.0-4.0-2.5	
	9.0	2.0-4.0-3.0	
	10.0	2.0-4.0-3.0	
	11.0	2.0-4.0-3.0	
	12.0	2.0-4.0-4.0	
	13.0	2.0-4.0-4.0	
	14.0	2.0-4.5-4.0	
	15.0	2.5-5.0-3.5	
	16.0	2.5-5.0-3.5	
	17.0	2.5-5.0-3.6	
	18.0	2.0-5.5-3.0	
	19.0	2.5-5.5-3.0	
	20.0	2.0-5.5-3.0	2.0-5.5-3.0
7.	8.0	4.0-4.0-1.0	
	9.0	4.0-4.0-1.0	
	10.0	4.0-4.0-1.0	
	11.0	4.0-4.0-1.5	
	12.0	3.5-4.5-2.0	
	13.0	3.5-4.5-2.0	
	14.0	3.5-5.0-1.5	
	15.0	3.0-5.5-1.5	
	16.0	2.5-5.5-1.5	
	17.0	2.5-5.5-2.5	
	18.0	2.5-5.5-1.5	
	19.0	2.5-5.5-1.5	3.5-5.5-2.0
8.	5.0	6.0-3.5-0.5	
	6.0	4.5-3.5-1.0	
	7.0	3.5-4.0-2.5	
	8.0	3.5-4.0-2.5	
	9.0	3.5-4.0-2.5	
	10.0	3.5-4.0-2.5	
	11.0	3.5-4.0-2.5	
	12.0	3.0-4.0-3.0	
	13.0	3.0-4.0-3.0	
	14.0	3.0-4.0-4.0	
	15.0	3.0-4.0-3.0	
	16.0	3.5-4.5-3.0	
	17.0	3.5-4.5-3.0	
	18.0	3.5-4.5-2.5	
	20.0	3.5-4.5-2.5	3.5-5.5-2.0
9.	6.0	5.0-3.5-1.0	
	7.0	5.0-3.5-1.0	
	8.0	4.0-3.5-2.5	
	9.0	3.5-3.5-2.5	
	10.0	3.0-3.5-3.5	
	11.0	3.0-3.5-3.0	

(*continued*)

Appendix III.1. (*Continued*)

No.	Age	Heath–Carter somatotype[a]	Tanner somatotype[b]
	12.0	3.0-3.5-3.5	
	13.0	3.0-3.5-4.0	
	14.0	3.0-3.5-4.0	
	15.0	3.0-4.0-3.0	
	16.0	3.0-4.0-3.5	
	17.0	3.0-4.0-3.5	
	19.0	3.0-4.5-3.5	
	20.0	3.0-4.5-3.5	3.0-5.0-4.0
10.	4.0	7.0-3.0-0.5	
	5.0	6.5-3.0-0.5	
	6.0	5.0-3.0-1.0	
	7.0	4.5-3.0-1.5	
	8.0	3.5-3.0-2.0	
	9.0	3.5-3.0-2.5	
	10.0	3.0-3.0-2.5	
	11.0	3.0-3.0-2.5	
	11.5	3.0-3.0-2.5	
	13.0	3.0-3.5-3.0	
	14.0	3.0-4.0-2.5	
	15.0	3.0-4.0-2.5	
	16.0	3.0-4.5-2.5	
	17.0	4.0-4.5-2.5	
	18.0	4.0-4.5-2.0	4.0-5.0-2.0
11.	6.5	4.5-3.0-1.0	
	7.0	4.5-3.0-1.0	
	8.0	4.0-3.5-1.5	
	9.0	3.5-3.5-3.0	
	10.0	3.0-3.5-3.0	
	11.0	3.0-3.5-3.0	
	12.0	2.5-3.5-3.0	
	13.0	2.5-3.5-3.0	
	14.0	2.5-3.5-3.0	
	15.0	3.0-3.5-3.0	
	16.0	3.0-4.0-3.5	
	17.0	3.0-4.0-3.5	
	18.0	3.0-4.0-3.0	
	19.0	3.0-4.5-2.5	
	20.0	3.5-4.5-3.5	3.0-4.5-3.0
12.	4.5	8.0-4.0-0.5	
	5.0	7.0-4.0-0.5	
	6.0	6.0-4.0-0.5	
	7.0	6.0-4.0-1.0	
	8.0	5.0-4.0-2.0	
	9.0	4.0-4.0-2.0	
	10.0	4.0-4.0-2.5	
	11.0	3.5-4.0-2.5	
	12.0	3.5-4.0-2.5	
	13.0	3.0-4.0-3.0	
	14.0	2.5-4.5-3.0	

Appendix III.1. (*Continued*)

No.	Age	Heath–Carter somatotype[a]	Tanner somatotype[b]
	15.0	2.5-5.0-3.0	
	16.0	3.0-5.0-2.5	
	17.0	3.0-5.0-3.0	
	18.0	3.0-5.0-3.0	
	19.0	3.0-5.0-2.5	3.0-4.5-2.5
13.	6.0	3.5-3.0-1.5	
	7.0	3.5-3.0-2.0	
	8.0	3.0-3.0-2.0	
	9.0	3.0-3.0-2.5	
	10.0	3.0-3.0-2.5	
	11.0	3.0-3.5-3.0	
	12.0	3.0-4.0-2.5	
	13.0	3.0-4.0-2.5	
	14.0	3.0-4.0-3.0	
	15.0	2.5-4.5-3.0	
	16.0	3.0-4.5-2.0	
	17.0	3.0-4.5-2.0	
	18.0	3.0-4.5-2.0	
	19.0	3.5-4.5-2.5	
	20.0	3.5-4.5-2.5	3.5-4.5-2.0
14.	8.0	3.5-3.5-2.0	
	9.0	3.0-3.5-2.0	
	10.0	3.0-3.0-3.0	
	11.0	3.0-3.0-3.0	
	12.0	3.0-3.5-3.5	
	13.0	3.0-3.5-3.5	
	14.0	3.0-4.0-4.0	
	15.0	3.0-4.0-3.5	
	16.0	2.5-5.0-3.5	
	17.0	3.0-5.0-3.0	
	18.0	3.0-5.0-2.0	
	19.0	2.5-5.0-2.5	
	20.0	3.0-5.0-2.5	3.5-4.5-3.0
15.	8.5	2.5-2.5-5.0	
	9.0	2.5-2.5-5.0	
	10.0	2.0-2.0-5.5	
	11.0	2.0-2.0-6.0	
	12.0	2.0-2.0-6.0	
	13.0	2.0-2.0-6.0	
	14.0	2.0-2.0-6.0	
	15.0	2.0-2.0-6.0	
	16.0	2.0-2.0-7.0	
	17.0	2.0-2.0-7.0	1.5-1.5-7.0
16.	5.0	4.0-3.0-1.0	
	6.0	3.0-3.0-1.5	
	7.0	3.0-3.0-3.0	

(*continued*)

Appendix III.1. (*Continued*)

No.	Age	Heath–Carter somatotype[a]	Tanner somatotype[b]
	8.0	3.0-3.0-4.0	
	9.0	2.5-3.0-4.0	
	10.0	2.0-3.0-5.0	
	11.0	2.0-2.5-5.0	
	12.0	2.0-2.5-5.0	
	13.0	2.5-2.5-5.5	
	14.0	2.5-2.5-5.5	
	15.0	2.0-2.5-6.0	
	16.0	2.0-2.5-6.0	
	17.0	2.0-2.5-6.0	
	18.0	2.0-2.5-6.0	
	19.0	2.0-2.5-5.5	2.0-2.0-6.0
17.	5.0	3.5-2.5-2.0	
	6.0	3.0-2.5-2.5	
	7.0	2.5-2.5-4.0	
	8.0	2.5-2.5-5.0	
	9.0	2.5-2.5-5.0	
	10.0	2.5-2.5-5.0	
	11.0	2.0-2.5-5.0	
	12.0	2.0-2.5-5.5	
	13.0	2.0-2.5-5.5	
	14.0	2.0-2.5-5.0	
	15.0	2.0-3.0-5.0	
	16.0	2.0-3.0-5.0	
	17.0	2.0-3.0-5.5	
	18.0	2.5-3.0-5.5	2.0-3.0-6.0
18.	3.5	8.0-3.0-0.5	
	4.0	7.0-3.0-0.5	
	5.0	6.0-3.0-1.0	
	6.0	4.5-3.0-1.5	
	7.0	4.5-3.0-1.5	
	8.0	3.0-3.0-2.0	
	9.0	3.0-3.0-3.5	
	10.0	2.5-3.0-3.5	
	11.0	2.5-3.0-4.0	
	12.0	2.0-3.0-4.0	
	13.0	2.0-3.0-4.0	
	14.0	2.0-3.0-4.0	
	15.0	2.0-3.0-4.0	
	15.5	1.5-3.0-4.0	
	16.0	3.0-3.0-4.0	
	17.0	2.0-3.0-4.0	
	18.0	2.0-3.0-4.0	2.0-3.0-6.0
19.	6.0	3.5-3.5-3.0	
	7.0	3.0-3.5-3.5	
	8.0	3.0-3.5-3.5	
	9.0	2.5-3.5-4.0	
	10.0	2.5-3.5-4.5	

Appendix III.1. (*Continued*)

No.	Age	Heath–Carter somatotype[a]	Tanner somatotype[b]
	11.0	2.0-3.5-4.5	
	12.0	2.0-3.5-5.0	
	13.0	2.0-3.5-5.5	
	14.0	2.0-3.5-5.5	
	15.0	2.0-4.0-5.0	
	16.0	2.0-4.0-5.0	
	17.0	2.0-4.0-5.0	
	18.0	2.5-4.0-5.0	2.0-4.0-5.0
20.	3.0	6.0-3.5-0.5	
	4.0	4.5-3.5-1.5	
	5.0	4.5-3.0-2.5	
	6.0	3.0-3.0-3.0	
	7.0	3.0-3.0-3.0	
	8.0	2.5-3.0-4.0	
	9.0	2.5-3.0-4.5	
	10.0	2.5-3.0-4.5	
	11.0	2.5-3.0-5.0	
	12.0	2.5-3.0-4.5	
	13.0	2.0-3.0-5.0	
	14.0	2.0-3.5-5.0	
	15.0	2.0-3.5-5.5	
	16.0	2.5-3.5-5.0	
	17.0	2.5-3.5-5.0	
	18.0	3.5-3.5-5.0	3.0-3.5-5.0
21.	5.0	6.5-3.0-0.5	
	6.0	4.0-3.5-1.5	
	7.0	4.0-3.0-2.0	
	8.0	3.0-3.0-2.0	
	9.0	2.5-3.0-3.0	
	10.0	2.5-3.0-3.0	
	11.0	2.5-3.5-4.0	
	12.0	2.0-3.5-4.0	
	13.0	2.5-3.5-4.0	
	14.0	2.0-3.5-4.0	
	15.0	2.0-4.0-4.0	
	16.0	2.0-4.0-4.0	
	17.0	2.0-4.0-4.0	
	18.0	2.0-4.0-4.0	
	20.0	2.0-4.0-4.0	2.0-4.0-4.5
22.	5.0	5.5-3.0-1.5	
	6.0	4.5-3.0-1.5	
	7.0	4.0-3.0-2.0	
	8.0	3.0-3.0-3.0	
	9.0	2.5-3.0-4.0	
	10.0	2.0-3.0-4.5	
	11.0	2.0-3.0-4.5	

(*continued*)

Appendix III.1. (*Continued*)

No.	Age	Heath–Carter somatotype[a]	Tanner somatotype[b]
	12.0	2.0-3.0-4.5	
	13.0	2.0-3.5-5.0	
	14.0	2.0-3.5-4.5	
	15.0	2.0-4.0-4.5	
	16.0	2.0-4.0-4.5	
	17.0	2.0-4.0-4.0	
	19.0	3.0-4.0-4.0	3.0-4.0-4.5
23.	4.0	5.0-3.0-1.0	
	5.0	4.0-3.0-1.0	
	6.0	3.0-3.0-2.0	
	7.0	3.0-3.0-3.0	
	8.0	2.5-3.0-3.0	
	9.0	2.0-3.0-3.5	
	10.0	2.0-3.0-4.0	
	11.0	2.0-3.0-4.0	
	12.0	2.5-3.0-4.5	
	13.0	2.0-3.0-4.5	
	14.0	2.0-3.5-5.0	
	15.0	2.0-3.5-4.5	
	16.0	2.0-3.5-5.0	
	17.0	2.5-3.5-5.0	
	18.0	2.5-3.5-5.0	3.0-4.0-4.5
Female somatotypes			
24.	9.5	4.5-3.5-1.5	
	10.0	3.5-3.5-3.0	
	11.0	3.5-3.5-3.0	
	12.0	3.0-3.5-4.4	
	13.0	4.0-3.5-3.5	
	14.0	4.0-3.5-3.5	
	15.0	3.0-3.5-4.5	
	16.0	3.0-3.5-4.0	
	17.0	4.0-3.5-4.0	
	18.0	5.0-3.5-2.5	
	19.0	6.5-3.5-1.5	
	20.0	6.5-3.5-1.0	6.0-2.5-2.5
25.	8.5	4.5-3.5-1.0	
	9.0	4.5-3.5-1.0	
	10.0	4.5-3.5-1.5	
	11.0	4.0-3.5-3.0	
	12.0	5.0-3.5-2.0	
	13.0	5.0-3.5-1.5	
	14.0	5.0-4.0-1.5	
	15.0	4.0-4.0-2.0	
	16.0	4.0-4.0-2.0	
	17.0	4.5-4.5-1.0	
	18.0	4.5-4.5-1.0	
	19.0	5.0-4.5-1.0	6.0-3.0-1.5

Appendix III.1. (*Continued*)

No.	Age	Heath–Carter somatotype[a]	Tanner somatotype[b]
26.	7.5	4.0-3.0-1.0	
	8.0	4.0-3.0-2.0	
	9.0	3.5-3.0-2.0	
	10.0	4.5-3.0-2.0	
	11.0	5.0-3.0-1.5	
	12.0	3.5-4.0-3.0	
	13.0	3.0-4.0-3.0	
	14.0	4.5-4.0-2.0	
	15.0	4.5-4.5-1.0	
	16.0	4.0-4.5-2.0	5.0-3.5-1.5
27.	9.0	3.0-3.5-1.5	
	10.0	3.0-3.5-2.0	
	11.0	3.0-3.5-2.5	
	12.0	3.5-3.5-2.0	
	13.0	3.0-3.5-2.5	
	14.0	4.0-4.0-3.0	
	15.0	3.5-4.0-3.0	
	15.5	3.0-4.0-3.0	
	17.0	4.5-4.5-1.5	
	18.0	4.5-4.5-1.5	
	19.0	4.5-4.5-1.5	
	20.0	5.5-4.5-1.0	4.5-4.5-1.5
28.	3.0	8.0-3.0-0.5	
	4.0	7.0-3.0-0.5	
	5.0	6.0-3.0-1.0	
	6.0	5.0-3.0-1.0	
	7.0	5.0-3.0-1.0	
	8.5	4.5-3.0-3.0	
	9.0	4.5-3.5-3.0	
	10.0	4.0-3.5-3.0	
	11.0	4.0-3.5-3.0	
	12.0	3.5-3.5-3.0	
	13.0	3.0-3.5-4.0	
	14.0	3.0-3.5-4.0	
	15.0	3.0-3.5-4.0	
	16.0	3.5-3.5-4.0	
	17.0	4.0-3.5-4.0	4.0-3.0-4.0
29.	5.0	5.0-3.0-1.0	
	6.0	4.0-3.0-2.0	
	7.0	3.5-3.0-3.0	
	8.0	3.5-3.0-3.0	
	9.0	3.0-3.0-3.0	
	10.0	2.5-3.0-4.0	
	11.0	2.5-3.0-4.0	
	12.0	2.0-3.0-4.0	
	13.0	2.5-3.0-4.0	

(*continued*)

Appendix III.1. (*Continued*)

No.	Age	Heath–Carter somatotype[a]	Tanner somatotype[b]
	14.0	3.0-3.0-4.0	
	15.0	2.5-3.0-4.5	
	16.0	3.0-3.0-4.5	
	17.0	3.5-3.0-4.0	
	18.0	3.5-3.0-4.0	4.0-2.0-4.0
30.	7.5	4.0-3.0-2.0	
	8.0	3.5-3.0-2.0	
	9.0	2.5-3.0-3.0	
	10.0	2.5-3.0-3.0	
	11.0	2.5-3.0-3.5	
	12.0	3.0-3.0-4.0	
	13.0	3.0-3.0-4.0	
	14.0	3.0-3.0-4.0	
	15.0	2.5-3.5-4.0	
	16.0	3.5-3.5-4.0	
	17.0	3.5-4.0-4.0	
	18.0	3.0-4.0-4.0	3.0-4.0-4.0
31.	5.0	6.0-3.5-0.5	
	6.0	4.5-3.5-1.0	
	7.0	3.5-3.5-2.0	
	8.0	3.0-3.5-2.5	
	9.0	3.0-3.5-3.5	
	10.0	3.0-3.5-3.0	
	11.0	3.0-3.5-3.5	
	12.0	3.0-3.5-3.5	
	13.0	3.0-3.5-4.0	
	14.0	3.5-3.5-4.0	
	15.0	3.5-3.5-4.0	
	16.0	4.0-4.0-3.5	
	17.0	3.0-4.0-4.0	
	19.0	3.0-4.0-4.0	4.0-4.0-3.0
32.	5.0	6.5-3.0-1.0	
	6.0	6.0-3.0-1.0	
	7.0	5.0-3.5-1.5	
	8.0	4.0-3.5-2.0	
	9.0	4.0-3.5-2.5	
	10.0	4.0-3.5-3.0	
	11.0	3.5-3.5-3.5	
	12.0	3.0-3.5-3.5	
	13.0	3.5-3.5-3.5	
	14.0	3.0-3.5-3.5	
	15.0	3.0-3.5-4.0	
	16.0	2.5-3.5-4.5	
	17.0	2.5-3.5-4.5	
	18.0	2.5-3.5-4.0	
	19.0	3.0-3.5-4.0	3.0-3.5-4.0
33.	5.0	5.5-3.0-1.0	
	6.0	4.5-3.0-2.0	

Appendix III.1. (*Continued*)

No.	Age	Heath–Carter somatotype[a]	Tanner somatotype[b]
	7.0	3.5-3.0-2.0	
	8.0	3.0-3.0-3.0	
	9.0	3.0-3.0-3.5	
	10.0	3.0-3.0-3.5	
	11.0	2.5-3.0-4.0	
	12.0	2.5-3.0-4.0	
	13.0	2.0-3.0-4.0	
	14.0	2.0-3.0-4.0	
	15.0	2.0-3.5-4.5	
	16.0	2.5-3.5-4.0	
	17.0	2.0-3.5-4.5	
	18.0	3.0-3.5-4.0	
	19.0	2.5-4.0-4.0	3.0-3.5-4.0
34.	6.0	4.0-3.0-1.5	
	7.0	3.5-3.0-2.0	
	8.0	3.5-3.0-2.0	
	9.0	3.0-3.0-2.0	
	10.0	3.0-3.0-3.0	
	11.0	3.5-3.0-3.0	
	12.0	2.5-3.5-4.0	
	13.0	3.0-3.5-4.0	
	14.0	3.0-3.5-4.0	
	15.0	3.5-3.5-4.0	
	16.0	3.5-3.5-4.0	
	17.0	3.5-3.5-4.0	3.5-3.5-4.0
35.	5.0	6.5-3.0-0.5	
	6.0	5.0-3.0-1.5	
	7.0	4.0-3.0-2.0	
	8.0	4.0-3.0-2.0	
	9.0	4.0-3.0-2.5	
	10.0	4.0-3.0-2.5	
	11.0	3.5-3.0-3.0	
	12.0	4.0-3.0-3.0	
	13.0	4.0-3.5-2.5	
	14.0	4.0-3.5-2.5	
	15.0	4.0-4.0-1.5	
	16.0	4.5-4.0-1.5	
	17.0	4.5-4.0-1.5	
	18.0	4.0-4.0-2.5	
	19.0	4.5-4.0-2.0	
	20.0	5.0-4.0-1.5	4.0-3.5-3.0
36.	4.0	8.0-3.5-0.5	
	5.0	6.5-3.5-1.0	
	6.0	5.5-3.5-1.5	
	7.0	4.5-3.5-1.5	
	8.0	4.0-3.5-2.0	

(*continued*)

Appendix III.1. (*Continued*)

No.	Age	Heath–Carter somatotype[a]	Tanner somatotype[b]
	9.0	4.0-3.5-2.5	
	10.0	3.5-3.5-3.0	
	11.0	3.5-3.5-3.0	
	12.0	4.0-3.5-3.0	
	12.5	4.0-3.5-3.0	
	13.0	4.0-3.5-3.0	
	14.0	4.5-4.0-2.5	
	15.0	5.5-4.0-2.0	
	16.0	6.0-4.0-2.0	
	17.0	5.0-4.0-2.0	
	18.0	5.0-4.0-2.0	
	20.0	4.5-4.0-2.5	5.5-3.0-2.5
37.	5.5	5.0-3.0-1.0	
	6.5	4.0-3.0-1.0	
	7.5	3.0-3.0-2.0	
	8.5	3.0-3.0-2.0	
	9.5	3.0-2.5-3.0	
	10.5	2.5-2.5-3.0	
	11.5	2.5-2.5-4.5	
	12.5	3.0-3.0-4.0	
	13.5	3.0-3.0-4.0	
	14.5	4.0-3.5-3.5	
	15.5	4.0-3.0-3.5	
	17.0	3.5-3.0-3.5	
	19.0	3.5-3.0-4.0	3.5-3.5-4.0
38.	5.5	4.5-2.5-2.0	
	6.0	4.0-2.5-2.5	
	7.0	4.0-2.5-3.0	
	8.0	3.0-2.5-3.5	
	9.0	3.0-3.0-3.5	
	10.0	3.0-3.0-4.5	
	11.0	3.0-3.0-5.0	
	12.0	3.5-3.0-5.0	
	13.0	3.0-3.0-5.5	
	14.0	3.5-3.0-5.0	
	15.0	3.5-3.0-5.0	
	16.0	4.0-3.0-4.5	
	18.0	5.0-3.0-3.5	4.0-3.0-5.0
39.	4.0	7.0-3.0-0.5	
	5.0	6.0-3.0-1.0	
	6.0	6.0-3.0-1.0	
	7.0	5.0-3.0-2.0	
	8.0	4.0-3.0-3.0	
	9.0	4.0-3.0-3.0	
	10.0	4.0-3.0-3.5	
	11.0	4.0-3.0-3.5	
	12.0	4.0-3.0-3.5	
	13.0	4.0-3.0-5.0	

Appendix III.1. (*Continued*)

No.	Age	Heath–Carter somatotype[a]	Tanner somatotype[b]
	14.0	4.0-3.5-3.5	
	15.0	4.0-3.5-3.0	
	16.0	3.5-3.5-4.0	
	17.0	5.0-3.5-2.5	4.0-3.0-4.0
40.	5.5	5.0-2.5-2.0	
	6.5	4.5-3.0-2.0	
	7.0	4.0-3.0-2.0	
	7.5	4.0-3.0-3.0	
	8.5	4.0-3.0-3.0	
	9.5	4.0-3.0-4.0	
	10.5	4.0-3.0-4.0	
	11.5	4.5-3.5-3.0	
	12.5	4.0-3.0-3.5	
	13.5	5.0-3.0-2.5	
	14.5	5.5-3.0-2.0	
	16.0	4.5-2.5-3.0	
	17.0	4.5-2.5-3.0	5.0-1.5-4.0
41.	5.0	6.0-2.0-1.5	
	6.0	6.0-2.0-2.0	
	7.0	6.0-2.0-2.0	
	8.0	4.0-2.0-3.0	
	9.0	4.0-2.0-4.0	
	10.0	4.0-2.0-4.0	
	11.0	4.0-2.0-4.0	
	12.0	3.5-2.0-6.0	
	13.0	4.0-2.0-5.0	
	14.0	3.5-2.0-6.0	
	15.0	4.5-2.0-3.5	
	16.0	4.0-2.0-3.5	
	17.0	5.0-2.0-3.0	5.0-1.0-4.0
42.	3.5	8.5-3.0-0.5	
	4.5	6.0-3.0-1.0	
	5.5	5.5-3.0-1.5	
	6.5	5.0-3.0-3.0	
	7.5	5.0-3.0-3.0	
	8.5	5.0-3.0-2.5	
	9.5	5.5-3.5-1.5	
	10.5	6.0-3.0-1.5	
	11.5	6.0-3.5-1.5	
	12.5	5.5-3.5-1.0	
	13.5	7.5-3.5-1.0	
	14.0	8.0-3.5-0.5	
	16.0	6.0-4.0-1.5	
	17.0	6.0-4.0-2.0	
	18.0	5.5-4.0-2.0	6.0-2.0-2.0

(*continued*)

Appendix III.1. (*Continued*)

No.	Age	Heath–Carter somatotype[a]	Tanner somatotype[b]
43.	2.0	9.0-3.0-0.5	
	3.0	9.0-3.0-0.5	
	4.0	7.5-3.0-0.5	
	5.0	7.0-3.0-1.0	
	6.0	5.0-3.0-1.0	
	7.0	4.0-3.0-2.0	
	8.0	4.0-3.0-2.5	
	9.0	4.0-3.0-3.0	
	10.0	4.0-3.0-3.0	
	11.0	3.5-3.0-3.0	
	12.0	3.0-3.5-3.0	
	13.0	3.5-3.5-3.0	
	14.0	4.0-4.0-2.5	
	15.0	4.0-4.0-2.5	
	16.0	5.0-4.0-1.5	
	17.5	5.5-4.0-1.5	
	20.0	6.0-4.0-1.0	6.0-3.0-2.0
44.	5.0	9.5-3.0-0.5	
	6.0	8.5-3.0-0.5	
	7.0	7.0-3.0-1.0	
	8.0	6.0-3.0-1.0	
	9.0	6.0-3.0-1.0	
	10.0	5.5-3.0-1.0	
	11.0	5.5-3.0-1.5	
	12.0	5.0-3.0-1.5	
	13.0	5.0-3.0-1.5	
	13.8	5.0-3.0-2.0	
	15.0	5.0-3.0-2.0	
	16.0	5.0-3.0-2.0	
	17.0	5.5-3.0-1.0	
	18.0	5.5-3.0-1.5	6.0-1.0-2.0
45.	5.0	6.0-3.0-1.0	
	6.0	4.5-3.0-2.0	
	7.0	4.5-3.0-3.0	
	8.0	4.0-3.0-3.0	
	9.0	4.0-3.0-3.0	
	10.0	4.0-3.0-3.0	
	11.0	3.5-3.0-3.5	
	12.0	3.5-3.5-3.5	
	13.0	4.0-3.5-3.0	
	14.0	5.5-3.5-2.0	
	15.0	5.5-3.5-2.0	
	16.0	5.5-3.5-1.5	
	18.0	6.5-3.5-1.0	
	19.0	5.5-3.5-1.5	6.0-2.0-3.0
46.	7.0	4.0-3.0-2.0	
	8.0	4.0-3.0-2.5	
	9.0	4.0-3.0-3.0	

Appendix III.1. (*Continued*)

No.	Age	Heath–Carter somatotype[a]	Tanner somatotype[b]
	10.0	3.5-3.0-3.5	
	11.0	3.0-3.0-3.0	
	11.5	3.0-3.0-4.0	
	12.0	3.0-3.0-4.0	
	13.0	3.5-3.0-4.0	
	13.8	3.5-4.0-3.5	
	14.7	4.5-4.0-2.5	
	15.0	5.0-4.0-1.5	
	16.0	6.0-4.0-1.0	
	17.0	8.0-4.0-1.0	6.0-2.0-2.0
47.	3.5	9.5-3.0-0.5	
	4.5	7.0-3.0-1.0	
	5.5	6.0-3.0-1.0	
	6.5	5.5-3.0-1.0	
	7.5	5.5-3.0-1.5	
	8.5	5.0-3.0-2.0	
	9.5	5.0-3.0-2.0	
	10.5	5.0-3.0-3.0	
	11.5	5.0-3.0-2.5	
	12.5	5.0-3.0-2.5	
	13.5	5.0-3.0-2.5	
	14.5	5.5-3.0-1.5	
	16.0	5.5-3.0-2.0	
	17.0	5.5-3.0-2.0	
	18.0	6.0-3.5-1.5	5.0-2.5-2.0
48.	5.0	5.5-3.0-1.0	
	6.0	5.0-3.0-1.5	
	7.0	3.5-3.0-3.0	
	8.0	3.5-3.0-3.0	
	9.0	3.5-3.0-3.0	
	10.0	3.5-3.0-3.0	
	11.0	3.5-3.0-3.5	
	12.0	3.0-3.0-4.0	
	13.0	4.0-3.0-2.5	
	14.0	3.0-3.5-3.5	
	15.0	3.0-3.5-3.5	
	16.0	4.0-3.5-2.0	
	17.0	4.5-3.5-2.0	
	18.0	4.5-3.5-3.0	
	19.0	4.0-3.5-3.0	5.0-3.0-2.0
49.	6.0	5.0-3.0-1.0	
	7.0	5.0-3.0-1.5	
	7.4	4.5-3.0-1.5	
	8.0	4.5-3.0-1.5	
	9.0	4.5-3.0-3.0	
	10.0	4.5-3.0-3.0	

(*continued*)

Appendix III.1. (*Continued*)

No.	Age	Heath–Carter somatotype[a]	Tanner somatotype[b]
	11.0	4.0-3.0-3.0	
	12.0	3.5-3.0-3.0	
	13.0	3.0-3.0-3.0	
	13.5	3.5-3.5-3.5	
	14.0	4.5-3.5-3.0	
	15.0	4.0-3.5-3.0	
	16.0	4.5-3.5-2.0	
	17.0	5.0-3.5-1.5	
	19.0	4.5-3.5-2.5	5.0-2.5-2.0
50.	4.5	8.0-3.0-0.5	
	5.5	7.5-3.0-0.5	
	6.6	7.0-3.0-1.0	
	7.5	5.0-3.5-1.5	
	8.5	5.0-3.5-1.5	
	9.5	4.5-3.5-1.5	
	10.5	4.5-3.5-1.5	
	11.5	4.0-4.0-3.0	
	12.5	4.0-4.0-3.0	
	13.5	4.0-4.0-2.5	
	14.5	4.0-4.0-2.5	
	15.0	4.0-4.0-2.5	
	16.0	4.0-4.0-2.5	
	17.0	4.0-4.0-2.0	
	18.0	5.0-4.0-2.0	5.0-3.0-3.0
51.	5.0	4.5-3.0-1.5	
	6.0	4.5-3.0-2.0	
	7.0	3.5-3.0-2.5	
	8.0	3.0-3.0-3.0	
	9.0	3.0-3.0-3.0	
	10.0	3.0-3.0-3.5	
	11.0	3.0-3.0-3.5	
	12.0	3.5-3.5-4.0	
	13.0	3.5-3.5-4.0	
	14.0	4.0-3.5-3.5	
	15.0	4.0-3.5-3.0	
	16.0	5.0-3.5-3.0	
	17.0	4.5-3.5-3.0	4.5-3.0-3.0
52.	3.5	6.5-3.0-0.5	
	4.0	6.0-3.0-1.0	
	5.5	4.0-3.0-2.0	
	6.5	4.0-3.0-2.0	
	7.0	4.0-3.0-3.0	
	7.5	3.0-3.0-3.0	
	8.5	3.0-3.0-4.0	
	9.5	3.0-3.0-4.0	
	10.0	3.0-3.0-4.0	
	10.5	2.0-3.0-5.0	
	11.5	2.0-3.0-5.0	

Appendix III.1. (*Continued*)

No.	Age	Heath–Carter somatotype[a]	Tanner somatotype[b]
	12.2	3.0-3.0-5.0	
	13.5	4.0-3.0-4.0	
	14.5	5.0-3.0-3.0	4.5-3.0-4.0
53.	4.0	7.5-3.0-0.5	
	5.0	7.0-3.0-0.5	
	6.0	5.0-3.0-1.5	
	7.0	5.0-3.0-3.0	
	8.0	4.5-3.0-3.0	
	9.0	4.0-3.0-3.0	
	10.0	3.5-3.0-3.5	
	11.0	3.5-3.0-3.5	
	12.0	3.5-3.0-4.0	
	13.0	4.0-3.0-4.0	
	14.0	4.0-3.0-4.0	
	15.0	4.0-3.0-4.5	
	16.0	4.0-3.0-4.0	
	17.0	4.0-3.0-4.0	
	18.0	3.5-3.0-4.5	4.5-2.5-4.0
54.	7.0	3.5-3.0-2.5	
	8.0	2.5-2.5-3.5	
	9.0	2.5-2.5-3.5	
	10.0	2.5-2.5-4.0	
	11.0	2.5-2.5-4.5	
	12.0	3.0-2.5-4.0	
	13.0	3.0-3.0-5.0	
	14.0	3.0-3.0-4.0	
	15.0	3.0-3.5-4.0	
	16.0	3.0-3.5-4.0	
	17.0	3.0-3.5-4.0	
	18.0	3.0-3.5-4.5	
	19.0	4.5-3.5-4.0	
	20.0	3.5-3.5-3.5	5.0-3.0-4.0

Specially selected subjects, Heath ratings
(Tanner ratings given for nos. 73–77 only)

No.	Age	Heath–Carter somatotype[a]	Tanner somatotype[b]
61.	6.0	4.5-3.0-1.5	Female
	9.0	4.0-3.0-2.5	
	12.0	4.0-3.5-3.0	
	15.0	6.0-3.5-2.0	
62.	9.6	3.5-2.5-3.0	Female – monozygotic twin of no. 63
	13.0	3.5-2.5-4.5	
	14.5	3.5-3.0-4.5	
	16.0	4.5-3.0-4.0	
63.	9.6	4.5-2.5-3.0	
	13.0	4.0-3.0-3.5	
	14.5	4.0-3.0-3.0	
	16.0	4.5-3.5-3.0	

(*continued*)

Appendix III.1. (*Continued*)

No.	Age	Heath–Carter somatotype[a]	Tanner somatotype[b]
64.	2.0	9.0-3.0-0.5	Female – dizygotic twin of no. 65
	3.5	9.0-3.0-0.5	
	5.0	6.5-3.0-0.5	
	6.5	5.0-3.0-1.5	
	8.0	4.0-3.0-2.5	
	9.5	4.0-3.0-2.5	
	11.0	3.5-3.0-3.0	
	12.5	4.0-3.0-3.0	
	14.0	3.0-3.0-4.0	
	15.5	3.0-3.5-4.0	
	17.0	4.0-3.5-3.0	
	18.0	4.0-3.5-3.0	
65.	2.0	9.0-3.0-0.5	
	3.5	8.5-3.0-0.5	
	5.0	6.0-3.0-1.5	
	6.5	4.5-3.0-2.0	
	8.0	3.5-3.0-3.0	
	9.5	3.5-3.0-3.0	
	11.0	3.5-3.0-3.0	
	12.5	3.0-3.0-3.5	
	14.0	3.0-3.0-3.0	
	15.5	4.5-3.2-2.0	
	17.0	4.5-3.5-2.0	
	18.0	4.5-3.5-2.0	
66.	8.0	3.5-3.0-3.0	Female – sibling of no. 67
	9.5	3.5-3.0-3.0	
	11.5	3.0-3.0-4.0	
	12.5	3.0-3.0-4.0	
	14.0	3.5-3.5-4.0	
	15.5	4.0-4.0-2.0	
	17.0	4.0-4.0-2.0	
67.	6.5	5.0-3.0-1.0	Female – sibling of no. 66
	8.0	4.5-3.0-2.0	
	9.5	4.0-3.5-2.0	
	11.0	4.0-3.5-2.0	
	12.5	4.5-3.5-2.0	
	14.0	5.0-4.0-1.5	
	17.0	6.5-4.0-1.0	
68.	10.0	3.0-3.5-3.5	Male – sibling of nos. 69 and 18 (male) and
	11.0	2.5-3.5-3.5	49 and 70 (female)
	12.0	2.5-3.5-3.5	
	13.0	2.5-3.5-4.0	
	14.0	2.5-3.5-4.0	
	15.0	2.5-4.0-3.5	
	16.0	2.0-4.0-3.5	
	18.0	2.0-4.5-3.5	
69.	11.0	2.5-3.0-3.5	Male – see above
	12.0	2.5-3.0-3.5	
	13.0	2.0-3.0-4.0	

Appendix III.1. (*Continued*)

No.	Age	Heath–Carter somatotype[a]	Tanner somatotype[b]
	14.0	2.0-3.0-4.0	
	15.0	2.5-3.0-3.5	
	16.0	3.5-3.5-4.0	
	18.0	3.5-4.0-3.5	
70.	No photo		

Subjects 71–77 = 'Growth from 20 to 35 years'

No.	Age	Heath–Carter somatotype[a]	Tanner somatotype[b]
71.	20.0	2.5-5.0-2.0	Male
	25.0	2.0-5.0-2.0	
	30.0	2.5-5.0-2.0	
72.	20.0	3.0-4.0-3.0	Male
	25.0	3.0-4.0-3.0	
	30.0	3.5-4.0-2.5	
	35.0	3.5-4.0-2.0	
73.	20.0	3.5-4.5-2.0	Male
	25.0	4.5-4.5-1.5	4.0-4.5-2.0
	35.0	7.0-4.5-1.0	
75.	19.0	2.5-4.5-3.5	Male
	25.0	2.5-4.5-3.5	2.0-6.0-3.0
76.	20.0	4.0-4.0-2.5	Male
	25.0	5.5-4.0-1.5	4.5-3.5-2.5
77.	20.0	2.0-4.5-3.5	Male
	30.0	2.0-4.5-4.0	3.0-4.0-3.5
78.	4.0	8.5-3.5-0.5	Male
	5.0	8.5-3.5-0.5	True isolated growth hormone deficiency
	6.0	8.0-3.5-0.5	
	7.0	6.0-3.5-1.0	
	8.0	5.5-3.5-1.0	
	9.0	5.5-3.5-1.0	
	10.3	5.0-3.5-1.5	
	11.3	5.0-3.5-2.0	
	12.3	5.0-3.5-2.0	
	13.3	5.0-3.5-2.0	
	14.0	5.5-3.5-2.0	
	15.5	5.5-3.5-2.0	
	16.5	6.0-4.0-1.0	
	17.5	7.0-4.0-1.0	
	18.5	8.0-4.0-1.0	
79.	3.4	8.0-2.5-0.5	Male
	3.7	8.0-2.5-0.5	True isolated growth hormone deficiency
	4.7	7.0-2.5-1.0	
	5.8	6.5-2.5-1.0	
	6.8	6.0-2.5-1.0	
	7.8	5.0-2.5-1.0	
	8.8	5.5-2.5-1.0	
	9.9	4.5-2.5-1.5	

(*continued*)

Appendix III.1. (*Continued*)

No.	Age	Heath–Carter somatotype[a]	Tanner somatotype[b]
	10.9	3.5-2.5-2.0	
	11.8	3.5-3.0-2.0	
	12.8	3.5-3.0-3.0	
	13.8	3.5-3.0-3.0	
	14.8	3.5-3.0-2.5	
	16.8	3.5-3.0-3.0	
80.	6.0	6.0-3.0-1.0	Male
	6.5	6.5-3.0-1.0	Isolated growth hormone deficiency in one of two
	6.9	6.0-3.5-1.0	monozygotic twins
	7.5	5.0-3.5-1.0	
	8.0	4.0-3.5-1.5	
	8.5	4.0-3.5-1.5	
	9.0	4.0-3.5-1.0	
81.	6.0	4.5-3.5-1.5	Male
	6.5	4.0-3.5-1.5	Normal twin
	6.9	4.0-3.5-2.0	
	7.5	4.0-3.5-2.0	
	8.0	3.5-3.5-2.5	
	8.5	3.5-3.5-3.0	
	9.0	3.5-3.5-3.0	
82.	11.2	6.5-3.5-1.0	Female
	11.5	6.5-3.5-1.0	True isolated growth hormone deficiency
	12.6	7.5-3.5-1.0	
	13.6	7.0-3.5-1.0	
	14.6	7.5-3.5-1.0	
	15.6	6.5-3.5-1.5	
	16.6	6.0-3.5-2.0	
	18.5	6.0-4.0-2.0	
	20.6	6.0-4.0-2.0	
83.	9.5	8.5-3.0-0.5	Female
	10.5	8.5-3.0-0.5	Growth hormone deficiency plus gonadotrophin
	11.7	8.5-3.0-0.5	deficiency
	12.7	8.0-3.0-0.5	
	13.7	8.0-3.0-1.0	
	14.7	6.5-3.0-1.0	
	15.7	6.5-3.0-1.0	
	16.7	6.5-3.0-1.0	
	17.5	6.5-3.0-1.0	
	18.5	6.5-3.0-1.0	
	19.4	6.5-3.0-1.5	
	20.4	6.5-3.0-1.0	
	21.5	8.5-3.0-0.5	
	22.0	8.5-3.0-0.5	
84.	12.9	5.0-2.0-3.0	Male
	13.9	6.0-2.0-1.0	Growth hormone deficiency plus gonadotrophin
	14.9	6.0-2.0-2.0	deficiency
	16.0	5.0-2.0-3.0	
	17.0	4.5-2.0-3.5	
	18.0	4.5-2.0-3.5	

Appendix III.1. (*Continued*)

No.	Age	Heath–Carter somatotype[a]	Tanner somatotype[b]
	19.3	4.5-2.0-3.5	
	20.0	4.0-2.0-4.5	
	21.3	4.0-2.0-4.5	
	22.3	4.5-2.0-5.0	
	23.3	4.5-2.0-4.5	
	25.5	4.0-2.0-5.0	
85.	10.9	6.0-2.5-1.0	Male
	12.3	5.5-2.5-2.0	Multiple pituitary deficiencies due to
	13.8	4.0-2.5-3.0	craniopharyngiomas
	15.3	4.5-2.5-3.5	
	16.6	4.5-2.5-4.0	
	17.6	3.5-2.5-4.5	
86.	8.0	9.0-2.0-1.0	Female
	9.5	8.0-2.0-1.0	Multiple pituitary deficiencies due to
	11.8	8.0-2.0-1.0	craniopharyngiomas
	12.8	6.0-2.0-1.0	
	14.3	5.0-2.0-1.0	
	15.7	6.0-2.0-1.0	
87.	5.6	9.0-3.5-0.5	Female
	6.7	6.0-3.0-1.0	Primary hypothyroidism
	7.8	6.0-3.0-1.0	
	8.8	5.0-4.0-1.5	
	9.8	5.0-4.0-1.5	
	10.9	4.5-4.0-2.0	
	11.9	5.0-4.0-2.0	
	12.8	4.5-4.0-2.0	
	13.8	5.0-4.5-1.5	
	14.8	5.5-4.5-1.5	
	15.8	4.5-4.5-2.0	
88.	4.6	10.0-3.0-0.5	Female
	5.8	8.0-3.0-1.0	Primary hypothyroidism
	6.7	7.0-3.0-1.0	
	8.7	7.0-3.0-1.5	
	9.7	6.0-3.0-2.0	
	10.7	5.0-3.0-2.5	
	11.7	4.0-3.0-4.0	
	12.7	3.5-3.0-4.0	
	13.7	3.5-3.0-4.0	
	14.7	3.5-3.5-4.0	
	15.7	5.0-3.5-2.5	
	16.9	5.0-3.5-2.5	
89.	13.0	3.0-2.5-3.5	Female
	14.0	3.0-2.5-3.0	Constitutional growth delay
	14.5	3.5-2.5-4.0	
	15.0	3.5-2.5-4.0	
	16.4	4.0-2.5-3.0	

(*continued*)

Appendix III.1. (*Continued*)

No.	Age	Heath–Carter somatotype[a]	Tanner somatotype[b]
	16.9	3.0-2.5 4.0	
	17.5	2.5-2.0-4.5	
	18.0	2.5-2.0-4.5	
	19.0	2.5-2.5-5.0	
	20.0	3.0-2.5-5.0	
	21.0	3.5-2.5-5.0	
90.	10.0	3.5-3.0-2.0	Male
	11.4	4.5-3.0-2.0	Constitutional growth delay
	12.8	3.0-3.0-3.0	
	13.8	3.5-3.0-3.0	
	14.8	4.0-2.0-3.0	
	15.8	4.0-3.0-3.0	
	16.8	4.0-3.0-4.0	
	17.8	4.0-3.0-4.0	
91.	2.5	8.0-3.5-0.5	Male
	3.4	6.0-3.5-1.0	Constitutional growth delay
	4.4	5.5-3.5-1.0	
	5.4	5.0-3.5-1.0	
	6.4	4.0-3.5-1.5	
	7.4	3.5-3.5-2.0	
	8.4	3.5-3.5-2.5	
	9.6	3.5-3.5-2.5	
	10.6	3.0-3.5-3.0	
	11.6	3.5-3.5-3.0	
	12.6	3.5-3.5-3.0	
	14.6	4.0-3.5-2.0	
	16.6	4.5-4.0-2.0	
92.	2.1	10.0-3.0-0.5	Female
	3.1	10.0-3.0-0.5	Constitutional tallness
	4.1	9.0-3.0-0.5	
	5.4	6.5-3.0-1.0	
	6.4	6.0-3.0-1.0	
	7.5	5.5-3.0-1.0	
	8.5	6.0-3.0-1.0	
	9.5	5.5-3.0-2.0	
	10.0	5.0-3.0-2.5	
	11.0	5.0-3.0-2.5	
	11.6	5.0-3.0-2.5	
	12.5	5.5-3.0-2.0	
	13.2	5.0-3.0-3.0	
93.	5.8	6.0-3.0-1.0	Female
	6.7	5.0-3.0-1.0	Silver–Russell syndrome
	7.9	4.5-3.0-2.0	
	8.4	4.0-3.0-2.5	
	8.4	4.0-3.0-3.0	
	8.9	4.0-3.0-3.0	
	9.8	4.0-3.0-3.0	
	11.0	4.0-3.0-3.0	
	11.5	3.5-3.0-3.0	

Appendix III.1. (*Continued*)

No.	Age	Heath–Carter somatotype[a]	Tanner somatotype[b]
	12.0	4.0-3.5-3.0	
	13.0	4.0-3.5-3.0	
	14.0	4.0-3.5-3.0	
	15.6	4.0-4.0-2.0	
	16.0	4.0-4.5-1.5	
94.	1.8	10.0-2.0-1.0	Female
	4.0	8.0-2.0-1.0	Silver–Russell syndrome
	5.3	4.0-2.0-2.0	
	6.3	4.0-2.0-2.0	
	7.3	3.5-2.0-3.0	
	8.3	2.5-2.0-4.5	
	9.3	2.0-2.0-4.5	
	10.3	2.5-2.0-5.0	
	11.3	2.5-2.0-5.0	
	12.3	3.0-2.0-4.0	
	13.3	4.0-2.5-3.0	
	14.3	4.0-2.5-3.0	
95.	2.7	8.5-3.0-0.5	Female
	3.7	7.5-3.0-1.0	Turner's syndrome
	4.6	7.0-3.0-1.0	
	5.8	5.5-3.5-1.0	
	6.8	5.5-3.5-1.0	
	7.9	5.5-3.5-1.0	
	8.7	5.5-3.5-1.0	
	9.7	5.0-3.5-1.5	
	10.7	5.0-3.5-1.5	
	11.6	5.5-3.5-1.0	
	12.4	5.5-3.5-1.0	
	13.4	5.5-3.5-1.0	
	14.8	6.0-4.0-1.0	
	16.6	6.0-4.0-1.0	
96.	10.0	4.5-3.5-1.0	Female
	10.4	4.5-3.5-1.0	Turner's syndrome
	11.0	5.5-4.0-1.0	
	11.6	5.5-4.0-1.0	
	12.1	5.0-4.0-1.0	
	12.6	4.5-4.5-1.0	
	13.1	4.5-4.5-1.0	
	13.4	4.5-4.5-1.0	
	14.0	4.0-5.0-1.0	
	15.0	5.0-5.0-1.0	
	15.6	6.0-5.0-1.0	
	16.2	5.5-5.0-1.0	
	16.7	5.5-5.0-1.0	
	17.7	6.0-5.0-1.0	

[a] Photoscopic Heath–Carter rating by Barbara H. Heath at sequential ages for each subject.
[b] Photoscopic rating by James M. Tanner. A single somatotype rating was assigned when the subject reached young adulthood.

Appendix III.2. *Ratings of somatotype photographs from* Varieties of Delinquent Youth *(Sheldon, 1949)*
Heath–Carter photoscopic somatotypes of 199 males whose photographs are in *Varieties of Delinquent Youth* (Sheldon, 1949). Ratings were made in 1985 by Barbara Heath. The same photographs also appear in *Physique and Delinquent Behavior* (Hartl *et al.*, 1982), but the heights for some subjects are not the same in the two books. The ratings were made for comparisons with the Sheldon methods and these are reported in Chapter 2. (Subject 57 was excluded from the analysis because data provided seem to be in error.) A somatochart (Fig. III.1) follows the list of data.

No.	Age (yrs)	Heath–Carter photoscopic rating	No.	Age (yrs)	Heath–Carter photoscopic rating
1	16	4.5-5.0-1.0	42	19	3.5-4.0-3.0
2	20	4.0-4.0-3.0	43	16	4.0-4.0-2.5
3	17	3.0-5.5-2.5	44	18	3.0-4.0-4.0
4	18	3.5-4.0-4.5	45	16	4.0-3.5-3.5
5	18	2.0-4.0-4.5	46	16	3.5-4.0-3.5
6	19	3.5-4.5-2.0	47	20	4.5-5.0-2.0
7	16	3.5-5.0-1.5	48	18	4.5-5.0-1.0
8	18	3.0-4.5-3.0	49	19	4.5-4.0-1.5
9	19	1.5-4.5-3.0	50	18	6.0-4.0-1.5
10	19	4.0-3.0-4.0	51	16	3.5-4.0-3.5
11	19	3.0-5.5-2.5	52	16	3.5-4.0-2.5
12	21	4.5-5.0-1.5	53	16	2.0-5.5-2.0
13	18	4.5-5.0-1.5	54	17	5.5-3.5-2.5
14	21	4.5-3.0-3.0	55	17	2.5-4.0-3.0
15	16	4.5-3.5-2.5	56	19	2.5-4.0-4.0
16	17	2.5-5.0-2.5			
17	17	3.0-5.5-2.0	58	17	4.0-5.5-1.5
18	19	5.0-4.5-1.0	59	17	3.5-3.5-3.5
19	19	5.5-4.0-2.0	60	16	4.0-4.0-2.0
20	19	3.0-4.5-3.0	61	19	3.5-4.5-2.5
21	21	3.5-4.5-2.0	62	16	3.0-5.5-1.5
22	21	4.0-4.0-3.0	63	18	2.5-5.0-2.0
23	15	4.0-4.5-2.0	64	18	3.0-4.0-3.5
24	18	3.0-4.5-3.0	65	17	2.5-5.0-2.0
25	20	4.0-4.0-3.0	66	23	3.5-5.0-1.5
26	16	3.5-4.0-3.5	67	21	3.5-4.5-3.0
27	19	2.5-6.0-1.5	68	17	4.0-4.0-4.0
28	20	4.5-4.5-1.0	69	18	3.5-4.5-2.0
29	16	3.5-4.5-2.0	70	18	3.5-5.0-2.0
30	16	4.5-4.5-2.5	71	21	4.0-4.5-2.5
31	19	2.0-3.0-5.0	72	16	2.5-5.0-3.0
32	21	4.0-4.0-1.5	73	22	2.5-4.0-3.5
33	17	4.0-3.5-3.5	74	18	3.0-5.0-2.5
34	21	2.0-6.0-1.5	75	21	3.5-3.5-4.5
35	16	2.0-6.0-1.5	76	17	4.0-4.5-1.5
36	17	4.5-4.0-1.0	77	19	4.0-4.0-1.5
37	16	4.0-4.5-2.0	78	20	3.5-4.5-2.0
38	18	3.0-4.0-3.0	79	16	4.5-3.0-4.0
39	17	3.0-4.5-3.0	80	19	3.5-4.5-2.5
40	17	3.0-3.5-4.0	81	16	4.0-3.0-2.5
41	17	2.5-4.5-3.5	82	18	3.5-4.0-3.0

Appendix III.2. (*Continued*)

No.	Age (yrs)	Heath–Carter photoscopic rating	No.	Age (yrs)	Heath–Carter photoscopic rating
83	16	3.5-5.0-3.0	130	15	3.0-5.0-3.0
84	17	3.5-5.0-2.5	131	17	2.5-4.5-4.0
85	18	4.0-5.0-3.0	132	16	3.5-5.0-2.5
86	24	4.0-4.5-3.0	133	17	9.5-3.0-0.5
87	20	4.0-5.0-2.0	134	21	3.5-5.5-2.0
88	19	3.0-4.0-4.5	135	20	2.5-2.5-6.0
89	21	4.0-4.0-3.0	136	21	3.0-5.0-2.0
90	18	4.0-5.5-1.5	137	17	5.0-3.5-2.0
91	18	3.0-5.0-3.0	138	21	5.0-6.0-1.0
92	21	5.0-4.0-1.5	139	16	2.5-2.5-6.0
93	18	4.5-4.0-4.0	140	22	2.0-5.0-3.5
94	19	2.5-4.0-4.0	141	20	2.5-7.0-1.0
95	18	2.5-4.0-4.0	142	21	3.5-5.0-1.5
96	18	4.5-4.5-1.5	143	18	3.5-4.0-4.0
97	17	5.0-3.5-3.0	144	18	3.0-3.5-4.0
98	23	2.5-5.0-3.0	145	17	2.5-7.0-1.0
99	19	6.0-4.5-1.0	146	16	3.5-4.0-3.0
100	17	10.0-4.5-0.5	147	20	4.0-4.0-3.5
101	18	5.0-3.5-3.0	148	18	4.0-4.0-4.0
102	21	1.0-6.0-2.0	149	22	3.5-5.5-1.5
103	18	5.0-6.0-1.0	150	17	2.0-5.0-3.5
104	18	2.5-4.5-3.5	151	17	5.0-5.0-1.0
105	17	3.0-4.0-4.0	152	18	4.0-3.5-4.5
106	16	10.0-3.0-0.5	153	17	3.0-5.0-2.5
107	21	3.0-5.0-3.0	154	16	2.5-4.0-5.0
108	22	5.0-4.5-2.0	155	17	3.0-5.0-2.0
109	16	3.5-3.0-4.5	156	16	2.0-5.5-2.5
110	18	6.0-4.0-1.0	157	21	3.0-4.0-3.5
111	20	5.0-4.0-1.5	158	17	4.5-4.5-2.5
112	16	5.0-4.0-2.5	159	19	3.5-4.0-3.5
113	16	3.0-7.0-1.0	160	19	3.5-3.5-3.5
114	17	4.0-4.0-3.0	161	21	1.5-4.5-4.0
115	18	3.5-2.0-5.5	162	16	3.0-5.0-2.5
116	17	3.0-4.0-4.0	163	17	3.5-5.0-1.5
117	19	5.0-4.5-1.0	164	20	4.5-5.0-1.5
118	21	4.0-5.5-1.0	165	21	4.0-4.0-2.5
119	19	3.5-7.0-1.0	166	24	5.0-4.5-1.5
120	18	4.5-5.0-2.5	167	21	3.5-5.5-2.0
121	17	4.0-6.0-1.0	168	18	4.0-3.5-4.0
122	17	3.0-5.5-1.5	169	22	5.0-3.5-4.0
123	19	3.5-4.5-3.0	170	16	2.5-4.0-4.0
124	20	3.5-5.0-2.0	171	18	3.5-4.5-3.0
125	16	3.5-4.5-4.0	172	18	4.5-5.0-2.0
126	21	3.0-4.0-4.0	173	17	2.0-5.0-3.0
127	22	6.5-4.5-1.0	174	18	3.5-4.0-3.5
128	18	4.0-4.0-4.0	175	20	2.5-4.5-3.0
129	21	2.5-4.5-3.5	176	21	4.0-3.5-4.5

(*continued*)

Appendix III.2. (*Continued*)

No.	Age (yrs)	Heath–Carter photoscopic rating	No.	Age (yrs)	Heath–Carter photoscopic rating
177	19	3.0-5.5-2.5	189	16	3.5-6.5-1.0
178	21	2.5-3.5-4.0	190	20	4.0-4.0-3.0
179	18	4.5-4.0-2.0	191	16	1.5-6.5-2.0
180	23	4.0-4.0-3.5	192	15	4.5-5.0-1.5
181	17	4.5-3.5-4.0	193	19	4.5-5.0-1.5
182	17	4.0-4.0-3.0	194	18	2.0-6.0-2.0
183	16	4.0-4.0-3.5	195	19	4.0-5.5-1.0
184	22	4.5-4.0-3.0	196	17	4.0-5.0-1.5
185	17	3.0-5.5-1.5	197	17	3.0-6.5-1.0
186	19	4.0-4.0-2.5	198	19	4.0-4.5-2.0
187	18	3.5-5.0-2.5	199	17	3.0-5.0-2.0
188	20	3.5-5.0-2.0	200	21	4.0-5.0-1.5

Fig. III.1. Somatotype distribution of males in *Varieties of Delinquent Youth* (Sheldon, 1949). Photoscopic ratings Heath–Carter ratings by Heath. ▲ = mean somatotype.

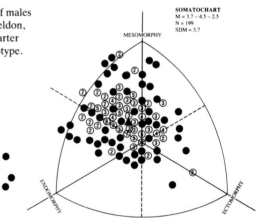

SOMATOCHART
M = 3.7 – 4.5 – 2.5
N = 199
SDM = 3.7

MESOMORPHY

ENDOMORPHY

ECTOMORPHY

Appendix III.3. *Ratings of somatotype photographs from* Atlas of Men

Heath–Carter photoscopic somatotypes of 256 subjects whose photographs are in *Atlas of Men* (Sheldon, 1954). Ratings were made in 1978 by Barbara Heath. The data are listed by four age groups. The last group is of 24 subjects whose HWR was less than 11.00 (36.37 metric). Further information is given in Chapter 2 (including somatocharts) and has been published in Carter (1985*a*).

No.	Age (yrs)	HWR	Heath–Carter photoscopic rating
Age = 18–29 years, n = 141			
AM0001	18	14.82	1.0-2.0-7.0
AM0008	19	14.50	1.5-2.5-6.5
AM0043	18	13.87	1.5-4.0-5.0
AM0104	19	12.60	2.0-6.5-1.5
AM0111	19	13.09	1.5-6.0-3.0
AM0127	19	12.49	2.0-7.0-1.5
AM0134	19	14.32	2.5-1.5-5.5
AM0141	19	13.98	3.5-1.5-5.0
AM0146	19	14.35	2.5-2.0-6.5
AM0154	18	14.02	2.0-2.0-5.0
AM0172	18	14.03	2.0-2.5-5.5
AM0204	18	13.76	2.5-3.0-4.5
AM0213	19	13.91	3.0-3.0-5.5
AM0219	18	13.73	3.0-3.5-5.0
AM0227	19	13.35	2.0-4.5-3.0
AM0256	18	13.72	2.0-4.0-4.5
AM0264	19	13.10	2.0-5.0-2.5
AM0326	19	12.47	2.0-6.5-1.0
AM0349	19	12.62	2.0-6.5-1.5
AM0384	18	13.96	3.5-2.0-5.5
AM0401	19	13.73	3.5-2.0-5.0
AM0408	19	13.62	4.0-3.0-5.0
AM0415	18	13.84	3.5-3.0-5.5
AM0422	19	13.65	3.0-3.0-4.0
AM0470	19	13.03	3.5-4.0-2.5
AM0537	19	12.48	4.0-5.0-1.5
AM0562	18	13.13	3.0-5.0-4.0
AM0576	19	12.46	3.0-6.0-1.5
AM0582	19	12.23	4.0-6.0-1.0
AM0628	18	13.46	4.0-2.0-3.5
AM0633	19	13.28	4.0-2.5-3.5
AM0639	18	13.23	4.5-2.5-3.0
AM0647	19	13.53	4.0-3.0-5.0
AM0715	18	12.99	4.0-3.5-2.5
AM0729	19	12.79	4.5-3.5-2.0
AM0755	18	12.89	4.5-4.0-3.0
AM0821	18	12.21	4.0-6.0-1.0
AM0848	19	11.94	4.5-6.5-1.0
AM0858	19	12.06	4.5-6.5-1.0
AM0876	19	13.22	5.0-2.5-3.5
AM0894	18	12.74	5.0-3.5-2.0

(continued)

Appendix III.3. (*Continued*)

No.	Age (yrs)	HWR	Heath–Carter photoscopic rating
AM0901	19	12.51	5.5-3.5-2.0
AM0907	19	13.01	5.0-3.0-3.0
AM0918	18	12.69	4.5-4.0-2.0
AM0940	19	12.29	5.5-4.5-1.5
AM0979	18	11.85	5.0-6.0-0.5
AM1006	19	12.01	5.5-5.5-1.0
AM1025	18	11.97	5.5-5.5-1.0
AM1033	19	11.66	5.5-6.5-1.0
AM1042	18	12.46	7.0-1.5-1.5
AM1059	18	12.30	7.5-2.5-2.0
AM1066	19	11.75	8.0-3.5-1.0
AM1076	19	12.26	6.5-4.0-1.5
AM1087	19	11.97	6.0-3.0-2.0
AM1095	18	11.88	6.0-5.0-1.0
AM1112	18	12.09	5.5-5.0-1.5
AM1119	19	11.48	8.0-4.5-1.0
AM1132	18	12.13	6.5-4.0-1.5
AM1137	19	11.39	6.5-6.0-0.5
AM1144	19	11.11	8.0-5.5-0.5
AM1149	18	11.35	9.5-3.0-0.5
AM1158	18	11.40	8.5-3.5-0.5
AM0017	20	14.37	1.5-2.5-6.5
AM0024	24	13.99	2.0-3.0-5.0
AM0031	26	14.33	2.0-2.5-6.5
AM0038	21	13.93	2.0-3.0-5.0
AM0052	22	13.65	2.0-4.0-4.5
AM0062	23	13.52	1.5-5.0-4.0
AM0075	21	13.40	2.0-5.0-4.0
AM0084	24	12.93	1.0-6.0-2.0
AM0093	23	12.98	2.0-5.5-2.5
AM0118	26	12.87	2.0-6.0-2.5
AM0166	24	13.61	2.5-3.5-4.0
AM0179	24	14.16	3.5-2.0-6.5
AM0184	28	13.92	3.0-3.0-6.0
AM0196	24	13.63	3.0-4.0-5.0
AM0240	20	13.32	3.0-4.0-3.5
AM0251	23	13.23	2.5-4.5-4.0
AM0277	28	12.83	3.0-4.5-2.0
AM0289	25	13.02	2.5-5.0-2.5
AM0303	22	13.00	2.5-5.0-2.5
AM0311	22	12.74	2.5-6.0-2.0
AM0318	29	12.92	2.0-5.5-3.0
AM0332	28	12.22	3.0-7.0-1.0
AM0340	21	12.65	2.5-6.0-2.0
AM0356	20	12.91	2.5-6.0-3.0
AM0361	22	12.93	2.5-5.5-3.0
AM0368	23	12.07	3.5-7.0-1.0
AM0375	20	12.21	3.0-7.0-1.0
AM0389	28	13.61	3.5-2.5-4.5

Appendix III.3. (*Continued*)

No.	Age (yrs)	HWR	Heath–Carter photoscopic rating
AM0433	23	13.33	3.5-3.0-3.5
AM0445	28	13.53	3.0-3.5-4.5
AM0457	23	13.13	3.5-4.0-3.0
AM0486	23	13.30	3.0-4.5-4.0
AM0495	20	13.07	3.5-4.5-3.0
AM0502	23	13.20	3.5-4.5-4.0
AM0511	20	12.82	2.5-5.5-2.0
AM0521	23	12.68	3.5-4.5-1.5
AM0553	22	12.78	3.0-6.0-3.0
AM0570	20	12.93	3.0-5.5-3.0
AM0588	28	11.93	4.0-7.0-1.0
AM0595	24	12.50	3.5-6.0-2.0
AM0603	23	12.21	4.0-6.5-1.5
AM0609	21	11.90	4.0-7.0-0.5
AM0616	24	11.87	4.0-7.0-1.0
AM0622	22	13.44	4.0-2.5-3.5
AM0652	23	13.22	4.5-3.0-4.0
AM0662	24	12.83	4.0-3.5-1.5
AM0677	24	12.80	4.0-4.0-2.0
AM0687	26	13.03	4.0-4.0-3.0
AM0695	22	12.99	4.0-4.0-4.0
AM0705	24	13.49	3.0-4.0-4.5
AM0740	27	12.33	5.0-4.5-1.5
AM0761	25	12.93	4.5-4.5-3.5
AM0770	23	12.83	4.5-4.5-3.0
AM0779	20	12.42	4.0-5.0-1.0
AM0791	20	12.32	4.5-5.0-1.0
AM0800	24	11.83	5.0-6.0-1.0
AM0811	20	12.60	4.0-5.0-2.0
AM0831	24	12.36	4.5-5.5-2.0
AM0838	22	12.33	4.5-5.5-2.0
AM0867	27	11.22	5.5-7.0-0.5
AM0882	22	12.63	5.0-3.0-2.0
AM0912	24	12.99	5.5-2.5-3.5
AM0924	29	12.01	5.0-5.5-1.0
AM0933	20	12.22	5.0-5.0-1.0
AM0948	20	12.68	5.0-3.5-2.0
AM0953	21	12.20	5.5-4.5-2.0
AM0959	27	12.30	6.5-3.5-3.0
AM0969	20	12.02	5.0-5.5-0.5
AM0989	25	11.98	5.5-5.0-1.5
AM0999	21	12.50	5.0-4.5-2.5
AM1016	25	11.37	5.5-6.5-1.0
AM1048	23	12.29	7.0-3.5-2.0
AM1053	21	12.45	8.0-2.0-2.5
AM1071	29	11.15	7.5-5.5-1.0
AM1081	20	12.40	6.5-3.5-2.0

(*continued*)

Appendix III.3. (*Continued*)

No.	Age (yrs)	HWR	Heath–Carter photoscopic rating
AM1104	24	11.27	8.0-4.5-0.5
AM1126	20	11.43	6.5-6.0-0.5
AM1164	26	10.55	11.5-3.0-0.5
AM1170	21	10.48	10.5-4.0-0.5
Age = 30–39 years, n = 32			
AM0006	34	14.34	2.0-1.5-5.5
AM0072	37	13.09	2.0-5.5-3.0
AM0152	38	13.70	3.5-2.0-4.0
AM0228	35	13.04	2.5-5.0-3.0
AM0247	37	13.17	4.0-3.5-4.0
AM0283	37	12.20	4.0-6.0-1.0
AM0309	34	12.82	3.5-4.5-2.0
AM0339	34	11.87	4.5-7.0-1.0
AM0378	32	12.03	4.0-7.0-1.0
AM0430	34	13.05	4.5-3.5-3.0
AM0453	38	12.71	4.5-4.0-2.0
AM0484	37	12.93	4.5-4.0-3.0
AM0513	32	12.45	5.0-4.0-1.5
AM0535	32	12.31	4.5-5.0-1.5
AM0551	36	12.20	4.5-6.0-1.5
AM0608	35	11.49	6.0-7.0-0.5
AM0657	33	12.71	4.5-3.5-2.0
AM0684	32	12.37	5.5-4.0-1.5
AM0711	33	12.75	4.5-4.0-3.0
AM0746	35	12.46	4.5-4.5-1.5
AM0777	39	11.63	6.0-5.5-0.5
AM0823	39	11.25	6.5-6.5-1.0
AM0843	34	11.35	6.0-7.0-1.0
AM0868	33	11.21	3.0-9.0-0.5
AM0900	30	11.94	7.0-4.0-1.0
AM0944	33	12.24	7.0-3.5-1.5
AM0984	37	11.61	6.0-5.5-1.0
AM1010	32	11.40	6.5-6.0-1.0
AM1029	39	11.17	8.0-5.0-0.5
AM1089	30	11.02	9.0-3.5-0.5
AM1111	34	11.65	8.0-4.0-1.0
AM1139	34	10.40	8.5-6.0-0.5
Age = 40–49 years, n = 25			
AM0003	40	14.76	1.0-1.0-7.0
AM0034	47	14.07	2.5-2.5-6.0
AM0054	40	13.42	1.5-4.0-4.5
AM0074	49	13.01	2.0-6.0-3.0
AM0101	49	12.41	3.0-6.5-1.5
AM0147	43	14.18	2.5-2.0-6.0
AM0177	44	13.52	4.0-2.5-4.0
AM0211	47	13.48	4.0-3.0-5.0
AM0275	41	12.55	3.5-5.0-1.5
AM0313	43	12.28	4.5-5.5-1.5

Appendix III.3. (*Continued*)

No.	Age (yrs)	HWR	Heath–Carter photoscopic rating
AM0362	48	12.43	5.0-5.0-2.0
AM0414	46	13.37	4.0-3.0-4.0
AM0449	42	13.19	3.5-4.0-4.0
AM0491	42	12.88	4.5-3.5-3.0
AM0552	40	12.08	5.5-5.0-1.5
AM0613	40	11.31	5.5-7.0-0.5
AM0663	46	12.20	6.0-4.0-1.0
AM0724	49	11.82	6.0-5.0-1.0
AM0751	40	12.40	5.5-4.0-1.5
AM0810	47	11.48	7.0-6.0-1.0
AM0850	44	10.70	8.5-6.0-0.5
AM0929	45	11.59	7.5-4.5-1.0
AM0985	44	11.46	7.5-5.0-0.5
AM1018	45	10.50	9.5-5.0-0.5
AM1092	44	11.07	8.5-5.0-0.5
Age = 50 and above, n = 34			
AM0016	50	14.11	3.0-2.0-5.5
AM0064	54	13.33	1.5-5.0-3.5
AM0088	54	12.39	2.0-6.5-1.0
AM0218	54	13.52	4.0-3.0-5.0
AM0236	52	13.01	4.5-3.5-2.5
AM0263	50	12.47	4.0-5.0-1.0
AM0281	51	12.25	5.0-5.0-1.0
AM0296	50	12.59	3.5-5.0-1.5
AM0403	51	13.11	4.0-2.5-4.0
AM0454	57	12.52	5.0-4.0-1.5
AM0569	51	12.44	4.0-5.5-2.0
AM0625	58	13.25	5.0-2.0-3.0
AM0675	54	12.05	7.5-3.5-1.0
AM0692	57	12.38	5.5-4.5-2.0
AM0782	56	11.53	6.0-6.0-1.0
AM0796	53	11.44	7.5-5.0-0.5
AM0885	51	12.22	6.5-3.0-1.0
AM0925	58	11.47	8.0-4.5-0.5
AM1012	55	11.08	7.5-6.0-0.5
AM0119	61	12.71	3.0-6.0-2.5
AM0161	66	13.71	3.0-1.5-4.5
AM0195	62	13.23	3.0-4.0-3.0
AM0427	66	12.81	4.5-3.5-3.0
AM0467	61	12.49	4.5-4.5-1.5
AM0485	61	12.71	4.5-4.0-2.0
AM0520	60	11.87	5.0-6.0-1.0
AM0528	64	11.73	6.5-5.0-0.5
AM0544	62	11.60	5.5-6.0-1.0
AM0714	61	11.98	5.5-5.0-1.0
AM0732	65	11.91	7.5-3.5-1.0
AM0749	65	12.17	6.0-4.0-1.0

(*continued*)

Appendix III.3. (*Continued*)

No.	Age (yrs)	HWR	Heath–Carter photoscopic rating
AM0814	63	11.66	7.5-4.5-1.0
AM0841	65	11.65	7.0-5.0-1.0
AM0974	60	11.28	8.0-5.0-1.0
Subjects whose HWR was less than 11.00, n = 24			
AM1018	45	10.50	9.5-5.0-0.5
AM1032	55	10.77	8.5-6.0-0.5
AM1036	37	10.55	6.5-8.0-0.5
AM1037	41	10.51	7.5-7.0-0.5
AM1038	52	10.00	7.0-9.0-1.0
AM1041	30	10.90	7.5-6.0-0.5
AM1118	34	10.77	9.5-5.0-0.5
AM1128	24	10.81	8.0-5.0-0.5
AM1131	53	10.82	9.0-5.5-1.0
AM1138	23	10.95	8.5-5.0-0.5
AM1139	34	10.40	8.5-6.0-0.5
AM1141	26	10.90	8.5-5.0-0.5
AM1142	43	10.18	9.5-6.0-0.5
AM1146	25	10.61	9.5-5.0-0.5
AM1155	39	9.10	13.0-4.0-0.5
AM1156	13	10.82	11.0-3.5-0.5
AM1159	24	10.91	9.5-4.0-0.5
AM1160	43	9.82	11.5-5.0-0.5
AM1161	20	10.90	8.5-5.0-0.5
AM1166	26	10.20	10.5-5.0-0.5
AM1167	39	9.70	12.5-4.0-0.5
AM1169	20	10.65	8.5-6.0-0.5
AM1174	23	10.18	11.0-4.0-0.5
AM1175	31	9.70	12.5-4.0-0.5

Glossary

anthropometry The measurement of dimensions of the human body.

component As used in reference to a somatotype it is one of the three components that make up the somatotype, namely endomorphy, mesomorphy and ectomorphy.

dysplasia Literally, bad shape or form. In somatotyping it refers to disharmony or uneven distribution of a component or components in different anatomical regions of the body.

ectomorphy The relative linearity or slenderness of a physique. The third number in the somatotype.

endomorphy The relative fatness of a physique. The first number in the somatotype.

height–weight ratio (HWR) Height divided by the cube root of weight. Closely related to ratings in ectomorphy.

mesomorphy The musculoskeletal robustness relative to height of a physique. The second number in the somatotype.

photoscopy Visual observation of the physical characteristics of the somatotype as observed and rated in a somatotype photograph. (Sometimes used interchangeably with **anthroposcopy**, which has a broader meaning in physical anthropology.)

physique A general word for the whole body. There are many methods of classifying physiques, of which somatotyping is one.

rating The process of assigning values to each of the three components that make up the somatotype.

somatochart A two-dimensional projection of the relationships among somatotypes.

somatoplot A projection of a somatotype location in three-dimensional space (i.e. somatopoint) onto a two-dimensional grid or somatochart.

somatopoint A point in three-dimensional space determined from the somatotype which is represented by a triad of x, y and z coordinates for the three components. The scales on the coordinate axes are component units with the hypothetical somatotype of 0-0-0 at the origin of the three axes.

somatotype (S) A quantification of the present shape and composition of the human body. It is expressed in a three-number rating representing endomorphy, mesomorphy and ectomorphy respectively and always in the same order.

somatotype attitudinal distance (SAD) The distance in three dimensions between any two somatopoints. Calculated in component units.

somatotype attitudinal mean (SAM) The average of the SADs of each somatopoint from the mean somatopoint ($\bar{S}$) of a sample.

somatotype dispersion distance (SDD) The distance in two dimensions between any two somatoplots. Calculated in the y-units of the two-dimensional somatochart.

453

somatotype dispersion mean (SDM)   The average of the SDDs of each somatoplot from the mean somatoplot ($\bar{S}$) of a sample.

somatotyping   The process of obtaining or rating the somatotype. A specific method of classifying physique.

References

Acheson, R.M. & Dupertuis, C.W. (1957). The relationship between physique and rate of skeletal maturation in boys. *Human Biology*, **29**, 167–93.

Adcock, G.J. (1948). A factorial examination of Sheldon's types. *Journal of Personality*, **16**, 312–19.

Adelson, D. & Turner, A. (1963). A note on the relation of somatotype to outcome after phenothiazine therapy in chronic schizophrenic patients. *Journal of Nervous and Mental Disease*, **137**, 242–5.

Aitken, A., Clarke, P.T., Brown, I.J. & Kay, D. (1980). The relationship between body shape (somatotype) and body satisfaction (cathexis) in male physical educators. *Scottish Journal of Physical Education*, **8**(3), 30–3.

Alexander, M.J.L. (1976). The relationship of somatotype and selected anthropometric measures to basketball performance in highly skilled females. *Research Quarterly of the American Association for Health, Physical Education and Recreation*, **47**, 575–85.

Allen, N. (1965). A factor analysis of selected college football ability test items. PhD Thesis, University of Oregon, Eugene (Microcard PE700).

Alonso, R.F. (1986). Estudio del somatotipo de los atletas de 12 años de la EIDE occidentales de Cuba. *Boletin de Trabajos de Anthropologia*, [April], 3–18.

Alt, P.M. (1953). Relationship of physique and temperament. *The School Review*, **61**, 267–76.

Amador, M., Rodríguez, C. & Gonzales, M.E. (1983). Somatotyping as a tool for nutritional assessment in preschool children. *Anthropologiai Közlemények*, **27**, 109–18.

Anastasi, A. (1958). *Differential Psychology*. New York: Macmillan.

Anderson, L.P. (1985). Somatotype and skinfolds of Down's syndrome adolescents. MA Thesis, San Diego State University.

Angel, J.L. (1949). Constitution in female obesity. *American Journal of Physical Anthropology*, **7**, 443–71.

Ansley, H.R., Sheldon, W.H. & Elderkin, R.D. (1957). Erythrocytes in schizophrenic patients. *Diseases of the Nervous System*, **18**, 444–5.

Ansley, H.R., Lawrenson, S. & Ansley, S. (1963). Internal structural correlates with somatotypes. I. Red blood cells, small veins and viscera. *Journal of the Mount Sinai Hospital*, **30**, 199–216.

Araújo, C.G.S. (1978). Somatotyping of top swimmers by the Heath–Carter method. In *Swimming Medicine IV*, ed. B. Erickson & B. Furberg, pp. 188–98. Baltimore: University Park Press.

Araújo, C.G.S. (1979). Comparison of somatotypes on different age groups of Brazilian swimmers. *Medicine and Science in Sports*, **11**, 103 (abstract).

Araújo, C.G.S. & Moutinho, M.F.C. (1978). Somatotype and body composition of adolescent Olympic gymnasts. *Caderno Artus de Medicina Desportiva*, **1**, 39–42.

Araújo, C.G.S., Gomes, P.S.C. & Moutinho, M.F.C. (1978*a*). Compogram: a new method to plot somatotypes. *Caderno Artus de Medicina Desportiva*, **1**, 43–6.

Araújo, C.G.S., Gomes, P.S.C. & Novaes, E.V. (1978*b*). The somatotype of high level Brazilian judokas. *Caderno Artus de Medicina Desportiva*, **1**, 21–30.

Araújo, C.G.S., Pavel, R.C. & Gomes, P.S.C. (1978*c*). Comparison of somatotype and speed in competitive swimming at different phases of training. In *Swimming Medicine III*, ed. J. Terauds & E.W. Beddingfield, pp. 329–37. Baltimore: University Park Press.

Araújo, C.G.S., Clarys, J.P. & Duarte, M.F. (1986). Somatotypes of male physical education students – a new approach. In *Celafiscs – Dez Anos de Contribuição às Ciências do Esporte*, p. 138. Laboratorio de Aptidao Fisica de Sao Caetano do Sul, SP, Brasil. (Abstract in Portuguese).

Arnot, R.B. & Gaines, C.L. (1984). *Sportselection*. New York: Viking Press.

Arraj, T. & Arraj, J. (1985). *A Tool for Understanding Human Differences: How to Discover and Develop your Type According to Dr C.G. Jung and Dr William Sheldon*. Chiloquin, Oregon: Tools for Inner Growth.

Atchley-Carlson, B.J. (1981). Effects of a five-month training program on anthropometric dimensions of female and male competitive swimmers. MA Thesis, San Diego State University.

Bagnall, K.M. & Kellett, D.W. (1977). A study of potential Olympic swimmers I. The starting point. *British Journal of Sports Medicine*, **11**, 127–32.

Bahamondes, A., Oyarzun, F. & Matte, I. (1952). Determinacion del somatotipo de 100 alcoholicos mediante el metodo de Sheldon. *Revista de Psiquiatria*, **17**, 47–68.

Bailey, D.A., Carter, J.E.L. & Mirwald, R. (1982). Somatotypes of Canadian men and women. *Human Biology*, **54**, 813–28.

Bailey, S.M. (1985). Human physique and susceptibility to noninfectious disease. *Yearbook of Physical Anthropology*, **28**, 149–73.

Bainbridge, D.R. & Roberts, D.F. (1966). Dysplasia in Nilotic physique. *Human Biology*, **38**, 251–78.

Bakker, H.K. & Struikenkamp, R.S. (1977). Biological variability and lean body mass estimates. *Human Biology*, **49**, 187–202.

Bale, P. (1969). Somatotyping and body physique. *Physical Education*, **61**, 75–82.

Bale, P. (1979). The relationship between physique and basic motor performance in a group of female physical education students. *Carnegie Research Papers*, **1**, 26–32.

Bale, P. (1980). The relationship of physique and body composition to strength in a group of physical education students. *British Journal of Sports Medicine*, **14**, 193–8.

Bale, P. (1981). Body composition and somatotype characteristics of sportswomen. In *The Female Athlete*, ed. J. Borms, M. Hebbelinck & A. Venerando, pp. 157–67. Basel: Karger.

Bale, P. (1983). *The Somatotypes of Sportsmen and Sportswomen*. Eastbourne: Brighton Polytechnic.

Bale, P. (1986). The relationship of somatotype and body composition to strength in a group of men and women sport science students. In *Perspectives in Kinanthropometry*, ed. J.A.P. Day, pp. 187–97. Champaign, Illinois: Human Kinetics.

Bale, P., Colley, E. & Mayhew, J. (1984). Size and somatotype correlates of strength and physiological performance in adult male students. *Australian Journal of Science and Medicine in Sport*, **16**, 2–6.

Bale, P., Colley, E. & Mayhew, J.L. (1985*a*). Relationships among physique, strength, and performance in women students. *Journal of Sports Medicine and Physical Fitness*, **25**, 98–103.

Bale, P., Rowell, S. & Colley, E. (1985*b*). Anthropometric and training characteristics of female marathon runners as determinants of distance running performance. *Journal of Sports Sciences*, **3**, 115–26.

Barrell, G.V. & Cooper, P.J. (1982). Somatotype characteristics of international orienteers. *Perceptual and Motor Skills*, **53**, 767–70.

Barton, W.H. & Hunt, E.E., Jr (1962). Somatotype and adolescence in boys: a longitudinal study. *Human Biology*, **34**, 254–70.

Bayley, N. & Bayer, L.M. (1946). The assessment of somatic androgeny. *American Journal of Physical Anthropology*, **4**, 433–61.

Behnke, A.R. (1953). The relation of lean body weight to metabolism and some consequent systematizations. *Annals of the New York Academy of Sciences*, **56**, 1095–142.

Behnke, A.R. & Wilmore, J.H. (1974). *Evaluation and Regulation of Body Build and Composition*. Englewood Cliffs, New Jersey: Prentice–Hall.

Behnke, A.R., Guttentag, O.E. & Brodsky, C. (1957). *Quantifications of Body Configuration in Geometrical Terms*. Research and Development Technical Report USNRDL-TR-204, NS 080-001 US.

Berry, J.N. (1972). Somatotype distribution in male college students in northern India. *American Journal of Physical Anthropology*, **36**, 85–94.

Berry, J.N. & Deshfukh, P.Y. (1964). Somatotypes of male college students in Nagpur, India. *Human Biology*, **36**, 157–76.

Beulen, A. (1956). Sheldon: vers une psychologie constitutionelle. *Revue d'Education Physique*, **178**, 313–21.

Beunen, G. (1973–4). Somatotype and skeletal maturity in boys 12 through 14. *Hermes (Leuven)*, **8**, 411–22.

Beunen, G. & Van Hellemont, A. (1980). Body structure and somatotype in physical education students. *Anthropologiai Közlemények*, **24**, 15–21.

Beunen, G., Ostyn, M., Renson, R., Simons, J., Swalus, P. & Van Gerven, D. (1977). Somatotype and physical fitness in fourteen-year-old boys. In *Frontiers of Activity and Child Health*, ed. H. Lavallee & R.J. Shephard, pp. 115–23. Quebéc: Pelican.

Beunen, G., Claessens, A., Ostyn, M., Renson, R., Simons, J. & Van Gerven, D. (1981*a*). Skeletal maturation (TW2) and somatotype. In *Child Growth and Development*, ed. H. Lavallee & R.J. Shephard, pp. 245–53. Université du Quebéc a Trois-Rivières.

Beunen, G., Claessens, A. & van Esser, M. (1981*b*). Somatic and motor characteristics of female gymnasts. In *The Female Athlete*, ed. J. Borms, M. Hebbelinck & A. Venerando, pp. 176–85. Basel: Karger.

Beunen, G., Claessens, A., Lefevre, J., Renson, R., Simons, J. & Van Gerven, D. (1986). Somatotype as related to age at peak velocity in height, weight, and static strength. In *Kinanthropometry II*, ed. T. Reilly, J. Watkins & J. Borms, pp. 68–72. London: Spon.

Bevans, M.T. (1977). A comparison of somatotypes in female gymnasts and distance runners. MS Thesis, Springfield College. (University of Oregon, Microform PE1991f).

Bláha, P. & Seifertová, V. (1981). Dílčí výsledky z longitudinálního sledování tělesného rozvoje dětí sportovních tříd (Partial results of longitudinal follow-up study of body development of children attending sport classes). In *Telesná kultúra v živote dieťaťa*. Materiály z celoštátnej konferencie usporiadanej pri príležitosti Medzinárodného roka dieťaťa. Bratislava. 22–23 November, 1979. Sborník vědecké rady ÚV ČSTV, 12, pp. 265–8.

Bodel, J.K., Jr (1950). Distribution and permanence of body build in boys. PhD Thesis, Harvard University.

Bodzsar, E.B. (1982). The indices of the physique and the socio-economic factors based on a growth study in Bakony girls. *Anthropologiai Közlemények*, **26**, 129–34.

Boennec, P., Prevot, M. & Ginet, J. (1980). Somatotype de sportif de haut niveau. Résultats dans huit disciplines différentes. *Médecine du Sport*, **54**, 309–18.

Bok, V. (1974). A comparison of selected illustrations of creative works from the point of view of constitutional typology. *Acta Universitatis Carolinae (Gymnica)*, **10**, 79–91.

Bok, V. (1976). Comparison of somatotypes in certain works of art with the view to the beauty of the living human body. In *International Conference on Physical Education, AIESEP*, ed. R. Linc, pp. 191–6. Prague: Universita Karlova.

Bok, V. (1981) Příspěvek k poznání dědičné podmíněnosti somatotypu a některých antropometrických charakteristík (Contribution to the knowledge of genetical determination of somatotype and of some anthropometric characteristics). PhD Dissertation, Universita Karlova, Prague.

Bok, V. (1983). The comparison of Adam's and Eve's depiction in selected style periods from the point of view of somatotype. *Acta Universitatis Carolinae (Gymnica)*, **19**, 73–84.

Bok, V. & Tlapáková, E. (1982). New ways of somatotyping and their application. *Acta Universitatis Carolinae (Gymnica)*, **18**, 5–19.

Borms, J. & Hebbelinck, M. (1984). Review of studies on Olympic athletes. In *Physical Structure of Olympic Athletes. Part II: Kinanthropometry of Olympic Athletes*, ed. J.E.L. Carter, pp. 7–27. Basel: Karger.

Borms, J., Hebbelinck, M. & Ross, W.D. (1972). Somatotype and skeletal maturity in 12 year old boys. *Pediatric Work Physiology, Proceedings of the Fourth International Symposium*, ed. O. G. Bar-Or, pp. 85–91. Israel: Wingate Institute.

Borms, J., Hebbelinck, M. & Schraepen, D. (1976). Anthropometric assessment of the self-portraits of eleven and twelve year old boys. In *Physical Education, Sports and the Sciences*, ed. J. Broekhoff, pp. 307–20. University of Oregon, Eugene: Microform Publications.

Borms, J., Hebbelinck, M. & Van Gheluwe, B. (1977). Early and late maturity in Belgian boys, 6 to 13 years of age and its relation to body type. In *Growth and Development: Physique*, ed. O. Eiben, pp. 399–406. Budapest: Akadémiai Kiadó (Hungarian Academy of Sciences).

Borms, J., Van Der Meer, J., De Schepper, P. & Hebbelinck, M. (1978). A multivariate comparison of the physical and socio-psychological profile of Belgian physical educators and general subject teachers. In *Proceedings of Twentieth World Congress of Physical Education, Health and Recreation*, July 1977, Mexico City, vol. 2, pp. 627–39. Mexico City: SODIFEF.

Borms, J., Ross, W.D., Duquet, W. & Carter, J.E.L. (1986). Somatotypes of world class body builders. In *Perspectives in Kinanthropometry*, ed. J.A.P. Day, pp. 81–90. Champaign, Illinois: Human Kinetics.

Bösze, P., Eiben, O.G. & Laszlo, J. (1982). Somatotype analysis of patients with hypergonadotropic primary amenorrhoea. (In Hungarian). *Humanbiologia Budapestinensis*, **13**, 61–3.

Bouchard, C. (1977). Univariate and multivariate genetic analysis of anthropometric and physique characteristics of French Canadian families. PhD Thesis, University of Texas, Austin.

Bouchard, C., Landry, E., LeBlanc, C. & Mondor, J. (1974). Quelques-unes des caracteristiques physiques et physiologiques des jouers de hockey et leurs relations avec la performance. *Mouvement*, **9**, 95–110.

Bouchard, C., Demirjian, A. & Malina, R.M. (1980a). Heritability estimates of somatotype components based upon familial data. *Human Heredity*, **30**, 112–18.

Bouchard, C., Demirjian, A. & Malina, R.M. (1980b). Path analysis of family resemblance in physique. *Studies in Physical Anthropology*, **6**, 61–70.

Bridges, P.K. & Jones, M.T. (1973). Relationship between some psychological assessments, body-build, and physiological stress responses. *Journal of Neurology, Neurosurgery, and Psychiatry*, **36**, 839–45.

Brief, F.K. (1986). Somatotipo y caracteristicas antropometricas de los atletas Bolivarianos. Universidad Central de Venezuela, Caracas.

Broekhoff, J. (1966). Relationships between physical, socio-psychological, and mental characteristics of thirteen-year-old boys. PhD Thesis, University of Oregon, Eugene (Microcard PSY248).

Broekhoff, J. (1976). Physique types and perceived physical characteristics of elementary school children with low and high social status. In *Physical Education, Sports and the Sciences*, ed. J. Broekhoff, pp. 332–5. University of Oregon, Eugene: Microform Publications.

Broekhoff, J., Santomier, J. & Bennett, J. (1978). Somatotypes and personality profiles of male and female physical education students. In *Proceedings of the Twentieth World Congress of Physical Education, Health and Recreation*, July 1977, Mexico City, vol. 2, pp. 706–18. Mexico City: SODIFEF.

Broekhoff, J., Nadgir, A. & Pieter, W. (1986). Morphological differences between young gymnasts and non-athletes matched for age and gender. In *Kinanthropometry III*, ed. T. Reilly, J. Watkins, & J. Borms, pp. 204–10, London: Spon.

Brouwer, D. (1957). Somatotypes and psychosomatic diseases. *Journal of Psychosomatic Research*, **2**, 23–34.

Brown, G.M. (1960). Relationship between body types and static posture of young women. *Research Quarterly*, **31**, 403–8.

Brown, G.M. & Mott, P.H. (1951). Somatotyping: two technical modifications of the Sheldon procedure. *Medical Radiography and Photography*, **27**, 20–2.

Brožek, J. (1961). Body measurements, including skinfold thickness, as indicators of body composition. In *Techniques for Measuring Body Composition*, ed. J. Brožek & A. Henschel, pp. 3–35. Washington, D.C.: National Academy of Sciences.

Brožek, J. (1965). *Human Body Composition*. New York: Pergamon Press.

Brožek, J. and Keys, A. (1952). Body-build and body composition. *Science*, **116**, 140–2.

Buday, J. & Eiben, O.G. (1982). Somatotype of adult Down's patients. *Anthropologiai Közlemények*, **26**, 71–7.

Bulbulian, R. (1984). The influence of somatotype on anthropometric prediction of body composition in young women. *Medicine and Science in Sports and Exercise*, **16**, 4, 389–97.

Bullen, A.K. (1945). A cross-cultural approach to the problem of stuttering. *Child Development*, **16**, 1–88.

Bullen, A.K. (1952). Some problems in the practical application of somatotyping. *Florida Anthropologist*, **5**, 17–20.

Bullen, A.K. & Hardy, H.L. (1946). Analysis of body build photographs of 175 college women. *American Journal of Physical Anthropology*, **4**, 37–65.

Burdick, J.A. & Tess, D. (1983). A factor analytic study based on the Atlas of Men. *Psychological Reports*, **52**, 511–16.

Burian, R.J. (1969). A study of the relationship between female body physique and a number of psycho-social correlates. PhD Thesis, Arizona State University, Tempe.

Burt, C. (1944). The factorial study of physical types. *Man*, **44**, 82–6.

Butts, N.K. (1982). Physiological profile of high school female cross-country runners. *The Physician and Sportsmedicine*, **10**, 11, 103–11.

Caldeira, S., Matsudo, V.K.R., Vívolo, M.A. & Sessa, M. (1986a). The somatotype characteristics of South American volleyball players. In *Celafiscs – Dez Anos de Contribuição às Ciências do Esporte*, p. 136. Laboratorio de Aptidao Fisica de Sao Caetano do Sul, SP, Brasil. (Abstract in Portuguese.)

Caldeira, S., Stanziola, L. & Matsudo, V.K.R. (1986b). Relationships between somatotype and agility among top athletes. In *Celafiscs – Dez Anos de Contribuição às Ciências do Esporte*, p. 139. Laboratorio de Aptidao Fisica de Sao Caetano do Sul, SP, Brasil. (Abstract in Portuguese.)

Caldeira, S., Vívolo, M.A. & Matsudo, V.K.R. (1986c). Somatotype among Brasilian volleyball players. In *Celafiscs – Dez Anos de Contribuição às Ciências do Esporte*, pp. 116–9. Laboratorio de Aptidao Fisica de Sao Caetano do Sul, SP, Brasil. (Abstract in Portuguese.)

Caldeira, S., Vívolo, M.A. & Matsudo, V.K.R. (1986d). Somatotype characteristics from college volleyball players from different Brazilian regions. In *Celafiscs – Dez Anos de Contribuição às Ciências do Esporte*, p. 137. Laboratorio de Aptidao Fisica de Sao Caetano do Sul, SP, Brasil. (Abstract in Portuguese.)

Caldeira, S., Vívolo, M.A. & Matsudo, V.K.R. (1986e). Somatotype of athletes at different sports in Olympic center of training and research. In *Celafiscs – Dez Anos de Contribuição às Ciências do Esporte*, p. 137. Laboratorio de Aptidao Fisica de Sao Caetano do Sul, SP, Brasil. (Abstract in Portuguese.)

Carruth, W.A. (1952). The relationships of constitutional build and motor ability. PhD Thesis, New York University.

Carter, J.E.L. (1958). An analysis of somatotypes of boys aged twelve to seventeen years. MA Thesis, State University, Iowa, Iowa City.

Carter, J.E.L. (1964). The physiques of male physical education teachers in training. *Journal of the Physical Education Association of Great Britain and Northern Ireland*, **56**, 66–76.

Carter, J.E.L. (1965). The physiques of female physical education teachers in training. *Journal of the Physical Education Association of Great Britain and Northern Ireland*, **57**, 6–16.

Carter, J.E.L. (1966). The somatotypes of swimmers. *Swimming Technique*, **3**, 76–9.

Carter, J.E.L. (1968). Somatotypes of college football players. *Research Quarterly*, **39**, 476–81.

Carter J.E.L. (1970). The somatotypes of athletes: a review. *Human Biology*, **42**, 535–69.

Carter, J.E.L. (1971). Somatotype characteristics of champion athletes. In *Anthropological Congress dedicated to Aleš Hrdlička*, ed. V.V. Novotny, pp. 242–52. Czechoslovak Academy of Sciences, Prague: Academia.

Carter, J.E.L. (1974). Somatotype, growth and physical performance. In *The Regulation of the Adipose Tissue Mass*, ed. J. Vague & J. Boyer, pp. 254–64. Amsterdam: Excerpta Medica.

Carter, J.E.L. (1978). The prediction of outstanding athletic ability – the structural perspective. In *Exercise Physiology*, ed. F. Landry & W.A.R. Orban, pp. 29–42. Miami: Symposia Specialists.

Carter, J.E.L. (1980a). *The Heath–Carter Somatotype Method*, 3rd edn. San Diego: San Diego State University Syllabus Service.

Carter, J.E.L. (1980b). The contributions of somatotyping to kinanthropometry. In *Kinanthropometry II*, ed. M. Ostyn, A. Beunen & J. Simons, pp. 411–24. Baltimore: University Park Press.

Carter, J.E.L. (1981). Somatotypes of female athletes. In *The Female Athlete*, ed. J. Borms, M. Hebbelinck & A. Venerando, pp. 85–116. Basel: Karger.

Carter, J.E.L. (ed.) (1984a). *Physical Structure of Olympic Athletes. Part II. Kinanthropometry of Olympic Athletes*. Basel: Karger.

Carter, J.E.L. (1984b). Somatotypes of Olympic athletes from 1948 to 1976. In *Physical Structure of Olympic Athletes. Part II. Kinanthropometry of Olympic Athletes*, ed. J.E.L. Carter, pp. 80–109. Basel: Karger.

Carter, J.E.L. (1985a). A comparison of ratings by Heath and Sheldon of somatotypes in 'Atlas of Men'. In *Physique and Body Composition*, vol. 16, ed. O.G. Eiben, pp. 13–22. Budapest: Humanbiologia Budapestinensis.

Carter, J.E.L. (1985b). Morphological factors limiting human performance. In *The Limits of Human Performance, The American Academy of Physical Education Papers*, No. 18, ed. H.M. Eckert & D.H. Clarke, pp. 106–17. Champaign, Illinois: Human Kinetics.

Carter, J.E.L. (1988). Somatotypes of children in sport. In *Children and Sport*, ed. R.M. Malina, pp. 153–65. Champaign, Illinois: Human Kinetics.

Carter, J.E.L. & Brallier, R.M. (1988). Physiques of specially selected female gymnasts. In *Children and Sport*, ed. R.M. Malina, pp. 167–75. Champaign, Illinois: Human Kinetics.

Carter, J.E.L. & Heath, B.H. (1971). Somatotype methodology and kinesiology research. In *Kinesiology Review*, pp. 10–19. Washington D.C.: American Association for Health, Physical Education and Recreation.

Carter, J.E.L. & Heath, B.H. (1986). Comparison of somatotypes of young adults by two methods. In *Kinanthropometry III*, ed. T. Reilly, J. Watkins & J. Borms, pp. 63–7. London: Spon.

Carter, J.E.L. & Lucio, F.D. (1986). Body size, skinfolds, and somatotypes of high school and Olympic wrestlers. In *Perspectives in Kinanthropometry*, ed. J.A.P. Day, pp. 171–80. Champaign, Illinois: Human Kinetics.

Carter, J.E.L. & Pařízková, J. (1978). Changes in somatotypes of European males between 17 and 24 years. *American Journal of Physical Anthropology*, **48**, 251–4.

Carter, J.E.L. & Phillips, W.H. (1969). Structural changes in exercising middle-aged males during a two-year period. *Journal of Applied Physiology*, **27**, 787–94.

Carter, J.E.L. & Rahe, R.H. (1975). Effects of stressful underwater demolition training on body structure. *Medicine and Science in Sports*, **7**, 304–8.

Carter, J.E.L. & Rendle, M.L. (1965). *The Physiques of Royal New Zealand Air Force Men*. Report to the R.N.Z.A.F., Air Department, Wellington, New Zealand. (University of Oregon, Microcard PSY188.)

Carter, J.E.L., Ross, W.D., Kasch, F.W. & Phillips, W.H. (1965). Body types of middle-aged males in training. *Journal of the Association for Physical and Mental Rehabilitation*, **19**, 148–52.

Carter, J.E.L., Sleet, D.A. & Martin, G.N. (1971). Somatotypes of male gymnasts. *Journal of Sports Medicine and Physical Fitness*, **11**, 2–11.

Carter, J.E.L., Štěpnička, J. & Clarys, J.P. (1973). Somatotypes of male physical education majors in four countries. *Research Quarterly*, **44**, 361–71.

Carter, J.E.L., Štěpnička, J. & Climie, J.F. (1978). Somatotypes of female physical education majors in three countries. In *Proceedings of the Twentieth World Congress of Physical Education, Health, and Recreation*. July 1977, Mexico City, vol. 2, pp. 698–705. Mexico City: SODIFEF.

Carter, J.E.L., Rendle, M.L. & Gayton, P.H. (1981). Size and somatotype of Olympic male field hockey players. *New Zealand Journal of Sports Medicine*, **9**, 8–13.

Carter, J.E.L., Aubry, S.P. & Sleet, D.A. (1982). Somatotypes of Montreal Olympic athletes. In *Physical Structure of Olympic Athletes. Part I. The Montreal Olympic Games Anthropological Project*, ed. J.E.L. Carter, pp. 53–80. Basel: Karger.

Carter, J.E.L., Ross, W.D., Duquet, W. & Aubry, S.P. (1983). Advances in somatotype methodology and analysis. *Yearbook of Physical Anthropology*, **26**, 193–213.

Chen, K.P., Damon, A. & Elliot, O. (1963). Body form, composition, and some physiological functions of Chinese on Taiwan. *Annals of the New York Academy of Sciences*, **110**, 760–77.

Chernilo, B.B., Soto, J. & Fernández, A.A.V. (1979). *Composición Corporal y Somatotipo en Judokas. Juegos Panamericanos, Puerto Rico, 1979*. Santiago: Unidad de Salud, Comité Olímpico de Chile.

Child, I.L. (1950). The relation of somatotype to self ratings on Sheldon's temperamental traits. *Journal of Personality*, **18**, 440–3.

Child, I.L. & Sheldon, W.H. (1941). The correlation between components of physique and scores on certain psychological tests. *Character and Personality*, **10**, 23–34.

Chovanová, E. (1976*a*). The physique of the Czechoslovak top ice-hockey players. *Acta Facultatis Rerum Naturalium Universitatis Comenianae Anthropologia*, **22**, 115–18.

Chovanová, E. (1976*b*). Somatotypes of top-class skiers. (Jumpers, cross-country runners, combined events competitors and downhill runners.) *Acta Facultatis Rerum Naturalium Universitatis Comenianae Anthropologia*, **24**, 63–77.

Chovanová, E. (1981). The problems of the selection of talented downhill skiers from the somatic point of view. *Teorie a Praxe Tělesné Výchovy*, **29**(4), 212–18.

Chovanová, E. & Pataki, L. (1982). Physique of young throwers and its relation to the inter-individual variability of sports performance. *Humanbiologia Budapestinensis*, **13**, 27–40.

Chovanová, E. & Zapletalová, L. (1980). Size, shape and body proportion of young Czechoslovak basketball players. *Anthropologiai Közlemények*, **24**, 39–44.

Chovanová, E. & Zrubák, A. (1972). Somatotypes of prominent Czechoslovak ice-hockey and football players. *Acta Facultatis Rerum Naturalium Universitatis Comenianae Anthropologia*, **21**, 59–62.

Chovanová, E., Bergman, P. & Štukovský, R. (1981). The share of heredity on forming a somatotype. Sborník VR UV CSTV, 'Telesná Kultúra v živote dieťaťa'. *Sport (Bratislava)*, **12**, 218–21.

Chovanová, E., Bergman, P. & Štukovský, R. (1982a). Genetic aspects of somatotypes in twins. *Modern Man Anthropos* (Brno), **22**, 5–12.

Chovanová, E., Drobńy, I., Bednarcikova, M. & Stefko, P. (1982b). Body build, somatotypes and muscular efficiency of children brought up in children's homes and mentally retarded children from special boarding schools. *IInd Anthropological Congress of Aleš Hrdlička, Universitas Carolina Pragensis*, pp. 295–8.

Chovanová, E., Pataki, L. & Vavrovic, D. (1983a). Somatotypological characteristics of young weight lifters. *Teorie a Praxe Tělesné Výchovy*, **31**(1), 32–5.

Chovanová, E., Slamka, M. & Pataki, L. (1983b). Utilization of the ascertainment of body structure and composition in the training process. *Teorie a Praxe Tělesné Výchovy*, **31**(3), 102–6.

Chytráčková, J. (1979). The relation of somatotypes and performance in women. *Teorie a Praxe Tělesné Výchovy*, **27**(3), 161–6.

Claessens, A. (1981). Stability of the body structure and of the somatotype. Follow-up study on Belgian boys aged 13 to 18 years. PhD Dissertation, Katholieke Universiteit Leuven, Belgium.

Claessens, A., Beunen, G., Lefevre, J., Martens, G. & Wellens, R. (1986a). Body structure, somatotype, and motor fitness of top-class Belgian judoists and karateka: a comparative study. In *Kinanthropometry III*, ed. T. Reilly, J. Watkins, & J. Borms, pp. 53–7. London: Spon.

Claessens, A., Beunen, G., Simons, J.M., Wellens, R., Geldof, D. & Nuyts, M.N. (1986b). Body structure, somatotype, and motor fitness of top-class Belgian judoists. In *Perspectives in Kinanthropometry*, ed. J.A.P. Day, pp. 155–64. Champaign, Illinois: Human Kinetics.

Clarke, H.H. (1967). *Application of Measurement of Health and Physical Education*, 4th edn. Englewood Cliffs, NJ: Prentice Hall.

Clarke, H.H. (1971). *Physical and Motor Tests in the Medford Boys Growth Study*. Englewood Cliffs, NJ: Prentice–Hall.

Clarys, J.P. & Borms, J. (1971). Typologische studie van waterpolospelers en gymnasten (Typological study of waterpoloplayers and gymnasts). *Geneeskunde en sport*, **4**(1), 2–8.

Clarys, J.P., Borms, J. & Hebbelinck, M. (1970). Somatotypologie bij kinderen uit het lager onderwijs (Somatotypology of children from primary education). *Belgisch Archief van Sociale Geneeskunde en Hygiene*, **6**, 427–40.

Clauser, C.E., McConville, J.H. & Young, J.W. (1969). *Weight, Volume, and Center of Mass of Segments of the Human Body*. AMRL-TR-69-70, Aerospace Medical Research Laboratory, Air Force Systems Command, Wright-Patterson Air Force Base, Ohio.

Cockeram, B.W. (1975). Physique of arts and physical education students. *New Zealand Journal of Health, Physical Education and Recreation*, **8**(2), 62–6.

Conrad, K. (1941). *Die Konstitutionstypen als Genetisches Problem*. Berlin: Springer–Verlag.

Conrad, K. (1963). *Der Konstitutionstypus*, 2nd edn. Berlin: Springer–Verlag.

Copley, B.B. (1980a). An anthropometric, somatotypological and physiological study of tennis players with special reference to the effects of training. PhD Thesis, University of the Witwatersrand, Johannesburg.

Copley, B.B. (1980*b*). A morphological and physiological study of tennis players with special reference to the effects of training. *South African Journal for Research in Sports, Physical Education and Recreation*, **3**(2), 33–44.

Cortes, J.B. (1961). Physique, need for achievement, and delinquency. PhD Thesis, Harvard University, Cambridge.

Cortes, J.B. & Gatti, F.M. (1965). Physique and self-description of temperament. *Journal of Consulting Psychology*, **29**, 432–9.

Cortes, J.B. & Gatti, F.M. (1966). Physique and motivation. *Journal of Consulting Psychology*, **30**, 408–14.

Cortes, J.B. & Gatti, F.M. (1970). Physique and propensity. *Psychology Today*, October, 42–4, 82–4, 99.

Cortes, J.B. & Gatti, F.M. (1972). *Delinquency and Crime: A Biopsychosocial Approach*. New York: Seminar Press.

Craig, L.S. & Bayer, L.M. (1967). Androgenic phenotypes in obese women. *American Journal of Physical Anthropology*, **26**, 23–34.

Cressie, N.A.C., Withers, R.T. & Craig, N.P. (1986). The statistical analysis of somatotype data. *Yearbook of Physical Anthropology*, **29**, 197–208.

Cross, J.A. (1968). Relationships between selected physical characteristics of boys at twelve and fifteen years of age and their personality characteristics at eighteen years of age. EdD Thesis, University of Oregon, Eugene. (Microcard PSY343.)

Cureton, T.K. (1947). *Physical Fitness, Appraisal and Guidance*. London: Kimpton.

Cureton, T.K. (1951). *Physical Fitness of Champion Athletes*. Urbana, Illinois: University of Illinois Press.

Cureton, T.K. & Hunsicker, P. (1941). Body build as a framework of reference for interpreting physical fitness and athletic performance. *Research Quarterly*, **12**, 301–30.

Damon, A. (1942). A note on the estimation of dysplasia in human physiques: Sheldon's method and the analysis of variance. *Human Biology*, **14**, 110–12.

Damon, A. (1955). Physique and success in military flying. *American Journal of Physical Anthropology*, **13**, 217–52.

Damon, A. (1957). Constitutional factors in acne vulgaris. *Archives of Dermatology*, **76**, 172–8.

Damon, A. (1960). Host factors in cancer of the breast and uterine cervix and corpus. *Journal National Cancer Institute*, **24**, 483–516.

Damon, A. (1961). Constitution and smoking. *Science*, **134**, 339–41.

Damon, A. (1963). Constitution and alcohol consumption: physique. *Journal of Chronic Diseases*, **16**, 1237–50.

Damon, A. (1965*a*). Delineation of the body build variables associated with cardiovascular diseases. *Annals of the New York Academy of Sciences*, **126**, 711–27.

Damon, A. (1965*b*). Discrepancies between findings of longitudinal and cross-sectional studies in adult life, physique and physiology, *Human Development*, **8**, 16–22.

Damon, A. (1970). Constitutional medicine. In *Anthropology and the Behavioural and Health Sciences*, ed. O. Von Mering & L. Kasdan, pp. 179–95. University of Pittsburgh Press, Pittsburgh.

Damon, A. & McFarland, R.A. (1955). The physique of bus and truck drivers, with a review of occupational anthropology. *American Journal of Physical Anthropology*, **13**, 711–42.

Damon, A. & Polednak, A.P. (1967a). Constitution, genetics, and body form in peptic ulcer: a review. *Journal of Chronic Diseases*, **20**, 787–802.

Damon, A. & Polednak, A.P. (1967b). Physique and serum pepsinogen. *Human Biology*, **39**, 355–67.

Damon, A., Fowler, E.P., Jr. & Sheldon, W.H. (1955). Constitutional factors in otosclerosis and Meniere's disease. *Transactions of American Academy of Ophthalmology and Otolaryngology*, **59**, 444–58.

Damon, A., Bleibtreu, K., Elliot, O. & Giles, E. (1962). Predicting somatotype from body measurements. *American Journal of Physical Anthropology*, **20**, 461–74.

Danby, P.M. (1953). A study of the physique of some native East Africans. *Journal of the Royal Anthropological Institute*, **83**, 194–214.

Davidson, M.A., Lee, D., Parnell, R.W. & Spencer, S.J.C. (1955). The detection of psychological vulnerability in students. *Journal of Mental Science*, **101**, 810–25.

Davidson, M.A., McInnes, R.G. & Parnell, R.W. (1957). The distribution of personality traits in seven year old children: A combined psychological, psychiatric and somatotype study. *British Journal of Educational Psychology*, **27**, 48–61.

Day, J.A.P., Duquet, W. and Meerseman, G. (1977). Anthropometry and physique type of female middle and long distance runners, in relation to speciality and level of performance. In *Growth and Development: Physique*, ed. O. Eiben, pp. 385–97. Budapest: Académiai Kiadó (Hungarian Academy of Sciences).

De Garay, A.L., Levine, L. & Carter, J.E.L. (1974). *Genetic and Anthropological Studies of Olympic Athletes*. New York: Academic Press.

De Pauw, D. & Vrijens, J. (1972). Physique, muscle strength and cardiovascular fitness of weight-lifters. *Journal of Sports Medicine and Physical Fitness*, **12**, 192–200.

De Rose, E.H., Lampert, A. & Oliveira, J.L. (1979). Avaliação funcional dos arbritos de futebol da di visao especial da federação gaúcha de futebol. *Revista Brasileira de Ciências do Esporte*, **1**(1), 40 (abstract).

De Rose, E.H. & Guimarães, A.C.S. (1980). A model for optimization of somatotype in young athletes. In *Kinanthropometry II*, ed. M. Ostyn, G. Beunen & J. Simons, p. 222. Baltimore: University Park Press.

De Woskin, S.F. (1967). Somatotypes of women in a fitness program. MA Thesis, San Diego State University. (University of Oregon, Microcard PE10007.)

Deabler, H.L., Hartl, E.M. & Willis, C.A. (1973). Physique and personality somatotype and the 16 personality factor questionnaire. *Perceptual and Motor Skills*, **36**(3), 927–33.

Dempster, W.T. (1955). Space requirements of the seated operator. (WADC Tech. Report 55-159) U.S.A.F. Wright Patterson Air Force Base, Ohio, 245 pp.

Deutsch, M. & Ross, W.D. (1978). The impact of CAN/SDI on the emergence of a new multidisciplinary field: Kinanthropometry. In *Biomechanics of Sports and Kinanthropometry*, vol. 6, ed. F. Landry & W.A.R. Orban, pp. 465–74. Symposia Specialists: Miami.

Dibiase, W.J. & Hjelle, L.A. (1968). Body-image stereotypes and body-type preference among male college students. *Perceptual and Motor Skills*, **27**, 1143–6.

Dobzhansky, T. (1941). *Genetics and Origin of the Species*. New York: Columbia University Press.

Dobzhansky, T. (1962). *Mankind Evolving*. New Haven: Yale University Press.

Dobzhansky, T. (1967). On types, genotypes, and the genetic diversity in populations. In *Genetic Diversity and Human Behaviour*, ed. J.N. Spuhler, pp. 1–18. Viking Fund Publications in Anthropology, No. 45. New York: Wenner–Gren Foundation for Anthropological Research.

Docherty, D., Eckerson, J.D. & Hayward, J.S. (1986). Physique and thermoregulation in prepubertal males during exercise in a warm, humid environment. *American Journal of Physical Anthropology*, **70**, 19–24.

Dolan, K. (1987). Physical characteristics of elite male triathletes. MA Thesis, San Diego State University.

Donnan, F. de S. (1959). Physique and hazardous occupations: a comparison of 160 Royal Marines. *British Medical Journal*, **2**(5154), 728–31.

Draper, G. (1924). *Human Constitution: A Consideration of its Relationship to Disease*. Philadelphia: Saunders.

Draper, G., Dupertuis, C.W. & Caughey, J.L., Jr. (1944). *Human Constitution in Clinical Medicine*, pp. 132–43. New York: Hoeber.

Drobńy, I., Chovanová, E., Hlatká, M., Bednarčíkova, M., Štefko, P. & Danihelová, Z. (1980). Physique and muscle strength of children brought up in children's homes and of handicapped (mentally retarded) children from special boarding schools. *V Zborníku, Sociálna a pracovná integrácia postihnutého jedinca*. MŠ SSR, 15 stran.

Drowatzky, J.N. (1965). Mental, social, maturity, and physical characteristics of boys underaged and normal-aged in elementary school grades. EdD Thesis, University of Oregon, Eugene (Microcard PSY217).

Dupertuis, C.W. (1950). Anthropometry of extreme somatotypes. *American Journal of Physical Anthropology*, **8**, 367–83.

Dupertuis, C.W. (1963). A preliminary somatotype description of Turkish, Greek and Italian military personnel. In *Anthropometric Survey of Turkey, Greece, and Italy*, ed. H.T.E. Hertzberg, E. Churchill, C.W. Dupertuis, R.M. White & A. Damon, pp. 35–60. New York: Macmillan.

Dupertuis, C.W. & Emanuel, I. (1956). *A Statistical Comparison of the Body Typing Methods of Hooton and Sheldon*. U.S. Air Force, Wright Air Development Center. WACD Technical Report 56-366, (ASTIA Document AD 97205).

Dupertuis, C.W. & Michael, N.B. (1953). Comparison of growth in height and weight between ectomorphic and mesomorphic boys. *Child Development*, **24**, 203–214.

Dupertuis, C.W. & Tanner, J.M. (1950). The pose of the subject for photogrammetric anthropometry, with special reference to somatotyping. *American Journal of Physical Anthropology*, **8**, 27–47.

Dupertuis, C.W., Pitts, G.C., Osserman, E.F., Welham, W.C. & Behnke, A.R. (1951*a*). Relation of specific gravity to body build in a group of healthy men. *Journal of Applied Physiology*, **3**, 676–80.

Dupertuis, C.W., Pitts., G.C., Osserman, E.F., Wellham, W.C. & Behnke, A.R. (1951*b*). Relation of body water content to body build in a group of healthy men. *Journal of Applied Physiology*, **4**, 364–7.

Duquet, W. (1980). Studie van de toepasbaarheid van de Heath & Carter somatotype-methode op kinderen van 6 tot 13 jaar (Applicability of the Heath–Carter somatotype method to 6 to 13 year old children). PhD Dissertation, Vrije Universiteit Brussel, Belgium.

Duquet, W. & Hebbelinck, M. (1977). Applications of the somatotype attitudinal distance to the study of group and individual somatotype status and relations. In

Growth and Development: Physique, ed. O. Eiben, pp. 377–84. Budapest: Akadémiai Kiadó (Hungarian Academy of Sciences).

Duquet, W., Hebbelinck, M. & Borms, J. (1975). Somatotype distributions of primary school boys and girls. In *Proceedings of the 18th International Congress of the International Council on Health, Physical Education and Recreation*, ed. D. Schmull, pp. 326–34. Zeist, The Netherlands: The Jan Luiting Foundation.

Eiben, O.G. (1980). Recent data on variability in physique: some aspects of proportionality. In *Kinanthropometry II*, ed. M. Ostyn, G. Beunen & J. Simons, pp. 69–77. Baltimore: University Park Press.

Eiben, O.G. (1981). Physique of female athletes – anthropological and proportional analysis. In *The Female Athlete*, ed. J. Borms, M. Hebbelinck & A. Venerando, pp. 127–41. Basel: Karger.

Eiben, O. G. (1982). The Körmend Growth Study: Body measurements. *Anthropologiai Közlemények*, **26**, 181–210.

Eiben, O.G. (1985). The Körmend Growth Study: Somatotypes. In *Physique and Body Composition*, vol. 16, ed. O.G. Eiben, pp. 37–52. Budapest: Humanbiologia Budapestinensis.

Eiben, O.G. & Eiben, E. (1979). The physique of European table-tennis players. *Collegium Anthropologist*, **3**, 1, 67–76.

Eiben, O.G., Kelemen, A., Pethö, B. & Felsövályi, Á. (1980). Physique of endogenous psychotic female patients. *Anthropologiai Közlemények*, **24**, 77–82.

Eiben, O.G., Bösze, P., László, J., Buday, J. & Gaál, M. (1985). Somatotype of patients with streak gonad syndrome. In *Physique and Body Composition*, vol. 16, ed. O.G. Eiben, pp. 53–64. Budapest: Humanbiologia Budapestinensis.

Ekman, G. (1951). On typological and dimensional systems of reference in describing personality. *Acta Psychologica*, **8**, 1–24.

Epps, P. & Parnell, R.W. (1952). Physique and temperament of women delinquents compared with women undergraduates. *British Journal of Medical Psychology*, **25**, 249–55.

Eränkö, O. & Karvonen, M.J. (1955). Body types of Finnish champion lumberjacks. *American Journal of Physical Anthropology*, **13**, 331–43.

Everett, P.W. & Sills, F.D. (1952). The relationship of grip strength to stature, somatotype components, and anthropometric measurements of the hand. *Research Quarterly*, **23**, 161–6.

Falls, H.B. & Humphrey, L.D. (1978). Body type and composition differences between placers and non-placers in an AIAW gymnastics meet. *Research Quarterly*, **49**, 38–43.

Farmosi, I. (1978). Telesna grada igračica (The physique of dancers). *Glasnik Antropološkog Društva Jugoslavije*, **15**, 147–50.

Farmosi, I. (1980*a*). Body-composition, somatotype and some motor performance of judoists. *The Journal of Sports Medicine and Physical Fitness*, **20**(4), 431–4.

Farmosi, I. (1980*b*). A sportoló nők alkati es morotikus vizsgalatanak tapasztalatai (Experiences at the morphologic and motor examinations of sportswomen). *Tanulmányok a TFKI kutatásaiból*, 105–17.

Farmosi, I. (1982). Results of constitutional and motor examinations of male athletes. *Glasnik Antropološkog Društva Jugoslavije*, **19**, 35–51.

Farmosi, I., Horvath, B. & Semjen, S. (1985). Okolvivok testosszetetele es szomatotipusa (Body composition and somatotype of boxers). *Testnevelés és Sporttudomány*, **4**, 35–8.

Faulkner, R.A. (1976). Physique characteristics of Canadian figure skaters. MSc Thesis, Simon Fraser University, Burnaby. (University of Oregon, Microcard PE1841f.)

Felker, D.W. (1972). Social stereotyping of male and female body types with differing facial expressions by elementary school age boys and girls. *Journal of Psychology*, **82**(1), 151–4.

Fernandes, E. & Corazza, S. (1986). Tennis player's somatotype in both sexes. In *Celafiscs – Dez Anos de Contribuição às Ciências do Esporte*, p. 439. Laboratorio de Aptidao Fisica de Sao Caetano do Sul, SP, Brasil. (Abstract in Portuguese.)

Finlay, S.E. (1956). The physique of students. *University of Leeds Medical Journal*, **5**, 76–9.

Fisher, R. (1975). Grandmothers and granddaughters: A descriptive and comparative study of selected measurements. PhD Thesis, University of Southern California, Los Angeles.

Fiske, P.W. (1944). A study of relationships to somatotype. *Journal of Applied Psychology*, **28**, 504–19.

Fleischmann, J., Linc, R., Štěpnička, J., Hošek, V. & Vaněk, M. (1977). Somatotypen und ihre biochemische und psychische korrelation. (The somatotypes and their biochemical and psychic correlations.) *Acta Universitatis Carolinae, Gymnica*, **13**, 83–8.

Fredman, M. (1972). Somatotypes in a group of Tamil diabetics. *South African Medical Journal*, **46**, 1836–7.

Fredman, M. (1974). Body constitution and blood glucose and serum insulin levels in a group of Tamil Indians. In *The Regulation of the Adipose Tissue Mass*, ed. J. Vague & J. Boyer, pp. 194–7. Amsterdam: Exerpta Medica.

Garn, S.M. & Gertler, M.M. (1950). An association between type of work and physique in an industrial group. *American Journal of Physical Anthropology*, **8**, 387–97.

Garn, S.M., Gertler, M.M. & Sprague, H.B. (1950). Somatotype and serum-cholesterol. *Circulation*, **2**, 380–91.

Gayton, P.H. (1975). The physique of New Zealand surf life savers. *New Zealand Journal of Health, Physical Education and Recreation*, **8**(3), 114–20.

George, S.C. (1985). Physique and body composition of type II diabetic women. MA Thesis, San Diego State University.

Gertler, M.M. (1950). The chemotype and the somatotype. Proceedings of the 19th Annual Meeting, A.A.P.A. *American Journal of Physical Anthropology*, **8**, 261–2.

Gertler, M.M. (1967). Ischemic heart disease, heredity and body build as affected by exercise. *Canadian Medical Association Journal*, **96**, 728–30.

Gertler, M.M. & White, P.D. (1954). *Coronary Heart Disease in Young Adults*. Cambridge, Mass: Harvard University Press.

Gertler, M.M., Garn, S.M. & Sprague, H.B. (1950). Cholesterol, cholesterol esters and phospholipids in health and in coronary artery disease. II. *Morphology and Serum Lipids in Man*, pp. 381–91.

Gertler, M.M., Garn, S.M. & Levine, S.A. (1951*a*). Serum uric acid in relation to age and physique in health and in coronary heart disease. *Annals of Internal Medicine*, **34**, 1421–31.

Gertler, M.M., Garn, S.M. & White, P.D. (1951*b*). Young candidates for coronary heart disease. *Journal of the American Medical Association*, **147**, 621–5.

Gibbens, T.C.N. (1963). *Psychiatric Studies of Borstal Lads*. New York: Oxford University Press.

Glueck, S. & Glueck, E. (1950). *Unravelling Juvenile Delinquency*. Cambridge, Mass: Harvard University Press.

Glueck, S. & Glueck, E. (1956). *Physique and Delinquency*. New York: Harper Brothers.

Glueck, S. & Glueck, E. (1970). *Toward a Typology of Juvenile Offenders*. New York: Grune and Stratton.

Goff, C.W. (1954). *Legg–Calvé–Perthes Syndrome. Chapter 5*. Constitutional Aspects. Springfield, Illinois: Thomas.

Gordon, E. (1984). Physique, lipids and kilojoule intake of a group of South African Caucasoid young adults. PhD Thesis, University of the Witwatersrand, Johannesburg.

Greene, W.H. (1961). Interrelations between measures of personal-social status and the relationships of these measures to selected physical factors of ten year old boys. MS Thesis, University of Oregon, Eugene (Microcard PWY139).

Greene, W.H. (1964). Peer status and level of aspiration of boys as related to their maturity, physique, structural, strength, and motor ability characteristics. EdD Thesis, University of Oregon, Eugene (Microcard PSY192).

Greenlee, G.A. (1986). The relationship of somatotype and isokinetic strength measures to lower extremity injuries in female athletes. In *Kinanthropometry III*, ed. T. Reilly, J. Watkins, & J. Borms, pp. 191–6. London: Spon.

Gregerson, M.I. & Nickerson, J.L. (1950). Relation of blood volume and cardiac output to body type. *Journal of Applied Physiology*, **3**, 329–41.

Guedes, D.P. (1983). Estudos antropometricos entre escolares. *Reviste Brazileire de Edução Fisica e Desportos*, **out/mar**, 12–17.

Guimarães, A. & De Rose, E.H. (1980). Somatotypes of Brazilian student track and field athletes of 1976. In *Kinanthropometry II*, ed. M. Ostyn, G. Beunen & J. Simons, pp. 231–8. Baltimore: University Park Press.

Gyenis, G. (1985). Body composition and socio-economic factors in male university students in Hungary. In *Physique and Body Composition*, vol. 16, ed. O.G. Eiben, pp. 65–70. Budapest: Humanbiologia Budapestinensis.

Gyenis, G., Héra, G., Endrödi, K., Kardos, I.L. & Eiben, O.G. (1980). Somatic and psychological characteristics of Hungarian female drivers. *Anthropologiai Közlemények*, **24**, 99–104.

Haley, J.S. (1974). The somatotypes of fifteen-year-old male basketball players, distance runners and sprinters. MA Thesis, San Diego State University.

Hall, R. (ed.) (1982). *Sexual Dimorphism in Homo Sapiens*. New York: Praeger.

Hammond, W.H. (1957). The status of physical types. *Human Biology*, **29**, 223–41.

Hanley, C. (1951). Physique and reputation of Junior High School boys. *Child Development*, **22**, 247–60.

Harlan, W.R., Osborne, R.K. & Graybiel, A. (1962). A longitudinal study of blood pressure. *Circulation*, **26**, 227–33.

Haronian, F. (1964). Physique and learning: an exploratory study. PhD Dissertation, New School for Social Research. (Dissertation Abstracts, pp. 7391–2.)

Haronian, F. & Sugarman, A.A. (1965). A comparison of Sheldon's and Parnell's methods for quantifying morphological differences. *American Journal of Physical Anthropology*, **23**, 135–41.

Hartl, E.M., Monnelly, E.P. & Elderkin, R.D. (1982). *Physique and Delinquent Behavior. A Thirty-Year Follow-Up of William H. Sheldon's Varieties of Delinquent Youth*. New York: Academic Press.

Hay, J.G. & Watson, J.M. (1970). The somatotypes of women track and field athletes. *New Zealand Journal of Health, Physical Education and Recreation*, **3**, 28–49.

Hayward, J.S., Lisson, P.A., Collis, M.L. & Eckerson, J.D. (1978). *Survival Suits for Accidental Immersion in Cold Water: Design Concepts and their Thermal Protection Performance*. Report. British Columbia: University of Victoria.

Hayward, J.S., Eckerson, J.D. & Dawson, B.T. (1986). Effect of mesomorphy on hyperthermia during exercise in a warm, humid environment. *American Journal of Physical Anthropology*, **70**, 11–18.

Heath, B.H. (1963). Need for modification of somatotype methodology. *American Journal of Physical Anthropology*, **21**, 227–33.

Heath, B.H. (1973). Somatotype patterns and variation within a Melanesian population *Proceedings IXth International Congress of Anthropological and Ethnological Sciences*, Chicago.

Heath, B.H. (1977). Applying the Heath–Carter somatotype method. In *Growth and Development: Physique*, ed. O. Eiben, pp. 335–47. Budapest: Akademiai Kiadó (Hungarian Academy of Sciences).

Heath, B.H. & Carter, J.E.L. (1966). A comparison of somatotype methods. *American Journal of Physical Anthropology*, **24**, 87–99.

Heath, B.H. & Carter, J.E.L. (1967). A modified somatotype method. *American Journal of Physical Anthropology*, **27**, 57–74.

Heath, B.H. & Carter, J.E.L. (1971). Growth and somatotype patterns of Manus children, Territory of Papua and New Guinea: application of a modified somatotype method to the study of growth patterns. *American Journal of Physical Anthropology*, **35**, 49–67.

Heath, B.H., Hopkins, C.E. & Miller, C.D. (1961). Physiques of Hawaii-born young men and women of Japanese ancestry. *American Journal of Physical Anthropology*, **19**, 173–84.

Heath, B.H., Mead, M. & Schwartz, T. (1968). A somatotype study of a Melanesian population. *Proceedings of the VIIIth Congress of Anthropological and Ethnographic Sciences*, Tokyo, vol I, pp. 9–11. Science Council of Japan.

Heath, C.W. (1954). Physique, temperament and sex ratio. *Human Biology*, **26**, 337–42.

Hebbelinck, M. & Borms, J. (1978). *Körperliches Wachstum und Leistungsfähigkeit bei Schulkindern*. Leipzig: Johann Ambrosius Barth.

Hebbelinck, M. & Postma, J.W. (1963). Anthropometric measurements, somatotype ratings, and certain motor fitness tests of physical education majors in South Africa. *Research Quarterly*, **34**, 327–34.

Hebbelinck, M. & Ross, W.D. (1974a). Body type and performance. In *Fitness, Health, and Work Capacity*, ed. L.A. Larsen, pp. 266–83. New York: Macmillan.

Hebbelinck, M. & Ross, W.D. (1974b). Kinanthropometry and biomechanics. In

Biomechanics IV, ed. R.C. Nelson & C.A. Morehouse, pp. 536–52. Baltimore: University Park Press.

Hebbelinck, M., Duquet, W. & Ross, W.D. (1973). A practical outline for the Heath–Carter somatotyping method applied to children. In *Pediatric Work Physiology Proceedings*, 4th International Symposium, 71–84. Wingate Institute, Israel.

Hebbelinck, M., Carter, L. & De Garay, A. (1975). Body build and somatotype of Olympic swimmers, divers, and water polo players. In *Swimming II*, ed. L. Lewillie & J.P. Clarys, pp. 285–305. Baltimore: University Park Press.

Hebbelinck, M., Ross, W.D., Carter, J.E.L. & Borms, J. (1980). Anthropometric characteristics of female Olympic rowers. *Canadian Journal of Applied Sport Sciences*, **5**, 255–62.

Hellerstein, H.K., Friedman, E., Brdar, P.J., Weiss, M. & Dupertuis, C.W. (1969). A comparison of the personality of adult subjects with rheumatic heart disease and with arteriosclerotic heart disease. In *Rehabilitation of Non-Coronary Heart Disease*, ed. H. Denolin, pp. 220–82. Symposium by the Council of Rehabilitation of the International Society of Cardiology, Hohenried.

Hertzberg, H.T.E., Churchill, E., Dupertuis, C.W., White, R.M. & Damon, A. (1963). *Anthropometric Survey of Turkey, Greece, and Italy*. New York: Macmillan.

Hirata, K-I. & Kaku, K. (1964). *Method of Physique and Physical Fitness and its Practical Application*. Gifu City: Hirata Institute of Health.

Hoit, J.D. & Hixon, T.J. (1986). Body type and speech breathing. *Journal of Speech and Hearing Research*, **29**, 313–24.

Holopainen, S., Lumiaho-Häkkinen, P. & Telama, R. (1984). Level and rate of development of motor fitness, motor abilities and skills by somatotype. *Scandinavian Journal of Sports Sciences*, **6**(2), 67–75.

Holzer, E., Hauser, G. & Prokop, L. (1984). The physique of Austrian table-tennis players, sociodemographic and ergometric aspects. *Hungarian Review of Sports Medicine*, 25, 27–41.

Hooton, E.A. (1946). *Up from the Ape*. New York: Macmillan.

Hooton, E.A. (1948). *Body Build in Relation to Military Function in a Sample of the U.S. Army*. Cambridge, Mass: Harvard University, Department of Anthropology.

Hooton, E.A. (1951). *Handbook of Body Types in the United States Army*. Cambridge, Mass: Harvard University, Department of Anthropology.

Hooton, E.A. (1959). *Body Build in a Sample of the U.S. Army*. Quartermaster Research and Engineering Command, Technical Report EP-102, Natick, Mass: U.S. Army.

Howard, R. & Gertler, M.M. (1952). Axis deviation and body build. *American Heart Journal*, **44**, 35.

Howells, W.W. (1952). A factorial study of constitutional types. *American Journal of Physical Anthropology*, **10**, 91–118.

Hrčka, J. & Zrubák, A. (1975). Vybrané charakteristiky posluchačov FTVŠ UK z hľadiska somatometrie, somatotypie a sily (Selected characteristics of students from the Faculty of Physical Education and Sport of Comenius University from the point of view of somatometry, somatotypology and strength). *Acta Facultatis Educationis Physicae Universitatis Comenianae*, **16**, 191–208.

Hulse, F.S. (1971). *The Human Species*. New York: Random House.

Hulse, F.S. (1981). Habits, habitats, and heredity: a brief history of studies in human plasticity. *American Journal of Physical Anthropology*, **56**, 495–501.

Humphreys, L.G. (1957). Characteristics of type concepts with special reference to Sheldon's typology. *Psychological Bulletin*, **54**, 218–28.

Hunt, E.E., Jr. (1949). A note on growth, somatotype, and temperament. *American Journal of Physical Anthropology*, **7**, 79–89.

Hunt, E.E., Jr. (1951). Physique, social class and crime among the Yap islanders. (Proceedings of 20th Annual Meeting of A.A.P.A.) *American Journal of Physical Anthropology*, **9**, 241–2 (abstract).

Hunt, E.E., Jr. (1952). Human constitution: an appraisal. *American Journal of Physical Anthropology*, **10**, 55–73.

Hunt, E.E., Jr. (1961). Measures of adiposity and muscularity in man: some comparisons by factor analysis. In *Techniques for Measuring Body Composition*, ed. J. Brožek & A. Henschel, pp. 192–211. Washington, D.C.: National Academy of Sciences.

Hunt, E.E., Jr. (1981). The old physical anthropology. *American Journal of Physical Anthropology*, **56**, 334–46.

Hunt, E.E., Jr. & Barton, W.H. (1959). The inconstancy of physique in adolescent boys and other limitations of somatotyping. *American Journal of Physical Anthropology*, **17**, 27–36.

Hunt, E.E., Jr., Cocke, G. & Gallagher, J.R. (1958). Somatotype and sexual maturation in boys: a method of developmental analysis. *Human Biology*, **30**, 73–91.

Imlay, R.C. (1966). The physiques of college baseball players in San Diego, California. MA Thesis, San Diego State University (University of Oregon, Microcard PE930).

Janoff, I.S., Beck, L.N. & Child, I.L. (1950). The relationship of somatotype to reaction time, resistance to pain, and expressive movement. *Journal of Personality*, **18**, 454–60.

Jennett, C.W. (1959). An investigation of tests of agility. PhD Thesis, University of Iowa, Iowa City.

Jensen, R.K. (1981). The effect of a 12-month growth period on the body moments of inertia of children. *Medicine and Science in Sports and Exercise*, **13**, 4, 238–42.

Johnson, G.A. (1969). Somatotypes of dancers and nondancers. MA Thesis, San Diego State University.

Johnston, R.E. & Watson, J.M. (1968). A comparison of the phenotypes of women basketball and hockey players. *New Zealand Journal of Health, Physical Education and Recreation*, **3**, 48–54.

Jones, P.R.M. & Stone, P.G. (1964). An advance in somatotype photography. *American Journal of Physical Anthropology*, **22**, 259–64.

Jones, P.R.M., Worth, W.J.C., Stone, P.G., Ellis, M.J. & Jeffery, J.A. (1965). *The Influence of Somatotype and Anthropometric Measures on Heart Rate During Work, in Students and Special Sporting Groups*. Loughborough: Physical Education and Industrial Fitness Unit.

Jordan, D.B. (1964). A longitudinal analysis of the mental health of boys age fifteen to seventeen years. MA Thesis, University of Oregon, Eugene (Microcard PSY203.)

Kalenda, L.M. (1964). Relationships of body alignment with somatotype and centre of gravity in college women, a pilot study. MA Thesis, Louisiana State University, Alexandria.

Kane, J. (1961). Body shapes and personalities of athletes. *Sport and Recreation*, **2**, 11–13.

Kane, J. (1964). Psychological correlates of physique and physical abilities. In *International Research in Sport and Physical Education*, ed. E. Jokl & E. Simon, pp. 85–94. Springfield, Illinois: Thomas.

Kane, J. (1972). Personality, body concept and performance. In *Psychological Aspects of Physical Education and Sport*, ed. J. Kane, pp. 91–127. London: Routledge and Kegan Paul.

Kansal, D.K. (1981). A study of age changes in physique and body composition in males of two communities of Punjab. PhD Thesis, Punjabi University, Patiala.

Kansal, D.K., Gupta, N. & Gupta, A.K. (1986). A study of intrasport differences in the physique of Indian university football players. In *Perspectives in Kinanthropometry*, ed. J.A.P. Day, pp. 143–54. Champaign, Illinois: Human Kinetics.

Katch, F.I., Behnke, A.R. & Katch, V.L. (1979). Estimation of body fat from skinfolds and surface area. *Human Biology*, **51**, 411–24.

Khasigian, H.A., Evanski, P.M. & Waugh, T.R. (1978). Body type and rotational laxity of the knee. *Clinical Orthopaedics and Related Research*, **130**, 228–32.

Kiker, V.L. & Miller, A.L. (1967). Perceptual judgement of physiques as a factor in social image. *Perceptual and Motor Skills*, **24**, 1013–14.

Kline, N.S. & Tenney, A.M. (1950). Constitutional factors in the prognosis of schizophrenia. *American Journal of Psychiatry*, **107**, 434–41.

Knoll, W. (1928). *Die sportärztlichen Ergebnisse der II. Olympischen Winterspiele in St. Moritz 1928*. Bern: Haupt.

Kohlrausch, W. (1930). Zusammenhänge von Körperform und Leistung. Ergebnisse der anthropometrischen Messungen an den Athleten der Amsterdamer Olympiade. *Arbeitsphysiologie*, **2**, 187–204.

Kovaleski, J.E., Parr, R.B., Hornak, J.E. & Roitman, J.L. (1980). Athletic profile of women college volleyball players. *The Physician and Sportsmedicine*, **8**, 112–18.

Kovář, R. (1977). Somatotype of twins. *Acta Universitatis Carolinae, Gymnica*, **13**(2), 49–59.

Kozel, K. (1978). Motor activity of endomorphic children. *Teorie a Praxe Tělesné Výchovy*, **26**(3), 149–60.

Kraus, B.S. (1951). Male somatotypes among the Japanese of Northern Honshu. *American Journal of Physical Anthropology*, **9**, 347–66.

Kretschmer, E. (1921). *Körperbau und Charakter*. Berlin: Springer–Verlag.

Kroll, W. (1954). An anthropometrical study of some Big Ten varsity wrestlers. *Research Quarterly*, **25**, 307–12.

Kurimoto, E. (1963). Longitudinal analysis of maturity, structural, strength and motor development of boys fifteen through eighteen years of age. PhD Thesis, University of Oregon, Eugene (Microcard PSY167).

Landers, D.M., Boutcher, S.H. & Wang, M.Q. (1986). A psychobiological study of archery performance. *Research Quarterly for Exercise and Sport*, **57**, 236–44.

Lapiccirella, R., Marrama, P. & Bonati, B. (1961). Hormones, excretion, diet, and physique in Somalis. *Lancet*, **281**, 24–5.

Larivière, G., Dulac, S. & Bonlay, M. (1978). Tests de patinage visant à mesurer la condition physique de joueurs de hockey sur glace. In *Ice Hockey. Research, Development and New Concepts*, ed. F. Landry & W.A.R. Orban, pp. 47–53. Miami: Symposium Specialists.

Lasker, G.W. (1947). The effects of partial starvation on somatotype: an analysis of material from the Minnesota starvation experiment. *American Journal of Physical Anthropology*, **5**, 323–41.

Lasker, G.W. (1952). Note on the nutritional factor in Howell's study of constitutional type. *American Journal of Physical Anthropology*, **1**, 375–9.

Laubach, L.L. (1969). Body composition in relation to muscle strength and range of joint motion. *Journal of Sports Medicine and Physical Fitness*, **9**, 89–97.

Laubach, L.L. & Marshall, M.E. (1970). A computer program for calculating Parnell's anthropometric phenotype. *Journal of Sports Medicine and Physical Fitness*, **10**, 217–24.

Laubach, L.L. & McConville, J.T. (1966a). Relationships between flexibility, anthropometry, and the somatotype of college men. *Research Quarterly*, **37**, 241–51.

Laubach, L.L. & McConville, J.T. (1966b). Muscle strength, flexibility, and body size of adult males. *Research Quarterly*, **37**, 384–92.

Laubach, L.L. & McConville, J.T. (1969). The relationship of strength to body size and typology. *Medicine and Science in Sport*, **1**, 189–94.

Laubach, L.L., Hollering, B.L. & Goulding, D.V. (1971). Relationships between two measures of cardiovascular fitness and selected body measurements of college men. *Journal of Sports Medicine and Physical Fitness*, **11**, 222–6.

Lavoie, J.M. & Lèbe-Néron, R.M. (1982). Physiological effects of training in professional and recreational jazz dancers. *Journal of Sports Medicine and Physical Fitness*, **22**, 231–6.

Leake, C.N. (1987). Body composition and body shape of trained female triathletes. MA Thesis, San Diego State University.

Lebedeff, A. (1980). Body structure of female intercollegiate and junior tennis players. MA Thesis, San Diego State University.

Leek, G.M. (1968). The physiques of New Zealand water polo players. *New Zealand Journal of Health, Physical Education and Recreation*, **3**, 39–47.

Leek, G.M. (1969). The physique of swimming champions. *New Zealand Journal of Health, Physical Education and Recreation*, **2**(3), 30–41.

Leek, G.M. (1970). The physique of voluntary Antarctic personnel. *New Zealand Journal of Health, Physical Education and Recreation*, **3**, 50–60.

Lerner, R.M. (1969). The development of stereotyped expectancies of body build behavior relations. *Child Development*, **40**, 137–41.

Lerner, R.M. & Korn, S.J. (1972). Development of body build stereotypes in males. *Child Development*, **43**(3), 908–20.

Lerner, R.M., Karabenick, S.A. & Meisels, M. (1975). One-year stability of children's personal space schemata towards body build. *Journal Genetic Psychology*, **127**, 151–2.

Lewis, A.S. (1966). The physique of New Zealand basketball players. *New Zealand Journal of Physical Education*, **39**, 25–6.

Lewis, A.S. (1969). Physique as a determinant of success in sport with particular reference to Olympic oarsmen. *New Zealand Journal of Health, Physical Education and Recreation*, **2**(3), 5–21.

Lindegard, B. (1953). Variations in human body-build. *Acta Psychiatrica et Neurologica, Supplementum*, **86**, 1–163.

Lindzey, G. (1967). Behaviour and morphological variation. In *Genetic Diversity and Human Behaviour*, ed. J.N. Spuhler, pp. 227–40. Chicago: Aldine.

Lister, J. & Tanner, J.M. (1955). The physique of diabetics. *Lancet*, **269**, 1002–4.

Livson, N. & McNeill, D. (1962). Physique and maturation rate in male adolescents. *Child Development*, **33**, 145–52.

Lohman, T.G. (1981). Skinfolds and body density and their relation to body fatness: A review. *Human Biology*, **53**, 181–225.

Lopez, A., Rojas, J. & Garcia, E. (1979). Somatotype et composition du corps chez les gymnastes de haut niveau. *Cinésiologie*, **72**, 608–18.

Lowdon, B.J. (1980). The somatotype of international male and female surfboard riders. *Australian Journal of Sports Medicine*, **12**, 34–9.

Lubin, A. (1950). A note on Sheldon's table of correlation between temperamental traits. *British Journal of Psychology*, **3**, 186–9.

Lynde, R.E. (1968). Longitudinal analysis of interest scores of boys ten to twelve years of age as related to selected physical measure. MS Thesis, University of Oregon, Eugene (Microcard PE1025).

Lynde, R.E. (1969). Longitudinal analysis of occupational interest scores of boys fifteen through seventeen years of age as related to various physical characteristics. EdD Thesis, University of Oregon, Eugene.

Majors, P.A. (1982). Effects of two training programs on physique and physical fitness in weight control classes for women. MA Thesis, San Diego State University.

Malina, R.M. & Shoup, R.F. (1985). Anthropometric and physique characteristics of female volleyball players at three competitive levels. In *Physique and Body Composition*, vol. 16, ed. O.G. Eiben, pp. 105–12. Budapest: Humanbiologia Budapestinensis.

Marcotte, G. & Herminston, R. (1978). Ice hockey. In *The Scientific Aspects of Sports Training*, ed. A.W. Taylor, pp. 222–9. Springfield, Illinois: Thomas.

Martin, A.D., Ross, W.D., Drinkwater, D.T. & Clarys, J.P. (1985). Prediction of body fat by skinfold caliper: assumptions and cadaver evidence. *International Journal of Obesity*, **9**, Suppl. 1, 31–9.

Martin, A.D., Drinkwater, D.T., Clarys, J.P. & Ross, W.D. (1986). The inconstancy of fat-free mass: a reappraisal with implications for densitometry. In *Kinanthropometry* III, ed. T. Reilly, J. Watkins & J. Borms, pp. 92–7. London: Spon.

Matsudo, V.K.R. (1986). Effects of soccer training on adolescents and adults physical fitness characteristics. (In Portuguese). In *Celafiscs – Dez Anos de Contribuição às Ciências do Esporte*, pp. 298–304. Laboratorio de Aptidao Fisica de Sao Caetano do Sul, SP, Brasil.

McFarland, R.A. (1953). *Human Factors in Air Transportation*. New York: McGraw–Hill.

McFarland, R.A. & Franzen, R. (1944). *The Pensacola study of naval aviators*. Washington, D.C.: U.S. Department of Commerce, Civil Aeronautics Administration.

McLure, C.C. (1967). The physiques of professional and amateur women golfers. MA Thesis, San Diego State University (University of Oregon, Microcard PE1027).

Mechikoff, R.A. & Francis, L. (1984). Social and demographic influences on the physique of the Olympic athlete. In *Physical Structure of Olympic Athletes. Part II: Kinanthropometry of Olympic Athletes*, ed. J.E.L. Carter, pp. 39–52. Basel: Karger.

Medeková, H. & Havlíček, I. (1982). Genetic contingency of variability of somatic traits in parents and children. *Modern Man Anthropos (Brno)*, **22**, 13–19.

Meleski, B.W., Shoup, R.F. & Malina, R.M. (1982). Size, physique, and body composition of competitive female swimmers 11 through 20 years of age. *Human Biology*, **54**, 609–25.

Meredith, H.V. (1940). Comments on 'Varieties of Human Physique'. *Child Development*, **11**, 301–9.

Mészáros, J. & Mohácsi, J. (1982a). The somatotype of Hungarian male and female class I paddlers and rowers. *Anthropologiai Közlemények*, **26**, 175–9.

Mészáros, J. & Mohácsi, J. (1982b). An anthropometric study of top level athletes in view of the changes that took place in the style of some ball games. *Humanbiologia Budapestinensis*, **13**, 15–20.

Miller, A.R. and Stewart, R.A. (1968). Perception of female physiques. *Perceptual and Motor Skills*, **27**, 721–2.

Miller, A.R., Kiker, V.L., Watson, R.A., Frauchiger, R.A. & Moreland, D. (1968). Experimental analysis of physiques as social stimuli. Part 2. *Perceptual and Motor Skills*, **27**, 355–9.

Miller, J.O. (1967). Longitudinal analysis of the relationship between self-differentiation and social interaction and selected physical variables in boys twelve to seventeen years of age. EdD Thesis, University of Oregon, Eugene (Microcard PSY305).

Mittleman, K.D. (1982). Etiological factors associated with menstrual irregularities in distance runners. MA Thesis, San Diego State University.

Mohácsi, J. & Mészáros, J. (1982). Serdulokoru üszök testalkati es fiziologiai vizsgalata (Physical and physiological examinations of young swimmers). *Humanbiologia Budapestinensis*, **13**, 21–5.

Monnelly, E.P., Hartl, E.M. & Elderkin, R. (1983). Constitutional factors predictive of alcoholism in a follow-up of delinquent boys. *Journal of Studies of Alcoholism*, **44**, 530–7.

Morris, P.C. (1960). A comparative study of physical measures of women athletes and unselected college women. EdD Thesis, Temple University, Philadelphia.

Morris, R.W. & Jacobs, M.L. (1950). On the application of somatotyping to the study of constitution in disease. *South African Journal of Clinical Science*, **1**, 347–70.

Morton, A.R. (1967). Comparison of Sheldon's trunk-index and anthroposcopic methods of somatotyping and their relationships to the maturity, structure, and motor ability of the same boys, nine through sixteen years of age. EdD Thesis, University of Oregon, Eugene (Microcard PE1032).

Munroe, R.A., Clarke, H.H. & Heath, B.H. (1969). A somatotype method for young boys. *American Journal of Physical Anthropology*, **30**, 195–201.

Murphy, G.B. (1972). The somatotypes of New Zealand table tennis players. *New Zealand Journal of Health, Physical Education and Recreation*, **5**, 6–8.

Murphy, S.J. (1975). A somatotype comparison of PCAA sprint and distance freestyle swimmers. MA Thesis, San Diego State University.

Muthiah, C.M. & Sodhi, H.S. (1980). The effect of training on some morphological parameters of top-ranking Indian basketball players. *The Journal of Sports Medicine and Physical Fitness*, **20**(4), 405–12.

Náprstková, J. (1973). Typology of women. *Teorie a Praxe Tělsné Výchovy*, **21**, 544–51.

Newman, R.W. (1952). Age changes in body build. *American Journal of Physical Anthropology*, **10**, 75–90.

Newton, R.M. (1978). A kinanthropometric study of thirteen to fifteen year-old ice hockey players. MA Thesis, San Diego State University.

Niederman, J.C., Spiro, H.L. & Sheldon, W.H. (1964). Blood pepsin as a marker of susceptibility to duodenal disease. *Archives of Environmental Health*, **8**, 540–6.

Novak, L.P., Woodward, W.A., Bestit, C. & Mellerowitz, H. (1977). Working capacity, body composition, and anthropometry of Olympic female athletes. *Journal of Sports Medicine and Physical Fitness*, **17**, 275–83.

Novak, L.P., Mellerowicz, H., Bestit, C. & Woodward, W.A. (1978). Body composition of Olympic male swimmers. *Journal of Sports Medicine and Physical Fitness*, **18**, 139–51.

Olgun, P. & Gürses, C. (1986). Relationships between somatotypes and untrained physical abilities. In *Perspectives in Kinanthropometry*, ed. J.A.P. Day, pp. 115–21. Champaign, Illinois: Human Kinetics.

Oliveira, G.N., Soares, J. & Vívolo, M.A. (1986). Somatotype determination of German handball players from different regions of Brazil. In *Celfiscs – Dez Anos de Contribuição às Ciências do Esporte*, p. 138. Laboratorio de Aptidao Fisica de Sao Caetano do Sul, SP, Brasil. (Abstract in Portuguese.)

Olson, A.L. (1960). Characteristics of fifteen year old boys classified as outstanding athletes, scientists, fine artists, leaders, scholars, or as poor students or delinquents. PhD Thesis, University of Oregon, Eugene (Microcard PSY129).

Orczykowska-Światkowska, Z., Rogucka, E. & Welon, Z. (1978). Genetic and environmental determinants of body components. *Studies in Physical Anthropology*, **4**, 11–19.

Orvanová, E. (1984). Body build, heredity and sport achievements. In *Genetics of Psychomotor Traits in Man*. pp. 111–23. Warsaw: International Society of Sport Genetics and Somatology.

Orvanová, E. (1986). Comparison of body size, shape and composition between three age groups of elite weight-lifters and non-athletes. In *Kinanthropometry III*, ed. T. Reilly, J. Watkins & J. Borms, pp. 73–80. London: Spon.

Orvanová, E., Uher, L., Slamka, M., Pataki, L. & Ramacsay, L. (1984). Body size, shape and composition analysis of weightlifters and variables discriminating them according to performance and age. In *Human Growth and Development*, ed. J. Borms, R. Hauspie, A. Sand, C. Susanne & M. Hebbelinck, pp. 511–23. London: Plenum Press.

Osborne, R.H. & De George, F.V. (1959). *Genetic Basis of Morphological Variation: An Evaluation and Application of the Twin Study Method*. Cambridge, Mass: Harvard University Press.

Ostyn, M., Simons, J., Beunen, G., Renson, R. & Van Gerven, D. (1980). *Somatic and motor development of Belgian secondary schoolboys*. Leuven: Leuven University Press.

Oyarzun, F. (1952). Somatotipos de 76 adictos alcoholicos. *Revista de Psiquiatria*, **17**, 35–46.

Palát, M., Štukovský, R. & Chovanová, E. (1982). *Somatometric aspects of juvenile hypertension*. IInd Anthropological Congress of Aleš Hrdlička, pp. 275–7. Prague: Universitas Carolina Pragensis.

Pallulat, D.M.A. (1984). Physiques of female professional tennis players. MA Thesis, San Diego State University.

Pancorbo, A.E. & Rodríquez, C. (1986). Somatotype of high performance junior female swimmers. *Boletin Científici-técnico, Inder Cuba*, 1/2, 30–5.

Papai, J. (1980). Variations of physique in female college students. *Anthropologiai Közlemények*, 24, 173–8.

Pařízková, J. & Carter, J.E.L. (1976). Influence of physical activity on stability of somatotypes in boys. *American Journal of Physical Anthropology*, 44, 327–40.

Pařízková, J., Adamec, A., Berdychová, J., Čermák, J., Horná, J. & Teplý, Z. (1984). *Growth, Fitness and Nutrition in Preschool Children*. Prague: Charles University.

Parnell, R.W. (1953). Physique and choice of faculty. *British Medical Journal*, 2, 472–5.

Parnell, R.W. (1954). Somatotyping by physical anthropometry. *American Journal of Physical Anthropology*, 12, 209–39.

Parnell, R.W. (1956). Physique and individual alcohol consumption. *International Journal of Alcohol and Alcoholism*, 1, 127–33.

Parnell, R.W. (1957). Physique and mental breakdown in young adults. *British Medical Journal*, 1, 1485–90.

Parnell, R.W. (1958). *Behaviour and Physique*. London: Edward Arnold.

Parnell, R.W. (1959). Etiology of coronary artery disease. *British Medical Journal*, 1 (24 Jan), 232.

Parnell, R.W. (1984). *Family physique and fortune*. Sutton Coldfield: Parnell Publications.

Parsons, J.M. (1973). Prediction of athletic performance through physique classification. *British Journal of Physical Education*, **July**, 21–4.

Perbix, J.A. (1954). Relationship between somatotype and motor fitness in women. *Research Quarterly*, 25, 84–90.

Pérez, B. (1977). Somatotypes of male and female Venezuelan swimmers. In *Growth and Development: Physique*, ed. O. Eiben, pp. 349–55. Budapest: Akademiai Kiadó (Hungarian Academy of Sciences).

Pérez, B. (1981). *Los Atletas Venezolanos, su tipo físico*. Caracas: Universidad Central de Venezuela.

Pérez, B., Castillo, T.L. & Brief, F.K. (1985). *Caracteristicas somatotipicas asociadas con la edad sexo en un grupo de escolares Venezolanos*. Caracas: Instituto de Investigaciones Economicas y Sociales. FACES Universidad Central de Venezuela.

Petersen, G. (1967). *Atlas for Somatotyping Children*. Assen, The Netherlands: Royal Vangorcum and Springfield, Illinois: C.C. Thomas.

Pinto, J.R. (1978). *The position and specificity of somatotype in professional occupations*. Rio de Janeiro: Faculdades Integradas Castello Branco, Educação Fisicà, Ladebio.

Pirie, L. (1974). The somatotype of hockey players. MA Thesis, University of Western Ontario, London.

Pomerantz, H.Z. (1962). Relationship between coronary heart disease and certain physical characteristics. *Journal of the Canadian Medical Association*, 86, 57–60.

Porter, A.M. (1958). The physiques of explorers. *American Journal of Physical Anthropology*, **16**, 485–7.

Powell, G.E., Tutton, S.J. & Stewart, R.A. (1974). The differential stereotyping of similar physiques. *British Journal of Sociological and Clinical Psychology*, **13**, 421–3.

Preston, T.A. & Singh, M. (1972). Redintegrated somatotyping. *Ergonomics*, **15**(6), 693–700.

Pugh, L.G.C.E., Edholm. O.G., Fox, R.H., Wolff, H.S., Hervey, G., Hammond, W.H., Tanner, J.M. & Whitehouse, R.H. (1960). A physiological study of channel swimming. *Clinical Science*, **19**, 257–73.

Rahe, R.H. & Carter, J.E.L. (1976). Middle-aged male competitive swimmers. Background and body structure characteristics. *Journal of Sports Medicine and Physical Fitness*, **16**, 309–18.

Rangan, S.C.B. (1982). Validity of age, socio-economic belonging and dietary type as somatotype determinants in boys of secondary schools. PhD Thesis, Bangalore University, Bangalore, India.

Reilly, T. & Bretherton, S. (1986). Multivariate analysis of fitness of female field hockey players. In *Perspectives in Kinanthropometry*, ed. J.A.P. Day, pp. 135–43. Champaign, Illinois: Human Kinetics.

Reilly, T. & Hardiker, R. (1981). Somatotype and injuries in adult student rugby football. *The Journal of Sports Medicine and Physical Fitness*, **21**, 2, 186–91.

Renson, R. & Swalus, P. (1970). Motor aptitude and biotypological data of climbers. *Sport (Belgium)*, **13**, 4, 210–20.

Renson, R. & Van Gerven, C. (1968–69). Afkoelingsreakties bij zwemmen in koud zeewater (Cooling down reactions while swimming in cold seawater). *Hermes (Leuven)*, **3**, 191–218.

Reynolds, R.M. (1965). Responses on the Davidson Adjective Check List as related to maturity, physical, and mental characteristics of thirteen year old boys. PhD Thesis, University of Oregon, Eugene (Microcard PSY227).

Riegrová, J. (1978). Typologická studie studentu tělesné výchovy Univerzity Palackého (Typological study of physical education students of Palacký University). *Teorie a Praxe Tělesné Výchovy*, **26**, 7, 421–9.

Rivera, M.A., Albarrán, M.A., Malavé, R.D. & Frontera, W.R. (1986). The Puerto Rican athlete kinanthropometry project: age group and senior wrestlers. In *Perspectives in Kinanthropometry*, ed. J.A.P. Day, pp. 181–6. Champaign, Illinois: Human Kinetics.

Roberts, D.F. (1977). Physique and environment in the northern Nilotes. *Mitt. unt Ges. Wien,*, **107**, 161–8.

Roberts, D.F. & Bainbridge, D.R. (1963). Nilotic physique. *American Journal of Physical Anthropology*, **21**, 341–70.

Robson, H.E. (1974). Physique variations of women in physical education teaching and recreational sport. *Proceedings of International Congress of Sports Medicine, Melbourne*, pp. 80–85.

Rocha, M.L., Gomes, P.S.C., Gil, C., de Freitas, J. & Villasboas, L.F.P. (1977*a*). O somatotipo dos candidatos à escola de educação física e desportos da UFRJ. *Arquivos de Anatomia e Antropologia, Instituto de Antropologia, (Professor Souza Marques)*, **2**(2), 203–8.

Rocha, M.L., Araújo, C.G.S., de Freitas, J. & Villasboas, L.F.P. (1977*b*). Antropometria dinâmica da Natação. *Revista de Educação Física, Brasil*, **102**, 46–54.

Rodríguez, C., Sánchez, G., García, E., Martínez, M. & Cabrera, T. (1986).

Contribution to the study of the morphological profile of highly competitive male Cuban athletes. *Boletin Científico-técnico, Inder Cuba*, **1/2**, 6–24.

Rodríguez, F.A. (1986). Physical structure of international lightweight rowers. In *Kinanthropometry III*, ed. T. Reilly, J. Watkins & J. Borms, pp. 255–61. London: Spon.

Roldan, A. (1967). *Personality Types and Holiness*. (Translated from the Spanish by G.M. McCaskey. Originally published by Editorial Razon y Fo, S.A., Madrid.) Staten Island, New York: Alba House.

Ross, W.D. (1976). Metaphorical models for the study of human shape and proportionality. In *Physical Education, Sports and the Sciences*, ed. J. Broekhoff, pp. 285–304. University of Oregon, Eugene: Microform Publications.

Ross, W.D. & Day, J.A.P. (1972). Physique and performance of young skiers. *Journal of Sports Medicine and Physical Fitness*, **12**, 30–7.

Ross, W.D. & Ward, R. (1982). Human proportionality and sexual dimorphism. In *Sexual Dimorphism in Homo Sapiens*, ed. R. Hall, pp. 317–61. New York: Praeger.

Ross, W.D. & Wilson, B.D. (1973). A somatotype dispersion index. *Research Quarterly*, **44**, 372–4.

Ross, W.D. & Wilson, N.C. (1974). A stratagem for proportional growth assessment. *Acta Paediatrica Belgica*, **28** (Supplement), 169–82.

Ross, W.D., Hebbelinck, M. & Wilson, B.D. (1974). Somatotype in sport and the performing arts. *Medicina Dello Sport*, **20**, 314–26.

Ross, W.D., McKim, D.R. & Wilson, B.D. (1976). Kinanthropometry and young skiers. In *Application of Science and Medicine in Sport*, ed. A. Taylor, pp. 257–77. Springfield, Illinois: Thomas.

Ross, W.D., Brown, S.R. & Faulkner, R.A. (1977*a*). Age of menarche in Canadian skaters and skiers. *Canadian Journal of Applied Sports Sciences*, **1**, 191–3.

Ross, W.D., Brown, S.R., Yu, J.W. & Faulkner, R.A. (1977*b*). Somatotypes of Canadian figure skaters. *Journal of Sports Medicine and Physical Fitness*, **17**, 195–205.

Ross, W.D., Carter, J.E.L., Rasmussen, R.L. & Taylor, J. (1978). Anthropometric and photoscopic somatotyping of children. In *Physical Fitness Assessment: Principles, Practices and Application*, ed. R.J. Shephard & H. LaVallee, pp. 257–62. Springfield, Illinois: Thomas.

Ross, W.D., Marfell-Jones, M.J. & Stirling, D.R. (1982). Prospects in kinanthropometry. In *The Sport Sciences. Physical Education Series Number 4*, ed. J.J. Jackson & H.A. Wenger, pp. 134–50. Victoria, Canada: University of Victoria.

Ryan, B.T. (1980). Influences of family history on coronary heart disease risk factor identification in 10–15 year old boys. MA Thesis, San Diego State University.

Salmela, J.H. (1979). Growth patterns of elite French-Canadian female gymnasts. *Canadian Journal of Applied Sports Science*, **4**, 219–22.

Salokun, S.O. & Toriola, A.L. (1985). Perceived somatotype and stereotypes of physique among Nigerian school children. *The Journal of Psychology*, **119**, 587–94.

Sanford, R.N. (1953). Physical and physiological correlates of personality structure. In *Personality*, 2nd edn, ed. C. Kluckhohn & H.A. Murray, pp. 100–3. New York: Knopf.

Sargent, D.A. (1887). The physical characteristics of the athlete. *Scribners*, **2**, 541–61.

Schori, T.R. & Thomas, C.B. (1973). Rorschach factors and somatotype. *Journal of Clinical Psychology*, **29**(4), 491–2.

Schreiber, M.L. (1973). Anaerobic capacity as a function of somatotype and participation in varsity athletics. *Research Quarterly*, **44**, 197–205.

Seltzer, C.C. (1950). A comparative study of the morphological characteristics of delinquents and non-delinquents. In *Unravelling Juvenile Delinquency*, ed. S. Glueck & E. Glueck, pp. 307–50. New York: Commonwealth Fund.

Seltzer, C.C. (1951). Constitutional aspects of juvenile delinquency. *Cold Springs Harbor Symposium on Quantitative Biology*, **15**, 361–72.

Seltzer, C.C. (1966). Some re-evaluations of the build and blood pressure study, 1959, as related to ponderal index, somatotype, and mortality. *New England Journal of Medicine*, **274**, 254–9.

Seltzer, C.C. & Mayer, J. (1964). Body build and obesity – who are the obese? *Journal of the American Medical Association*, **189**(9), 677–84.

Seltzer, C.C. & Mayer, J. (1969). Body build (somatotype) distinctiveness in obese women. *Journal of the American Dietetic Association*, **55**, 454–8.

Seltzer, C.C., Wells, F.L. & McTernan, E.B. (1948). A relationship between Sheldonian somatotype and psychotype. *Journal of Personality*, **16**, 431–6.

Sharma, S.S. & Dixit, N.K. (1985). Somatotype of athletes and their performance. *International Journal of Sports Medicine*, **6**, 161–2.

Sheldon, W.H. (with the collaboration of S.S. Stevens and W.B. Tucker) (1940). *The Varieties of Human Physique*. New York: Harper and Brothers.

Sheldon, W.H. (with the collaboration of S.S. Stevens) (1942). *The Varieties of Temperament*. New York: Harper and Brothers.

Sheldon, W.H. (1943*a*). *A Basic Classification Applied to Aviation Cadets*. A.A.F. School of Aviation Medicine, Report No. 1. (Project No. 127).

Sheldon, W.H. (1943*b*). *Use of the Somatotype in Standardizing and Objectifying the Adaptability Rating for Military Aeronautics*. A.A.F. School of Aviation Medicine, Report No. 2. (Project No. 127).

Sheldon, W.H. (1944*a*). *Validity of the Somatotype and Anthropometric Variables in Flying Training*. Aviation Psychology Abstract Series (A.A.F. Air Surgeon's Office). Abstract No. 127.

Sheldon, W.H. (1944*b*). Constitutional factors in personality. In *Personality and the Behavior Disorders*, ed. J.M. Hunt, pp. 526–49. New York: Ronald Press.

Sheldon, W.H. (with the collaboration of E.E. Hartl and E. McDermott) (1949). *Varieties of Delinquent Youth*. New York: Harper and Brothers.

Sheldon, W.H. (1951). The somatotype, the morphophenotype, and the morphogenotype. *Cold Springs Harbor Symposia on Quantitative Biology*, **15**, 373–82.

Sheldon, W.H. (1952). Frontiers in human physique studies. *Professional Contributions of the American Academy of Physical Education*, **2**, 67–75.

Sheldon, W.H. (with the collaboration of C.W. Dupertuis and E. McDermott) (1954). *Atlas of Men*. New York: Harper and Brothers.

Sheldon, W.H. (1961). New developments in somatotyping technique. Lecture delivered at Childrens Hospital, March 13, Boston.

Sheldon, W.H. (1963). Constitutional variation and mental health. In *Encyclopedia of Mental Health*, vol. 2, pp. 355–66. New York: Franklin Watts.

Sheldon, W.H. (1965). A brief communication on somatotyping, psychiatyping and other Sheldonian delinquencies. Paper delivered at the Royal Society of Medicine, May 13, London.

Sheldon, W.H. (1971). The New York study of physical constitution and psychotic pattern. *Journal of the History of Behavioral Sciences*, **7**, 115–26.

Sheldon, W.H. & Tucker, W.B. (1938). The anthrotyping technique; a method for studying physical variation. Unpublished manuscript, Tozzer Library, Harvard University, Cambridge, Mass.

Sheldon, W.H., Lewis, N.D.C. & Tenney, A.M. (1969). Psychotic patterns and physical constitution. In *Schizophrenia, Current Concepts and Research*, ed. D.V. Siva Sanker, pp. 839–911. Hicksville, New York: PJD Publications.

Shin, S-G. (1985). A research on athlete's somatotype, body composition and maximum oxygen uptake ability. MA Thesis, Dong-A University, Pusan, Korea.

Shoup, R.F. (1978). Anthropometric and physique characteristics of Black, Mexican-American and White female high school athletes in three sports. MA Thesis, University of Texas, Austin.

Shoup, R.F. (1987). Growth and aging in the Manus of Pere village, Papua New Guinea: a mixed-longitudinal and secular perspective. PhD Thesis, University of Texas, Austin.

Shoup, R.F. & Malina, R.M. (1985). Anthropometric and physique characteristics of female high school varsity athletes in three sports. In *Physique and Body Composition*, vol. 16, ed. O.G. Eiben, pp. 153–77. Budapest: Humanbiologia Budapestinensis.

Sidhu, L.S. & Kansal, D.K. (1974). Comparative study of body composition of Jat-Sikhs and Banias of Punjab (India). *Zeitschrift fur Morphologie und Anthropologie*, **65**, 276–84.

Sidhu, L.S. & Wadhan, S.P.S. (1975). A study of somatotype distribution of sportsmen specializing in different events. *Sports Medicine*, **4**, 13–19.

Sidhu, L.S., Singal, P. & Kaur, S. (1982). Physique and body composition of Jat-Sikh and Bania girl students of Punjab. *Zeitschrift fur Morphologie und Anthropologie*, **73**, 51–8.

Sills, F.D. (1950). A factor analysis of somatotypes and of their relationship to achievement in motor skills. *Research Quarterly*, **21**, 424–37.

Sills, F.D. & Everett, P.W. (1953). The relationship of extreme somatotypes to performance in motor and strength tests. *Research Quarterly*, **24**, 223–8.

Sills, F.D. & Mitchem, J. (1957). Prediction of performance on physical fitness tests by means of somatotype ratings. *Research Quarterly*, **28**, 64–71.

Sinclair, G.D. (1966). Stability of physique types of boys nine through twelve years of age. MS Thesis, University of Oregon, Eugene (Microcard PE833.)

Sinclair, G.D. (1969). Stability of somatotype components of boys ages twelve through seventeen years and their relationships to selected physical and motor factors. PhD Thesis, University of Oregon, Eugene (Microcard PE1042.)

Singal, P. & Sidhu, L.S. (1984). Age changes and comparison of somatotypes during 20 to 80 years in Jat-Sikh and Bania females of Punjab (India). *Anthropologia Anzeiger*, **42**, 281–9.

Singh, R. (1976). A longitudinal study of the growth of trunk surface area measured by planimeter on standard somatotype photographs. *Annals of Human Biology*, **3**, 181–6.

Singh, S.P. (1978). Growth patterns from 4 to 20 years of male Rajput Gaddis – a Himalayan Tribe. PhD thesis, Punjabi University, Patiala.

Singh, S.P. (1981). Body morphology and anthropometric somatotypes of Rajput and Brahmin Gaddis of Dhaula Dhar Range, Himalayas. *Zeitschrift Fur Morphologie und Anthropologie*, **72**, 315–23.

Singh, S.P. & Malhotra, P. (1986). Morphology, body composition and somatotype of Indian national cyclists. In *Kinanthropometry*, III, ed. T. Reilly, J. Watkins & J. Borms, p. 82. London: Spon.

Singh, S.P. & Sidhu, L.S. (1980). Changes in somatotypes during 4 to 20 years in Gaddi Rajput boys. *Zeitschrift Fur Morphologie und Anthropologie*, **71**, 285–93.

Singh, S.P. & Sidhu, L.S. (1982). Physique and morphology of Jat-Sikh cyclists of Punjab. *Journal of Sports Medicine and Physical Fitness*, **22**, 185–90.

Singh, S.P., Sidhu, L.S. & Malhotra, P. (1985). Body morphology of high altitude Spitians of North West Himalayas. Unpublished paper, Human Biology, Punjabi University, Patiala, India.

Sinning, W.E. (1978). Anthropometric estimation of body density, fat and lean body weight in women gymnasts. *Medicine and Science in Sports*, **10**, 243–9.

Sinning, W.E., Wilensky, N.F. & Meyers, E.J. (1976). Post season body composition changes and weight estimation in high-school wrestlers. In *Physical Education, Sports and the Sciences*, ed. J. Broekhoff, pp. 137–53. University of Oregon, Eugene: Microform Publications.

Sinning, W.E., Cunningham, L.N., Racaniello, A.P. & Sholes, J.L. (1977). Body composition and somatotype of male and female nordic skiers. *Research Quarterly*, **48**, 741–9.

Skibińska, A. & Sklad, M. (1979). Genetyczene uwarunkowania somatotypn Sheldona. (Genetic conditioning of the Sheldonian somatotype.) *Wychowania fizyczne i sport*, **2**, 3–12.

Skottowe, I. & Parnell, R.W. (1962). The significance of somatotype and other signs in psychiatric prognosis and treatment. *Royal Society of Medicine Proceedings*, **55**, 707–16.

Slamka, M., Chovanová, E. & Pataki, L. (1983). Somatic parameters differentiating the weight lifters according to performance. *Teorie a Praxe Tělesné Výchovy*, **31**, 4, 232–6.

Slaughter, M.H. (1970). An analysis of the relationship between somatotype and personality traits of college women. *Research Quarterly*, **41**, 569–75.

Slaughter, M.H. & Lohman, T.G. (1976). Relationship of body composition to somatotype. *American Journal of Physical Anthropology*, **44**, 237–44.

Slaughter, M.H. & Lohman, T.G. (1977). Relationship of body composition to somatotype in boys, ages 7 to 12 years. *Research Quarterly*, **48**, 750–8.

Slaughter, M.H., Lohman, T.G. & Boileau, R.A. (1977a). Relationship of Heath and Carter's second component to lean body mass and height in college women. *Research Quarterly*, **48**, 759–68.

Slaughter, M.H., Lohman, T.G. & Misner, J.E. (1977b). Relationship of somatotype and body composition to physical performance in 7–12 year old boys. *Research Quarterly*, **48**, 159–68.

Slaughter, M.H., Lohman, T.G. & Misner, J.E. (1980). Association of somatotype and body composition to physical performance in 7–12 year-old-girls. *Journal of Sports Medicine and Physical Fitness*, **20**, 2, 189–98.

Sleet, D.A. (1968). Somatotype and social image. MA Thesis, San Diego State University. (University of Oregon, Microcard PSY339.)

Sleet, D.A. (1969). Physique and social image. *Perceptual and Motor Skills*, 28, 295–9.

Sleet, D.A. (1982). Typecasting – does your body type affect your personality? *Shape*, 2(1), 40–2, 90–1.

Smit, P.J., Daehne, H.O. & Burger, E. (1979a). Somatotypes of South African rugby players. In *Sport and Somatology in Ischaemic Heart Disease*, ed. P.J. Smit, pp. 15–21. South Africa: University of Pretoria.

Smit, P.J., Daehne, H.O., Halhuber, M.H. & Stocksmeier, U. (1979b). Somatotypes of cardiac infarction patients. In *Sport and Somatology in Ischaemic Heart Disease*, ed. P.J. Smit, pp. 1–14. South Africa: University of Pretoria.

Smith, H.C. (1949). Psychometric checks on hypotheses, derived from Sheldon's work on physique and temperament. *Journal of Personality*, 17, 310–20.

Smith, H.C. & Boyarsky, S. (1943). Relationship between physique and simple reaction time. *Character and Personality*, 12, 46–53.

Smithells, P.A. (1949). *Physique and Temperament in Relation to Physical Education*. Dunedin, New Zealand: Coulls Somerville Wilkie.

Smithells, P.A. & Cameron, P.E. (1962). *Principles of Evaluation in Physical Education*. New York: Harper and Brothers.

Sobral, F., Brito, A.P., Alves, J., Fragoso, M.I. & Rodríguez, M.A. (1986). Physique, personality and strength as related with menarcheal age in college women. In *Kinanthropometry III*, ed. T. Reilly, J. Watkins & J. Borms, pp. 181–4. London: Spon.

Sodhi, H.S. (1976). The physique and body composition of Indian athletes and sportsmen of selected physical activity. PhD Thesis, Punjabi University, Patiala.

Sodhi, H.S. (1980). A study of morphology and body composition of Indian basketball players. *The Journal of Sports Medicine and Physical Fitness*, 20(4), 413–22.

Sodhi, H.S. & Sidhu, L.S. (1984). *Physique and Selection of Sportsmen: A Kinanthropometric Study*. Patiala: Punjab Publishing House.

Somerset, H.C.A. (1953). Some investigation into dimensions of physique and their relationships with Rorschach responses. MA Thesis, Victoria University College, Wellington, New Zealand.

Song, T.M.K. (1982). Relationship of physiological characteristics to skiing performance. *The Physician and Sportsmedicine*, 10(12), 97–102.

Spain, D.M., Bradess, V.A. & Huss, G. (1953). Observations on atherosclerosis of the coronary arteries in males under age 46: a necropsy study with special reference to somatotypes. *Annals of Internal Medicine*, 38, 254–77.

Spain, D.M., Bradess, V.A. & Greenblatt, I.J. (1955). Postmortem studies on coronary atherosclerosis, serum beta lipoprotein, and somatotypes. *American Journal of Medical Sciences*, 229, 294–301.

Spain, D.M., Nathan, D.J. & Gellis, M. (1963). Weight, body type, and the prevalence of coronary atherosclerotic heart disease in males. *American Journal of Medicine and Science*, 245, 63–8.

Staffieri, J.R. (1967). A study of social stereotype of body image in children. *Journal of Personality and Social Psychology*, 7, 101–4.

Stejskal, F. & Náprstková, J. (1975). Somatotypes of track cyclists. *Teorie a Praxe Tělesné Výchovy*, 23, 8, 483–7.

Stephens, W.G.S. & Taylor, J.H. (1962). The schematic two-dimensional plotting of the spatial relationship among somatotypes. *American Journal of Physical Anthropology*, **20**, 395–8.

Štěpnička, J. (1970). Somatotypes of Czechoslovak superior sportsmen. (In Czech.) *Teorie a Praxe Tělesné Výchovy*, **18**, 722–31.

Štěpnička, J. (1972). *Typological and motor characteristics of athletes and university students.* (In Czech.) Prague: Charles University.

Štěpnička, J. (1974a). Typology of sportsmen. *Acta Universitatis Carolinae, Gymnica*, **1**, 67–90.

Štěpnička, J. (1974b). Somatotype and basic physical performance of male physical education majors, Charles University. *Acta Universitatis Palackianae Olomucensis Facultatis Medicae*, **48**, 185–90.

Štěpnička, J. (1976a). The aesthetics of human physique. In *International Conference of Physical Education*, A.I.E.S.E.P., ed. R. Linc, pp. 211–6. Prague: Universita Karlova.

Štěpnička, J. (1976b). Somatotypes of Bohemian and Moravian youth. *Acta Facultatis Medicae Universitatis Brunensis*, **57**, 233–42.

Štěpnička, J. (1976c). Somatotype, body posture, motor level and motor activity of youth. *Acta Universitatis Carolinae, Gymnica*, **12**, 1–93.

Štěpnička, J. (1977). Somatotypes of Czechoslovak athletes. In *Growth and Development: Physique*, ed. O. Eiben, pp. 357–64. Budapest: Akadémiai Kiadó (Hungarian Academy of Sciences).

Štěpnička, J. (1979). Methods of somatotyping in youth. In *Methods of Functional Anthropometry*, ed. V. Novotny & S. Titlbachová, pp. 155–202. Prague: Universitas Carolina Pragensis.

Štěpnička, J. (1983). The possibilities of influencing human figure beauty by means of physical exercises. (In Czech.) *Acta Universitatis Carolinae (Gymnica)*, **19**(1), 47–55.

Štěpnička, J. (1986). Somatotype in relation to physical performance, sports and body posture. In *Kinanthropometry III*, ed. T. Reilly, J. Watkins & J. Borms, pp. 39–52. London: Spon.

Štěpnička, J. & Broda, T. (1977). Somatotypes of young downhill skiers. *Teorie a Praxe Tělesné Výchovy*, **25**, 166–9.

Štěpnička, J. & Potměšil, J. (1970). Somatotypes of cross-country skiers. *Lyzarstvi*, **56**, 7–8,15–16.

Štěpnička, J., Chytráčková, J. & Kasalická, V. (1976). A somatotypic characteristic of the Czechoslovak superior downhill skiers, wrestlers, and road cyclists. *Teorie a Praxe Tělesné Výchovy*, **24**(3), 156–60.

Štěpnička, J., Chytráčková, J. & Kasalická, V. (1977). Somatotype and motor abilities in pupils of the basic nine year school. *Teorie a Praxe Tělesné Výchovy*, **25**(9), 551–9.

Štěpnička, J., Chytráčková, J., Kasalická, V. & Kubrychtová, I. (1979a). *Somatic preconditions for study of physical education.* (In Czech.) Prague: Universita Karlova.

Štěpnička, J., Táborský, F. & Kasalická, V. (1979b). The somatic prerequisites of women handball players. *Teorie a Praxe Tělesné Výchovy*, **27**, 746–55.

Štěpnička, J., Chytráčková, J., Kasalická, V., Kohoutek, M., Kovář, R. & Linc, R. (1981). Somatic and motor viewpoints of physical education process effectivity. *Acta Universitatis Carolinae (Gymnica)*, **17**, 2, 5–43.

Sterner, T.G. & Burke, E.J. (1986). A comparison of visual estimation and skinfold techniques. *The Physician and Sportsmedicine*, **14**, 101–7.

Stewart, H. (1980). Body type, personality temperament, and psychotherapeutic treatment of male adolescents. *Adolescence*, **15**, 927–32.

Stewart, R., Tutton, S.J. & Steele, R.E. (1973). Stereotyping and personality. 1. Sex differences in perception of female physiques. *Perceptual and Motor Skills*, **36**(3), 811–14.

Strong, M.L. (1980). Somatotypes and body composition of junior female gymnasts. MA Thesis, San Diego State University.

Strongman, K.T. & Hart, C.J. (1968). Stereotyped reactions to body build, *Psychological Reports*, **23**, 1175–8.

Štukovský, R., Palát, M. & Chovanová, E. (1983). Somatotype and body proportions in juvenile hypertension. *Acta Facultatis Rerum Naturalium Universitatis Comenianae, Anthropologia*, **XXVIII–XXIX**, 103–11.

Sucec, M.J. (1979). A study of the relationship between bioenergetic character types and somatotypes. MA Thesis, United States International University, San Diego.

Sugarman, A.A. & Haronian, F. (1964). Body type and sophistication of body concept. *Journal of Personality*, **32**, 380–94.

Sutherland, E.H. (1951). Critique of Sheldon's Varieties of Delinquent Youth. *American Sociological Review*, **16**, 10–13.

Sutorius, G.C. (1969). Somatotypes of United States rowers. MA Thesis, San Diego State University. (University of Oregon, Microcard PE985.)

Sutorius, G.C. & Carter, J.E.L. (1967). Somatotypes of college rowers. Unpublished study, San Diego State University.

Swalus, P. (1967–68a). Contribution a l'étude des relations entre le maintain d'une part, le somatotype, la souplesse et la force d'autre part. I. *Hermes*, **11**, 102–24.

Swalus, P. (1967–68b). Contribution a l'étude des relations entre le maintain d'une part, le somatotype, la souplesse et la force d'autre part. II. *Hermes*, **11**, 157–84.

Swalus, P. (1969). Etude des relations entre le somatotype et différent facteurs de l'aptitude motrice chez les garcons de 12 à 19 ans. *Kinanthropologie*, **1**, 3–14.

Swalus, P. & Van der Maren, B. (1968–69). Etude des relations entre le somatotype et une série de mesures anthropométriques. *Hermes (Louvain)*, **3**, 41–51.

Swalus, P., Beunen, G., Ostyn, M., Renson, R., Simons, J. & Van Gerven, D. (1970). Comparison des méthodes de Sheldon, Parnell et Heath–Carter pour la détermination du somatotype ou du phénotype. *Kinanthropologie*, **2**, 31–42.

Szmodis, I. (1977). Physique and growth estimated by Conrad's and Heath–Carter's somatocharts in athletic children. In *Growth and Development: Physique*, ed. O. Eiben, pp. 407–15. Budapest: Akadémiai Kiadó (Hungarian Academy of Sciences).

Tanner, J.M. (1947). The morphological level of personality. *Proceedings of the Royal Society of Medicine*, **40**(6), 301–8.

Tanner, J.M. (1951a). Current advances in the study of physique. Photogrammetric anthropometry and an androgeny scale. *Lancet*, **1**, 574–9.

Tanner, J.M. (1951b). The relationship between serum cholesterol and physique in healthy young men. *Journal of Physiology*, **115**, 371–90.

Tanner, J.M. (1952). The physique of students. An experiment at Oxford. *Lancet*, **2**, 405–9.

Tanner, J.M. (1954a). Reliability of anthroposcopic somatotyping. *American Journal of Physical Anthropology*, **12**, 257–65.

Tanner, J.M. (1954b). Physique and choice of careers. *The Eugenics Review*, **406**, 149–57.

Tanner, J.M. (1956). Physique, character and disease. A contemporary appraisal. *Lancet*, **271**, 635–7.

Tanner, J.M. (1964). *The Physique of the Olympic Athlete*. London: George Allen and Unwin.

Tanner, J.M. & Weiner, J.S. (1949). The reliability of the photogrammetric method of anthropometry, with a description of a miniature camera technique. *American Journal of Physical Anthropology*, **7**, 145–86.

Tanner, J.M. & Whitehouse, R.H. (1982). *Atlas of Children's Growth*. New York: Academic Press.

Tanner, J.M., Healy, M.J.R., Whitehouse, R.H. & Edgson, A.C. (1959). The relation of body build to the excretion of 17 ketosteroids and 17 ketogenic steroids in healthy young men. *Journal of Endocrinology*, **19**, 87–101.

Tappen, N.C. (1950). An anthropometric and constitutional study of championship weight lifters. *American Journal of Physical Anthropology*, **8**, 49–64.

Temesi, Z. & Szmodis, I. (1982). History of sports activity and variation of physique in female junior handball players. (In Hungarian.) *Humanbiologia Budapestinensis*, **13**, 47–52.

Thompson, J.C. (1952). An analysis of the factors affecting the achievement of undergraduate men majoring in physical education, at the State University of Iowa. *Research Quarterly*, **23**, 417–27.

Thorland, W.G., Johnson, G.O., Fagot, T.G., Tharp, G.D. & Hammer, R.W. (1981). Body composition and somatotype characteristics of junior Olympic athletes. *Medicine and Science in Sports and Exercise*, **13**(5), 332–8.

Thorland, W.G., Johnson, G.O., Housh, T.J. & Refsell, M.J. (1983). Anthropometric characteristics of elite adolescent competitive swimmers. *Human Biology*, **55**(4), 735–48.

Tittel, K. & Wutscherk, H. (1972). *Sportanthropometrie*. Leipzig: Johann Ambrosius Barth.

Tobias, P.V. (1972). Physique and body composition in Southern Africa. *Journal of Human Evolution*, **1**, 339–43.

Tomášová, I. (1977). The longitudinal study of female students of physical education and sport, Charles University. MA Thesis, FTVS, Charles University, Prague.

Toriola, A.L. & Igbokwe, N.U. (1985). Relationship between perceived physique and somatotype characteristics of 10 to 18 year old boys and girls. *Perceptual and Motor Skills*, **60**, 878.

Toriola, A.L., Salokun, S.O. & Mathur, D.N. (1985). Somatotype characteristics of male sprinters, basketball, soccer, and field hockey players. *International Journal of Sports Medicine*, **6**, 344–6.

Toteva, M. (1986). Somatotype characteristics of children at primary school age. In *Kinanthropometry III*, ed. T. Reilly, J. Watkins & J. Borms, pp. 58–62. London: Spon.

Toteva, M. & Sumanov, B. (1984). Sravnitelna somatotipna charakteristika na sstezateli ot razlicnite ski-disciplini (Comparative somatotypological characteristics of competitors in various ski-disciplines). *Naucni trudove*, **25**, Sofia: ECIPKFKS – VIF 'Georgi Dimitrov', 13–20.

Travill, A.L. (1984). Physique differences in male Olympic high, long and triple jumpers. MA thesis, San Diego State University.

Tucker, L.A. (1982). Relationship between perceived somatotype and body cathexis of college males. *Psychological Reports*, **50**, 983–9.

Tucker, L.A. (1983*a*). Self-concept: a function of self-perceived somatotype. *Journal of Psychology*, **113**, 123–33.

Tucker, L.A. (1983*b*). Muscular strength: a predictor of personality in males. *Journal of Sports Medicine and Physical Fitness*, **23**, 213–20.

Tucker, L.A. (1983*c*). Effect of weight training on self-concept: a profile of those influenced most. *Research Quarterly for Exercise and Sport*, **54**, 389–97.

Tucker, W.B. & Lessa, W.A. (1940*a*). Man: A constitutional investigation. *The Quarterly Review of Biology*, **15**, 411–55.

Tucker, W.B. & Lessa, W.A. (1940*b*). Man: A constitutional investigation. *The Quarterly Review of Biology*, **15**, 265–89.

Vaccaro, P., Clarke, D.H. & Wrenn, J.P. (1979). Physiological profiles of elite women basketball players. *Journal of Sports Medicine and Physical Fitness*, **19**, 45–54.

Vaccaro, P., Gray, P.R., Clarke, D.H. & Morris, A.F. (1984). Physiological characteristics of world class white-water slalom paddlers. *Research Quarterly for Exercise and Sport*, **55**(2), 206–10.

Vargas, L.A., Casillas, L.E. & Luján, J.M. (1975). Morfología externa de un grupo de jóvenes mexicanos. *Anales de Antropologia*, **XII**, 85–101.

Verdonk, P.F. (1972). *Lichaamshouw en gedrag (Body Size and Behaviour)*. Haarlem: Uitgeverijde De Toorts.

Vervaeke, H. & Persyn, U. (1981). Some differences between men and women in various factors which determine swimming performance. In *The Female Athlete*, ed. J. Borms, M. Hebbelinck & A. Venerando, pp. 150–6. Basel: Karger.

Villanueva, M. (1976). Comparación de cuatro téchnicas somatotipológicas. *Anales de Antropologia*, **13**, 289–303.

Villaneuva, M. (1979). *Manual de técnicas somatotipológicas*. Instituto do Investigaciones Antropológicas, Universidad Nacional Autonoma de Mexica.

Vívolo, M.A., Caldeira, S. & Matsudo, V.K.R. (1986*a*). Anthropometric study of the Japanese volleyball female national team according to Heath–Carter somatotype method. (In Portuguese.) In *Celafiscs – Dez Anos de Contribuição às Ciências do Esporte*, pp. 120–4. Laboratorio de Aptidao Fisica de Sao Caetano do Sul, SP, Brasil.

Vívolo, M.A., Matsudo, V.K.R. & Caldeira, S. (1986*b*). Somatotype of top level gymnastics. In *Celafiscs – Dez Anos do Contribuição às Ciências do Esporte*, p. 137. Laboratorio de Aptidao Fisica de Sao Caetano do Sul, SP, Brasil.

Votto, M.J. (1976). Somatotype and physical performance characteristics of major college football players. MS Thesis, University of Oklahoma.

Walker, R.N. (1962). Body build and behaviour in young children. 1. Body build and nursery school teacher's ratings. *Monographs of the Society for Research in Child Development*, **27**(3), 1–94.

Walker, R.N. (1963). Body build and behaviour in young children. 2. Body build and parent's ratings. *Child Development*, **34**, 1–24.

Walker, R.N. (1974*a*). Standards for somatotyping children. 1. Prediction of young adult height from children's growth data. *Annals of Human Biology*, **1**(2), 149–58.

Walker, R.N. (1974*b*). Standards for somatotyping children. 2. Prediction of somatotyping ponderal index from children's growth data. *Annals of Human Biology*, **1**(3), 289–99.

Walker, R.N. (1978). Pre-school physique and late-adolescent somatotype. *Annals of Human Biology*, **5**(2), 113–29.

Walker, R.N. (1979). Sheldon's trunk index and the growth of the thoracic and lumbar trunk. *Annals of Human Biology*, **6**(4), 315–36.

Walker, R.N. & Tanner, J.M. (1980). Prediction of adult Sheldon somatotypes I and II from ratings and measurements at childhood ages. *Annals of Human Biology*, **7**, 213–24.

Watson, C.G. (1972). Psychopathological correlates of anthropometric types in male schizophrenics. *Journal of Clinical Psychology*, **28**(4) 474–8.

Weese, C.H., Ross, W.D. & Bailey, D.A. (1975). Computer prospectus for somatotype analysis in longitudinal growth studies. (Second Canadian Symposium on Child Growth and Development, 17–18 November, 1972.) *Na'páo*, **5**, 33–5.

Weiner, J.S. & Lourie, J.A. (1969). *Human Biology, A Guide to Field Methods*. IBP Handbook No. 9. Oxford: Blackwell.

Wells, W.D. & Siegel, B. (1961). Stereotyped somatotypes. *Psychological Reports*, **8**, 77–8.

Westlake, D.J. (1967). Somatotypes of female track and field competitors. MA Thesis, San Diego State University. (University of Oregon, Microcard PE1050).

White, J.A., Quinn, G., Al-Dawalibi, M. & Mulhall, J. (1982*a*). Seasonal changes in cyclists' performance. Part I, the British Olympic road race squad. *British Journal of Sports Medicine*, **16**(1), 4–12.

White, J.A., Quinn, G., Al-Dawalibi, M. & Mulhall, J. (1982*b*). Seasonal changes in cyclists' performance. Part II, the British Olympic track squad. *British Journal of Sports Medicine*, **16**(1), 13–21.

Wichary, J.J. (1984). Physique differences between women runners and other active exercisers of 40 years of age. MA Thesis, San Diego State University.

Willgoose, C.E. (1952). Educational implication of constitution psychology. *Education*, **73**, 1–8.

Willgoose, C.E. (1961). *Evaluation in Health, Education, and Physical Education*. New York: McGraw–Hill.

Williams, K.M. (1984). Size, somatotypes and skinfold patterns of women rugby players. MA Thesis, San Diego State University.

Williams, L.R.T. (1977). The psychobiological model and multiple discriminant function analysis of high-calibre oarsmen. *Medicine and Science in Sport*, **9**(3), 178–84.

Wilmore, J.H. (1970). Validation of the first and second components of the Heath–Carter modified somatotype method. *American Journal of Physical Anthropology*, **32**, 369–72.

Wilmore, J.H., Parr, R., Haskell, W., Costill, D. & Milburn, L.J. (1976*a*). Athletic profile of professional football players. In *Physical Education, Sports and the Sciences*, ed. J. Broekhoff, pp. 155–68. University of Oregon, Eugene: Microform Publications.

Wilmore, J.H., Parr, R.B., Haskell, W.L., Costill, D.L. & Milburn, L.J. (1976*b*). Football pros' strengths – and CV weakness – charted. *The Physician and Sportsmedicine*, **4**(10), 45–54.

Wilmore, J.H., Brown, C.H. & Davis, J.A. (1977). Body physique and composition of the female distance runner. *Annals of New York Academy of Sciences*, **301**, 764–76.

Winthrop, H. (1957). The consistency of attitude patterns as a function of body type. *Journal of Personality*, **25**, 372–82.

Withers, R.T., Craig, N.P. & Norton, K.I. (1986). Somatotypes of South Australian male athletes. *Human Biology*, **58**, 337–56.

Withers, R.T., Whittingham, N.O., Norton, K.I. & Dutton, M. (1987). Somatotypes of South Australian female games players. *Human Biology*, **59**, 575–87.

Withers, R.F.J. (1964). Problems of genetics in obesity. *Eugenics Review*, **56**, 81–90.

Wittman, P.M., Sheldon, W.H. & Katz, C.J. (1948). A study of the relationships between constitutional variations and fundamental psychotic behavior reactions. *Journal of Nervous and Mental Disease*, **108**, 470–6.

Yates, J. & Taylor, J. (1978). Stereotypes for somatotypes: shared beliefs about Sheldon's physiques. *Psychological Reports*, **43**, 777–8.

Yuhasz, M.S., Eynon, R.B. & MacDonald, S.B. (1980). The body composition, fat pattern and somatotype of young female gymnasts and swimmers. *Anthropologiai Közlemények*, **24**, 283–9.

Zeng, L. (1985). The morphological characteristics of elite Chinese athletes who participated in gymnastics, swimming, weightlifting and track and field events. Master's Thesis, State University of New York, Cortland.

Zerssen, D.V. (1965). A biometric examination of Sheldon's theories concerning the connection between body build and temperament. *Zeitschrift fur Experimentelle und Angewandte Psychologie*, **12**, 521–48.

Zerssen, D.V. (1969). Comparative studies in the psychomorphological constitution of schizophrenics and other groups: A survey. In *Schizophrenia: Current Concepts and Research*, ed. D.V. Siva Sankar, pp. 913–25. Hicksville, New York: PJD Publications.

Znášik, A. (1979). Sledovanie motorických schopností a somatotypov lyžiarov-zjazdárov a žiakov ZDŠ v Bratislave (Follow-up of motoric abilities and somatotypes of alpine-skiers and of the children from elementary schools in Bratislava). MA Thesis, Comenius University, Bratislava.

Zrubák, A. & Hrčka, J. (1976). Body composition and somatotypes of body builders, football players, and fencers. *Acta Facultatis Rerum Naturalium Universitatis Comenianae Anthropologia*, **22**, 321–6.

Zrubák, A., Varga, G. & Hatiar. B. (1981). Somatická charakteristika žiakov v športovej gymnastike (Somatic characteristics of young gymnasts). In *Telesná Kultúra v Živote Dieťaťa*. Materiály z celoštátnej konferencie usporiadanej pri príležitosti Medzinárodného roka dieťaťa. Bratislava, 22–23 November, 1979. Sbornik vědecké rady ÚV ČSTV. 12, pp. 277–82.

Zubin, J. & Taback, M. (1941). A note on Sheldon's method for estimating dysplasia. *Human Biology*, **13**, 405–10.

Zuk, G.H. (1958). The plasticity of the physique from early adolescence through adulthood. *Journal of Genetic Psychology*, **92**, 205–14.

Author index

(This index also contains names of people referred to in the text who are not cited in the references.)

491

Subject index

497